The Effects of Low Temperatures on Biological Systems

The Effects of Low Temperatures on Biological Systems

Edited by

B.W.W. Grout

Department of Biological Sciences
Plymouth Polytechnic

G.J. Morris

Cell Systems Limited
Cambridge Science Park
Cambridge
Formerly of Institute of Terrestrial Ecology
(Natural Environment Research Council)
Cambridge

CAMBRIDGE
UNIVERSITY PRESS

CAMBRIDGE UNIVERSITY PRESS
Cambridge, New York, Melbourne, Madrid, Cape Town, Singapore, São Paulo, Delhi

Cambridge University Press
The Edinburgh Building, Cambridge CB2 8RU, UK

Published in the United States of America by Cambridge University Press, New York

www.cambridge.org
Information on this title: www.cambridge.org/9780521105767

First published by Edward Arnold (Publishers) Ltd 1987
This digitally printed version by Cambridge University Press 2009

A catalogue record for this publication is available from the British Library

ISBN 978-0-521-41749-5 hardback
ISBN 978-0-521-10576-7 paperback

Contributors

A. Clarke
British Antarctic Survey (Natural Environment Research Council), High Cross, Madingley Road, Cambridge CB3 0ET

B.J. Fuller
Academic Department of Surgery, Royal Free Hospital School of Medicine, Pond Street, London NW3 2Q9

C.J. Green
Surgical Research Group, Division of Comparative Medicine, MRC Clinical Research Centre, Northwick Park, Middlesex

B.W.W. Grout
Department of Biological Sciences, Plymouth Polytechnic, Drake Circus, Plymouth PL4 8AA

E. James
Department of Ophthalmology, Medical University of South Carolina, 171 Ashley Avenue, Charleston SC29425, USA

J.J. McGrath
Director, Bioengineering Transport Processes Laboratory, Associate Professor, Mechanical Engineering Department, Michigan State University, East Lansing, Michigan 48824, USA

G.J. Morris
Cell Systems Limited, Cambridge Science Park, Milton Road, Cambridge CB4 4FY
Formerly of Institute of Terrestrial Ecology (Natural Environment Research Council), Culture Centre of Algae and Protozoa, 36 Storey's Way, Cambridge CB3 0DT

D.S. Reid
Department of Food Science and Technology, University of California, Davis, California 95616, USA

H. le B. Skaer
ARC Unit of Insect Neurophysiology and Pharmacology, Department of Zoology, University of Cambridge, Downing Street, Cambridge CB2 3EJ

M.J. Taylor
MRC Medical Cryobiology Group, University Department of Surgery, Douglas House, Trumpington Road, Cambridge CB2 2AH

L.C.H. Wang
Department of Zoology, University of Alberta, Edmonton, Alberta, Canada T6G 2E9

J.M. Wilson
School of Plant Biology, University College of North Wales, Bangor, Gwynedd LL57 2UW, UK

L.A. Withers
Department of Agriculture and Horticulture, University of Nottingham, School of Agriculture, Sutton Bonington, Loughborough LE12 5RD

Preface

The effects of low temperature, in the natural environment or laboratory, are likely to attract the attentions of most biologists at some point in their studies. These effects are diverse, complex and can be studied at many levels. For example, the distribution, behaviour and reproductive strategies of whole organisms are commonly moderated by low temperature, influencing the ecology of both animal and plant populations. Reduced temperature and its effects on the productivity and survival of plants is also an area of major concern, particularly where food crops are involved. In these instances low temperatures are typically restrictive, although they may have important developmental functions, particularly in plants. The diversity of effects is increased following freezing when temperature *per se* is separated, as it must be for a full understanding of the system, from the presence of ice.

In the laboratory the influences of low temperature can be studied through to the cellular and molecular levels to give insight into basic biological mechanisms, and a more complete understanding of natural systems at all levels. The practical benefits of preservation of biological systems at low and ultra-low temperatures can also be manipulated, ranging from maintenance of activity in isolated biological macromolecules to keeping a wide range of microorganisms, higher organism cells and tissues viable whilst frozen. Beneficial aspects of the application of low temperatures are also exploited, both in fixation and observation, in light and electron microscopy.

Each of these areas of study has already accumulated a large body of literature to occupy much of the time of the investigator, increasingly restricting the opportunities to stray into other areas of cryobiology. This restraint is effectively strengthened as most workers are involved in a specific area of biology firstly, and with the effects of low temperature subsequently. Rarely does one find a genuine investigator of low temperature effects who is bold enough to range widely across the biological spectrum.

This book was begun as an attempt to encourage excursions by interested biologists, at all levels from undergraduate finalists, into areas of biology possibly unfamiliar to them to look further at low temperature effects. It is hoped that they will see signposts in these areas indicating similarities, parallels and divergences from their own experience that will help in the approach to their specific interests. It is not intended as a comprehensive text but as a guidebook to foreign parts.

We would like to acknowledge the willingness of the publishers to take on this task, and the skill and care with which each of the contributors has used their specific knowledge to develop the general theme of effects of low temperature on biological systems.

1986

BWWG
GJM

Contents

Section 1

Fundamental principles

1

Physico-chemical principles in low temperature biology

M.J. Taylor
MRC Medical Cryobiology Group
Cambridge

Introduction

Much of our current understanding of the effects of reduced temperatures on biological systems has evolved from studies on a large variety of biological samples from both the animal and plant kingdoms. Furthermore, many evaluations of the influence of low temperatures have been based solely upon some measurement of survival. This approach has provided a considerable degree of understanding of the factors and variables which govern whether a biological system either withstands decrease in temperature successfully or succumbs to injury. Nevertheless, the exact mechanism of cryoprotection or cryoinjury at the molecular and macromolecular level is ill-understood. A knowledge of the physico-chemical aspects of low temperatures is fundamental to a complete understanding of the nature of resistance to low temperature or cryoinjury and cell death.

The major effect of reduced temperatures on any system is the reduction of molecular motion which, for biological systems, has important consequences. The inhibitory effects of low temperatures on chemical and physical processes, including the biochemical reactions of living tissue, provide the basic means for achieving long-term preservation of cells, tissues and organs. In the extreme, temperature reduction to absolute zero (−273.16°C) brings to a halt all molecular motion and causes all reactions to cease. In practice, however, no changes of biological importance occur at temperatures below −150°C (Mazur, 1964) and it is only under these conditions that a state of true 'suspended animation' can exist. However, cooling below 0°C brings about dramatic changes in biological systems, especially if the water in the intra- and extracellular solutions freezes and separates as ice, giving rise to physico-chemical changes which may be lethal to cells. In nature some organisms have adapted to low temperature environments to avoid injury and in the laboratory, by careful manipulation of cooling and rewarming conditions and application of certain protective compounds, injury may also be avoided to allow the long-term preservation of a wide variety of biological cells and some tissues.

Methods of inhibiting cellular activity by temperature reduction are determined by principles of thermodynamics and heat transfer. These physical methods are complemented by the use of added cryoprotectant compounds whose mode of action is based on physiological and pharmacological principles. The purpose of this chapter is to summarize the physical aspects of reduced temperature and its effects on water and aqueous solutions as a basis for discussion of the effects of low temperatures on biological systems. The treatment of many of the concepts in this review will, of necessity, be brief and in some cases simplified, so the reader requiring greater detail will be directed to more comprehensive articles in the literature.

Heat energy and temperature

Heat is a particular form of energy, and the science of thermodynamics, which arose from interest in processes involving work, heat and material transformations, can be defined as the study of changes in energy of bulk matter and radiation. However, since it is conjectured that the total energy of the universe as a whole is constant (First Law of Thermodynamics), it is necessary to limit consideration to a special isolated part of the universe, a 'system'. The term 'surroundings' thus denotes the rest of the universe. An 'equilibrium' condition is said to apply when the macroscopic properties (volume, temperature, pressure, concentration, energy, etc.) of a system are constant with time. A 'state' is defined as an equilibrium condition for a system. Thermodynamics thus enables a number of useful predictions of what must happen in going from one state to another, as in a chemical reaction, a phase change, translocation of solvent and/or ions across a semi-permeable membrane, flow of electric current, or change in temperature.

As with all forms of energy, heat is a product of an intensity factor (temperature) and a capacity factor (the heat capacity). When heat is removed from a substance, its temperature will fall. Just how much the temperature will fall depends on:

(1) the amount of heat removed;
(2) the amount of substance present;
(3) the chemical nature and physical state of the substance;
(4) the conditions under which heat is removed.

In general, the temperature change (ΔT) for a given amount of substance is directly proportional to the heat (q) removed or added:

$$\Delta T \propto q \qquad (1)$$

$$\Delta T = Cq \qquad (2)$$

where C, the proportionality constant, is the heat capacity. Heat capacities vary greatly from substance to substance. Liquid water has a large heat capacity (75.3 J $(\text{mol K})^{-1}$), so that it requires more heat to raise the temperature of 1 mole of water by 1 degree than it would for 1 mole of copper (24.47 J $(\text{mol K})^{-1}$) under similar conditions. Many different conditions under which heat is added or removed from a system can be realized in practice, but the two important cases are constant volume (Cv) and constant pressure (Cp). For liquids and solids, Cv and Cp are very nearly the same because the volume change with temperature is quite small. For gases, however, Cp is larger than Cv by the amount of heat necessary to expand the gas. Heat capacities are, in general, temperature-dependent and should be defined precisely only in terms of a differential heat flow (dq) and temperature change (dT) thus:

$$C = \frac{dq}{dT} \qquad (3)$$

The experimental values are usually summarized in the form of a polynomial in temperature; three items suffice to represent data over a considerable range of temperature:

$$C = a + bT + cT^2 \qquad (4)$$

Heat can be defined as that form of energy which passes from one body to another solely as a result of a difference of temperature and manifests itself as disordered motion of molecules. This motion can be resolved into:

(1) translation of molecules from one position to another;
(2) rotation of molecules and parts of molecules;
(3) vibration of parts of molecules in relation to each other.

Each type of molecular motion contributes to the heat capacity of a body. The energy content of a system is therefore a function of its temperature. However, if ice absorbs heat at 0°C or if two blocks of ice are rubbed together, melting, but not necessarily any change in temperature, occurs. Thus there is a difference in the energy content of ice and water. This is called a latent energy or latent heat. The text of Adamson (1979) is recommended as further reading for detailed accounts of the concepts of heat, energy and related thermodynamic topics.

Heat transfer

During cooling or freezing of a biological system the cold environment acts as a heat sink which absorbs the thermal energy given off by the biomaterial until that material is in equilibrium with the surroundings. The rate of heat transfer has important consequences during the cooling and warming of a biological system and is a major determinant of cell survival during cryopreservation (Mazur *et al.*, 1970; Taylor, 1984 and elsewhere in this volume). Within a solution, the number of ice nuclei that form after freezing has been initiated, the rate at which they grow and the size they attain are all dependent upon the rate at which heat is removed from the immediate environment. During warming, the rate at which ice crystals melt and the opportunity for crystal growth and recrystallization will depend on the rate at which heat is pumped into the system. These rates are influenced not only by temperature gradients outside the solution but also by gradients within it, as determined by the *thermal conductivity* and *specific heat* (heat capacity per unit mass of substance) of the solution both in the liquid and frozen state, by its latent heat of fusion and by the overall geometry of the system.

Heat transfer is usually categorized into three modes of transmission: conduction, convection and radiation.

Conduction is a process in which heat is transferred through a material (solid, liquid or gas), as a result of a temperature differential, or between different materials in direct physical contact. It is a convenient way of cooling and warming a material and involves the transmittance of energy by direct molecular communication without appreciable displacement of the molecules involved.

Convection, on the other hand, is a process of energy transfer which involves the actual movement of bulk material, as either liquid or gas, and it is the most important mechanism of heat transfer between a solid surface and a surrounding liquid or gas. This process of energy transfer has been described as combining the action of heat conduction, energy storage and mixing motion in a number of steps (Burdette, 1981). Initially, heat flows by conduction from the solid surface to adjacent molecules of fluid and this transferred energy causes an increase in the temperature and internal energy of adjacent fluid molecules. Next, these molecules move to a region of lower temperature in the liquid where mixing and energy transfer to other fluid molecules occurs. Thus, energy is stored in the fluid molecules and is carried as a result of their mass motion. Convection is the primary means of heat transfer during cooling and freezing of biological systems.

Heat transfer by radiation involves heat flow by emission of electromagnetic energy from a material at a high temperature to a material at a lower temperature, when the materials are separated in space (even by a vacuum). This mode of heat transfer is important at relatively high temperature differentials, such as that between liquid nitrogen used in cryogenic storage (−196°C) and room temperature.

The influence of temperature

The removal of heat from a system slows down both physical and chemical processes in proportion to the loss of heat and therefore to the fall in temperature. Physical phenomena such as osmotic pressure depend solely on the rate of molecular motion so that the decrease in the rate of the process is proportional to the fractional change in absolute temperature. Many chemical reactions, however, depend upon an energy of activation, which is the minimum energy required for molecules to react, and this results in a special relationship between the rate of reaction and temperature (Arrhenius, 1889). Arrhenius found that the log of the reaction rate was proportional to the reciprocal of the absolute temperature. This relationship can be expressed as:

$$\ln K = \text{a constant} - E/RT \qquad \mathbf{5}$$

From this it follows that a plot of log K against 1/T should be a straight line (the so-called Arrhenius plot) with a slope of $-E/2.3R$, where E is the activation energy and R is the gas constant. It is generally true that rate of reaction decreases with decreasing temperature and a plot of log K against 1/T yields a linear relation over a significant range in accordance with the Arrhenius equation. Such behaviour is not, however, always observed, for an Arrhenius plot can show curvature (and sometimes distinct breaks) probably reflecting changes in reaction mechanism. For example, when applied to the cooling of certain enzyme systems Arrhenius plots have a discontinuity of slope and approximate to two straight lines meeting at an angle, possibly indicating that there is a change in the value of activation energy at the transition temperature (Douzou, 1977).

Furthermore, the phenomenon of cooling retarding metabolism by reducing the number of activated molecules available to take part in physiological processes does not mean that all reaction rates are affected to the same extent or even in the same way by a reduction of temperature. For example, a disparate effect of temperature on the permeability of dog erythrocytes to cations has been reported (Elford and Solomon, 1974) where sodium flux was shown to increase during cooling from 37°C to 20°C and then decrease during subsequent cooling. Also, potassium flux demonstrated a minimum at 12°C and then increased during further cooling, whereas water transport decreased in accordance with a typical Arrhenius relationship. These observations suggest that sodium, potassium and water are transported across the dog red cell membrane by independent mechanisms that are affected differently by temperature. It is clear that the effect of cooling on integrated metabolizing systems is complex, and often unpredictable, such that reaction pathways may become uncoupled producing harmful consequences. Pegg (1981) has recently summarized, in general terms, the cellular mechanisms essential for cell survival and outlined some specific aspects of cooling which have practical implications for cell survival *in vitro*. These include the reduction in demand for metabolites, the effect on active ion transport and cell volume regulation, the consequences of these effects for the design of storage media for tissues to be held at low temperatures, and the phenomenon of thermal shock.

Water – structure and physical properties as liquid and as ice

Some physico-chemical properties of water and its role in life processes

The universal importance of water as a biological solvent for life processes is obvious, and all such processes are sensitively attuned to the physical properties of water (Franks, 1977). Many of the important chemical properties of macromolecules in biological systems, including conformational stability and biochemical specificity, depend on the interaction of their constituent groups with the surrounding solvent medium. These interactions are, in turn, markedly influenced by the structural features of liquid water and by structural changes in water caused by solutes and changes in temperature.

Cooling alters these interactions and freezing may be influenced by the presence of solutes. Although little is known specifically about solute interactions during the freezing of water, an understanding of this process undoubtedly requires an appraisal of the properties of aqueous solutions. With regard to our current understanding of the structure and properties of water, particularly at sub-zero temperatures, reference must be made to volume seven of the major reference exposition entitled *Water: A Comprehensive Treatise* (Franks, 1982a). The reviews on the properties of supercooled water (Angell, 1982) and on the properties of aqueous solutions at sub-zero temperatures (Franks, 1982a) are particularly useful.

Water has often been described as an 'unusual' solvent and there are many distinctive properties of water which allow it to play its important role in biological processes. Firstly, the *dielectric constant* or *permittivity* (ϵ), of water is one of the highest known (80 at 20°C). Since the force of attraction between ions varies inversely with ϵ, the attraction between ions is diminished when they are dissolved in water and, as a result, ionic compounds are very water-soluble. Secondly, water has an unusually high heat capacity, a major factor in temperature control in homeothermic animals, and is important in protecting plants from adverse effects of fluctuating temperature. Thirdly, the high heat of vaporization of water enables man and certain animals to effect surface cooling by evaporation of water from the skin, or from the surface of the tongue by panting. These, and other important properties including high surface tension, low viscosity, high melting and boiling points and higher density of the liquid state compared with the solid state (ice) are all ascribed to features of the unique structure of water and to the presence of hydrogen-bonding as described below.

Water structure in the solid and liquid state

The structure of water in the liquid phase is not nearly so clearly defined and understood as that of ice. It is therefore helpful to consider the structure of ice in the first instance.

Structure of ice

As shown in Fig. 1.1a, a water molecule exists in the form of an isosceles triangle with an O–H bond length between 0.96 and 1.02Å and an H–O–H

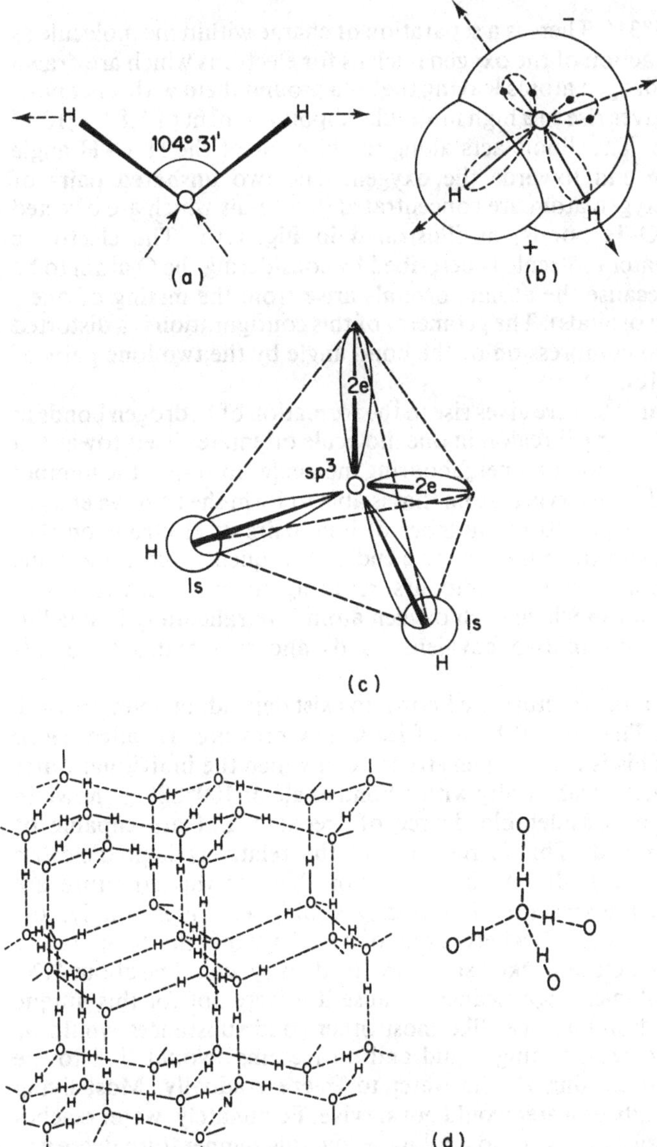

Fig. 1.1 The water molecule and the hydrogen-bonding arrangement in the structure of *Ice I*. Details are given in the text. (a) The bond angle between oxygen and hydrogen in the water molecule. (b) The relative location of the unshared electron pairs. (c) The overall geometry of the molecule is a tetrahedron. (d) When formed into ice, water molecules are hydrogen-bonded to each other. The typical hexagonal ice form at low pressures (*Ice Ih*) is shown here. The tetrahedral hydrogen-bond arrangement for a single water molecule is shown separately for clarity.

The schematic representation of the three-dimensional network of molecules in the crystal structure of ice (d) is adapted from Némethy and Scheraga, 1962. Covalent bonds are represented by solid lines and hydrogen bonds as broken lines.

bond angle of 104°31′. There is a separation of charge within the molecule as a result of the attraction of the oxygen nucleus for electrons which are drawn away from the hydrogen atoms leaving the area around them with a net positive charge. This gives rise to a high molecular dipole moment of 1.87×10^{-18} electrostatic units. The dipole acts along the bisector of the H–O–H angle with the negative end towards the oxygen. The two unshared pairs of electrons of the oxygen atom are concentrated in orbitals which are directed away from the O–H bonds, as illustrated in Fig. 1.1b. The electronic structure of the water molecule is described by considering the O atom to be sp^3 hybridized (because the atomic orbitals arise from the mixing of one s orbital and three p orbitals). The geometry of this configuration is a distorted tetrahedron due to compression of the bond angle by the two lone pairs of electrons (Fig. 1.1c).

This highly polar structure gives rise to the formation of hydrogen bonds in which a positively charged region in one molecule orientates itself towards a negatively charged region of a neighbouring molecule. In water, the number of protons around each oxygen atom that is able to form the positive ends of hydrogen bonds is equal to the number of lone pairs of electrons on that oxygen atom that can form the negative ends; consequently, each molecule tends to have four nearest neighbours resulting in an extensive three-dimensional network in which each oxygen atom is tetrahedrally bonded to four hydrogen atoms in two covalent bonds and two hydrogen bonds (Fig. 1.1d).

Several forms of ice structure are known to exist depending upon pressure and temperature. The normal form of ice at low pressures is called *ice Ih* (hexagonal ice). This is a very open structure in which the individual water molecules, arranged tetrahedrally with a bond angle of 109°28′ as shown in Fig. 1.1d, retain a considerable degree of freedom and are capable of rotation in the crystal. This is reflected in the relatively high dielectric constant of ice. The hydrogen bonds responsible for this structure are straight and thus the structure is virtually strain-free. Interestingly, the density of ice at 0°C is 0.917, whereas that of liquid water at this temperature is 0.999 (if ice was a close-packed structure, its density would be about 1.7). This fact is of ecological significance because if it were not for this unique type of hydrogen bonding, ice, like most other solid substances would be heavier than the corresponding liquid and on freezing, would sink to the bottom of a pond causing all the water to freeze gradually. Most living organisms in the body of water would not survive. Fortunately, water reaches its maximum density at +4°C so cooling below this temperature decreases the density, allowing it to rise to the surface where freezing occurs. The ice layer that forms on the pond surface does not sink and, more importantly, acts as a thermal insulator for the water below it.

At high pressures (>2000 atm) six different crystalline forms of ice can exist, each in a particular pressure and temperature range (Kamb, 1965; Eisenberg and Kauzmann, 1969; Fletcher, 1970; Franks, 1972) (see Fig. 1.3). Under these conditions the water molecules are fully hydrogen-bonded, but are arranged so that the space is more fully occupied, and often two interpenetrating networks are formed (Kamb, 1965). In such structures the hydrogen bonds are bent or the molecules compressed, however these forms of ice do not arise under the conditions of interest in low temperature biology.

At very low temperatures (below −140°C) a metastable form of ice I, cubic ice (*ice Ic*) can exist, but it can only form under special conditions, such as on condensation from the vapour (Bertie *et al.*, 1963). It cannot form by cooling ice Ih, although its hydrogen-bonding configuration is a modification of that of ordinary ice. It crystallizes in a cubical form instead of the hexagonal crystal structure and on warming it changes into ice Ih.

One other form of water crystal will be mentioned for completeness, although having restricted relevance in low temperature biology. *Clathrates* are inclusion compounds (or gas hydrates) that arise from the entrappment of a neutral 'guest' molecule such as Cl_2 or SO_2, which are accommodated in cavities of molecular size within the spatial network formed by the hydrogen-bonded water. No chemical bonds are formed between the two substances, although the crystals are stabilized by van de Waal's forces acting between the guest molecule and water molecules forming the cavity.

Among the various forms of ice, only ice I is produced during either slow or rapid freezing of water or of aqueous solutions at low pressures. The only other low pressure form, ice Ic, cannot form from the liquid.

Structure of liquid water

The structure of liquid water is not as well understood as that of solid ice; it is, however, generally agreed that the tetrahedral co-ordination of water molecules found in ice also exist in a more imperfect form in liquid water. (This configuration has been established by X-ray diffraction of ice, as well as by X-ray scattering of liquid water.) As we have seen, ice has a very open structure, and the increase in density on melting was at first explained as simply the result of the breakdown of the ice structure into individual water molecules, with perhaps some small hydrogen-bonded units present. However, this would not account for many of the properties of water, including its high critical temperature. Consequently, a number of models have been proposed to take account of the unique properties of this liquid. None of these models can account satisfactorily for all the observed properties of water, but progress has been made toward a detailed explanation of its structure and properties (e.g. Némethy and Scheraga, 1962; Némethy, 1968; Frank, 1970, 1972; Conway, 1981).

Bernal and Fowler (1933) suggested that the hydrogen-bonding found in ice also exists in water, although the regular three-dimensional network in the solid no longer exists – instead, some hydrogen bonds are broken and the coordination number increases above four. This explains why the volume decreases upon melting. The free monomeric water molecules, released as some of the hydrogen-bonds break, are able to occupy spaces in the remaining 'icelike' lattice; this explains why the density of water is greater than that of ice. As the temperature rises, so more hydrogen bonds are broken, but at the same time the kinetic energy of molecules increases. Higher temperature increases the density of water, while elevated kinetic energy decreases the density because each molecule will now occupy a greater volume. The net result is that a maximum density is reached at 4°C, above which the density decreases with increasing temperatures.

An alternative theory suggested that very few hydrogen bonds are actually broken when ice melts (Pople, 1951). Instead, the bonds are bent and distorted in liquid water. The relatively low heat of fusion for water (6.01 kJ mol^{-1}) offers some support for this model.

Pauling suggested that the structure of water may be similar to that of certain hydrocarbon-water clathrates, such as CH_4. $6H_2O$ and Cl_2. $8H_2O$ (Pauling, 1959). The hydrogen-bonded water molecules in these complexes form cages that have a characteristically open structure like ice, but with an overall arrangement which is looser than that of ice. The guest molecules, CH_4 or Cl_2, reside in the cavities of the ice-like structure. In the model for liquid water the guest molecule is replaced by another interstitial water molecule (Davidson, 1973). Pauling's model was considered to be too crystalline and rigid, and a similar model based on 'flickering clusters' was put forward by Frank and Wen (1957) who suggested that the clathrates were constantly breaking up and reforming with a mean lifetime of the order of 10^{-10} seconds (Fig. 1.2). This model was based upon the notion of co-operative hydrogen-bonding, which suggests that it is more favourable energetically for a water molecule to form several hydrogen bonds rather than

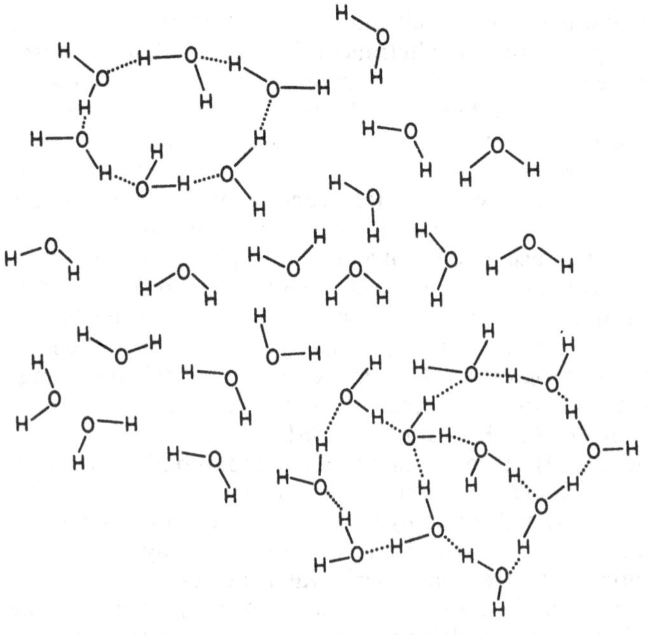

Fig. 1.2 Schematic representation of the 'flickering cluster' model for the structure of liquid water in which the tetrahedrally co-ordinated structured clusters and the non-structured regions of unbonded molecules form and disintegrate very quickly. (Redrawn and adapted from Némethy and Scheraga, 1962).

only one. Hydrogen bonds in water have a certain amount of covalent character that arises because of resonance between the various hydrogen-bonded structures. Once a hydrogen-bond forms it becomes easier for other water molecules to attach to the resonating structure through additional hydrogen-bonds. Similarly, as one hydrogen-bond is broken so the entire structure tends to disintegrate. This co-operative element gives rise to the formation, and breakdown, of three-dimensional structures which have been termed 'flickering clusters'. These may be of various sizes and shapes, and they are brought about by localized energy fluctuations. This cluster model was developed further in terms of a statistical thermodynamic calculation. This involved summing up all the energetic interactions of all the species postulated to exist in liquid water in order to derive equations which can predict the internal energy, free energy, entropy, heat capacity, molar volume, etc., of liquid water (Némethy and Scheraga, 1962). The results were in moderately good agreement with several properties of water, and the structures illustrated schematically in Fig. 1.2 are generally accepted as correct in principle if not in detail. The flickering-cluster aspect of the model permits a flexible structure that has the low viscosity characteristics of liquid water and yet provides an acceptable explanation for many of the thermodynamic, and other, properties of water which are known to be those of a highly structured medium as verified by techniques such as X-ray scattering.

The structure of aqueous solutions

The preceding discussion of the structure of pure water is somewhat simplistic in the context of the situation prevailing in biological samples, since it is the properties of aqueous solutions containing a variety of solutes and their interactions with the surrounding water that influences the events which constitute life processes. Depending on their functional groups, solute molecules interact in a variety of ways with the water surrounding them. These interactions with water molecules, either specific or non-specific, can alter the structure of water in their immediate vicinity and consequently alter the properties of the solution (Kavanau, 1964; Coetzee and Ritchie, 1969; Conway, 1981). The hydrogen-bonding between water molecules, and also between water and other molecules, is extremely sensitive to changes in temperature and this property will influence the ability (or inability) of a complex organism to cope with changes in temperature. Furthermore, the presence of non-aqueous organic molecules such as glycerol, dimethylsulphoxide (DMSO) and methyl alcohol, commonly used as cryoprotectants, in a biological system gives rise to mixed solvent systems, the structure and properties of which can be very different from an entirely aqueous medium.

The number of possible combinations of mixed solvent is vast, and their physical properties have therefore not been studied extensively except in a few cases at selected volume-ratios and temperatures. Some of the properties of mixed solvents have been summarized (Douzou, 1977). Their structure becomes complex in the presence of solutes because of the selective attraction of one of the components of the mixed solvent for the solutes. Molecules of water, as the more polar component, tend to cluster around the solutes, particularly ions and dipolar ions, forcing away the molecules of the less polar component. Consequently, the composition of the solvent in the vicinity of charged solutes does not correspond to that in the bulk of the solution. Preferential solvation of charged solutes by the more polar solvent molecules will have an important effect on the physical and chemical properties of mixed solvents as well as on the behaviour of solutes.

Solubility

Changes in water structure often have a more influential effect in determining the solubility of many substances, and the properties of their aqueous solutions, than the direct solvent–solute interactions themselves. Depending on the nature of the forces of interaction, and on the resultant changes in water structure, three classes of solute can be distinguished (Frank, 1963):

(i) electrolytes;
(ii) polar, but non-ionized, solutes usually capable of hydrogen bonding;
(iii) non-polar (chemically inert) solutes.

In addition, many substances of both low and high molecular weight contain different functional groups belonging to two or all of these categories. These are often referred to as polyfunctional solutes and may be considered as a fourth class of solute (Frank, 1963).

Simple electrostatic considerations indicate that ions must exert a strong effect on surrounding water molecules, and the concept that ions are hydrated (i.e. surrounded by orientated and immobilized water molecules) has been accepted for a long time. A scheme in which the structure of water around ions is considered to consist of three characteristic concentric regions has been supported by experimental studies of many properties of aqueous ionic solutions, and has served as a working hypothesis to later developments in theory (Frank and Wen, 1957; Kavanau, 1964).

Un-ionized, polar solutes are expected to have the simplest interactions with water because they can hydrogen-bond to it as either donors or acceptors, or both, and each interaction would be expected to cause smaller structural effects than in the case of other solutes. However, although these substances can hydrogen-bond to water molecules, in many cases their geometry is not compatible with that of the clusters and they can only interact with that water which is not part of the clusters. Consequently, the cluster–unstructured water equilibrium is shifted and these solutes can be considered as 'structure breakers'. Similarly, these solutes can also be expected to have an inhibitory effect upon freezing because water molecules hydrogen-bonded to them are orientated incorrectly for ice formation. These hydrogen-bonds have to be broken before the water molecules can crystallize.

Non-polar solutes have low aqueous solubility; this is thought to be related to some unusual structural changes in water. Generally, low solubility occurs when the intermolecular forces between solute and solvent are weaker than those between the pure components themselves. In such cases energy has to be supplied during the solution process, which is therefore endothermic. Typically, the intermolecular attraction is strongest between substances having similar structures. In water the observed situation is very different; hydrocarbons and non-polar solutes (rare gases, O_2, N_2 etc) dissolve exothermically compared with their solution in inert, hydrocarbon solvents. The low solubility has been ascribed to a negative entropy change corresponding to an increase in order. As small solutes cannot contribute much to the entropy, it has been concluded that the increase in ordering must arise from structural changes in the solvent (Frank and Evans, 1945; Némethy and Scheraga, 1962).

For larger solutes such as macromolecules, the non-polar groups tend to pack together in such a manner that the amount of their contact with water is minimized unless such packing is prevented or rendered unfavourable by some dominant structural restriction. The packing of non-polar groups is a consequence of the low solubility of non-polar materials. Their solution process is thermodynamically unfavourable and therefore its partial reversal, the aggregation of non-polar groups, is favoured. In macromolecules, the aggregation of non-polar groups is often referred to as *hydrophobic bonding*. It is of great importance in proteins, where it is believed to make a major contribution to the stability of compact conformations.

Most solutes of biochemical interest are polyfunctional in the sense of the classifications outlined above (i.e. they have both a non-polar part which can participate in hydrophobic bond formation and a polar, or ionizable, part). Many different functional groups are present in a macromolecule, and the question arises as to whether the effect of each functional group upon water structure, or upon stabilization of macromolecular structure, can be treated independently. In many cases the effects of the separate 'functions' seem to be additive (Némethy, 1968).

Current understanding of the interactions of these various types of solute with the variety of solvents of interest in low temperature biology, especially water, is still sketchy and an extended discussion of the ideas, models and experimental evidence available is beyond the scope of this chapter. The reader is directed towards articles by Frank (1963), Kavanau (1964), and Coetzee and Ritchie (1969) for further information.

Most of the quantitative, theoretical information available on the properties of aqueous solutions refers to very dilute solutions, whereas biology at low temperatures is often concerned with phenomena in much more concentrated solutions. The stresses imposed by reduced water activity are also of significance, particularly when freezing removes water from the system as ice and concentrates the solutes in the residual liquid phase (Franks, 1982b). At multimolar solute concentrations, and in semi-dry systems, where there is insufficient water to form the usual cluster structures the hydrogen-bonding tendency of water will still be a dominant feature, but theories of solution structure based on dilute solutions can no longer be applied exactly. Indeed, very few studies have been carried out on aqueous systems at high solute concentrations, and no theoretical treatment is available for them, which is unfortunate in view of the practical importance of systems with low water activity. Many aspects of the behaviour of water, and of dilute solutions, have yet to be clarified before concentrated solutions can be treated quantitatively, because interpretation of data on systems containing small amounts of water may require rules which are different from those applied to dilute solutions.

Anomolous water

Some years ago it was thought that another form of liquid water existed. A number of laboratories reported the existence of a new molecular form of liquid water which was more stable than ordinary water. If their claims had been true, it would have introduced another phase boundary in the liquid region of Fig. 1.3. Unfortunately, it has been shown that the experimenters were misled by impurities and 'polywater', a mixture of colloidal and molecular dissolved impurities, has now been dismissed as an illusion. A mass of scientific and anecdotal detail contributes to the story of polywater and the history of these events provides an outstanding example of how science reacts to an announcement of great discovery. The problem was not one of data, which were real, but one of interpretation. The whole story of the polywater phenomenon, which provides a salutary reminder of the pitfalls of modern science, has been entertainingly recounted by Franks (1981b).

Water in biological systems

The role of water as an essential component of life has many facets. Macroscopically, water is the substrate that transports essential nutrients and waste products within living organisms. Each cell, and organelle within a cell, requires a critical balance of water so that it can function properly. The flow of water into, and out of, cells across the plasma membrane accompanies many phenomena, for example a constant flow of water across cell membranes is coupled to the various ion fluxes. At the molecular level discrete water molecules or small aggregates are implicated in the maintenance of conformation of biologically active macromolecules such as nucleotides, proteins and carbohydrates. In metabolism, water is the necessary reagent in any process that involves hydrolysis, condensation and redox reactions (i.e. most biochemical reactions).

In the preceding sections the importance of hydrogen-bonding in the structure and properties of water, both as liquid and as ice, was emphasized. The ability of water to form and accept hydrogen-bonds has important consequences in low temperature biology, notably in its ability to bind both to cryoprotectant molecules and to the surface of macromolecules, contributing to their tertiary and quaternary structures.

Macromolecular conformation and solvent interaction

It is widely accepted that solvent interactions play a major role in the conformational stability and activity of proteins, although little is understood of the specific details of such effects. Concepts such as 'water structure' (spatial arrangement of water molecules with respect to each other and with respect to macromolecular surfaces), hydrogen-bonding and hydrophobic-bonding are all variously invoked as the basis of such interactions (see reviews by Kuntz and Kauzmann, 1974; Cooke and Kuntz, 1974; Franks, 1979; Finney, 1979; Pain, 1982).

Hydrophobic interactions

Folding, structural stability and dynamics of biologically significant macromolecules (proteins, polysaccharides and nucleic acids) are thought to be extensively controlled by solvent interactions with general emphasis upon the poorly understood, so-called 'hydrophobic' or 'apolar' interactions. The hydrogen-bond, as a structural element, had achieved a pre-eminence that had gone essentially unchallenged until, with the acquisition of more specific information on the conformation of these macromolecules it emerged that the hydrophobic, or apolar, bond has an important role in macromolecular structure. For globular proteins particularly, increasing experimental evidence favours an elliptically-shaped model with ionic and polar groups on its exterior and alkyl side-chains folded toward the interior, having little or no contact with water. In the interior, in a hydrocarbon-like atmosphere, the apolar side-chains are a major factor in holding the tertiary structure together through van der Waal's forces of attraction. It may be that a complete description of protein structure and function involves a variety of intermolecular forces including electrostatic ones, hydrogen-bonding, van der Waal's forces (dipole–dipole, ion-induced dipole, dipole-induced dipole, and dispersion forces), and the forces arising from hydrophobic interactions. The result is a complex energy balance responsible for the conformational stability of native proteins. Such forces are also invoked to explain the energetics of enzyme–substrate binding, the binding of a hormone to its receptor, and protein–protein interactions. The structural integrity of membranes, whose major components are amphiphilic lipids, also depends upon solvent interactions.

Vicinal water

The term 'vicinal water' is often encountered in the literature; it describes interfacial water, usually near a solid surface, having properties which differ from the corresponding bulk properties due to structural differences induced by proximity to the surface. The solvent close to the surface of a protein, the so-called 'hydration shell', is composed of several types of water which have properties that differ significantly from those of the bulk water, although how far this perturbation extends from the surface is still debatable (Cooke and Kuntz, 1974; Drost-Hansen, 1982).

A recent summary of solvent-mediated influences on the conformation and activity of proteins indicates that three types of water can be distinguished (Franks, 1979). Firstly, there is internal water within the protein as part of the structure which is irrotationally bound, and is intimately involved in the function of proteins. This water may be active as a co-enzyme and it cannot readily be resolved by methods other than diffraction as its motions are determined by those of the protein itself. Secondly, there are water molecules bound at various peripheral sites on the surface of the protein, which can also be distinguished by diffraction but may be quite mobile. Thirdly, there is an undefined region where the influence of the protein competes with the normal tetrahedral-like order in the water. The hydration helps to maintain the structural stability of the protein and influences both the mechanical and hydrodynamic properties of the protein as well as the transport behaviour of the protein and of substrates or ions towards it.

A number of experimental techniques have been used to investigate the structure, energetics and dynamics of protein solvation (Kuntz and Kauzmann, 1974; Cooke and Kuntz, 1974; Franks, 1979). Diffraction techniques undoubtedly provide the most important means of exploring structural parameters such as distances and angles. Raman and infra-red spectroscopy can, in principle, probe bond vibrations, but nuclear magnetic resonance (NMR) spectroscopy is probably the most potent technique for gaining information about structure and dynamics since it is possible to home-in on one particular nucleus of the several available in proteins. Many studies to probe solvation interactions and energetics have involved sorption and desorption of water vapour in dry proteins; such investigations may, however, be of limited value for it is not yet established that almost-dry protein has the same conformation as native protein. Calorimetry has also been used in the study of the solution thermodynamics of proteins, and heat capacity calorimetry, in particular, has provided much valuable information. In addition mention must also be made of 'unfreezable water'. This is the fraction of water in a solution which does not freeze even beyond its equilibrium freezing point (see below). Unfreezable water should not exist, for ice should separate out and, at a eutectic temperature, the protein should precipitate or crystallize. In practice, there is always a small fraction of the water that will not freeze, and this extraosmotic or extracolligative phenomenon can be detected either by scanning calorimetry or by high resolution NMR (Kuntz and Kauzmann, 1974). Finally, the techniques for studying the motions of water and of protein residues rely upon sedimentation and viscosity, (i.e. hydrodynamic properties), self-diffusion and rotational relaxation (dielectric relaxation and nuclear magnetic relaxation).

'Bound' or 'unfreezable' water

The techniques listed above have shown that water molecules in the vicinity of, and interacting strongly with, macromolecular surfaces have properties that differ from those of the bulk medium, for example they exhibit a lower vapour pressure, lower mobility and greatly reduced freezing point (Kunzt and Kauzmann, 1974; Cooke and Kuntz, 1974). Water demonstrating such properties is commonly referred to as bound water, but an operational definition for this term, based upon any of the observed properties, is subject to limitations and debate. Different methods measure different properties, and some water molecules may be designated as bound by some methods but not bound by others. Furthermore, the measured properties may change continuously with increasing distance from the macromolecular surface, rendering a quantification of bound water somewhat arbitrary. Despite such complications the water adjacent to macromolecules has sufficiently different properties to make the distinction between bound water and the remaining bulk water useful.

Calorimetric as well as spectroscopic studies have indicated, in a variety of macromolecular species, that an appreciable amount of water (0.3–0.5g H_2O/g polymer) will not freeze (Kuntz and Kauzmann, 1974; Cooke and Kuntz, 1974). This indicates that the properties of some of the water in such preparations are sufficiently different that the water cannot be transformed into ice (unfreezable water). This phenomenon might be due to a kinetic barrier, resulting in the supercooling of the water, but a greatly suppressed freezing point is thought to be a more likely explanation. The enthalpy of fusion of the water falls steadily as the water content of collagen decreases and approaches zero at water contents below 0.3g H_2O/g protein (Haly and Snaith, 1971). It has also been proposed that the amount of unfrozen water, which is reported to be independent of temperature below the eutectic point of the bulk solution, should be taken as the amount of bound water since it is easily quantified by both calorimetric and spectroscopic methods. However, the poor control of solute concentrations in frozen solutions and the possibility of low temperature denaturation complicate this approach.

On a less quantitative basis there have also been reports that some of the water in cellular systems is unfreezable. For example, in a study of NMR relaxation times of water protons in frozen striated muscle samples, approximately 20% of the water in muscle was found to be of the unfreezable type (Belton *et al.*, 1972). This equates approximately to 0.8g H_2O/g dry weight, which is considerably larger than the hydration values for proteins and nucleic acids as determined by calorimetry. However, in cells of the stratum corneum the amount of unfrozen water has been estimated as 0.37g H_2O/g dry weight (Hansen and Yellin, 1972).

Nuclear magnetic resonance is one of the most extensively employed techniques for studying the properties of water in biological systems. The resonances of aqueous ^1H, ^2H, and ^{17}O have been measured in whole cells and in protein solutions; the parameters usually studied are the line width, the spin-lattice relaxation time (T_1) and the spin-spin relaxation time (T_2). In examination of biopolymers in solutions below −20°C, under conditions when the bulk water has been frozen out, the amount of water remaining unfrozen can be determined readily by the area of the high-resolution NMR signal. The water line width is temperature dependent with a value of about 1000 Hz at −35°C, which implies a mobility at this temperature greater than that of ice but less than that of supercooled water. The NMR technique is in agreement with calorimetric studies indicating that about 0.3–0.5g H_2O/g polymer does not freeze.

The details of the measurement and properties of bound water have been reviewed with particular emphasis upon NMR data (Derbyshire, 1982; Franks and Mathias, 1982) and a consensus view of the state of bound water on proteins as perceived by the NMR spectroscopist has also been presented (Bryant and Halle, 1982).

The terms 'bound' or 'non-freezable' water are often used to explain the phenomenon in which the progressive freezing of a solution of a polar compound below the equilibrium freezing point does not usually result in the separation of the two solids below the eutectic point. Instead, the solution becomes supersaturated and supercools until it eventually undergoes a glass transition. This indicates that the proximity of chemical groups which are capable of forming hydrogen-bonds can prevent water molecules from diffusing to the ice front and participating in the crystallization process.

Osmotically inactive water

The ability of intracellular water to act as a solvent for the cellular constituents is of particular biological importance, and it can be measured indirectly from the osmotic behaviour of cells. The volume of water (V) associated with *n* moles of solute in a cell is related to the osmotic pressure by the van't Hoff equation ($\pi V = n RT$). If a cell behaved exactly according to this relationship, it could be described as a perfect osmometer. However, the observed osmotic response of cells does not conform to this equation until V is replaced by a term rV, where r represents the fraction of intracellular water that is able to participate in the osmotic equilibrium (Ponder, 1948). Values for r are generally approximately 0.8, implying that 20% of the intracellular water is osmotically inactive (Cooke and Kuntz, 1974).

Cold denaturation

The stability of macromolecular conformation relies upon a complex energy balance in which the free energy of stabilization is the sum of a variety of energetic elements. These separate interactions vary in different ways with temperature, such that stability of proteins might be expected only within a restricted temperature range. Indeed, it is well-known that most macromolecules exist in a biologically active state only up to certain temperatures which are often characteristic of the particular molecular species and environmental conditions such as ionic strength and pH. Thermal denaturation of proteins by raised temperatures is believed to involve a monomolecular transition from a compactly-folded native protein to a highly-unfolded, denatured protein. This unfolding exposes many of the normally shielded groups and side-chains which can then become solvated. Under appropriate solution conditions this process may be reversible, but is often hindered by aggregation of the denatured protein (Brandts, 1964).

Consideration of the temperature-dependence of the energy balance responsible for the conformational stability of macromolecules suggests that denaturation might also be induced by reduced temperatures. Some of the most detailed studies of the effects of low temperatures on protein conformational stability have been carried out using the enzyme chymotrypsinogen. Analysis of the denaturation reaction in the temperature range +80°C to −80°C indicates that the thermodynamic properties of this protein are continuous over the entire temperature range and that the observed behaviour at low temperature is in part predictable on the basis of observations at high temperatures (Brandts *et al.*, 1970). However, detailed investigations below the freezing point of the solution suggest that the protein, in frozen solutions, is distributed in more than a single phase. These studies also indicate that the distribution depends upon temperature, the concentration of small solutes and possibly on the concentration of proteins, and that the denaturation process is unable to occur in at least one of these phases. The protein denaturation that is evident in frozen solutions has been suggested as an unfolding process, essentially the same as that which takes place above the normal freezing point, but occurring in unfrozen pools (Brandts *et al.*, 1970). Furthermore, it is believed that denaturation is brought about by a combination of factors and that lowering of temperature *per se* leads to a general instability of native proteins which renders them susceptible to the influence of other factors. Apart from temperature, these factors include changes in the concentration of small solutes and protein, changes in distribution of protein into multiple phases and, possibly, more specific effects resulting from the dehydration of protein molecules. The dehydration of protein molecules is inevitable as the vapour pressure of ice is reduced with a fall in temperature.

The loss of activity of another enzyme, phosphofructokinase, at reduced temperatures has been correlated with changes in the conformational stability of the protein as determined by heat capacity measurements (Dixon *et al.*, 1981). Using differential scanning calorimetry an exothermic dissociation of the macromolecule was demonstrated during cooling between +20 and −80°C, representing the weakening of hydrophobic interactions responsible for the stability of the active configuration of the enzyme at ordinary temperatures. This results in the spontaneous dissociation of the protein into inactive subunits. Furthermore, heat capacity measurements do not show a quantitative reversal of the dissociation during warming, indicating that the denaturation may not be fully reversible as the system is restored to physiological temperatures.

A notable feature of these studies on the temperature dependence of denaturation is the very large heat capacity involved. This has been attributed to the exposure of apolar side-chains during the denaturation process and to the additional energy that has to be introduced into the system to break down the ordered water structures which are thought to form around such groups in aqueous solution at low temperatures. The denatured form of a protein, with its hydrophobic side-chains exposed to aqueous solvent, would be expected to have a larger heat capacity than the native form and thus give rise to the unusual temperature effects associated with denaturation reactions.

Water stress and the protective influence of bound water

Physiological stress may be defined as a change in the physical or chemical conditions necessay for the optimum function of living organisms. In many cases it can be related to water stress resulting from changes in the intermolecular nature of water and/or solute hydration induced by changes in temperature, pressure or solute concentration (Franks, 1982b). Stress conditions produced by low temperatures are amongst the most widespread in the biosphere, and organisms must in some way either resist or adapt to these stresses in order to survive. A number of mechanisms have evolved to resist water stress during chilling and freezing that involve a diversity of physical and chemical principles such as the inhibition of ice nucleation (undercooling), the promotion of efficient ice nucleation, the synthesis of freezing point depressants or substances capable of disturbing the diffusional motions of water (water-binding), and possibly the redistribution of cell water to areas of the tissue where freezing is harmless. Examples of these mechanisms in the avoidance, or tolerance, of water stress at low temperatures are summarized elsewhere (Franks, 1982a; 1982b) and will be discussed more fully in the ensuing chapters of this volume.

Under conditions of water stress a number of species of plants and microorganisms are able to exploit the phenomenon of bound or non-freezable water by synthesizing and accumulating molecules that bind relatively large quantities of water. In this way these organisms are able to resist dehydration whether caused by heat, cold or salinity (Gould and Measures, 1977; Pollard and Wyn Jones, 1979). Water-binding substances employed in this way are often referred to as 'compatible solutes' (Brown and Simpson, 1972) as they do not give rise to enzyme inactivation or to other toxic symptoms when synthesized or accumulated at concentration (raised by several orders of magnitude) in the cells of protected organisms (Gould and Measures, 1977; Pollard and Wyn Jones, 1979). Examples of solutes which possess this benign feature are the amino acids proline and trimethylglycine (glycine betaine) and polyols such as arabitol and glycerol. Such solutes are thought to provide the basis of an osmoregulatory mechanism which has evolved to prevent the loss of intracellular water that would otherwise occur as a result of an increase in external osmotic pressure. However, it is felt that this effect cannot have an osmotic origin but is mediated through a large kinetic barrier that prevents the establishment of osmotic equilibrium (Franks, 1982b).

An impressive example of adaptation to low water activity by such a mechanism is demonstrated by the bacterium *Bacillus subtilis* which responds to the addition of 1M NaCl in the external medium by increasing its intracellular proline concentration by more than 100-fold and recommencing its growth process under these conditions (Gould and Measures, 1977).

Despite it not being a highly charged molecule, proline possesses a surprisingly high capacity to bind water (Kuntz, 1971). Using high-resolution proton NMR it was determined that 3 moles of water per mole of proline are prevented from freezing when the aqueous macromolecular solution is rapidly cooled to temperatures as low as −35°C. Furthermore, it has been pointed out that these observations agree favourably with studies on aqueous solutions of polyvinylpyrrolidone (PVP) in which the pyrrolidone side-chain is chemically almost identical to proline (Franks *et al.*, 1977).

Low temperature and freezing

It has recently been emphasized that low temperature is a relative term dependant upon the field of interest (Franks, 1982a). Temperatures in the region of −270°C are regarded as low by those interested in low temperature physics whereas consideration of biological phenomena rarely involves temperatures lower than −70°C, although liquid nitrogen, as the most commonly used refrigerant may be considered as being at the lowest temperature of practical importance (−196°C). In the life sciences, relatively high sub-zero temperatures are typically regarded as 'low' because, in water-based systems, these are often equated with freezing. The highest temperature at which ice can possibly exist, 0°C, is generally considered to provide a convenient division between normal and low temperatures. This is, however, somewhat arbitrary because a variety of organisms show symptoms of chill injury, resulting from temperature reduction, at temperatures above 0°C (Wilson, Fuller, Morris – this volume). Furthermore, it should be appreciated that lowering temperature will not result in a discontinuity of properties such as reaction rates and transport phenomena as the temperature passes through 0°C, provided freezing is avoided and the system undercools.

The process of freezing is described in thermodynamic terms as a first-order phase transition; this means that the normal freezing point/melting point is the only temperature at which the solid and liquid can coexist in stable equilibrium at a given pressure and that the chemical potentials of the coexisting solid and liquid phases are equal. The process itself is associated with a latent heat, and all thermodynamic and transport properties of the system demonstrate discontinuities. The effects produced by the separate conditions of temperature reduction and freezing are quite independent but the combined effects of low temperatures and freezing produce a complex array of superimposed processes, which include changes in the rates of molecular motion, changes in the positions of chemical and biochemical equilibria, changes in the concentrations of the components in a mixture as a result of phase separations, and the possibility of irreversible aggregation or dissociation events involving macromolecules. The most obvious consequence of reduced temperatures in aqueous systems is freezing itself. When an aqueous solution freezes, water separates out as pure solid and the concentration of all the solutes in the residual liquid phase increases. These concentration effects are generally regarded as constituting the most important single damaging factor associated with freezing. Assumptions that freezing injury was predominantly of a physical nature arising from mechanical effects associated with the shape and size of ice crystals have been shown to be largely misfounded.

Interactive physico-chemical events at low temperatures are complex and poorly understood. No attempt will be made here to give a comprehensive account of current understanding of equilibrium and non-equilibrium processes and conditions at low temperatures. In line with the previous sections of this chapter the synopsis will be limited to an outline of principles and some of the techniques used to study them. Reviews by MacKenzie (1977) and Franks (1982a) are recommended for a more detailed treatment of the freezing behaviour of aqueous systems.

Phase changes – principles of heterogeneous and homogeneous equilibria

The process of freezing is a specific example of change of phase without change of chemical composition. In thermodynamic terms the underlying principle is the tendency of a system to move toward a state of lower chemical potential, as that corresponds to increasing overall entropy (at constant temperature and pressure). Entropy is the concept derived from the second law of thermodynamics which is a statement recognizing that all systems change spontaneously in such a manner as to decrease their capacity for change (i.e. to approach a condition of equilibrium). The entropy of a system is a measure of the degree to which it is removed from equilibrium, but in an inverse sense. If a system has a high capacity for change, its entropy is low; if a system is at equilibrium, its entropy is relatively high. Entropy has also been described as a measure of the chaotic dispersal of energy, and the natural tendency of spontaneous change is towards states of higher entropy. If, at a particular temperature and pressure, a solid has a lower chemical potential than the liquid, the solid is the stable phase under those conditions, and the liquid will have a natural tendency to freeze. Furthermore, for a state of equilibrium to exist the chemical potential of a sample of matter must be uniform, and this principle of uniform chemical potential applies however many phases are in equilibrium.

Depicting phase equilibria: phase diagrams

A 'phase diagram' describes quantitatively the behaviour of a system which is not homogeneous, but which contains homogeneous regions, known as *phases*, separated from one another by observable boundaries, sometimes called surfaces of discontinuity (Fig. 1.3). Any single phase is homogeneous, but it does not have to be continuous. For example, ice constitutes a phase whether it is a solid block or finely divided. The number of phases is the number of separate physical states in a system; there may, however, be more than one liquid or solid phase (e.g. ice and solid salt). The independent chemical individuals which must be specified in order to describe the chemical nature of the system are called the *components*. The phase diagram for water (Fig. 1.3) describes a one-component system. The solid–liquid boundary shows how the freezing point depends on the temperature and indicates that enormous pressures are needed to effect any significant change (very high pressures cause the formation of unique forms of ice as discussed earlier). Notice that the line slopes backwards (the gradient is negative), indicating that the freezing point drops as the pressure is raised. The thermodynamic reason for this is that ice contracts on melting and, as mentioned earlier, the molecular reason for the decrease in volume on melting is the very open structure of the ice crystal which 'partially collapses' on melting.

The phase rule

Salt (e.g. NaCl) dissolved in water constitutes a two-component system; a solution of two proteins in a buffer would constitute a five-component system – protein A, protein B, water, acid and the salt of the acid. The study of such heterogeneous systems has been greatly simplified by a generalization known as the Gibbs Phase Rule which may be stated simply:

$$f = c - p + 2 \qquad \textbf{6}$$

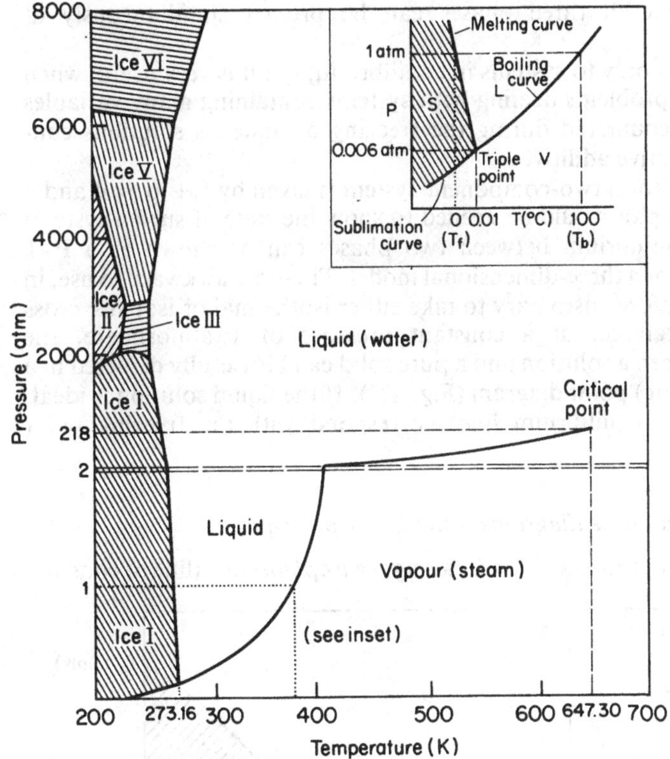

Fig. 1.3 The phase diagram for water – a single component system. The liquid–vapour boundary line represents the vapour pressure at any temperature and the boiling point at any pressure. The solid–liquid line shows how the freezing point depends on the temperature and indicates that enormous pressures are needed to bring about any significant change. (Details are given in the text see p. 24.)

where c is the number of components, p the number of phases, and f the number of degrees of freedom (i.e. the number of independent intensive variables such as temperature, pressure and concentration which must be specified in order to describe the system completely). For example, in the case of a single-component system such as for water (Fig. 1.3), c = 1 and there will be f = 2, 1 or 0 degrees of freedom corresponding to p = 1, 2 or 3 phases (Solid; Liquid or Vapour). Application of the phase rule shows that in a pure phase area (S, L or V), f = 2, meaning that the pressure can be varied independently of temperature. Along the boundary lines, however, p = 2 and f = 1 indicating that for a particular vapour pressure, the other degree of freedom (temperature) is constrained to the value given by the line. In other words, for a given pressure, water has a well-defined freezing point. Finally, at the *triple point*, p = 3 and f = 0 and under these conditions the system is totally fixed such that there is only one unique value of temperature and pressure at which all three phases can be present simultaneously at equilibrium.

The rule applies only to systems in equilibrium, but it is very useful when solving practical problems dealing with systems containing many variables such as those encountered during the freezing of aqueous solutions containing cryoprotective additives.

The phase rule for a two-component system is given by $f + p = 4$ and a four-dimensional plot would be needed to show the state of such a system. However, the equilibrium between two phases can be shown as a P–T composition plot or a three-dimensional model. These are awkward to use, in practice it is therefore customary to take either isothermal or isobaric cross sections. For example, at a constant pressure of 1 atmosphere, the equilibrium between a solution and a pure solid can be usefully depicted in a freezing (or melting) point diagram (Fig. 1.4). If the liquid solution is ideal, the solution–solid equilibrium lines correspond with the freezing point depression curves.

Information from phase diagrams – the Lever principle

Cooling a solution of composition C_1 along the *isopleth* C_1C_1' (line of constant

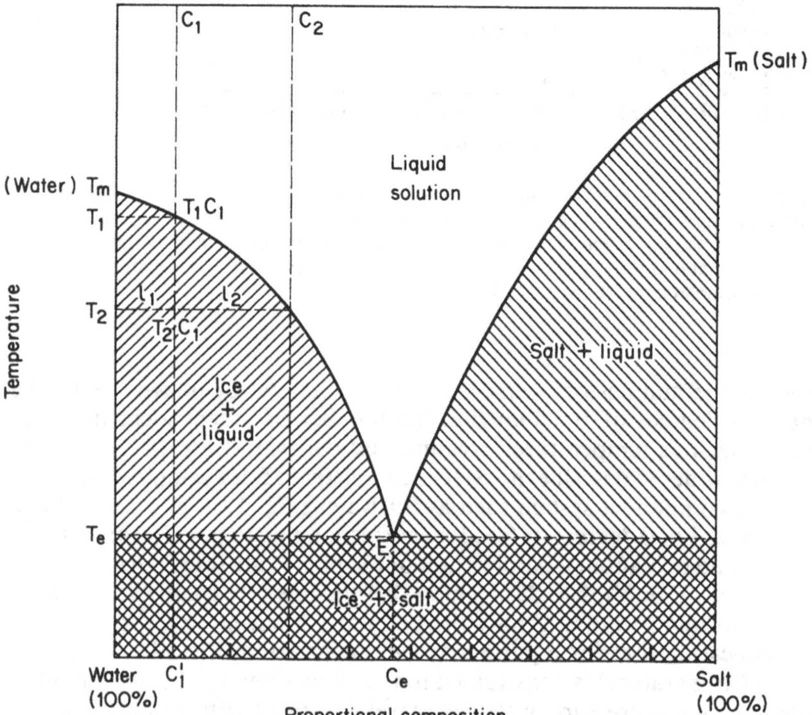

Fig. 1.4 A schematic freezing point or eutectic diagram for a simple two-component system such as an aqueous salt solution. (See text p. 26 for a description).

composition) initially leads to a phase change at temperature T_1 (Fig. 1.4). That is, at point T_1C_1, pure ice begins to separate from the solution which in turn becomes progressively richer in the other component, salt; the solution's composition is given by points along the line T_1C_1–E. At any point, T_2C_1, in the two-phase region the system consists of ice and liquid of composition C_2 in stable coexistence; the proportions of the two being given by the *Lever principle*. The broken line passing horizontally through a typical point T_2C_1 in the two-phase region is called a *tie-line*, and it connects the solid and liquid compositions that are in equilibrium. The lengths l_1 and l_2 of the two sections of the tie-line at T_2 are in the same proportion as the amounts of solution and of ice. If composition is plotted as mole fraction, the ratio will be of the number of moles of solution to the number of moles of ice, and if composition is in weight fraction or per cent, the ratio will be that of weight of solution to weight of ice.

At point E, the *eutectic point*, solution of composition C_e is simultaneously in equilibrium with pure ice and pure, solid salt. The eutectic temperature (T_e) is seen to be the lowest temperature at which liquid can exist in equilibrium with the solid phase. Further cooling beyond T_e leads to the disappearance of the liquid phase and the simultaneous freezing out of ice and salt in relative amounts corresponding to C_e. A eutectic involving ice as one of the solid phases is often called a *cryohydric point*. In biological systems, however, eutectic separation of solid phases is not often observed because of the multicomponent and complex nature of the aqueous phase.

A significant example of the usefulness and application of these principles is demonstrated by recent studies of the relative contributions of the fraction of unfrozen water and of salt concentration to the survival of frozen cells (Mazur *et al.*, 1981).

Phase diagrams of practical benefit to studies in low temperature biology are naturally far more complex than the simplified version used here to illustrate the fundamental principles. In the first instance the incorporation of cryoprotective compounds such as glycerol or dimethylsulphoxide into aqueous solutions demands consideration of tertiary composition phase diagrams. The derivation of phase diagrams for binary and ternary phase systems and their application in cryobiology have been presented in detail (Cocks and Brower, 1974; Shepard, *et al.*, 1976).

Phase transitions in biological membranes

A significant discovery from contemporary molecular biology has been the recognition that biological membranes are not static structures but have a liquid nature or fluidity in their interior. The degree of fluidity is of particular importance in many biological functions, for example transport across the membrane. Biological membranes are, in fact, multicomponent systems but even with as few as two components the phase rule predicts that when two such phases (e.g. solid and liquid) are present, there will still be two degrees of freedom. That is, a two-component system will not show a single sharp melting temperature since it is possible to vary the temperature of the system and still have both phases present. The biological value of this feature is clear, for, with a multicomponent composition, the membrane can gradually change its degree of fluidity over a wide temperature range, thus affording fine 'control' of cell function according to changes in the temperature of the surroundings. If biological membranes were composed of just one kind of lipid or component, the membrane would have a definite melting point and the organism could respond to temperature variation only by an 'all-or-none' change in membrane fluidity and thus in membrane function. The influence of low temperatures on structure and function of biological membranes is discussed in detail elsewhere in this volume.

Nucleation and the growth of ice - homogeneous and heterogeneous nucleation

The above description of phase transitions during the cooling of an aqueous solution might leave the impression that ice formation will automatically commence at the equilibrium freezing point (T_m). In the early 19th century observers recognized that the onset of crystallization during cooling is a highly unpredictable and sample-dependent event. In other words, there is a common tendency for solutions to cool to varying degrees beyond T_m without undergoing a phase change. This phenomenon is commonly referred to as *supercooling* but is better described as *undercooling* (Franks, 1982a).

The reason why liquids are relatively easily supercooled, and freezing does not take place spontaneously at the equilibrium freezing point, is that any crystallization process must be preceded by a nucleation event in which a solid-to-liquid interface is created. Freezing, or the growth of ice crystals, then proceeds by the transference of molecules from the liquid to this interface. A nucleus is a critical group of water molecules which cluster into a configuration which can be identified by other molecules as 'ice-like', and acts as the interface for condensation of additional molecules, with a corresponding reduction in their chemical potential. It is essential that these 'ice embryos', formed spontaneously by random molecular motion producing localized fluctuations in density and orientation, exist for sufficient time to enable diffusion and condensation to take place. Nucleation is therefore a statistical phenomenon.

The lifetime of these water molecule clusters is inversely proportional to the temperature and directly proportional to the size of the critical nucleus. Consequently, in the region of 0°C, the number of molecules necessary for the formation of an effective nucleating cluster is very large. The probability that such a grouping exists in a given sample of liquid is very low, and if it was present, its lifetime would be very short. However, as the temperature is reduced so the number of molecules necessary to provide a critical nucleus is decreased, and the probability of nucleation thus increases. The spontaneous nucleation of ice by such events is called *homogeneous nucleation*, and the absolute lower temperature limit at which the liquid state can exist is the homogeneous nucleation temperature (T_h).

The rate of formation of ice nuclei can, however, be markedly enhanced by other particulate matter which can provide a suitable interface or nucleating cluster. In this case the process is known as *heterogeneous nucleation*. The theory underlying heterogeneous nucleation is not nearly so well understood as that of homogeneous nucleation (Franks, 1982a), mainly because of the influence of extraneous factors such as electric fields and pressure waves on the phenomenon. Practically, however, heterogeneous nucleation is effected only by macroscopic particles, the molecular geometry of which is important. The effectiveness of a heteronucleator depends on the interfacial free energy at the solid–ice interface and, to be effective, the particle must have a diameter greater than 10nm. Although effective nucleators are often crystallographically similar to ice, little is known about the actual crystal structures that constitute efficient nucleators. It is thought that molecularly dispersed solutes cannot facilitate the nucleation of ice (Fletcher, 1970). It is also believed that certain microorganisms are effective ice nuclei, especially at high sub-zero temperatures (Maki *et al.*, 1974).

In practice, ice formation in aqueous solutions is usually initiated by heterogeneous nucleation, but if this has not occurred, freezing will occur by homogeneous nucleation at −38.5°C (the homogeneous nucleation temperature of pure water [Hobbs, 1974]).

Growth and propogation of ice

After a nucleation event has occurred, the liquid–solid phase transition is completed by the growth of the nuclei into crystals of variable size and shape. Factors which influence the morphology and crystal dimensions include the extent of supercooling, the rate of cooling, and the nature and concentrations of any dissolved solutes (Franks, 1982a). Whereas the rate of nucleation increases rapidly with decreasing temperature, the opposite is true for the rate of crystal growth. The propogation of the ice front is a function of the diffusion of water molecules to the liquid–solid interface and is governed by the activation energy. In water cooled slowly below 0°C, ice crystals form around the limited number of early-developing nuclei and, as the temperature falls, these crystals grow larger rather than new crystals forming at other nucleating sites. It is energetically more demanding for a new crystal to be initiated than for the growth of existing crystals to be promoted. Furthermore, very small ice crystals will have high surface energies which render them thermodynamically less stable than larger crystals. Therefore, unless water is frozen very slowly with a minimum of supercooling, the smaller ice crystals that form will be metastable and will fuse with other crystals, increasing their size, and thereby minimizing surface energy. This is the process known as *recrystallization*.

In the frozen food industry the process by which smaller ice crystals decrease in size during storage of a frozen product while larger ice crystals grow in size, even at constant temperatures, is called *maturation*. The number of ice crystals therefore decreases with time while their average size increases. The presence of solutes can affect the kinetics of this maturation process and can therefore affect the storage stability of the frozen product (Reid, 1983).

Generally, slow cooling (<1°C min^{-1}) produces a small number of very large crystals while rapid cooling (>1°C sec^{-1}) produces conditions in which ice nuclei are more plentiful but for which there is insufficient time for crystal growth; the result will be a multitude of very small crystals. For examples of the complexity and diversity of ice crystal types and the observable growth patterns of ice from slightly supercooled water [see Hobbs (1974) and the section on ice crystal types below]. In considering ice growth in multi-component aqueous solutions rather than pure water it must be stressed that a comprehensive theoretical analysis is not yet available. Such analysis is complicated by factors such as the build up of concentration gradients ahead of the advancing ice front (giving rise to a localized lowering of the freezing point), the possibility of incorporation of the solute onto the growing ice phase, the superimposition of a temperature gradient due to the dissipation of latent heat, and, in solutions of electrolytes, additional electrostatic effects brought about by the non-uniform distribution of ions. These latter may give rise to freezing potentials - the Workman–Reynolds effect (Gross, 1972).

Ice nucleation and propagation in biological cells and tissues

Ice nucleation and crystal growth are influenced by the rate of cooling. In the natural environment the rates of cooling are dictated by climatic fluctuations and are low. In the laboratory, however, the cooling rate imposed upon biological cells and tissue may vary over orders of magnitude.

The equilibrium freezing point for a wide range of cell types lies between 0 and -1°C. Despite this, such cells frequently demonstrate an ability to undercool by as much as 15°C before the nucleation of intracellular ice. Thus, when cells are cooled slowly, ice formation is typically initiated in the extracellular medium while the cells themselves remain unfrozen. (For this brief discussion plant cell walls should be considered as part of the extra-cellular medium and the protoplast synonymous with cell.) From such observations it has been stated that the plasma membrane is an effective barrier to ice propagation (Luyet *et al.*, 1964).

When the extracellular fluid is frozen, and the cell is not, a gradient of water potential is established such that water flows from the cell to the exterior, where it subsequently freezes, in an attempt to achieve an osmotic equilibrium (Grout and Morris, this volume). As a direct consequence there is a reduction in cell volume. If cooling is slow enough to permit this process to be prolonged, the cells could become so extensively dehydrated that no intra-cellular freezing would take place. Intracellular freezing is widely believed to be lethal to cells, but this has been questioned by some recent studies (Rall *et al.*, 1980) and it must be emphasized that the injurious process may not be the freezing itself but the subsequent recrystallization, either during further cooling or during slow thawing.

A number of reasons have been put forward for the preferential occurrence of extracellular nucleation. Firstly, as heat is removed by conduction from the external surface of a specimen, the coldest region will always be in the extracellular compartment. Secondly, the probability of ice nucleation occurring in any given compartment is directly related to size, and since the extracellular fluid forms one very large compartment compared with the many small compartments of the intracellular space, nucleation will occur in the extracellular fluid before a significant number of cells has frozen internally. Once nucleated, ice will propogate throughout a compartment until equilibrium is re-established. If nucleation occurs inside a cell, ice will grow throughout that cell but may not be able to seed any other intracellular compartment. Consequently, even if a limited number of cells in a tissue freeze internally before extracellular nucleation occurs, once extracellular crystallization has been initiated ice will continue to grow throughout that compartment.

The question of whether ice nucleation in biological cells and tissues is initiated by homo- or heteronucleation has not been unequivocally answered by the limited number of studies conducted to date. Despite some earlier reports that yeast cells and human erythrocytes, confined in droplet emulsions, freeze by homogeneous nucleation (Rasmussen *et al.*, 1975) it has been shown for several plant and animal cells cooled under conditions of maximal supercooling (20–30°C) that freezing is initiated by heterogeneous nucleation in the absence of extracellular ice (Franks and Bray, 1980; Franks *et al.*, 1983). It has also been shown that the body fluids of Antarctic insects freeze by heterogeneous nucleation (Block and Young, 1979). The influence of cryoprotective, non-electrolyte compounds on the intracellular nucleation of mammalian embryos has also been studied (Rall *et al.*, 1983). For mouse embryos cooled rapidly (15–20°C min^{-1}) it was concluded that intracellular ice-nucleation in the absence of cryoprotective additive is initiated heterogeneously, but in the presence of increasing concentrations of either glycerol or DMSO, the nucleation temperature is increasingly depressed until in the presence of 1.5 and 2M concentrations it drops close to the expected homogeneous nucleation temperature.

Rapid cooling promotes intracellular ice formation because there is insufficient time for supercooled water to escape from the cells, in response to the osmotic gradient established by extracellular ice, before the contained cell water freezes. As the temperature falls and, if freezing has not already been initiated by heterogeneous nucleation, any remaining supercooled cellular water may crystallize by homogeneous nucleation at *c.* −40°C (see above). There is, however, one further mechanism by which intracellular ice nucleation may be initiated, and that is a special case of heterogeneous nucleation in which extracellular ice grows through 'pores' or altered regions of the plasma membrane (Mazur, 1965). The occurrence of such trans-membrane nucleation may exist, however, only after injury has been sustained (Luyet *et al.*, 1964).

Factors, such as cell volume, water permeability and its temperature coefficient, surface area, rate of temperature change and intracellular nucleation temperature provide the basis of a physico-chemical model for a quantitative description of the movement of intracellular water during cooling. This enables calculation of the extent to which the cytoplasm of a particular type of cell is supercooled at any temperature and thus permits prediction of the likelihood of ice forming inside the cell during cooling at any specific rate (Mazur, 1963, 1966). The general validity of such calculations has been supported experimentally by measuring cell volumes and by demonstrating the presence or absence of intracellular ice (e.g. Leibo, 1977).

The importance and implications of these calculations for the cryopreservation of cells and tissues is presented elsewhere (Mazur, 1981).

The influence of cooling rate and temperature on the size of ice crystals in cells and tissues is also of particular importance for freeze-fixation techniques in ultrastructural research. Such techniques aim to minimize the size of ice crystals and even to approach the vitrified state using quenching techniques. The principle involved is that such rapid cooling will restrict the size of ice crystals to dimensions which are beyond the resolution of the electron microscope. It is hoped that freeze-fixation will provide a more accurate representation of the *in vivo* state than the more conventional chemical fixation methods (Skaer, this volume).

Finally, mention should be made of the exploitation of ice nucleation, or its specific inhibition in the plant kingdom, as a means of adaptation and protection during exposure to low temperatures (Burke *et al.*, 1976). Nucleation may actually enhance resistance, by way of thermal buffering, in Afro-alpine plants where specific nucleating agent-induced freezing occurs. *Lobelia telekii* grows at high altitudes in temperatures that cycle between +10°C during the day and −10°C at night, and employs ice nucleating substances (polysaccharides) to ensure freezing when the temperature drops to −0.5°C. In so doing the latent heat evolved helps to maintain the temperature of the plant close to 0°C (Krog *et al.*, 1979). Thus the plant can exist expressing both growth and resistance, whereas many resistant plants in Arctic and temperate regions suspend growth during the cold season.

An alternative and much more common approach, however, seems to be the exploitation of nucleation inhibitors as a means of achieving deep supercooling and, therefore, protection by freeze-avoidance. For example, in fully acclimated hickory it appears that supercooling to temperatures below −40°C can be achieved. Resistance mechanisms in animals generally rely on freeze avoidance rather than freeze tolerance and, although the mechanisms by which ice nucleation is inhibited are not fully understood, one possibility is that heteronuclei are rendered inactive by a 'chemical poisoning' effect, enabling the fluid to supercool to a limited extent. Such a process has been likened to the inactivation of enzymes by the irreversible adsorption of particular molecules at their active sites. It has been suggested that a possible mode of action for the so-called antifreeze proteins and glycoproteins isolated from the sera of polar fish species might originate from a direct effect of the peptides on nucleation and/or crystal growth rather than by a straightforward colligative effect of freezing point depression (De Vries, 1980).

Metastability – the non-equilibrium freezing behaviour of aqueous systems

When considering the physico-chemical and biophysical responses of biological systems to low temperatures, it is important to be aware that events rarely take place under true equilibrium conditions. For example, the phase change phenomena depicted schematically by phase boundaries (Fig. 1.4) would hold true only for a simple binary system such as NaCl–H_2O in which supercooling was avoided by ensuring nucleation occurred at the equilibrium freezing/melting point (T_m) during cooling at a sufficiently slow rate to prevent appreciable temperature gradients. It is well recognized that many interdependent factors determine whether an aqueous system, such as a biological system, approaches the thermodynamic state of lowest free energy during cooling. Metastability is thus often unavoidable, especially in concentrated systems. Many of these non-equilibrium states are, however, sufficiently reproducible and permanent to have been described as pseudo-equilibrium states. Conversion of such metastable thermodynamic states to more stable forms may be subject to large kinetic barriers. The prevalence of 'unfreezable' or 'bound' water (see above) is a prime example, where the expected path of thermodynamic stabilization by way of crystallization is prevented by large kinetic restraints. The fraction of total water that is unable to freeze is often regarded, for practical purposes, as water of hydration of the solute. It has already been mentioned that the thermodynamic equilibrium state of eutectic separation of solid phases is not often observed during the deep cooling of multicomponent biological systems.

A clear understanding of the occurrence and effects of metastable states during the cooling of compartmentalized living systems, is complicated by the interaction of thermodynamic and kinetic factors which are difficult to separate. Furthermore, these complexities are compounded when such systems are cooled rapidly to low sub-zero temperatures. Nevertheless, some basic principles have been established based upon studies using model systems; for example, aqueous solutions of ethyleneglycol, sucrose, glycerol, polyvinylpyrrolidone (PVP) and other macromolecules that interact with water by hydrogen bonding (MacKenzie, 1977). The knowledge from studies on model macromolecular systems can, qualitatively, permit some interpretation of the non-equilibrium phase behaviour of the fluids in cells and tissues during cooling. Data for phase transitions such as nucleation, glass transitions, devitrification and recrystallization as well as melting behaviour can be predicted from the known composition and concentration of many cell solutions. Such inferences must, however, be made with caution because the course of events during the cooling and warming of cells and tissues often appears to depend upon their compartmentalization into domains bounded by semi-permeable membranes. Further complications may arise from the presence of a variety of insoluble species which interfere with or are affected by the freezing/melting behaviour of the system (MacKenzie, 1977). It is clear that an understanding of the non-equilibrium freezing of cells and tissues requires the direct determination of phase behaviour in these systems; the results of such studies have only recently begun to emerge (Franks and Bray, 1980; MacKenzie, 1981; Franks *et al.*, 1983).

The principal non-equilibrium phenomena observed during cooling and attempted freezing of representative aqueous systems will be summarized here.

Experimental techniques

Studies of low temperature metastable states and transitions in aqueous systems have involved the use of a variety of experimental techniques. These have included light and electron microscopic techniques to examine the morphology and growth of ice crystals, and thermal techniques to reveal first and second order phase transitions. The thermal approach has been based specifically on two techniques, differential thermal analysis (DTA) and differential scanning calorimetry (DSC). DTA monitors the temperature difference between the experimental test sample and a reference sample for an experimental period during which heat is extracted from both samples at the same rate. DSC, however, maintains the samples at the same temperature as each other throughout the scanning sequence by differentially supplying or extracting heat and measures this differential rate of supply of energy. As the rate of temperature scanning is constant, the actual quantity measured is therefore the specific heat difference between the sample and the control. DSC therefore allows thermodynamic quantities such as enthalpy to be measured directly, which is not the case for DTA. This has been of particular benefit in the quantitative estimation of nucleation rates (Franks, 1981a).

The droplet emulsion technique is of value in the study of phase transitions and metastable states in aqueous solutions, especially homogeneous nucleation. In this technique samples of pure water or aqueous solutions are dispersed into droplets (0.5–100 μm diameter) forming an emulsion in an inert dispersant such as n-heptane or silicone oil (Rasmussen and MacKenzie, 1973; Franks *et al.*, 1983). The majority of droplets are free of extraneous nucleants and do not solidify until the maximum supercooling temperature is reached. The measured homogeneous nucleation temperature (T_h) depends to some extent upon droplet diameter, but there now seems to be general agreement from these and other methods of determination that T_h for water is $-39° \pm 1°C$. Alternative methods of sampling, dependent upon the type of measurement being performed, are thin layer films of 1–2 μm thickness and fine column capilliary techniques (Angell, 1982).

Although single cells have been successfully incorporated into emulsified water droplets to study the freezing behaviour after deep supercooling (Rasmussen *et al.*, 1975; Franks *et al.*, 1983), sampling of biological tissues for direct measurement of phase transitions by thermal analysis techniques is problematic. Electrical resistance measurements in specially developed sample holders have, however, been used successfully to reveal phase transitions in animal tissues and a satisfactory correlation between the strictly thermal and electrical methods has been claimed (MacKenzie, 1981). Thermal measurements are particularly well suited for analytical investigations of state changes, but a number of other techniques such as dielectric permittivity measurements (Amrhein and Luyet, 1966), low-temperature X-ray diffraction (Dowell and Rinfret, 1960), and nuclear magnetic resonance (Derbyshire, 1982) can also yield useful information about the dynamics of metastable systems.

Solid–liquid states and transition temperatures

The basic knowledge of the non-equilibrium freezing behaviour of aqueous systems cannot be summarized without acknowledging the pioneering work of Father Basile Luyet and his small group of associates (Rapatz, MacKenzie and Rasmussen) in the 1960s and early 1970s. Their work, which introduced and utilized several of the techniques described in the previous section, has contributed much to the understanding of the effects of solutes and cooling conditions on the kinetics of freezing and ice morphology. Studies were conducted on aqueous solutions of compounds such as ethylene glycol, sucrose, glycerol, gelatin and PVP, as models of the more complex biological systems in which instability might play an important role in low temperature injury. The recognition that, in the temperature range from absolute zero to the melting point, the structure of a frozen solution is not stable and that there exist regions of instability and increased molecular mobility, has led to detailed studies on the interactive effects of four major contributory variables. These are temperature, cooling rate, the nature of the solute and its concentration.

Aqueous solutions of simple, low molecular weight solutes tend to supercool before ice crystallization is finally initiated. This tendency becomes more pronounced with increasing concentration and with increasing molecular complexity of the solute. Furthermore, in many cases the solute does not crystallize and there is no eutectic transition in the solid–liquid phase diagram. It is also recognized that fast rates of cooling prevent attainment of the phase-transition equilibrium such that complete crystallization is prevented or hindered, and amorphous solid states (glassy states) are produced. This *vitrification* of pure water by very rapid cooling, using quenching techniques, has proved difficult to achieve (Angell 1982), whereas aqueous solutions tend to undergo glass transitions with increasing ease as the concentration of dissolved solute increases. Even at slow cooling rates it can prove quite difficult, even impossible, to initiate nucleation and the growth of ice in some highly concentrated aqueous solutions. The entire solution will solidify in the amorphous state. In practice the freezing of low, and moderately concentrated, aqueous solutions invariably gives rise to partial crystallization so that a finite volume fraction of the liquid will crystallize and the crystals are embedded in the glassy matrix. A frozen solution thus consists of a framework of ice, often with very fine ramifications, bathed in a medium in which marked concentration gradients may exist for prolonged periods because of the slow attainment of equilibrium due to slow diffusion at low temperatures. The proportion of solution remaining unfrozen is often considerable and even during slow cooling a relatively large fraction of a solution may be in the vitreous state (Luyet, 1969).

When sufficient time is allowed for a freezing solution to reach the phase transition equilibrium, the amount of ice formed (which is determined by the original concentration and the lowest temperature reached), can be calculated from the melting curve of the phase diagram. The corresponding amount of non-frozen solution can also be calculated for each original concentration and each final temperature. Such calculations provide the basis of studies to determine the influence of the fraction of unfrozen water as a potential mechanism of injury during the cryopreservation of cells (Mazur *et al.*, 1981). In situations where equilibrium is not reached or maintained, however, the proportion of ice and of non-frozen material depends upon the extent to which the cooling rate prevents or restricts the formation of ice. This cannot be calculated on the basis of an established rule but has to be determined in each particular case.

For experimental investigation metastable, non-equilibrium states are invariably induced by employing high-solute concentrations and/or very rapid cooling to temperatures below the glass transition (T_g). In this way, ice crystallization can be either inhibited or modified so that thermolabile crystal modifications are produced which will undergo a variety of state transitions as the temperature is raised towards the equilibrium melting point.

Ice crystal types

The dedication of Luyet to observing, characterizing and recording crystallization events and ice growth patterns in frozen aqueous solutions (see Meryman's obituary of Luyet, 1975) has provided a valuable insight into the freezing behaviour of solutions. On the basis of freezing thin films of solution on the cold stage of a purposely built light microscope (cryomicroscope) Luyet and Rapatz (1958) identified four principal forms of ice separated from solutions of a variety of solutes. These are described as compact solid masses (Fig. 1.5c), regularly arranged dendrites, irregular dendrites and spherulites. The latter three types have been classified together as arborescences; they have essentially the same structure but differ in the size and number of branches (Fig. 1.5). In pure water ice always crystallizes in hexagonal dendritic forms, whereas aqueous solutions containing dissolved solutes produce a variety of forms of ice dependent on the combined action of four principal factors; temperature, cooling rate, nature of the solute and its concentration. Some detailed studies of the influence of temperature, cooling rate and solute concentration on the type of ice crystal units formed during freezing have been reported for solutions containing either glycerol or PVP (Rapatz and Luyet, 1966; 1968). In general, slow rates of cooling to temperatures just a few degrees below the freezing point give rise to crystal formations with hexagonal symmetry for all concentrations tested. This indicates that an essential requirement for arrangement of molecules into regular hexagonal crystals is slow growth. Faster cooling to intermediate temperature results in irregularly formed arborescent structures and rapid cooling produces spherulitic crystalline formations (either *coarse* or *evanescent* spherulites, depending upon the freezing temperature and solute concentration). Under the light microscope, spherulites are seen to result from the uniform radial growth of numerous ice spears, apparently devoid of branches, arising from centres of nucleation. Evanescent spherulites, produced by very rapid cooling rates, appear as faint, almost transparent discs which usually require polarized light to reveal their outlines. The term 'evanescent' has been applied to those units having radial fibres which are so thin that they are not visible in ordinary light. The rate of cooling influences not only the morphological features of the crystallization units formed but also the completeness of crystallization. Evidence from X-ray diffraction data, in particular, has indicated that evanescent spherulites represent incompletely frozen material as the very high cooling velocities involved prevent the molecules from completing the process of crystallization. Before this was

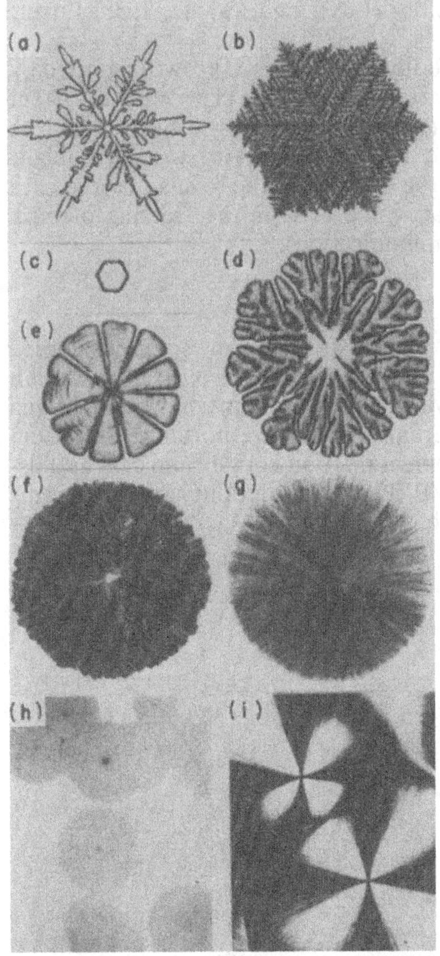

Fig. 1.5 The principal types of ice crystallization observed in aqueous solutions at various temperatures, cooling rates and solute concentrations: (a) and (b) regular dendritic hexagonal forms; (c) compact solid hexagonal form; (d) and (e) irregular dendritic rosettes; (f) transition form; (g) plain (coarse) spherulite; (h) evanescent spherulite (non-polarized); (i) evanescent spherulite (polarized light). (Reproduced from Luyet and Rapatz, 1958 by permission).

recognized, many rapidly cooled solutions were regarded as having vitrified when in actual fact they contained evanescent spherulites. Furthermore, it is now known that these incipient or incomplete forms of crystallization can be completed during warming; this is known as recrystallization.

Many crystalline formations which are intermediate between these main types can occur; some of these are recognized as *transition forms*. The irregular dendritic forms which develop in solutions of all concentrations at intermediate temperatures are themselves regarded as a transitional stage between hexagonal units and spherulites (Luyet and Rapatz, 1958; Luyet, 1965; Luyet, 1966).

It is possible that the detailed morphology of ice crystal growth reported by Luyet *et al.* (1964) may be subject to artefact due to the experimental system of freezing thin layers sandwiched between glass cover slips. Factors such as unnatural nucleation events initiated by the surfaces of the glass might influence the crystal morphology to the extent that the forms observed are not truly representative of those that would arise in the freezing of bulk solutions (Franks, 1984, personal communication).

Supplemented phase diagrams

A synopsis of the solid–liquid states that exist in aqueous solutions cooled under non-equilibrium conditions and of the characteristic transition temperatures identified by slowly rewarming such systems may best be given by specifically prepared 'phase diagrams'. These are normal equilibrium phase diagrams enhanced with supplementary, non-equilibrium data as illustrated for glycerol (Fig. 1.6) and PVP (Fig. 1.7) solutions. Such diagrams have been termed supplemented phase diagrams (MacKenzie, 1977) or solid–liquid state diagrams (Franks, 1982a).

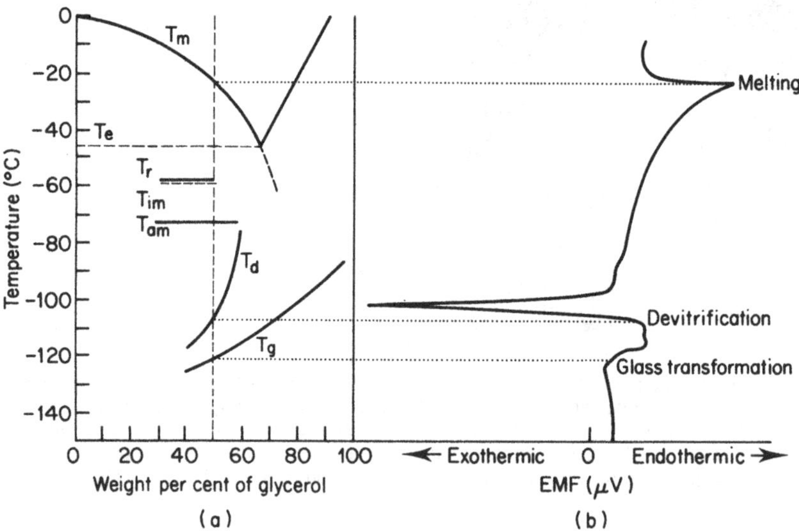

Fig. 1.6 (a) Supplemented phase diagram for aqueous solutions of glycerol showing the transition temperatures for the glass transformation (T_g), devitrification (T_d), ante-melting (T_{am}), incipient melting (T_{im}), recrystallization (T_r), eutectic separation (T_e) and melting (T_m) at various concentrations of glycerol. (Adapted from Luyet and Rasmussen, 1968).

(b) A differential thermogram obtained from DTA of a 50% (by weight) solution of glycerol in water. The thermogram has been redrawn (from Luyet and Rasmussen, 1968) so that the temperature scale corresponds to that of the supplemented phase diagram in (a), illustrating the correlation between events in the thermal analysis and points plotted on the supplemented phase diagram.

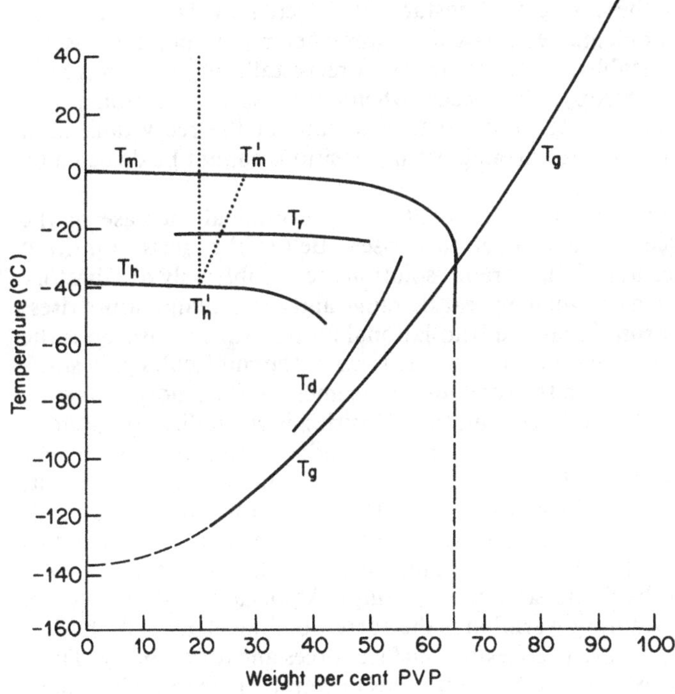

Fig. 1.7 Supplemented phase diagram for the system water–polyvinylpyrrolidone, illustrating solid–liquid states and the following transitions: melting (T_m), recrystallization (T_r), homogeneous nucleation (T_h), devitrification (T_d) and the glass transformation (T_g). The diagram has been redrawn from the compiled DTA data of MacKenzie and Rasmussen (1972) and the more recent DSC data of Franks *et al.* (1977).

It was mentioned previously that much supplementary non-equilibrium data has been derived from differential thermal analysis (DTA) measurements of the temperatures of instability in a variety of frozen aqueous solutions. One approach has been to cool the preparations very rapidly in order to limit the amount of ice formed and maximize the amount of amorphous material; another approach has been to cool samples slowly so that the crystallization of water is not hindered. In this way the transitions that both the amorphous and the crystalline phases undergo during the rewarming of the solidified solutions could be recorded for a variety of solutes over a range of concentrations. Supplemented phase diagrams can then be constructed to show the temperatures of transitions or instabilities in partially frozen solutions in terms of their concentration. Five principal transitions are identified (Fig. 1.6) of which three are well authenticated phase transitions, namely, *glass transformations* (labelled T_g), *devitrification* (T_d) and *melting* (T_m). In addition, two further thermal events have been identified a few degrees below the commencement of melting. These minor endothermic events, termed *ante-melting* (T_{am}) and *incipient melting* (T_{im}), seem to be well established experimentally, but little is known about their nature.

In addition to these stages of instability detected by DTA and DSC (Franks, 1982a) supplemented phase diagrams often represent temperatures at which the well-established phenomenon of recrystallization (T_r) has been determined by microscopy. This occurs when a transparent solution at low temperature is suddenly observed to turn opaque at the recrystallization temperature during slow rewarming (such transitions cannot be detected by DTA).

The nature of these transitions relates to the gradual increase in the mobility of molecules as temperature rises. Below the glass transition temperature molecules of the vitreous solution are capable only of vibratory motion, but within a certain temperature range and as the temperature rises, they acquire both rotational and translational motion (this transition is the glass transformation). As the temperature rises further molecules gain sufficient mobility to pass from the random arrangement of the amorphous state to the ordered structure of a crystal; this transition is crystallization or devitrification. At even higher temperatures under conditions of higher molecular activity the smaller ice crystals become unstable and, as they melt, their molecules are transferred to large crystals. The growth of large crystals at the expense of small crystals during rewarming is the phenomenon known as recrystallization. Finally, the crystalline material disappears as the solid phase changes to the liquid state (i.e. melting). Associated with increase in molecular motility during warming is the corresponding change in physical properties such as viscosity, cohesion, surface forces and diffusibility. These changes would be expected to be continuous and gradual except when interrupted at specific transitions, when they might be discontinuous and even abrupt. The glass transformation is accompanied by a marked change in physical properties such as specific heat and refractive index. The increase in specific heat of such a preparation results in a greater uptake of heat per unit rise in temperature thus allowing the transition temperature to be detected by thermal techniques such as DTA and DSC. Devitrification, being a crystallization event, is an exothermic process generally revealed by a high peak in the differential thermogram. Melting involves the absorption of the latent heat of crystallization and, as a marked endothermic process, also gives rise to a sharp event on the thermogram. These thermal events can be correlated with points on a supplemented phase diagram (Fig. 1.6). It is also apparent that a preparation which has been cooled rapidly demonstrates a devitrification peak but no 'ante-melting' endotherm (Fig. 1.6b).

Glass transition and melting are generally observed in both rapidly and slowly cooled samples, whereas devitrification occurs only in rapidly frozen material and the ante-melting transition occurs only in slowly cooled samples. It has also been recorded that, in slowly cooled preparations, when maximum crystallization was achieved, the beginning of the melting endotherm (T_{im}) could be detected well below the eutectic temperature (T_e) (Luyet and Rasmussen, 1968). It therefore seems that both rapidly and slowly frozen specimens contain amorphous and crystalline materials. The relative proportions of ice and of water that remain in the noncrystalline state when samples of solutions of various concentrations are frozen under a variety of conditions is of importance with regard to the survival of biological materials exposed to low temperatures. Luyet, reviewing some of the early studies on the amount of water remaining amorphous in frozen solutions, recorded his surprise at the high values obtained after slow freezing, and expressed the sentiment that 'cryobiologists, who are generally inclined to consider the amorphous component as negligible, are apparently unaware of these high figures' (Luyet, 1969). Quantitative data presented for aqueous glycerol solutions show that under equilibrium conditions of slow cooling, at temperatures below −58°C when no more water can crystallize as ice, significant proportions of the system remain amorphous, for example the unfrozen portion of a 2 molal solution of glycerol will amount to more than 20% of the system under these conditions.

Prediction of the amount of a tertiary system which remains unfrozen at low temperatures can be made using ternary phase diagram data derived from DTA measurements. For example, the addition of glycerol to H_2O-NaCl binary solutions in the weight ratio 77.5:22.5 depresses the eutectic point from −21.2°C in the H_2O-NaCl binary system to about −80°C with a eutectic composition of H_2O(28%)-NaCl(9%)-glycerol(63%) (Shepard *et al.*, 1976). These calculations are, however, based on equilibrium phase relationships, and therefore do not take into account the possible formation of vitreous phases. In practice it has been found that even with cooling as low as 5°C min^{-1} many samples contain metastable, glassy phases and the predicted formation of the ternary eutectic for H_2O–NaCl–glycerol in the region of −80°C was unattainable even after repeated cooling and warming to eliminate the glassy phases by devitrification (Shepard *et al.*, 1976). Studies on ternary systems involving the hydrophilic polymer hydroxyethyl starch (HES) as one of the components have stressed the importance of absorbed unfreezable water as a possible explanation for the mechanism of cryoprotection afforded by such compounds (Korber and Scheiwe, 1980; Korber *et al.*, 1982).

In addition to cooling rate, solute concentration is an important determinant of the temperatures of instability in solutions. However, it appears that not all transitions are dependent upon the initial concentration of solute prior to cooling. This is true particularly for the temperatures of antemelting, incipient melting and recrystallization (Fig. 1.6). In general, glass transformations were recorded at high concentrations; devitrification, antemelting, incipient melting and also the observations of recrystallization were made at moderate concentrations; and melting occurred at low and intermediate concentrations.

Supplemented phase diagrams of varying degrees of completeness have been prepared for a variety of aqueous mixtures. These include dimethyl sulphoxide (Rasmussen and MacKenzie, 1968); glycerol, ethylene glycol and glucose (Luyet and Rasmussen, 1968; Rasmussen and Luyet, 1969); sucrose (MacKenzie, 1977); ribose (Reid, 1979); 1,2-propanediol (Boutron and Kaufmann, 1979b); hydroxyethyl starch (Franks *et al.*, 1977); and polyvinyl pyrrolidine (PVP) (MacKenzie, 1977; Franks *et al.*, 1977). Studies using water-soluble polymers (e.g. PVP) are of particular interest because they may, to a greater or lesser extent, be representative of the behaviour of biological material. A particularly important observation is the apparent inability of the polymer to crystallize under any circumstances. The organic solute component of most aqueous solutions tends to retain a certain amount of water and demonstrate a readiness to form a supersaturated solution, and it is only the aqueous component that crystallizes with ease (Franks, 1982a). Qualitatively, the behaviour of polymer solutions is similar to that of small molecules, although the high viscosities of polymer solutions facilitate vitrification and make it much easier to initiate and maintain metastable states. In other words, the tendency for aqueous solutions of simple, low molecular weight solutes to supersaturate and supercool before ice finally crystallizes becomes more evident with increasing concentration and with increasing molecular complexity. Furthermore, there is the possibility that the solute does not crystallize and therefore fails to give rise to a eutectic break in the solid–liquid phase diagram (Fig. 1.7). Figure 1.7 illustrates the changes that could be expected when, for example a 20 per cent w/w solution of PVP is cooled from ambient temperatures. Unless freezing is induced at a particular temperature (T_m) by seeding with an ice crystal the solution will undercool and, if heterogeneous nucleation is avoided, undercooling can be maintained to the homogeneous nucleation temperature, T_h'. Assuming cooling is sufficiently slow, part of the latent heat evolved would return the system to the melting point curve at some point, T_m'. Further freezing would then follow the T_m curve with the concentration (and viscosity) of the residual liquid phase increasing until the glass transition temperature (T_g) is reached. At this point where the T_m curve becomes vertical (~ −30°C), no more ice will freeze out and the whole system solidifies with 65 per cent of the water remaining unfrozen. There are no indications of the hydrated polymer phase crystallizing, and further cooling produces solidification of the supersaturated solution as a glass. This phenomenon exposes the non-equilibrium, metastable nature of such solutions and, as discussed earlier, the usual interpretation of such unfreezable water is in terms of water molecules hindered from diffusing to the face of ice crystals by specific binding to polar groups on the polymer.

The general view of solid–liquid phase transitions as depicted in supplemented phase diagrams is that phase transitions encountered during rapid cooling and subsequent rewarming are fundamentally the same transitions that take place during slow freezing and rewarming. Rapid cooling rates merely hinder some stages of the process and rewarming allows resumption and completion of the hindered stages. The main factors responsible for delaying the motion of water molecules, and therefore hindering some stages of phase transitions such as crystallization, are low temperature and the presence of molecules of solute.

Vitrification (glass transformation)

Amorphous, solid states can be stabilized by employing fast rates of cooling, and the amorphous solid so obtained crystallizes (devitrifies) when rewarmed to a temperature T_d, unless rewarmed very rapidly. When a liquid is cooled so that it assumes the vitreous state, its volume decreases according to a contraction coefficient until the temperature is lowered to the glass transition point (T_g), below which the volume decreases according to a much smaller coefficient. During the first stages of contraction, above the glass transition, the molecules are separated from each other by appreciable distances and are capable of translational and rotational motion, and will move closer together. Eventually a stage is reached when the intermolecular distance is so reduced that the molecules are no longer capable of translational motion and they form a rigid framework for which the coefficient of contraction will be abruptly and markedly reduced. The translational diffusion is so slow that the metastable state can be maintained for many years. The physical behaviour of such a system is similar in several respects to that of crystals, except that the molecules which are essentially now only capable of vibratory motion do not possess an ordered arrangement. When a vitrified substance passes through the glass transformation stage it undergoes an abrupt change in several physical properties such as specific heat, expansion coefficient, refractive index and, to a lesser extent, dielectric permittivity. The glass transition temperature (T_g) is therefore defined as the temperature of abrupt change in physical properties without change of order. Below T_g, the substance is in the glassy (vitreous) state and above T_g it is a supercooled liquid (though very viscous). Both below and above T_g the substance is amorphous.

The viscosity of a supercooled liquid, which is related to rotational and translational molecular diffusion, is known to rise more sharply with decreasing temperature than would be predicted from the simple Arrhenius relationship. This marked increase in viscosity with falling temperature continues to the limit at the glass transition. The glassy state of a substance is therefore often defined in terms of its response to an applied shear stress (i.e. shear viscosity). Although the dividing line between liquid and solid is a little vague, it is often taken to be $\sim 10^{15}$ poise (Franks, 1982a).

A variety of techniques to secure the vitrification of pure water by ultra-rapid cooling show poor agreement in their estimates of the glass transition temperature (T_g) for water. Theoretical considerations suggest that water should be difficult, but not impossible, to vitrify but there appears to be some doubt that complete vitrification has yet been achieved in practice (Angell, 1982). Information about the glass transition of water has been provided mainly by quench-cooling aqueous solutions of glass-forming compounds such as a polyhydroxy compounds or polymers. Extrapolation of the T_g curves for such solutions to zero concentration have indicated that the T_g for water is about $-137°C$. Actual values are, however, dependent upon the warming rates used during the DTA analyses (MacKenzie, 1977).

Attempts to completely prevent crystallization in 20% gelatin solutions were unsuccessful even when estimated cooling rates of the order of 1×10^5 to 5×10^5 °C sec^{-1} were achieved (MacKenzie and Luyet, 1962). The recognition that partial crystallization seems unavoidable in such quench-cooled aqueous solutions has led to a definition of vitrification as the state where crystallites are smaller than 10nm (Riehle, 1968). The ultrastructural appearance of polymer–water mixtures in the vitrified state has generally been shown to have a microspherical texture consisting of close-packed spheres of nanometer dimensions. The microspheres are interpreted as representing aggregates of the hydrated polymer which have minimized their interfacial free energy by taking up a spherical conformation. It has been speculated that this might resemble the state of aggregation of the hydrated polymer in solution at normal temperatures. Such detail was achieved by rapidly cooling samples prior to freeze-fracture at a temperature below T_g (Franks *et al.*, 1977).

Devitrification

The crystallization phase transition invariably observed during the rewarming of a rapidly cooled solution is often referred to as devitrification. Associated with this transition is a marked exothermic displacement in the differential thermogram which represents the movement of water molecules from positions in the disorderly amorphous state to new positions in orderly arranged crystals. This occurs when the temperature is high enough to impart sufficient freedom in molecular mobility to effect the phase change.

When crystallization takes place during rewarming it is important to distinguish two cases: one consists of the growth of pre-existing crystallites arising from nucleation during cooling; the other case (devitrification) consists of the formation of new crystals involving both nucleation and crystal growth. In both cases the material passes from the amorphous to the crystalline state.

Stability of the amorphous state

Several mechanisms have been proposed for the protection of biological cells against freezing injury by cryoprotective additives (Shlafer, 1981). Most attention has been given to cryoprotection mediated through the colligative effect of freezing point depression by which the amount of ice that forms at a given sub-zero temperature is reduced, together with a reduction in the build-up of salt concentration. Other mechanisms have, however, been implicated; for example partial osmotic dehydration of the cells by the cryoprotectant, the retention of bound water and vitrification of cell water as a result of cryoprotectant-related inhibition of crystalline ice formation. A comprehensive study of the stability of the amorphous state in a number of systems containing a variety of cryoprotectant compounds has been reported (Boutron and Kaufmann, 1978, 1979a, 1979b). Such investigations attempt to provide an understanding of the factors governing the stabilization of the amorphous state in aqueous cryoprotectant mixtures in the hope that better cryoprotection might be achieved at low temperatures. In the extreme case when a solution is maintained entirely in the amorphous state even during slow cooling or warming rates, all the cells should be protected (in the absence of cold shock or osmotic stresses prior to cooling) because there would be no phase transition during cooling and warming and the injuries associated with the co-existence of two phases would be avoided.

The stability of the amorphous state has been defined empirically in terms of the *critical heating rate*, V_{cr}, above which an amorphous sample has insufficient time to crystallize, even partially, before the melting temperature is reached. The smaller the value of V_{cr} the more stable the amorphous state.

The dependence of T_d on the rate of warming can be measured, and the difference $T_m - T_d$, corresponding to a given warming rate, can be used to define the stability of the amorphous state (Boutron and Kaufmann, 1979b). The warming rate for which $T_m - T_d$ is zero is defined as the critical heating rate V_{cr} for which the supercooled mixture does not devitrify or recrystallize. In general $T_m - T_d$ was found to vary linearly with the logarithm of the warming rate. Interestingly, it has been shown on the basis of these considerations that the stability of the wholly amorphous state of aqueous solutions of 1,2-propanediol and its glass-formation tendency are much greater, for the same water contents, than for all of the other solutions studied including those of the more commonly used cryoprotectants such as glycerol, dimethylsulphoxide, ethylene glycol and ethanol (Boutron and Kaufmann, 1979b; Boutron *et al.*, 1982). On the basis of cryoprotection relating to the attainment and stability of the amorphous state, 1,2-propanediol should prove to be a better cryoprotectant than all of the other compounds. Indeed, 1,2-propanediol affords greater protection to red blood cells than equivalent concentrations of glycerol (15% and 20% w/w) at low cooling rates and also at high cooling rates using concentrations of 30% w/w (Boutron *et al.*, 1982).

The difficulties in relating these fundamental studies to practical cryopreservation concern the high concentrations of solute necessary (40–50%) to ensure total vitrification during cooling, and the high heating rates generally required to avoid crystallization during rewarming. Nevertheless, these studies have shown that some mixtures can readily be vitrified and returned to the liquid state without the intervention of crystallization.

Some recent studies aimed at avoiding freezing damage in whole isolated organs by achieving vitrification are relevant here. The approach has been to reduce the concentration of cryoprotectants required for vitrification by employing high hydrostatic pressures. High pressures (1000–2000 atm) significantly lower T_m, T_h and slightly elevate T_g making it easier (at least theoretically) to achieve vitrification with lower concentrations of cryoprotectant and possibly lower cooling rates. These pressures are themselves damaging, but initial studies have also revealed a baroprotective effect of the additives which may counteract the harmful effects of hyperbaria at low temperatures (Fahy and Hirsch, 1982; Fahy *et al.*, 1984).

Recrystallization

Recrystallization is a general phenomenon of systems that have been cooled under extreme non-equilibrium conditions. The phenomenon occurs because very small ice crystals are thermodynamically metastable due to their high surface energies and during warming they minimize their surface to volume ratios by fusing or growing into larger, more stable ice forms. Recrystallization generally occurs at higher sub-zero temperatures as indicated by T_r lines in the supplemented phase diagrams, but it has been detected as low as −130°C (Dowell and Rinfret, 1960). T_r is determined mostly by the nature of the solute; concentration has little effect on T_r.

Several types of recrystallization have been distinguished (Luyet, 1965).

Irruptive recrystallization describes the event in which evanescent spherulites formed during the rapid cooling of aqueous solutions resume their inhibited crystalline growth when the temperature reaches a specific, narrow range T_r during slow rewarming. This event is characterized at the macroscopic level by an abrupt transition of a transparent preparation under ordinary light to one that becomes intensely opaque. The opacity is due to the formation of a cloud of ice particles which have grown large enough to render the cloud visible. The term 'irruptive recrystallization' was coined to reflect the rather sudden increase in the rate of crystalline growth in a given temperature range resulting in the abrupt onset of opacity.

Migratory recrystallization, by constrast, describes the gradual growth of large crystals in a population at the expense of smaller ones and has been designated as such because the phenomenon involves the migration of molecules from small to large crystals as the specimen is warmed gradually from T_r to T_m. The rate of migration increases with rising temperature to the extent that the growth of ice masses can be observed under the microscope as the melting point is approached.

Spontaneous recrystallization is a phenomenon that can occur during rapid cooling. The latent heat released during freezing (formation of spherulites) is not dissipated sufficiently quickly to prevent localized rises in temperature and this, in turn, can give rise to recrystallization.

It should be appreciated that the nature of each of these types of recrystallization is not different from each other and that they all reflect the highly unstable nature of quickly cooled solutions. At a particular temperature the system strives to reduce its total surface area and achieve its equilibrium state which is also the state that would have resulted from very slow cooling with the initiation of freezing at T_m.

Melting

Systems which have previously been cooled slowly to maximize crystallization become unstable during warming at temperatures below the lowest point in the melting curve of the phase diagram (the eutectic point). In particular, two minor endothermic thermal events are detected at T_{am} (ante-melting) and T_{im} (incipient melting). In glycerol solutions the temperature at which the melting endotherm begins (T_{im}) is more than 10°C below the eutetic, and ante-melting instabilities occur approximately 15°C below T_{im} (Fig. 1.6).

The nature of the ante-melting and incipient melting transitions has not been explained, but it is thought that the instability of the system far below the eutectic is probably due to the few layers of molecules which separate the ice near its melting point from the highly concentrated solution. The absence of an ante-melting transition in the cases of both pure ice and of highly concentrated solutions without ice, suggests that the transition requires the presence of the two phases and that it might represent activity at the ice-solution interface (Rasmussen and Luyet, 1969). Ante-melting may reflect a relaxation at this interface similar to the molecular relaxation in the bulk of an amorphous solution undergoing the glass transition.

Both T_{im} and T_{am} appear to be independent of the initial composition of the solution. For example, with glycerol concentrations of between 20 and 50 per cent, the temperatures at which ante-melting and incipient melting are detected are independent of the concentration (Fig. 1.6). It appears that the highly concentrated solutions in contact with the ice phase, not the original solutions, are responsible for ante-melting and incipient melting (Luyet and Rasmussen, 1968). On this basis, it has been shown that T_{im} plotted at the highest concentration at which ice is still formed corresponds to an extrapolation of the melting curve (Fig. 1.6). T_{im} might therefore reflect partial melting of ice in areas of contact with the highest solute concentrations that still contain freezable water (Luyet and Rasmussen, 1968).

The final event during the warming of a frozen solution is the process of melting in which the latent heat of crystallization is absorbed as the crystalline solid phase disappears in its transformation to the liquid phase; this is represented by a pronounced endothermic peak in the differential thermogram.

Chemical and biochemical effects in solutions at low temperatures

All chemical properties of macromolecules and multimolecular structures in biological systems, including conformational stability and biochemical specificity, are dependent upon the interaction of their constituent groups with the surrounding solvent medium. These interactions are, in turn, influenced by the structural features of liquid water and by the structural changes in water caused by solutes. Furthermore, sensitivity of these interactions to changes of temperature gives rise to a complex interdependence of the physical and chemical processes which, in turn, governs the response of living systems to low temperatures and freezing. As we have seen, the process of freezing in biological tissues is complex because of their heterogeneous nature, being made of discreet regions bounded by semi-permeable membranes. Low temperature injury to biological tissues may be related to any of a wide variety of effects including: supercooling, nucleation, ice crystal morphology and growth, osmotic gradients leading to redistribution of water and osmotic stresses, solute-related stresses, thermal gradients and problems associated with recrystallization during rewarming. Some of the problems related to ice formation have been partly overcome by the addition of water-miscible protective substances (e.g. glycerol and dimethylsulphoxide) as cryoprotectants (Shlafer, 1981). In the context of chemical and biochemical effects in solution at low temperatures, the incorporation of multimolar concentrations of non-electrolyte, organic additives into aqueous solutions demands consideration of the influence of mixed solvent systems in combination with the effects of decreasing temperature and increasing ionic strength.

The principal properties of mixed solvents in relation to biochemistry at low temperatures are discussed elsewhere (Douzou, 1977) and will therefore not be repeated here. However, a summary of the general physico-chemical and biochemical effects of low molecular weight cryoprotectants on biological systems, both before and during freezing, is presented below (based on Franks, 1978).

Interdependent physico-chemical and biochemical effects of mixed solvents at low temperatures

Depression of the equilibrium freezing point (T_m) of substrate and cytoplasm on a colligative basis

This is the basis of a widely accepted explanation, originally proposed by Lovelock (1953; 1954), for the mechanism of cryoprotection afforded by permeating compounds such as glycerol and dimethylsulphoxide. The addition of these compounds to aqueous systems depresses T_m and, once freezing has begun, reduces the mole fraction of other solutes (primarily electrolytes) remaining in the non-frozen phase. At the same time the amount of ice that forms at a given sub-zero temperature is reduced. Cryoprotection of cells may therefore be mediated through one or more of the following mechanisms:

(a) reduction of salt concentration;
(b) reduction of cell shrinkage at a given temperature;
(c) reduction of the fraction of solution frozen at a given temperature.

Depression of the homogeneous nucleation temperature (T_h)

It has been shown that the lowering of T_h by a wide range of solutes (including soluble polymers) is a linear function of the colligative lowering of T_m (Rasmussen and MacKenzie, 1972). This means that the available range of supercooling is significantly extended; for example, in 30% ethylene glycol $T_h = -70°C$, compared with $-38°C$ in pure water.

Lowering of the critical cooling rate, above which freezing is controlled by homogeneous nucleation

Ice crystal size distribution depends on cooling rate and, consequently, is of practical importance in the development of rapid quench-cooling techniques to limit the size of ice crystals in tissues prepared for ultrastructural examination. Glycerol, for example, has been shown to reduce the critical cooling rate for water by almost an order of magnitude to values that are attainable in practice (Riehle, 1968).

Effects upon the rate of ice crystal growth

The presence of both electrolytes and non-electrolytes in aqueous solutions is known to influence the rate of ice crystal growth. A number of mechanisms have been proposed including:

(a) changes in the viscosity and thermal diffusivity of the solution;
(b) accumulation of rejected solute at the advancing ice–solution inter-

face, resulting in a lowering of the liquidus temperature, referred to as *constitutional supercooling*;

(c) absorption of solute at the growing crystal surface which is also responsible for the generation of so-called freezing potentials; and

(d) changes in water mobility due to specific water binding by the solute molecules or ions.

Effect upon various eutectic formations and T_e

The depression of binary eutectic points in aqueous solutions by the incorporation of organic cosolvents has already been mentioned emphasizing that the eutectic properties of multicomponent, cytoplasmic solutions are complex. Hydrogen-bonding between non-electrolyte, additive molecules and water inhibits simple crystalline eutectic separation. Instead, such solutions become very viscous and glassy at low temperatures.

Solubility changes of electrolytes including buffers

The solubility of salts in water is markedly affected by the presence of mixed solvents (Gordon, 1975), giving rise to the so-called *salting-in* and *salting-out* effects (Conway, 1981). Reduced temperatures, and freezing, aggravate the risks of salt precipitation. One important consequence of these differential solubility changes is the precipitation of buffer salts giving rise to grave shifts in pH which are inevitably reflected in biochemical events.

An extensive investigation of the pH of freezing aqueous salt solutions has been reported (Van den Berg, 1959; Van den Berg and Rose, 1959). Solutions of sodium and potassium phosphate both with and without the addition of sodium and potassium chloride were studied. From these phase diagrams it is possible to predict the sequence of eutectics and associated pH changes for any combination of these salts. It is evident that extremely wide fluctuations in pH can occur especially in the presence of sodium chloride which greatly reduces the pH of the eutectic. In contrast, potassium chloride has little effect on eutectic pH. It is clear from such studies that buffering the original medium is no guarantee that changes in pH sufficient to induce damage to biological material during freezing will not occur. A study of the factors affecting aggregation of lipoproteins during freezing has provided direct evidence for the importance of pH changes as a contributory factor to freezing damage (Soliman and Van den Berg, 1971b). This study examined the effects of changes in temperature, pH, salt and lipoprotein content on egg yolk lipoprotein aggregation during freezing in aqueous solutions with and without prior addition of sucrose, glycerol or dimethylsulphoxide. By changing only one environmental factor or one ingredient at a time, it was possible to determine the effect of each factor, including the presence of ice and cryoprotectants, without interference from the others which normally changed at the same time during freezing. The technique employs a knowledge of composition changes in the unfrozen phase of dilute solutions during freezing (Van den Berg and Soliman, 1969). The pH (measured at 25°C) was varied between 5.5 and 7.5 by altering the ratio of dibasic to monobasic phosphate, and in some instances, by replacing the monobasic phosphate, by citric acid.

From these studies, it was concluded that, as a result of freezing damage, the aggregation of low density egg yolk lipoprotein was dependent upon the pH, temperature and salt composition of the unfrozen phase but was largely independent of lipoprotein concentration. Maximum aggregation, equivalent to low temperature injury, occurred in the pH range 5.0–5.5 and decreased with decreasing temperature. Chlorides were shown to cause little freezing damage even at high concentrations, while phosphate, sulphate and nitrate in concentrations over 1M in the unfrozen phase caused extensive damage. Organic additives, such as sucrose, glycerol and DMSO reduced the damaging effect of phosphate by reducing the phosphate concentration.

This experimental approach has also been used to investigate which factors were responsible for the freezing damage of the important muscle enzyme, lactic dehydrogenase (Soliman and Van den Berg, 1971a). Inactivation of the enzyme occurred with decreasing temperature (+2 to −10°C) and decreasing pH (7.8–6.0) but with increasing salt concentration (up to 3.3M).

The lipoprotein of egg yolk and lactic dehydrogenase are both sensitive to decreasing pH in the 7–5 range and to high phosphate concentrations, but they differ with respect to the effects of temperature, high sodium chloride, glycerol, and protein concentrations (Soliman and Van den Berg, 1971a and b). Although lipoproteins and lactic dehydrogenase may not typify biological systems in their response to freezing, these results indicate the importance of pH and salt concentration changes in freezing injury and the necessity of taking both factors into account when studying such damage.

Ion solvation changes with associated changes of acidity (basicity) of the solvent reflected as apparent pH changes

The nature of a solvent, as well as temperature, has a profound effect upon the acidic or basic strength of a substance, for the solvent molecules themselves are involved in acid–base equilibria as acceptors or donors of protons. In addition, factors such as dielectric constant of the medium (a measure of the force acting between two electric charges, for example ions) and solute–solvent interactions (which can lead to complications in mixed solvents) such as the selective ordering of solvent molecules around ionic species (Douzou, 1977; Conway, 1981) can influence the acidic and basic strengths in solutions. Solute–solvent interactions, and in particular the question of proton activity in partially aqueous mixed solvents, involve the concept of the *medium effect*. This is a measure of the free energy change on transfer of one mole of a substance from the standard state in water to the standard state in another solvent medium (Bates 1969b, 1973). The medium effect reflects differences in the electrostatic and chemical interactions of the substance with the molecules of the two solvents (solute–solvent interactions). Of these interactions solvation is probably the most important when ions are transferred from one medium to another. The medium effect is closely allied to the activity coefficient which characterizes the departure from ideal behaviour as a result of interionic forces that are largely electrostatic (solute–solute interactions). The regularities of solvation in water, which is perfectly amphiprotic, may be less extensive in other media, or even absent. The distribution and orientation of water molecules, and therefore the physico-chemical properties of liquid water, are disturbed when a miscible, weakly protic, organic solvent is added and accompanied by lowering the temperature. A mixed solvent is not a continuous, structureless medium because molecules of the more polar solvent (usually water) tend to solvate the ions in accordance with electrostatics and thereby restrain the molecules of the less polar solvent from the vicinity of the ions (Douzou, 1977). Consequently, the composition of a mixed solvent around the ions is not the same as that of bulk solutions.

Effects of cellular components upon solubility, solvation, ion-binding equilibria and of acid and basic groups on the degree of dissociation (pK values)

Changes in the nature of the solvent medium, along the lines outlined above, coupled with a lowering of temperature are known to have a profound effect upon acid and base dissociation constants (usually expressed as pK_a ($-\log K_a$) and pK_b respectively). Considered in relation to biological activity of ionogenic residues of amino acids, proteins, lipids, polysaccharides and nucleotides such effects would be expected to have pronounced biochemical implications. However, little attention has been devoted to studying such matters which, inevitably, rely upon the need to make accurate pH measurements in mixed solvents at low temperatures (see below).

The thermodynamics of the ionization of water demonstrate that the ion product (K_w) has a significant temperature coefficient which will influence physico-chemical processes including acid–base equilibria (Harned and Owen, 1958; King, 1965; Hepler and Woolley, 1973). For example, the neutral point of water at 25°C is at pH 7.0, at 0°C it is at pH 7.47 (Robinson and Stokes, 1968) and at −35°C is estimated to be at pH 8.4 (Taylor, 1981). Some implications of these facts in relation to the meaning of pH at low temperatures have been considered (Taylor, 1981).

The effect of a number of organic cosolvents, including glycerol, on pK_w has been reported (Hepler and Woolley, 1973), but effects of dimethyl sulphoxide (DMSO), as a dipolar aprotic solvent, are most significant (Ritchie, 1969). DMSO does not donate protons in solution but functions only as a hydrogen-bond acceptor, and addition of DMSO to aqueous solutions will alter the dielectric constant of the medium with a corresponding influence upon ionic dissociation and acid–base equilibria in particular. Ionization constants, in the form of pK_a values, provide a convenient way of comparing the strength of acids and bases, and their determination can yield useful information about the effect of solvent composition upon acid–base equilibria (Ritchie, 1969). The pK_a values, together with their temperature coefficients, for a wide range of biological buffers in solutions containing DMSO at normal and sub-zero temperatures have been reported (Taylor and Pignat, 1982). A rise in pK_a with increasing organic content of solvent is expected, and generally observed, reflecting the increase in energy required to separate ions brought about by a lowering of the dielectric constant of the medium.

Perturbation of the intermolecular order and molecular dynamics characteristic of the aqueous medium

The role of 'water structure' in determining the solubility of many substances and, at the molecular level, its influence upon the maintenance of the conformations of biomacromolecules has been discussed above. The molecular order and interactions which are symptomatic of 'water structure' are perturbed by the presence of other polar molecules.

The interaction of DMSO in aqueous solutions is a particularly interesting example, not only because of the widespread use of DMSO as a cryoprotectant but also because studies of the physical properties of the binary system DMSO-H_2O have been undertaken to establish the changes in molecular structure and liquid motion that occur as a function both of composition and temperature (Packer and Tomlinson, 1971; Martin and Hauthal 1975). The thermodynamic excess functions (free energy, enthalpy and entropy of mixing) have been considered since these, more than any other properties, are determined by intermolecular forces. All the excess functions show minimum values in the concentration range 0.6–0.7 mole fraction of water, which together with observations that viscosities of the mixtures show a maximum in the same concentration range, indicate a significant deviation from regular behaviour caused by strong interactions (hydrogen-bonding through the $S \rightarrow O$ dipole, Fig. 1.8) between DMSO and water. All measurements indicate a minimum of molecular mobility, both rotational and translational, at approximately a mole fraction of water of 0.65 (70% DMSO-H_2O). This degree of association, which has been quoted as being 1.33 times greater than that in pure water (MacGregor, 1967), explains the often extraordinarily large mixing effects, for example freezing point depression. Furthermore there are indications that addition of small quantities of DMSO to water will produce a co-operative orientation of water molecules around each DMSO molecule. This has the effect of 'rigidifying' the water structure (Safford *et al.*, 1969). Implications of the perturbation of water structure on the solvation of ions and polar residues in mixed solvents is beyond the scope of this discussion, but details are available in the literature (Friedman and Krishnan, 1973; Martin and Hauthal, 1975). Moreover, the biological implications of DMSO based upon a review of its chemical properties has been reported (Rammler and Zaffaroni, 1967; Martin and Hauthal, 1975).

Interference with the conformational stabilities of both intracellular and extracellular macromolecules

The stability of the secondary and tertiary molecular structure of a protein is dependent upon the nature of its solvent environment together with factors such as temperature, ionic strength and pH. The conferment of increased conformational stability on globular proteins by glycerol and other

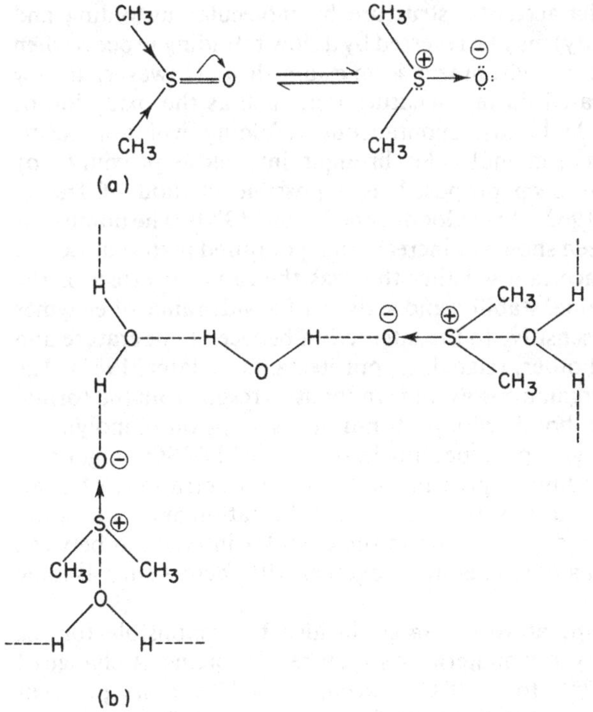

Fig. 1.8 Scheme for the association of dimethyl sulphoxide (DMSO) and water based upon hydrogen-bonding through the S→O dipole (a) Polarized form. (b) Hydrated form.

polyhydroxy compounds has been reported (Gerlsma, 1968). Such compounds are apparently able to substitute for water in their interactions with exposed polar sites on the macromolecule to induce a certain degree of stability.

Evidence has also been reviewed for the influence of DMSO and related dipolar, aprotic solvents on ionization equilibria, hydrogen-bonding and hydrophobic interactions of protein functional groups, stability of ground and transition states, chemical reactivities of protein functional groups, conformation of proteins and enzymatic activity (Friedman, 1968). Solvent effects on protein conformation and activity have been examined using enzymes (Henderson *et al.*, 1975). The conformational changes of trypsin in different solvents having a wide range of dielectric constants have been studied (Bettelheim and Senatore, 1964). A decrease in the proteolytic activity was observed with increasing DMSO concentration, and a complete inhibition of activity in a 70:30 weight per cent DMSO–H_2O medium (1:2 DMSO:H_2O ratio). Interestingly, this concentration of 70 per cent, in which maximum unfolding appears to occur, correlates with the concentration at which maximum influence of DMSO on 'water structure' was detected. DMSO has an inhibitory effect on the activity of several enzyme systems, and recovery of activity occurs upon its removal (Rammler, 1967). Denaturation of enzymes (loss of characteristic structure by molecular unfolding and subsequent loss of activity) may be reversed by a slow refolding process when the normal physiological environment is re-established. However, if new covalent bonds are created during denaturation, such as the oxidation of sulphydryl to disulphide bonds, spontaneous refolding will not occur. Bonding within or between molecules brought into close proximity by dehydration effects has been proposed as a possible method of freeze-induced injury (Levitt, 1962, 1966; Goodin and Levitt, 1970). The number of disulphide bonds has been shown to increase in cryoinjured material (Levitt, 1966) but it remains uncertain whether this was the cause or effect of the injury. The conformational stability and activity of a wide range of enzymes have been shown to be sensitive to the interaction between temperature and exposure to DMSO and other, related, cryoprotectants (Shlafer, 1981). The mode of action of the organic cosolvent is thought to result from the formation of strong hydrogen-bonds with proton-donor groups on biopolymers. The presence of methyl groups, especially in the case of DMSO, might also affect hydrophobic bonding in proteins at higher concentrations of cryoprotectant. At low temperatures, therefore, the stabilization or denaturation of biomacromolecules is dependent upon the complex interaction between cosolvent concentration and temperature together with factors such as ionic strength and pH.

The lowering of temperature *per se* could also be responsible for the denaturing of the tertiary and quaternary structure of proteins. A change of temperature from +20°C to −20°C is accompanied by a six per cent reduction in electrostatic interactions and free energy which, because of the extremely delicate stability of the higher levels of protein structure, may be quite significant (Franks, 1981a).

Ionic dissociation (with emphasis on the proton)

Lists of the factors responsible for the demise of biological systems during exposure to low temperatures invariably include 'pH changes' (Meryman, 1966); discussion of such effects is, however, rarely attempted because of the general lack of data. The universal recognition of the importance of pH effects at low temperatures on the one hand, and the clear lack of research devoted to its investigation on the other, is undoubtedly a result of the practical and theoretical difficulties encountered in attempts to make accurate pH measurements at low temperatures (Taylor, 1981). The importance of acidity and basicity in basic aspects of biological structure and function derives fundamentally from the role of ionic dissociation in such processes; this will now be discussed in brief with particular mention of the measurement and significance of pH at low temperatures.

Concept of an ideal solution

Vapour pressure is an important property of solutions as a measure of the tendency of a molecular species to escape from the solution into the vapour phase. This tendency is a direct reflection of the physical state of affairs within the solution. When a non-volatile solute is dissolved in a liquid the vapour pressure is lowered. The quantitative connection between lowering of vapour pressure and the composition of the solution is given by Raoult's Law which states that the relative lowering of vapour pressure is equal to the mole fraction of the solute in solution. Solutions which obey Raoult's Law at all concentrations and all temperatures are called *ideal solutions*, and are defined as having complete uniformity of cohesive forces between components. Genuinely ideal solutions are uncommon, but the deviation from ideal behaviour observed in dilute solutions is small. There are three important properties of solutions related to the variation of vapour pressure with solute concentration; these are: elevation of boiling point, depression of freezing point and osmotic pressure. They are known as colligative properties (*colligatus* - collected together) because they depend on the actual number of particles present in solution rather than on the size or nature of the molecules.

There are several reasons why a solution deviates from ideal behaviour, but by far the most important reason is the property of electrolytic dissociation. Arrhenius was able to correlate the apparently abnormal non-ideal behaviour of aqueous solutions of acids, bases and salts with their ability to conduct electricity. When an acid, base or salt is dissolved in water the polar properties of the solvent cause the forces of electrical attraction between the ions to be weakened and the ions move about in the solution as separate particles - solvated ions. Apart from endowing the solution with conducting properties the process of dissociation affects all the colligative properties of the solution.

The departure of a solution from ideal behaviour is described in terms of *activity*, which may be defined as the active or effective mass of a substance with non-ideal behaviour. The ratio of the activity (a) of a species to its molal concentration (m) is the activity coefficient (γ), thus:

$$a/m = \gamma \qquad \textbf{7}$$

At infinite dilution when $a = m$ then $\gamma = 1$.

Ionic activity coefficients are not amenable to measurement individually, but can be calculated from theoretical considerations at low concentrations. Under such conditions the interionic forces responsible for the departure from ideality depend primarily upon charge, radius and distribution of the ions and the dielectric constant of the medium rather than on chemical properties of the ions. On this basis Debye and Hückel developed an expression for the activity coefficient from electrostatic and statistical theory (Robinson and Stokes, 1968). The basic form of this expression is given as:

$$\log \gamma = - \frac{Az^2I^{1/2}}{1 + Ba^\circ I^{1/2}} \qquad \textbf{8}$$

where z is the valence, a° is the 'ion-size parameter' or mean distance of closest approach of ions (of the same order as ionic diameter), A and B are constants dependent upon temperature and dielectric constant of the solvent, and I is the ionic strength. Qualitatively, the numerator in expression **8** accounts for the effect of long-range coulombic forces while the denominator shows how these are modified by short-range interaction between ions.

It is clear from these fundamental principles that the complex interaction of temperature, ionic strength, mixed solvent composition and dielectric constant will govern basic processes such as solubility, ionic dissociation (pK_s) and acid–base equilibria (pH), and hence will be reflected in the stability, conformation and reactivity of macromolecules.

Definition and measurement of pH

The significance of pH is that it is an index of the chemical potential of protons. The unique influence of pH on biochemical reactions was first described, in terms of hydrogen ion concentration, in the classic papers of Sørensen in 1909. The application of pH measurements in industry and scientific research was made possible by the discovery of the hydrogen ion function of glass membranes and subsequent development of glass electrodes and pH meters. A practical, experimental method to supersede the electrometric determination of acidity using the glass electrode and a saturated calomel reference electrode has not yet been found.

The primary standard electrochemical cell against which all pH scales are defined consists of the hydrogen gas electrode in conjunction with a silver-silver chloride reference electrode represented as:

$$Pt; H_{2(g)}, HCl_{(m)}, AgCl; Ag_{(s)} \qquad \mathbf{9}$$

From thermodynamic energy considerations the electromotive force (E) of a galvanic cell is related to the chemical potential of the components in terms of their activities by the Nernst equation:

$$E = E^{\circ} - \frac{RT}{zF} \ln Q_a \qquad \mathbf{10}$$

where E = emf of the cell, E° = standard cell potential, Q_a = quotient of activities of the ions involved in the cell reaction, R = gas constant, F = Faraday charge, T = absolute temperature and z = number of equivalents of reactants converted to products in the cell reaction.

Written in terms of the activity of hydrogen ions only:

$$pH = -\log a_H = \frac{(E - E^{\circ})F}{2.303RT} \qquad \mathbf{11}$$

The determination of values for E° thus allows pH values to be derived from the emf of cell **9** or its counterpart with the glass electrode.

This procedure is, however, impractical (especially when glass electrodes are used) because the standard cell potential is often found to vary considerably between assemblies of apparently identical design and also with time. Furthermore, the measuring cell is also subject to certain extraneous potentials which lead to errors in the measured pH value. The most significant of these erroneous potentials are the liquid junction potential (LJP), the residual LJP and the asymmetry potential of the glass electrode (Bates, 1973; Taylor, 1981). In practice all these errors are minimized by calibrating the pH-cell regularly with a standard buffer whose pH is known accurately and designated pH(S). The universal procedure for measuring pH therefore relies upon an operational definition which is dependent upon a conventional standard pH unit, namely $pa_H = -\log a_H = -\log m_H \gamma_H$, and pH scales defined by standards, the pH values of which are assigned using cell **9** which does not have a liquid junction or an asymmetry potential.

pH measurement at low temperatures

The measurement of pH in biological systems at low temperatures is influenced by three principal factors:

(a) reduction of temperature;
(b) influence of partially aqueous mixed solvents due to the incorporation of organic co-solvents as cryoprotectants;
(c) increase in ionic strength.

Here, it is not convenient to discuss these factors as they are the subject of a recent synopsis of the meaning of pH at low temperatures (Taylor, 1981). The salient points may, however, be summarized for completeness.

(a) Temperature effects
Changes of temperature *per se* have an effect on several factors.

(i) pH scales – the establishment of numerical scales of activity requires a standard state to be assigned at a particular temperature. In the case of a conventional scale of hydrogen ion activity this means the standard potential $E°_H$ of the reversible hydrogen electrode is arbitrarily defined as zero at all temperatures. This unavoidably yields a different pH scale at each temperature, with the pH at one temperature having no quantitative meaning in the strictest sense relative to that at another. In practice, however, the apparent temperature coefficient of the hydrogen electrode potential is estimated to be *c.* -0.37 mV K^{-1} at pH 7 (Bates, 1973).

(ii) pa_H and pK – the process of ionization in solution is markedly affected by temperature as reflected in the temperature coefficients of acid dissociation constants (pKa), which can be regarded as indices of acid-base equilibria. Both pK_a and the temperature coefficient for some biological buffers in partially aqueous solutions at a range of concentrations and temperatures have been reported (Taylor, 1980b; Taylor and Pignat, 1982; Roy *et al.*, 1984). The effect of temperature on the pH of buffer solutions depends upon the temperature-dependence of the activity coefficient and upon the pK_a of the buffer species; the pK_a is usually more important so that the influence of temperature upon pH is not straightforward.

(iii) pK_w and 'neutrality' – the ion product of water, pK_w is of fundamental importance in considering the behaviour of acids and bases in solution, and its significant temperature coefficient and susceptibility to the influence of organic co-solvents have already been mentioned. The neutral point of supercooled water is predicted to rise significantly as temperature is lowered and will have a bearing on what is considered biologically 'normal' under the different conditions of solution chemistry that prevail during the cooling of biological systems (Taylor, 1981).

(iv) The pH-cell (adjustments of the pH meter) – since the standard potential of the pH-cell and the LJPs are temperature dependent it is imperative for the temperature of standardization to be the same as the temperature of measurement. The folly of not observing this requirement or of attempting to measure pH at low temperatures by keeping the reference electrode at temperatures above 0°C and coupling it with a modified glass electrode at sub-zero temperatures has been discussed elsewhere (Taylor, 1981). Furthermore, a method of temperature compensation for use when normal commercial laboratory pH meters are used for pH measurements at sub-zero temperatures has also been described (Taylor, 1978b; Taylor, 1980b).

(b) Mixed solvent effects

The influence of the nature and composition of a solvent on acid–base strength has been considered above. It is clear that the measurement and interpretation of pH values in partially aqueous mixed solvents becomes a problem of immense complexity, especially when the effects of a decrease in temperature are also involved (Bates, 1969b, 1973). At the present time, a universal pH scale relating proton activity uniformly to the aqueous standard reference state is not a practical possibility because the medium effect (defined earlier) for the proton and the difference in the LJPs between the two media cannot be measured physically. Nevertheless, separate scales for each medium can be derived using procedures analogous to those by which the practical aqueous pH scale is established (Bates, 1969a). In this way an operational scale in each medium of interest has to be established:

$$pH^*(X) = pH^*(S) + \frac{(E_X - E_S)F}{RT\ln 10} \qquad \textbf{12}$$

* indicates that standardization is made with reference to the standard state in a solvent medium other than pure water; X denotes the solution of unknown pH* and S denotes the standard reference solution of known or assigned pH*.

The preparation of hydrogen and silver–silver chloride electrodes for use in aqueous mixtures containing DMSO at normal and sub-zero temperatures has been described (Taylor, 1980a). These have been used to determine the standard cell potentials (E°) in some DMSO–H_2O mixtures at low temperatures (Taylor, 1978a). This primary standardizing cell has, in turn, been used to establish new pH* scales in 20 per cent and 30 per cent (w/w) DMSO–H_2O mixtures at temperatures between +25 and −12°C (Taylor, 1979). More recently, standard measurements have been extended to 40 and 50 per cent (w/w) DMSO–H_2O mixtures down to −20°C (Roy *et al.*, 1984), and data for other solvent mixtures such as ethylene glycol-water are currently being derived at sub-zero temperatures (Roy, personal communication).

The availability of these new standardizing buffers enables the calibration of conventional or modified electrodes for pH* measurements on an appropriately defined scale for these conditions of solvent composition and temperature. It has been possible using these standards to demonstrate a normal pH*-response of the glass-calomel pH-cell in 30 per cent (w/w) DMSO–H_2O at 25°C and −12°C (Taylor, 1978a). Moreover, these facilities have been used to clarify that the apparent rise in pH of physiological solutions (Elford and Walter, 1972) when DMSO is added is due to a real decrease in hydrogen ion activity as opposed to an artefactual increase in pH as a result of alteration of the variable cell characteristics such as LJP or asymmetry potentials. This leads to the question of whether this rise in pH* should be corrected by buffering-back to pH values which are regarded as normal for biological integrity. It is clear that such temptation should be resisted on the grounds that the optimum proton activity for biological integrity under these conditions may be very different from those that exist under physiological conditions. A specific illustration of this is given below. Furthermore, these considerations of the effect of mixed solvents on acid–base equilibria in solution should serve to warn against the practice of measuring pH in mixed aqueous solvents with conventional or modified electrodes which have been calibrated with pure aqueous standard buffer solutions.

(c) Ionic strength effects
The increase in ionic strength during freezing also imposes an influence upon hydrogen ion activity and its measurement, primarily by lowering the activity coefficients of ions and exceeding the limits of applicability of the Debye-Hückel law for estimating these coefficients. Salt-errors, due to increased ionic strength, can markedly influence the colour balance of pH-indicator equilibria making the interpretation of pH changes based on indicator colour changes extremely difficult, especially if the additional influences of mixed solvents and temperature change are also involved (Taylor, 1981; Taylor and Pignat, 1982).

The role of pH in the recovery of biological function after exposure to low temperatures*

The few attempts to study pH effects at low temperatures reported in the literature (see Taylor, 1977 for a brief review) have invariably involved pH measurements at or above 0°C or the use of inappropriately standardized electrodes, thus making any interpretation of the observed pH effects difficult and of dubious value. For example, pH and pK_a measurements carried out in mixed solvents on the basis of calibration with purely aqueous standard buffers lack thermodynamic significance, and the pH numbers taken from the instrument do not relate simply to the chemical equilibrium. However, some indications of the true effect of pH may be drawn from studies in which the accurate pH* scales discussed in the preceding section have been applied to determine the role of pH* and buffer capacity on the recovery of muscle tissue function after storage at sub-zero temperatures (Taylor, 1982).

The contractile recovery of mammalian smooth muscle after storage at −13°C in unfrozen medium containing 30 per cent (w/w) DMSO–H_2O was found to be markedly pH*-dependent with a clearly defined optimum at $pH^*_{-13} = 9.2$ (Fig. 1.9), which is surprisingly high compared with physiological pH.

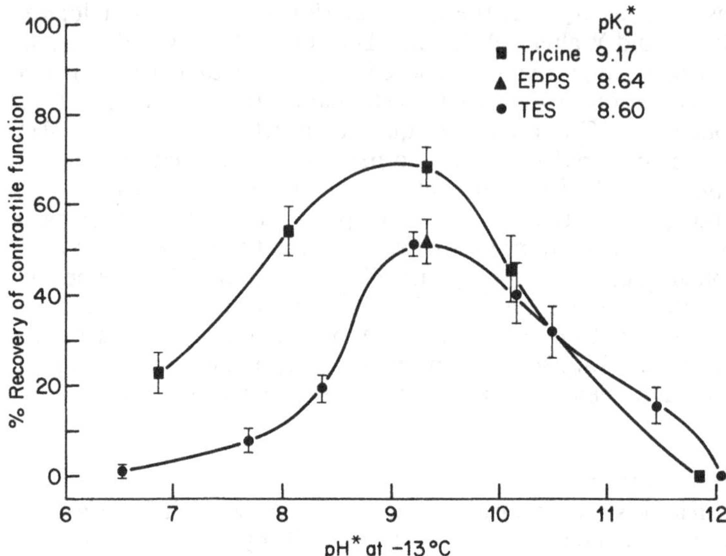

Fig. 1.9 The effect of pH^*_{-13} upon the functional recovery of smooth muscle after 65 hours storage at − 13°C in a potassium-rich bathing medium containing 3.84M DMSO and one of several different buffer compounds: Tricine (■): N-Tris (hydroxymethyl) methyl glycine; TES (●): N-Tris (hydroxymethyl) methyl-2-aminoethane sulphonic acid; EPPS (▲): N-2-Hydroxyethylpiperazine N[1]-3-propane sulphonic acid. (See Taylor, 1982 for details. Reproduced from Taylor, 1982 by permission of Academic Press).

In medium containing TES (N-Tris (hydroxymethyl) methyl-2-aminoethane sulphonic acid) buffer, which has a maximum buffer capacity at $pH^*_{-13} = 8.6$, the cooled muscles recover 50 per cent of their control contractibility, but in medium containing the buffer Tricine (N-tris hydroxymethyl)methylglycine) which has a maximum capacity at the optimum pH* for recovery, the contractile response upon rewarming improves to 70 per cent.

These observations indicate that the optimum pH for biological integrity under the conditions of low temperature preservation can be very different from those that exist under physiological conditions. It must also be emphasized that it is not possible to compare $pH_{37} = 7.4$ with $pH^*_{-13} = 9.2$ because these values are measured on two different scales of hydrogen-ion activity which are not related in any simple way (Bates, 1969b, 1973). In other words, differences in solvent acidity/basicity, dielectric constant, ion activities and mobilities mean that such scales are different for each medium and, as a result, solutions of different solvent composition may produce the same pH-meter reading but behave in different ways in acid–base reactions. Nevertheless, the accurate measurements of pH* in the study illustrated here have established that the optimum recovery of smooth muscle kept in unfrozen media at − 13°C is achieved using a solution which has a hydrogen ion concentration of 6.3×10^{-7} mM (pH* = 9.2). It has already been emphasized that the pK of any dissociable molecule or ion is strongly dependent on the solvent composition of the solution and its temperature, so that it is probable that the solvent-temperature effects on the pK of enzymes, substrates, contractile proteins, structural proteins and simpler ions such as buffers may cause optimal conditions to be shifted to a different pH*.

The precise nature of the pH* effect is unknown, and it is uncertain whether a similar effect would be a generally observed phenomenon for other cells or tissues. However, because most biological functions are known to be pH-dependent under normal physiological conditions, it may be that similar effects exist at low temperatures. The implications for those interested in biology at low temperatures are clear and, although the influence of pH changes has generally been neglected in the past, the principles and observations summarized here should make us wary of the role of pH changes in cryobiology.

Cryopreservation with the avoidance of freezing

It has been mentioned that future success in the cryopreservation of multicellular tissues and organs may come from approaches to avoid the damaging effects of extracellular ice formation. One approach aims to achieve solidification of the aqueous system at low temperatures by *vitrification* rather than by crystallization. Fahy *et al.* (1984) have recently described the principles of this approach to cryopreservation and Rall and Fahy (1985) have reported the successful cryopreservation of mouse embryos at $-196°C$ by vitrification. This was achieved using a solution containing 18.45% (w/v) dimethyl sulphoxide (DMSO), 13.95% (w/v) acetamide, 9% (w/v) propylene glycol and 5.4% (w/v) polyethylene glycol in modified Dulbecco's saline.

Successful vitrification (judging by the absence of visible ice formation) was achieved without the need for hyperbaric conditions and using either rapid cooling (approximately $2500°C\ min^{-1}$) or much slower cooling (approximately $20°C\ min^{-1}$) together with rapid warming ($2500°C\ min^{-1}$). Differential Scanning Calorimetry and cryomicroscopy have shown, however, that under these conditions the solution may only be partially vitrified (Reid *et al.*, 1985).

Extended pH* scales in mixed solvents at low temperatures

Since the preparation of this chapter, Roy *et al.* have published a number of papers extending the range of standard pH scales available for mixed solvents at low temperatures. These include both ethylene glycol/water mixtures and glycerol/water mixtures down to $-20°C$ (Roy *et al.*, 1985a and 1985b).

References

Adamson, A.W. (1979). *A Textbook of Physical Chemistry, 2nd edition.* Academic Press, New York.

Amrhein, E.M. and Luyet, B.J. (1966). Evidence for an Incipient Melting at the 'recrystallization' temperatures of aqueous solutions. *Biodynamica* **10**, 61-7.

Angell, C.A. (1982). Supercooled water. In *Water: A Comprehensive Treatise.* Vol. 7, pp. 1-81. Edited by Franks, F. Plenum Press, New York.

Arrhenius, S. (1889). Ueber die Reaktionsgeschwindigkeit bei der Inversion von Rohrzucker durch Sauren. *Physikalisch - Chemische Trenn und Nessenthoden* **4**, 226.

Bates, R.G. (1969a). Practical measurement of pH in amphiprotic and mixed solvents. *Pure and Applied Chemistry* **18**, 421-5.

Bates, R.G. (1969b). Medium effects and pH in nonaqueous solvents. In *Solute-Solvent Interactions*, pp. 45-96. Edited by Coetzee, J.F. and Ritchie, C.D. Marcel Dekker, New York and London.

Bates, R.G. (1973). *Determination of pH. Theory and Practice 2nd edition.* John Wiley and Sons Inc., New York.

Belton, P.S., Jackson, R.R. and Packer, K.J. (1972). Pulsed NMR studies of water in striated muscle I. Transverse nuclear spin relaxation times and freezing effects. *Biochimica et Biophysica Acta* **286**, 16-25.

Bernal, J.D. and Fowler, R.M. (1933). A theory of water and ionic solution, with particular reference to hydrogen and hydroxyl ions. *Journal of Chemical Physics* **1**, 515-48.

Bertie, J.E., Calvert, L.D. and Whalley, E. (1963). Transformations of Ice II, Ice III and Ice V at atmospheric pressure. *Journal of Chemical Physics* **38**, 840-46.

Bettelheim, F.A. and Senatore, P. (1964). Hydrophobic bond. Activity and conformation of trypsin in dimethylsufoxide - water systems. *J. Ch. Phys. Tome.* **61**, 105-10. (J. de Chimie Physique.)

Block, W. and Young, S.R. (1979). Measurement of Supercooling in small arthropods and water droplets. *CryoLetters* **1**, 85-91.

Boutron, P., Delaye, D., Roustit, B. and Körber, C. (1982). Ternary systems with 1, 2, propanediol - a new gain in the stability of the amorphous state in the system water - 1, 2-propanediol-1-propanol. *Cryobiology* **19**, 550-64.

Boutron, P. and Kaufmann, A. (1978). Stability of the amorphous state in the system water-glycerol-dimethylsulfoxide. *Cryobiology* **15**, 93-108.

Boutron, P. and Kaufmann, A. (1979a). Stability of the amorphous state in the system water-glycerol-ethylene glycol. *Cryobiology* **16**, 83-9.

Boutron, P. and Kaufmann, A. (1979b). Stability of the amorphous state in the system water - 1,2,-propanediol. *Cryobiology* **16**, 557-68.

Brandts, J.F. (1964). The thermodynamics of protein denaturation I. The denaturation of chymotrypsinogen. *Journal of the American Chemical Society* **86**, 4291-301.

Brandts, J.F., Fu, J. and Nordin, J.H. (1970). The low temperature demonstration of chymotrypsinogen in aqueous solutions and in frozen aqueous solutions. In *The Frozen Cell*, pp. 189–212. Ciba Foundation Symposium. Edited by Wolstenholme, G.E.W. and O'Connor, M. Churchill, London.

Brown, A.D. and Simpson, J.R. (1972). Water relations of sugar-tolerant yeasts: the role of intracellular polyols. *Journal of General Microbiology* **72**, 589–91.

Bryant, R.G. and Halle, B. (1982). NMR relaxation of water in heterogeneous systems – consensus views? In *Biophysics of Water*, pp. 389–93. Edited by Franks, F. and Mathias, S. John Wiley, Chichester.

Burdette, E.C. (1981). Engineering considerations in hypothermic and cryogenic preservation. In *Organ Preservation for Transplantation, 2nd edition*, pp. 213–59. Edited by Karow A.M. Jr and Pegg, D.E. Marcel Dekker, New York.

Burke, M.J., Gusta, L.V., Quamme, H.A., Weiser, C.J. and Li, P.H. (1976). Freezing and injury in plants. *Annual Review of Plant Physiology* **27**, 507–28.

Cocks, F.H. and Brower, W.E. (1974). Phase diagram relationships in cryobiology. *Cryobiology* **11**, 340–58.

Coetzee, J.F. and Ritchie, C.D. (1969). *Solute-Solvent Interactions*. Marcel Dekker, New York.

Conway, B.E. (1981). *Ionic Hydration in Chemistry and Biophysics*. Elsevier, Amsterdam.

Cooke, R. and Kuntz, I.D. (1974). The properties of water in biological systems. *Annual Review of Biophysics and Bioengineering* **3**, 95–126.

Davidson, D.W. (1973). Clathrate hydrates. In *Water: A Comprehensive Treatise*, Vol. 2, pp. 115–234. Edited by Franks, F. Plenum Press, New York.

Derbyshire, W. (1982). The dynamics of water in heterogeneous systems with emphasis on subzero temperatures. In *Water: A Comprehensive Treatise*, Vol. 7, pp. 339–430. Edited by Franks, F. Plenum Press, New York.

DeVries, A.L. (1980). Biological antifreezes and survival in freezing environments. In *Animals and Environmental Fitness*, pp. 583–607. Edited by Gilles, R. Pergamon Press, Oxford.

Dixon, W.L., Franks, F. and Rees, T. (1981). Cold-lability of phosphofructokinase from potato tubers. *Phytochemistry* **20**, 969–72.

Douzou, P. (1977). *Cryobiochemistry – An introduction*. Academic Press, London.

Dowell, L.G. and Rinfret, A.P. (1960). Low-temperature forms of ice as studied by X-ray diffraction. *Nature* **188**, 1144–8.

Drost-Hansen, W. (1982). The occurrence and extent of vicinal water. In *Biophysics of Water*, pp. 163–9. Edited by Franks, F. and Mathias, S. John Wiley, Chichester.

Eisenberg, D. and Kauzmann, W. (1969). *The structure and Properties of Water*, pp. 71–149. Oxford University Press, Oxford.

Elford, B.C. and Solomon, A.K. (1974). Temperature dependence of cation permeability of dog red cells. *Nature* **248**, 522–4.

Elford, B.C. and Walter, C.A. (1972). Effects of electrolyte composition and pH on the structure and function of smooth muscle cooled to – 79°C in unfrozen media. *Cryobiology* **9**, 82–100.

Fahy, G.M. and Hirsch, A. (1982). Prospect for Organ Preservation by Vitrification. In *Organ Preservation: Basic and Applied Aspects*, pp. 399–404. Edited by Pegg, D.E., Jacobsen, I.A. and Halasz, N.A. MTP Press, Lancaster.

Fahy, G.M., McFarlane, D.R., Angell, C.A. and Meryman, H.T. (1984). Vitrification as an approach to cryopreservation. *Cryobiology* **21**, 407–26.

Farrant, J. (1977). Water transport and cell survival in cryobiological procedures. *Philosophical Transactions of the Royal Society London B* **278**, 191–205.

Finney, J.L. (1979). The organization and function of water in protein crystals. In *Water: A Comprehensive Treatise*, Vol. 6, pp. 47–122. Edited by Franks. F. Plenum Press, New York.

Fletcher, N.H. (1970). *The Chemical Physics of Ice.* Cambridge University Press, Cambridge.

Frank, H.S. (1963). The structure of water. *Federation Proceedings* **24**, S1–S11.

Frank, H.S. (1970). The structure of ordinary water. *Science* **169**, 635–41.

Frank, H.S. (1972). Structural models. In *Water: A Comprehensive Treatise*, Vol. 1, pp. 515–43. Edited by Franks, F. Plenum Press, New York.

Frank, H.S. and Evans, M.W. (1945). Free volume and entropy in condensed systems III. Entropy in binary liquid mixtures; partial molal entropy in dilute solutions; structure and thermodynamics in aqueous electrolytes. *Journal of Chemical Physics* **13**, 507–32.

Frank, H.S. and Wen, W.Y. (1957). Structural aspects of ion–solvent interactions in aqueous solutions: A suggested picture of water structure. *Dissc. Farad. Soc.* **24**, 133–40.

Franks, F. (1972). The properties of ice. In *Water: A Comprehensive Treatise*, Vol. 1, pp. 115–49. Edited by Franks, F. Plenum Press, New York.

Franks, F. (1977). Solution and conformational effects in aqueous solutions of biopolymer analogues. *Philosophical Transactions of the Royal Society London B* **278**, 33–57.

Franks, F. (1978). Biological freezing and cryofixation. In *Low Temperature Biological Microscopy and Microanalysis*, pp. 3–16. The Royal Microscopial Society, Blackwell Scientific, Oxford.

Franks, F. (1979). Solvent mediated influences on conformation and activity of proteins. In *Characterisation of Protein Conformation and Function*, pp. 37–53. Edited by Franks, F. Symposium Press, London.

Franks, F. (1981a). Biophysics and biochemistry of low temperatures and freezing. In *Effects of Low Temperatures on Biological Membranes*, pp. 3–19. Edited by Morris, G.J. and Clarke, A. Academic Press, London and New York.

Franks, F. (1981b). *Polywater.* MIT Press, Cambridge, Massachusetts.

Franks, F. (1982a). The properties of aqueous solutions at subzero temperatures. In *Water: A Comprehensive Treatise*, Vol. 7, pp. 215–338. Edited by Franks, F. Plenum Press, New York.

Franks, F. (1982b). Physiological water stress. In *Biophysics of Water*, pp. 279–94. Edited by Franks, F. and Mathias, S. John Wiley, Chichester.

Franks, F., Asquith, M.H., Hammond, C.C., Skaer, H.le.B. and Echlin, P. (1977). Polymeric cryoprotectants in the preservation of biological ultrastructure: Low temperature states of aqueous solutions of hydrophilic polymers. *Journal of Microscopy* **110**, 223–38.

Franks, F. and Bray, M. (1980). Mechanism of ice nucleation in undercooled plant cells. *CryoLetters* **1**, 221–6.

Franks, F. and Mathias, S. (1982). *Biophysics of Water*. John Wiley, Chichester.

Franks, F., Mathias, F.S., Galfre, P., Webster, S.D. and Brown, D. (1983). Ice nucleation and freezing in undercooled cells. *Cryobiology* **20** 298–309.

Friedman, H.L. and Krishnan, C.V. (1973). Thermodynamics of ion hydration. In *Water: A Comprehensive Treatise*, Vol. 3, pp. 1–118. Edited by Franks, F. Plenum Press, New York.

Friedman, M. (1968). Solvent effects in reactions of protein functional groups. *Quarterly Report on Sulfur Chemistry* **3**, 125–44.

Gerlsma, S.Y. (1968). Reversible denaturation of ribonuclease in aqueous solutions as influenced by polyhydric alcohols and some other additives. *Journal of Biological Chemistry* **243**, 957–61.

Goodin, R. and Levitt, J. (1970). The cryoaggregation of bovine serum albumin. *Cryobiology* **6**, 333–8.

Gordon, J.E. (1975). *The Organic Chemistry of Electrolyte Solutions*. John Wiley and Sons Inc., New York.

Gould, G.W. and Measures, J.C. (1977). Water relations in single cells. *Philosophical Transactions of the Royal Society London B* **278**, 151–66.

Gross, G.W. (1972). Solute interference effects in freezing potentials of dilute electrolytes. In *Water Structure at the Water–Polymer Interface*, pp. 106–25. Edited by Jellinek, H.H.G. Plenum Press, New York and London.

Haly, A.R. and Snaith, J.W. (1971). Calorimetry of rat tail tendon collagen before and after denaturation: The heat of fusion of its absorbed water. *Biopolymers* **10**, 1681–99.

Hansen, J.R. and Yellin, W. (1972). NMR and infrared spectroscopic studies of stratum corneum hydration. In *Water Structure at the Water Polymer Interface*, pp. 19–28. Edited by Jellinek, H.H.G. Plenum Press, New York and London.

Harned, H.S. and Owen, B.B. (1958). *The Physical Chemistry of Electrolyte Solutions, 3rd edition*. Ch 15. Reinhold, New York.

Henderson, T.R., Henderson, R.F. and York, L.J. (1975). Effects of dimethyl sulfoxide on subunit proteins. *Annals of the New York Academy of Science* **243**, 38–53.

Hepler, L.G. and Woolley, E.M. (1973). Hydration effects and acid-base equilibria. In *Water: A Comprehensive Treatise*, Vol. 3, pp. 145–172. Edited by Franks, F. Plenum Press, New York.

Hobbs, P.V. (1974). *Ice Physics*. Oxford University Press, Oxford.

Kamb, B. (1965). Structure of ice VI. *Science* **150**, 205–9.

Kavanau, J.L. (1964). *Water and Solute-Water Interactions*. Holden-Day, San Francisco.

King, E.J. (1965). *Acid–Base Equilibria*. Ch 8. Pergamon Press, Oxford.

Körber, C. and Scheiwe, M.W. (1980). The cryoprotective properties of hydroxyethyl starch investigated by means of differential thermal analysis. *Cryobiology* **17**, 54–65.

Körber, C., Scheiwe, M.W., Boutron, P. and Rau, G. (1982). The influence of hydroxyethyl starch on ice formation in aqueous solutions. *Cryobiology* **19**, 478–92.

Krog, J.D., Zachariassen, K.E., Larsen, B. and Smidsrod, O. (1979). Thermal buffering in afro-alpine plants due to nucleating agent-induced water freezing. *Nature* **282**, 300–301.

Kuntz, I.D. (1971). Hydration of macromolecules. III hydration of polypeptides. *Journal of the American Chemical Society* **93**, 514–16.

Kuntz, I.D. Jr. and Kauzmann, W. (1974). Hydration of proteins and polypeptides. *Advances in Protein Chemistry* **28**, 239–345.

Leibo, S.P. (1977). Fundamental cryobiology of mouse ova and embryos. In *The Freezing of Mammalian Embryos*, pp. 69–96. Ciba Foundation Symposium, Elsevier - Excerpta Medica, North Holland, Amsterdam, Oxford and New York.

Leibo, S.P., McGrath, J.J. and Cravalho, E.G. (1978). Microscopic observations of intracellular ice formation in unfertilized mouse ova as a function of cooling rate. *Cryobiology* **15**, 257–71.

Levitt, J. (1962). A sulfhydryl–disulfide hypothesis of frost injury in plants. *Journal of Theoretical Biology* **3**, 355–91.

Levitt, J. (1966). Cryochemistry of plant tissue. Protein interactions. *Cryobiology* **3**, 243–51.

Lovelock, J.E. (1953). The mechanism of the protective action of glycerol against haemolysis by freezing and thawing. *Biochemica et Biophysica Acta* **11**, 28–36.

Lovelock, J.E. (1954). The protective action of neutral solutes against haemolysis by freezing and thawing. *Biochemical Journal* **56**, 265–70.

Luyet, B.J. (1965). Phase transitions encountered in the rapid freezing of aqueous solutions. *Annals of the New York Academy of Science* **125**, 502–21.

Luyet, B.J. (1966). Anatomy of the freezing process in physical systems. In *Cryobiology*, pp. 115–38. Edited by Meryman, H.T. Academic Press, New York.

Luyet, B. (1969). On the amount of water remaining amorphous in frozen aqueous solutions. *Biodynamica* **10**, 277–91.

Luyet, B. and Rapatz, G. (1958). Patterns of ice formation in some aqueous solutions. *Biodynamica* **8**, 1–68.

Luyet, B. and Rasmussen, D. (1968). Study by differential thermal analysis of the temperatures of instability of rapidly cooled solutions of glycerol, ethylene glycol, sucrose and glucose. *Biodynamica* **10**, 167–91.

Luyet, B.J., Williams, R.J. and Gehenio, P.M. (1964). Direct observation on the mode of invasion of living tissue by ice. *U.S. Defence Documentation Center. AAL-TDR-63*.

MacGregor, W.S. (1967). The Chemical and Physical Properties of DMSO. *Annals of the New York Academy of Science* **141**, 3–12.

MacKenzie, A.P. (1977). Non-equilibrium freezing behaviour of aqueous systems. *Philosophical Transactions of the Royal Society London B* **278**, 167–89.

MacKenzie, A.P. (1981). Complementary thermal and electrical studies on the nature of the frozen state. *Proceedings of the Royal Microscopic Society* **16**, 176–7.

MacKenzie, A.P. and Luyet, B.J. (1962). Electron microscope study of the structure of very rapidly frozen gelatin solutions. *Biodynamica* **9**, 47–69.

MacKenzie, A.P. and Rasmussen, D.H. (1972). Interactions in the water–polyvinylpyrrolidone system at low temperatures. In *Water Structure at the Water–Polymer Interface*, pp. 146–71. Edited by Jellinek, H.H.G. Plenum Press, New York.

Maki, L.R., Galyan, E.L., Chang-Chien, M., Caldwell, D.R. (1974). Ice Nucleation induced by *Pseudomonas syringae. Applied Microbiology* **28**, 456–9.

Martin, D. and Hauthal, H.G. (1975). *Dimethyl Sulphoxide*. Translated by Halberstadt, E.S. Van Nostrand, Reinhold, Wokingham.

Mazur, P. (1963). Kinetics of water loss from cells at subzero temperatures and the likelihood of intracellular freezing. *Journal of General Physiology* **47**, 347–69.

Mazur, P. (1964). Basic problems in cryobiology. In *Advances in Cryogenic Engineering*, Vol. 9, pp. 28–37. Edited by Timmerhaus, K.D. Plenum Press, New York.

Mazur, P. (1965). The role of cell membranes in the freezing of yeast and other single cells. *Annals of the New York Academy of Sciences* **125**, 658–76.

Mazur, P. (1966). Physical and chemical basis of injury in single-celled micro-organisms subjected to freezing and thawing. In *Cryobiology*, pp. 213–315. Edited by Meryman, H.T. Academic Press, London.

Mazur, P., (1981). Fundamental cryobiology and the preservation of organs by freezing. In *Organ Preservation for Transplantation, 2nd edition*, pp. 143–75. Edited by Karow, A.M. Jr. and Pegg, D.E. Marcel Dekker, New York.

Mazur, P., Leibo, S.P., Farrant, J., Chu, E.H.Y., Hanna, M.G. Jr. and Smith, L.H. (1970). Interactions of cooling rate, warming rate and protective additive on the survival of frozen mammalian cells. In *The Frozen Cell*, pp. 69–85, Ciba Foundation Symposium. Edited by Wolstenholme, G.E.W. and O'Connor, M. Churchill, London.

Mazur, P., Rall, W.F. and Rigopoulos, N. (1981). Relative contributions of the fraction of unfrozen water and of salt concentration to the survival of slowly frozen human erythrocytes. *Biophysical Journal* **36**, 653–75.

Meryman, H.T. (1966). Review of biological freezing. In *Cryobiology*, pp. 2–114. Edited by Meryman, H.T. Academic Press, London and New York.

Meryman, H.T. (1975). Basil. J. Luyet: In Memoriam. *Cryobiology* **12**, 285–92.

Morris, G.J. and Clarke, A. (1981). *Effects of Low Temperature on Biological Membranes.* Academic Press, London.

Némethy, G. (1968). The structure of water and of aqueous solutions. In *Low Temperature Biology of Foodstuffs, Recent Advances in Food Science*, Vol. 4, pp. 1–21. Edited by Hawthorn, J. and Rolfe, E.J. Pergamon Press, Oxford.

Némethy, G. and Scheraga, H.A. (1962). Structure of water and hydrophobic bonding in proteins. I. A model for the thermodynamic properties of liquid water. *Journal of Chemical Physics* **36**, 3382–400.

Packer, K.J. and Tomlinson, D.J. (1971). Nuclear spin relaxation and self-diffusion in the binary system, dimethylsulphoxide (DMSO) + water. *Transactions of the Faraday Society* **67**, 1302–14.

Pain, R.H. (1982). Molecular hydration and biological function. In *Biophysics of Water*, pp. 3–14. Edited by Franks, F. and Mathias, S. John Wiley, Chichester.

Parkes, A.S. (1964). Cryobiology. *Cryobiology* **1**, 3.

Pauling, L. (1959). The structure of water. In *Hydrogen Bonding*, pp. 1–6. Edited by Hadzi, D. Pergamon Press, London and New York.

Pegg, D.E. (1981). The biology of cell survival *in vitro*. In *Organ Preservation for Transplantation, 2nd edition*, pp. 31–52. Edited by Karow, A.M. Jr. and Pegg, D.E. Marcel Dekker, New York.

Pollard, A. and Wyn Jones, R.G. (1979). Enzyme activities in concentrated solutions of glycinebetaine and other solutes. *Planta* **144**, 291–8.

Ponder, E. (1948). *Hemolysis and Related Phenomena.* Churchill, London.

Pople, J.A. (1951). Molecular association in liquids II. A theory of the structure of water. *Proceedings of the Royal Society of London A* **205**, 163–78.

Rall, W.F. and Fahy, G.M. (1985). Ice-free cryopreservation of mouse embryos at $-196°C$ by vitrification. *Nature* **313**, 573–5.

Rall, W.F., Mazur, P. and McGrath, J.J. (1983). Depression of the ice-nucleation temperature of rapidly cooled mouse embryos by glycerol and dimethyl sulphoxide. *Biophysical Journal* **41**, 1–12.

Rall, W.F., Reid, D.S. and Farrant, J. (1980). Innocuous biological freezing during warming. *Nature (Lond.)* **286** 511–14.

Rammler, D.H. (1967). The effect of DMSO on several enzyme systems. *Annals of the New York Academy of Science* **141**, 291–9.

Rammler, D.H. and Zaffaroni, A. (1967). Biological implications of DMSO based on a review of its chemical properties. *Annals of the New York Academy of Science* **141**, 13–23.

Rapatz, G. and Luyet, B. (1966). Patterns of ice formation in aqueous solutions of glycerol. *Biodynamica* **10**, 69–80.

Rapatz, G. and Luyet, B. (1968). Patterns of ice formation in aqueous solutions of polyvinylpyrrolidone, and temperatures of instability of the frozen solutions. *Biodynamica* **10**, 149–66.

Rasmussen, D.H. and Luyet B. (1969). Complementary study of some non-equilibrium phase transitions in frozen solutions of glycerol, ethylene glycol, glucose and sucrose. *Biodynamica* **10**, 319–31.

Rasmussen, D.H., Macaulay, M.N. and MacKenzie. A.P. (1975). Supercooling and nucleation of ice in single cells. *Cryobiology* **12**, 328–39.

Rasmussen, D.H. and MacKenzie, A.P. (1968). Phase diagram for the system water–dimethylsulphoxide. *Nature* **220**, 1315–17.

Rasmussen, D.H. and MacKenzie, A.P. (1972). Effect of solute on ice–solution interfacial free energy; calculation from measured homogeneous nucleation temperatures. In *Water Structure at the Water-Polymer Interface*, pp. 126–45. Edited by Jellinek, H.H.G. Plenum Press, New York and London.

Rasmussen, D.H. and MacKenzie, A.P. (1973). Clustering in supercooled water. *Journal of Chemical Physics* **59**, 5003–13.

Reid, D.S. (1979). The low temperature phase behaviour of aqueous ribose. *CryoLetters* **1**, 35–8.

Reid, D.S. (1983). Fundamental physicochemical aspects of freezing. *Food Technology* **37**, 110–15.

Reid, D.S., Rall, W.F. and Fahy, G.M. (1985). DSC and cryomicroscope studies of the behaviour of VSI solutions at low temperatures *Cryobiology* **22**, 602.

Riehle, U. (1968). *The Vitrification of Dilute Aqueous Solutions* Ph.D. Thesis, Federal Technical University, Zurich, Switzerland.

Ritchie, C.D. (1969). Interactions in dipolar aprotic solvents. In *Solute-Solvent Interactions*, pp. 219–300. Edited by Coetzee, J.F. and Ritchie, C.D. Marcel Dekker, New York and London.

Robinson, R.A. and Stokes, R.H. (1968). *Electrolyte Solutions, 2nd edition*. Butterworth, London.

Roy, R.N., Gibbons, J.J., Baker, G. and Bates, R.G. (1984). Standard EMF of the H_2–AgCl; Ag cell in 30, 40 and 50 mass % dimethylsulfoxide/water from −20° to 25°C. pK_2 and pH values for a standard 'BICINE' buffer solution at subzero temperatures. *Cryobiology* **21**, 672–81.

Roy, R.N., Gibbons, J.J., and Pogue, R. (1985a). Acid dissociation constants and pH* values of the biological buffer "MOPS" in 50% (w/w) DMSO/water and ethylene glycol/water from 25°C to −20°C. *CryoLetters* **6**, 139–150.

Roy, R.N., Gibbons, J.J., McGinnis, T. and Woodmansee, R. (1985b). Electromotive force standard of the H_2–AgCl; Ag cell in 30, 40 and 50 mass % Glycerol/water from −20 to 25°C. pK_2 and pH values for a standard "MOPS" buffer in 50 mass % glycerol water. *Cryobiology* **22**, 578–88.

Safford, G.J., Schaffer, P.C., Leung, P.S., Doebbler, G.F., Brady, G.W. and Lyden, E.F.X. (1969). Neutron inelastic scattering and X-ray studies of aqueous solutions of dimethylsulphoxide and dimethylsulphone. *Journal of Chemical Physics* **50**, 2140–59.

Shepard, M.L., Goldston, C.S. and Cocks, F.H. (1976). The H_2O-NaCl-glycerol phase diagram and its application in cryobiology. *Cryobiology* **13**, 9–23.

Shlafer, M. (1981). Pharmacological considerations in cryopreservation. In *Organ Preservation for Transplantation, 2nd edition*, pp. 177–212. Edited by Karow, A.M. Jr. and Pegg, D.E. Marcel Dekker, New York.

Soliman, F.S. and Van den Berg, L. (1971a). Factors affecting freezing damage of lactic dehydrogenase. *Cryobiology* **8**, 73–8.

Soliman, F.S. and Van den Berg, L. (1971b). Factors affecting freeze aggregation of lipoprotein. *Cryobiology* **8**, 265–70.

Sørensen, S.P.L. (1909a). Enzymstudien. II Über die messung und die Bedentung der Wasserstaffionen Konzentration bei enzymatischen-Prozessen. *Biochem. Z* **21**, 131–304.

Sørensen, S.P.L. (1909b). Études enzymatiques. II sur la mesure et l'importance de la concentration des ions hydrogène dans les réactions enzymatiques. *C.R. Trav. Lab. Carlsberg* **8**, 1–168.

Taylor, M.J. (1977). pH measurements in dimethylsulphoxide-water mixtures: relevance to the integrity of smooth muscle exposed to low temperatures. Ph.D. Thesis. Council for National Academic Awards (CNAA), London.

Taylor, M.J. (1978a). Standard electromotive forces of the cell Pt; $H_{2(1atm)}$, $HCl_{(m)}$, AgCl: Ag containing mixtures of dimethylsulfoxide and water between +25 and −12°C. *Journal of Chemical Engineering Data* **23**, 308–13.

Taylor, M.J. (1978b). The response of the glass/calomel pH-cell in aqueous solutions containing dimethylsulfoxide at 25 and −12°C. *Cryobiology* **15**, 340–44.

Taylor, M.J. (1979). Assignment of standard pH values (pH*(S)) to buffers in 20 and 30% (w/w) dimethylsulfoxide/water mixtures at normal and subzero temperatures. *Journal of Chemical Engineering Data* **24**, 230–33.

Taylor, M.J. (1980a). Electrodes for the determination of thermodynamic quantities (including pH) in solutions containing dimethylsulphoxide at normal and subzero temperatures. *CryoLetters* **1**, 159–66.

Taylor, M.J. (1980b). Acid dissociation constants for some biological buffers in aqueous solutions containing dimethylsulphoxide at 25°C and −12°C. *CryoLetters* **1**, 449–60.

Taylor, M.J. (1981). The meaning of pH at low temperatures. *CryoLetters* **2**, 231–9.

Taylor, M.J. (1982). The role of pH* and buffer capacity in the recovery of function of smooth muscle cooled to −13°C in unfrozen media. *Cryobiology* **19**, 585–601.

Taylor, M.J. (1984). Subzero preservation and the prospect of long term storage of multicellular tissues and organs. In *Transplantation Immunology. Clinical and Experimental*, pp. 360–90. Edited by Calne, R.Y. Oxford University Press, Oxford.

Taylor, M.J. and Pignat, Y. (1982). Practical acid dissociation constants, temperature coefficients, and buffer capacities for some biological buffers in solutions containing dimethylsulfoxide between 25 and −12°C. *Cryobiology* **19**, 99–109.

Van den Berg, L. (1959). The effect of addition of sodium and potassium chloride to the reciprocal system: KH_2PO_4–Na_2HPO_4–H_2O on pH and composition during freezing. *Archives of Biochemistry and Biophysics* **84**, 305–15.

Van den Berg, L. and Rose, D. (1959). Effect of freezing on the pH and composition of sodium and potassium phosphate solutions: the reciprocal system KH_2PO_4–Na_2HPO_4–H_2O. *Archives of Biochemistry and Biophysics* **81**, 319–29.

Van den Berg, L. and Soliman, F.S. (1969). Effect of glycerol and dimethylsulphoxide on changes in composition and pH of buffer salt solutions during freezing. *Cryobiology* **6**, 93–97.

2

Cells at low temperatures

G.J. Morris
Cell Systems Limited
Cambridge
A. Clarke
British Antarctic Survey (Natural Environment Research Council)
Cambridge

Growth at low temperatures

The temperature range over which cells can grow varies from species to species, although the overall range in the biosphere extends from the freezing point to the boiling point of water. The temperature range over which cells may remain viable is even wider.

Growth of eukaryotic cells at temperatures above approximately 75°C (Brock, 1978) has not been reported, whereas prokaryotic growth appears to be limited not so much by temperature but by the presence of liquid water. Bacteria have been isolated from deep-sea hydrothermal vents (so-called 'black smokers') at temperatures over 300°C and have been found capable of chemolithotrophic growth at 250°C and 265 atmospheres (Baross and Deming, 1983; but also see criticism by Trent *et al.*, 1984). For any individual cell type, however, the temperature range over which growth will occur is approximately 30°C or less.

Traditionally, cells have been divided into three broad categories, thermophiles, mesophiles and psychrophiles, according to their ability to grow at high, medium and low temperatures respectively. The low temperature category has been further sub-divided on the basis of a combination of the range of temperatures over which growth can occur and the temperature of optimal (maximum) growth. Thus psychrophilic bacteria are defined as those with an optimal growth temperature less than 16°C and an upper temperature limit for growth of approximately 20°C. Psychrotrophic bacteria are those which will grow at temperatures as low as zero but which have an optimum temperature for growth approximately 20°C (Morita, 1975).

Simon (1981) has suggested that this classification of cells into distinct groups is misleading and that in reality there is a continuum of growth ranges. This raises an important theoretical consideration, namely the possibility that the same biochemical events might limit growth for a thermophile at a relatively high temperature and a psychrophile at a low temperature. Alternatively, those events which determine the lower temperature limit for growth may differ over various temperature ranges. All cell-types show an optimum temperature for growth and at both higher and lower temperatures the rate of cell division is reduced.

Two major observations can be made with regard to the cellular events which occur at sub-optimal temperatures.

(1) As the temperature is reduced, the rate of growth decreases, often in a non-linear manner. The factors responsible for this reduction in growth rate have not been elucidated for any cell-type.

(2) There is a minimum temperature below which cell division does not occur. This temperature may be determined with a precision of less than 1°C (Shaw *et al.*, 1971). It is not necessarily a property of the cell-type alone as it may be modified by environmental conditions such as nutrient levels, pH and water activity.

In some cell-types exposure to sub-optimal growth temperatures may induce resting stages, and for vegetative cells two major classes of response can be identified.

(1) The cells may survive for extended periods but still have the capacity to divide and grow when rewarmed. Obviously, the potential for integrated metabolism must be maintained so that essential functions (e.g. osmotic balance and turnover of essential components) can continue. In prokaryotes this has been termed maintenance (Pirt, 1965) and is comparable to the basal metabolic rate of higher organisms (Clarke, this volume).

(2) The cells slowly accumulate sub-lethal injuries to a level where the

potential for integrated metabolism can no longer be maintained, and so viability is lost.

Which of these alternatives prevails depends on the cell-type and how far below the minimum temperature for growth the cells are maintained. For example, the ciliated protozoan *Tetrahymena pyriformis* has a minimum growth temperature of about 15°C; at 10°C viability is retained for several days, but undercooling to −10°C causes complete loss of viability within 15 minutes (Morris *et al.*, 1984).

In this chapter we examine the effects of 'low' temperatures on the structure and activity of isolated cellular components, specifically membranes and proteins. The effects of temperatures below those optimal for growth on aspects of cell physiology will also be discussed. It must be stressed, however, that a reduction in temperature will simultaneously modify all aspects of cell biology. It is therefore perhaps an oversimplification to expect single determinants to explain the effects of low temperatures on integrated cellular processes such as transport or motility, even within a unicellular organism. Obviously, in multicellular systems such processes are more complex.

This discussion will be limited to the effects of slow rates of cooling, such as occur in the natural environment and in most experimental manipulations. The response to rapid cooling and the additional stresses associated with ice nucleation will be discussed elsewhere (Morris: Grout and Morris, this volume).

Cellular membranes

Biological membranes are complex structures whose composition and structure vary widely both between and within cells depending upon function. In order to discuss the effects of low temperatures on biological membranes it is first necessary to outline their composition and structure, and then to discuss the experimental work on simple model systems.

Membrane composition

The proportion of lipid in biological membranes varies between 20 and 80 per cent of the dry weight, and it is these lipids which give the membrane its unique structure and physical properties.

The dominant structural class of lipid in biological membranes are the glycerol phosphatides. These are usually fatty acyl esters of *sn* -glycerol-3-phosphate with a base of sugar head group covalently bonded to the phosphate group. Most fatty acyl chains are attached by acyl linkages, but in some organisms alkyl or alkyl-1-enyl linkages can be important. Alkyl-1-enyl linkages have, however, received very little attention, and interest has centered on the fatty acid linkages. This reflects the undoubted importance of fatty acid linkages, and also the relative ease with which they can be analysed.

The fatty acids of most biological membranes usually contain 16,18, 20 or 22 carbon atoms. Typically, they are either fully saturated or contain between one and six *cis* double bonds, usually methylene interrupted. There are normally two fatty acyl chains connected to a head group, often one is saturated, the other is not. Both *iso* -branched and *anteiso* -branched fatty acids may occur as components of membrane lipids, particularly in bacteria (Kaneda, 1977). Bacteria may also contain a number of cyclic or isoprenoid fatty acids in membrane lipids (Russell, 1984).

Membranes may also contain sterols, the predominant one in animal cells being cholesterol. Other cell-types contain sterols other than cholesterol in a proportion that may vary widely with cell-type.

In some organisms the membranes are of relatively simple composition. For example, the bacterium *Escherichia coli* contains only three phospholipids and only three different fatty acyl chains (Cronan and Gelman, 1975). Other cell-types contain a number of phospholipids together with a large variety of fatty acyl chains, and some cells may possess membranes that contain at least 200 lipid species.

Membrane structure

At temperatures above that of the gel to liquid crystalline phase transition, and in the presence of water, phospolipids form stable bilayers with the hydrophilic head groups exposed to the water and the hydrophobic hydrocarbon chains forming the core of the bilayer. This lamellar phase is the basic structure of biological membranes. It has been suggested that localized or transient non-bilayer lipid domains coexist within the bilayer structure, and some membrane functions may require the formation of such structures (de Kruijff *et al.*, 1980). At low water contents (below 0.25g H_2O/g lipid), other lipid phases may form (Grout and Morris, this volume).

In the lamellar phase, phospholipid head groups are exposed at both the intracellular and extracellular surfaces and so are able to interact with the appropriate aqueous phase. The fatty acyl chains are orientated largely at right angles to the plane of the membrane, with the terminal methyl groups in the interior of the bilayer. This arrangement presents a permeability barrier to a small polar molecule or ion since insertion into the hydrophobic interior requires a relatively large activation energy. The permeability properties of liposomes (lipid bilayer vesicles) to polar or ionic compounds are several orders of magnitude lower than those of cellular membranes (Fettiplace and Haydon, 1980). It is generally assumed that this difference is due to the presence of the membrane proteins, which are assumed to be responsible for much of the biological activity of cellular membranes.

The intramembraneous particles observed by freeze-fracture electron microscopy are generally assumed to be membrane proteins (Skaer, this volume), and the frequency of such particles is related to the metabolic activity of that membrane. Relatively inactive membranes such as the myelin sheath have very few particles, if any; erythrocytes have particle frequencies of 2000–3000 μm^{-2} while membranes across which there are very high rates of water transport (e.g. the descending loop of Henle in the mammalian kidney) have particle frequencies in excess of 4000 μm^{-2}.

Biological membranes may contain a large number of proteins of which the major classes are those involved in transmembrane transport, proteins specialized for recognition, such as receptors for hormones and neurotransmitters, and those which contribute to the mechanical structure of the membrane. The latter are often associated with the underlying cytoskeleton.

Membrane proteins may be associated with the surface, or penetrate to a greater or lesser extent into, the bilayer. A functional classification of membrane proteins may be produced according to the ease with which they may be removed from the membrane. Thus, peripheral proteins can be removed by mild procedures such as increasing the ionic strength of the external medium or by cold osmotic shock, while integral proteins require more severe treatments such as detergents to solubilize them.

It has been argued that integral proteins are surrounded by immobile boundary lipids and that this halo of lipids controls the activity of the proteins (Hesketh *et al.*, 1976). The existence of such rigidly associated boundary lipids is, however, disputed (Oldfield *et al.*, 1978, Kang *et al.*, 1979).

Membrane asymmetry

Membrane lipids are commonly distributed asymmetrically across the bilayer; the nature and extent of this asymmetry varies from one membrane to another. In the well-studied case of the human erythrocyte, phosphatidylserine and phosphatidylethanolamine are located primarily in the cytoplasmic half of the bilayer, while phosphatidylcholine and sphingomyelin are found mainly in the outer half (Bretscher, 1972). Protein distribution across the membrane is also asymmetric; proteins are associated more abundantly with the cytoplasmic side.

As well as the asymmetric distribution of components in the two halves of the bilayer, there may also be non-random distribution within the plane of the membrane. In highly polarized cells such as spermatozoa, domains of lipid have been demonstrated (Skaer, this volume). These specialized regions are actively maintained by the cells, and the membranes only become homogeneous with respect to lipid distribution upon cell death (Friend, 1982).

Membrane fluidity

Lipid bilayers may be considered as two-dimensional fluids surrounded by an environment with which they are immiscible. The motion of individual lipid molecules contributes the physical parameter described as bulk fluidity, the reciprocal of which is membrane viscosity. Bulk fluidity is composed of several distinct types of molecular motion including: lateral movements within the plane of the membrane, the rotation of molecules about a plane normal to the membrane, and the flexing of acyl chains by rotation about their carbon–carbon bonds. There is also a gradient of fluidity across the membrane, the interior being highly disorderd or fluid, the region close to the polar head groups at the bilayer surface is relatively rigid.

Several techniques have been used to measure the bulk fluidity of lipid bilayers and biological membranes; these include fluorescence depolarization, electron spin resonance (ESR) and nuclear magnetic resonance (NMR).

The measured fluidity of a biologial membrane is determined by its composition and the temperature of measurement (Cossins, 1981). Decreased fluidity would be expected to reduce the rates of many membrane processes which are dependent on diffusion or molecular collision for activity. A large part of the temperature dependence of enzyme activity is due to changes in the viscosity of the solvent environment (Gavish and Werber, 1979) and intrinsic membrane proteins are no exception to this. Alterations in membrane viscosity accompanying a reduction in temperature would have major effects on membrane-associated enzyme functions.

Thermotropic behaviour in model systems

It is a fundamental property of phospholipid bilayers that they exhibit thermotropism, that is a pure phospholipid in aqueous suspension undergoes an abrupt change from a disordered fluid (the liquid-crystalline state) to a highly ordered hexagonal lattice of fatty acyl chains (the gel state) over a specific temperature range. The temperature at the midpoint of this phase change is the transition temperature (T_c), and the change in state has been variously called the lipid phase transition, the gel-liquid crystalline transition or the order-disorder transition.

As lipid bilayers are essentially two-dimensional fluids the phase transition from liquid-crystalline to gel state is only a specialized case of the nucleation and phase behaviour of bulk aqueous samples (Taylor, this volume). To illustrate the processes occuring during such nucleation and growth of gel phase lipid we shall initially consider a simple phospholipid system and assume that nucleation is homogeneous.

At low rates of cooling the probability of substantial undercooling will be low and, consequently, there will be few nucleation sites within the bilayer. Under these conditions there is sufficient time for growth of gel phase lipid to occur with a minimum of structural imperfections or 'packing faults'. The two halves of the bilayer nucleate independently (Sillerud and Barnett, 1982) and, if both monolayers contain few 'packing faults', the probability of such structural imperfections being coincident between the monolayers is also low. Coincident packing faults would facilitate leakage of small molecular weight cellular components (Fig. 2.1). Therefore with bilayers composed of a single phospholipid cooled slowly to temperatures below the phase transition the formation of gel state lipid would not be expected to result in an increased permeability to small ions or molecules.

At the phase transition temperature itself there is a stable coexistence of gel and liquid phases. This is termed a phase separation and for a pure phospholipid it will occur over range of about 1 °C. The lipid molecules at the junction between the gel and liquid phases will be in a state of disturbed packing. In addition the molar volume of lipids in the gel state is smaller than in the liquid-crystalline phase. As a result a domain of gel phase might form in an environment of liquid-crystalline phase lipid and will be under tension. Conversely, a domain of liquid-crystalline lipid may be in a bulk gel phase and will be under compression. At the temperature of phase transition the bilayer

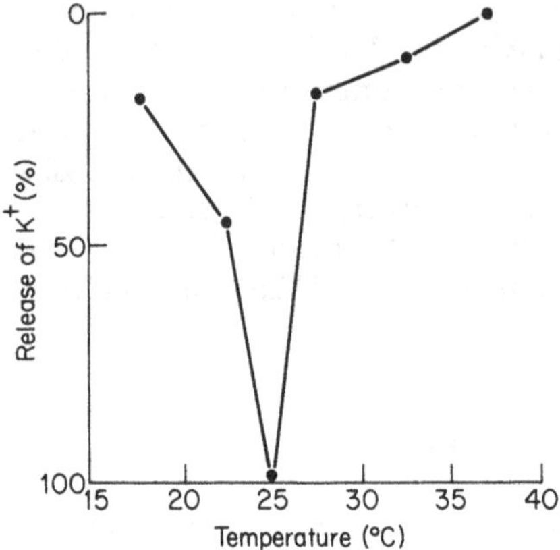

Fig. 2.1 Release of potassium (%) from liposomes prepared from dimyristoyl phosphatidylcholine following incubation for 30 seconds at different temperatures. The transition temperature (T_c) for dimyristoyl phosphatidylcholine is approximately 23°C. (Redrawn from Blok *et al.*, 1975.)

will thus exhibit an increase in the degree of lateral compressibility (Linden *et al.*, 1973; Marcelja and Wolfe, 1979).

Either or both these factors (i.e. the regions of molecular discontinuity between gel and liquid phases and the increased lateral compressibility) results in an increased permeability of the bilayer at the transition temperature (Fig. 2.1). This has been clearly demonstrated by a leakage of small molecules across the bilayer (Papahadjopoulos *et al.*, 1973; Blok *et al.*, 1975; Braganza *et al.*, 1983; Marsh *et al.*, 1976), hydrolysis by phospholipase A2 (Op den Kamp *et al.*, 1975) and increased uptake of membrane spin labels (Lee, 1977b).

The processes outlined above will depend both on the lipid composition of the membranes and on the presence of membrane proteins (Chapman *et al.*, 1977). In addition, membrane thermotropic behaviour would be influenced by the occurence (if any) of structures within the membrane which promote or depress nucleation.

The influence of bilayer composition on thermotropic behaviour

Studies on model systems (particularly liposomes) have provided much information on the relationship between the composition and thermotropic behaviour of lipid bilayers. These results have proved to be of great value in interpreting the thermotropic behaviour of biological membranes. The results may be summarized as follows.

(1) *Effect of fatty acyl chain length*

For any given phospholipid class the transition temperature (T_c) increases with acyl chain length (Fig. 2.2).

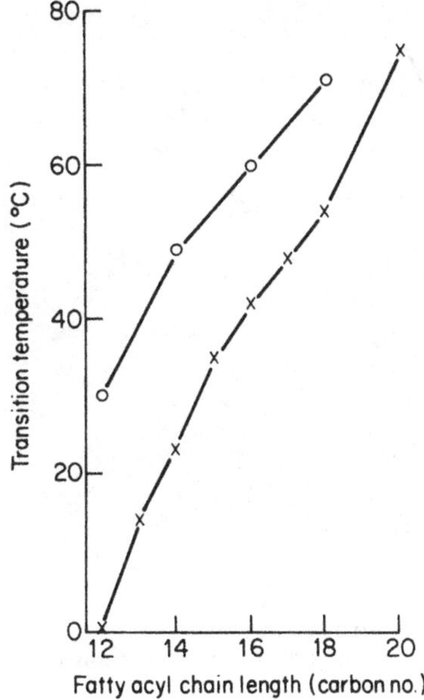

Fig. 2.2 Transition temperature (T_c) of saturated phosphatidylcholines (x) and phosphatidylethanolamines (o). (Redrawn from Seelig, 1981.)

(2) *Effect of unsaturation*

For any given phospholipid head group and fatty acid chain length, T_c decreases dramatically with the insertion of double (olefinic) bonds into the acyl chain. For example, the T_c for choline phosphoglyceride containing two stearic acids (both 18:0) is 54°C, whereas the T_c for the dioleyl form (18:1/18:1) is −22°C. Addition of a double bond to the fatty acid in the *sn*-2 position has a larger effect on T_c than when inserted in the *sn*-1 position (Table 2.1) and, in general, addition of further olefinic bonds into the same acyl chain has progressively smaller effect (Demel *et al.*, 1972).

The position of a single olefinic bond in the fatty acyl chain also has a

Table 2.1 Transition temperatures (T_c) for unsaturated 3-*sn*-phosphatidylcholines, all *cis* double bonds

Chain composition	T_c(°C)	Reference
18:1/18:1 (dioleoyl)	−22	Chapman, (1967)
18:0/18:1 1-stearoyl-2-oleoyl	+3	Phillips *et al.*, (1972)
18:1/18:0 1-oleoyl-2-stearoyl	+15	de Kruijff *et al.*, (1972)
16:0/18:1 1-palmitoyl-2-oleoyl	−5	de Kruijff *et al.*, (1973)

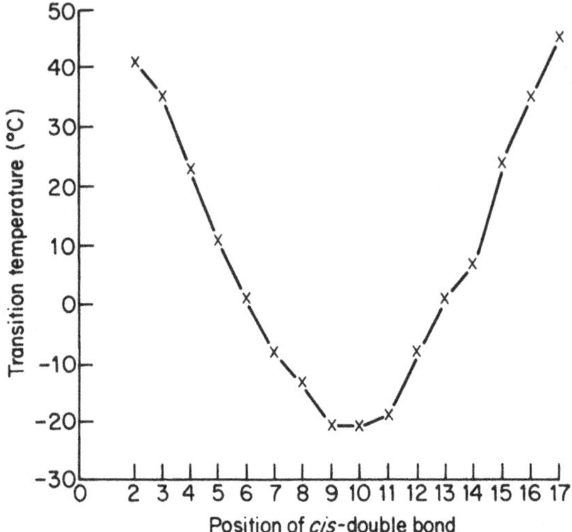

Fig. 2.3 Transition temperature (T_c) for fully hydrated 1,2-dioctadecenoyl-*sn* glycero-3-phosphocholines as a function of the position of the *cis* double bond. (Redrawn from Barton and Gunstone, 1975.)

pronounced effect (Fig. 2.3). The greatest effect on T_c occurs when the double bond is approximately in the middle of the hydrocarbon chain; the influence decreases if the double bond is sited closer to either the terminal methyl or carbonyl ends of the chain. Analyses of a wide variety of organisms has indicated that in monosaturated C16 and C18 fatty acids the single olefinic bond is almost always inserted close to the middle of the hydrocarbon chain. Thus in the 18:1 chain which is commonly a major component of membranes, over 90 per cent of the molecules will have a double bond at either position 7 or position 9.

(3) *Effects of branched chain structure*

The T_c of anteiso fatty acids is 25–35°C lower than the equivalent straight-chain fatty acid. The T_c of cyclopropane fatty acids is similar to that of the mono-unsaturated analogue (Cullen *et al.*, 1971). The effect of multi-branched chains (e.g. the isoprenoid fatty acyl chains of many thermophilic bacteria) does not appear to have been investigated.

(4) *Effect of polar head group*

For any saturated acyl chain the T_c is influenced by the phospholipid head group to which it is attached. For example, the transition temperatures of phosphatidylethanolamines are much higher than those for the correspon-ding phosphatidylcholines (Fig. 2.2). The transition temperatures of the other polar head groups are intermediate between these two extremes. This difference is, however, reduced when unsaturated fatty acids are present, for example the difference between phosphatidylcholine and phosphatidylethanolamine is only 2°C when both fatty acids are 18:1 (Seelig, 1981).

(5) *Effects of mixtures of phospholipids*

Biological membranes generally contain a variety of phospholipids, and much of our understanding of the effects of reduced temperatures on membranes derives from studies on bilayers of mixed phospholipids when variations in chain length, degree of unsaturation and head group have been controlled experimentally. Under equilibrium conditions (i.e. slow cooling) and using mixtures of only two phospholipids, three different physical situations are possible. These are:

(a) Complete miscibility in solid and liquid states – if two lipids have identical headgroups and fatty acyl chains of similar length, mixing is close to ideal in both the solid and liquid states (Fig. 2.4).

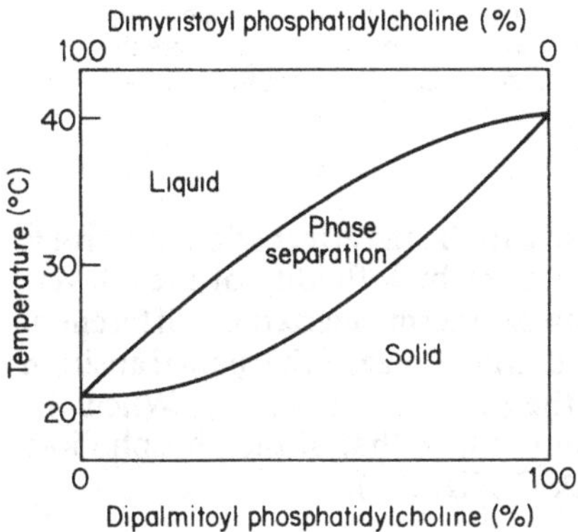

Fig. 2.4 Phase diagram for a binary mixture of dipalmitoyl phosphatidylcholine and dimyristoyl phosphatidylcholine, which have identical headgroups and possess acyl chains of similar length. They form ideal mixtures in both solid and fluid states and yield a simple phase diagram. (Redrawn from Shimschick and McConnell, 1973.)

(b) Solid phase immiscibility – with mixtures of lipids having the same headgroup but very different fatty acyl chains, packing in the solid state is different. Upon cooling, heterogeneous solid phase lipid is formed containing areas of the individual pure phase lipids together with solid solutions (as with some metal alloys). Solid phase immiscibility may also be displayed with mixtures of lipids with different headgroups.

(c) Fluid phase immiscibility – a limited fluid phase immiscibility has been demonstrated in mixtures of dielaidoyl phosphatidylcholine and dipalmitoyl phosphatidylethanolamine (Wu and McConnell, 1975).

(6) *Effect of sterol in phospholipid systems*

The effect of cholesterol on the thermotropic behaviour of phospholipid

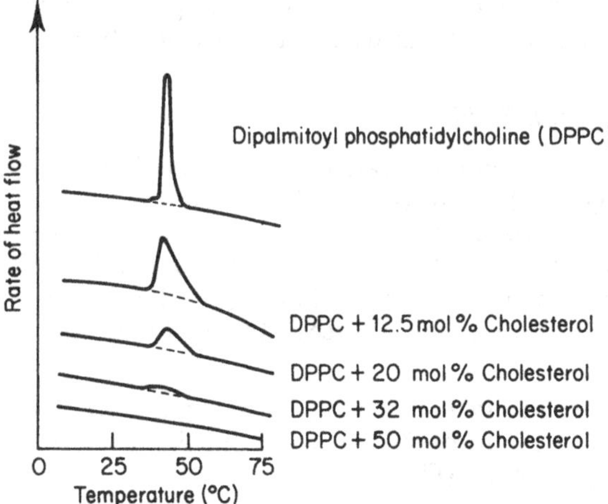

Fig. 2.5 Differential scanning calorimetry traces for dipalmitoyl phosphatidylcholine – cholesterol mixtures. By increasing the cholesterol content of phospholipid bilayers the endothermic phase transition is moved to lower temperatures and progressively broadened. (Redrawn from Ladbroke *et al.*, 1968.)

bilayers has been studied extensively. Above the T_c the effect of cholesterol is to order the acyl chains and thereby reduce the bulk fluidity of the bilayer. Below T_c cholesterol reduces the transition exotherm detected by differential scanning calorimetry (DSC) and the temperature range of the main transition is increased (i.e. the co-operativity of the process is reduced). When the cholesterol concentration becomes equimolar with that of the phospholipid no phase transition can be detected by DSC (Fig. 2.5).

(7) *Effect of proteins*

At high protein concentrations within lipid bilayers the co-operativity between phospholipid molecules should be reduced, and the incorporation of proteins will have an effect on thermotropic behaviour similar to that of cholesterol. For example, the addition of the polypeptide gramicidin A to dipalmitoyl phosphatidylcholine bilayers (Chapman *et al.*, 1977) or myelin apoprotein to dimyristoyl phosphatidylcholine bilayers (Curatolo *et al.*, 1978) causes a broadening of the phospholipid exotherm and, at high concentrations, a complete abolition of the phase transition.

At lower concentrations of protein within the bilayer, there is little effect on the thermotropic behaviour. However, following slow rates of cooling to temperatures below the phase transition, protein molecules are excluded from the growing phospholipid crystal lattice and are concentrated into regions of fluid lipid. Low temperatures can thus induce a lateral phase separation into regions of crystalline phospholipid and domains of high protein concentration containing trapped phospholipid (see below).

(8) *Other factors*

Several other factors have been shown to modify the thermotropic behaviour of phospholipid bilayers. These include: lateral pressure (Phillips and Chapman, 1968), the ionic composition of the surrounding aqueous environment (Lee, 1977a,) and the presence of compounds which interact with, or partition into, lipid bilayers. These include short-chain alcohols (Grisham and Barnett, 1973; Hui and Barton, 1973) dimethylsulphoxide (DMSO) (Lyman *et al.*, 1976), glycerol and trehalose (Crowe *et al.*, 1984a,b).

Thermotropic behaviour of biological membranes

Thermotropic events in biological membranes have been demonstrated by a variety of techniques. These include DSC, freeze-fracture electron microscopy, kinetics of membrane-associated enzymes and the behaviour of membrane probes. Since each of these processes involves different techniques and assumptions (and may even give different answers) they will be discussed separately.

Differential Scanning Calorimetry (DSC)

The least equivocal evidence of a change in state within biological membranes is provided by calorimetric studies. In addition to providing qualitative evidence of thermotropism, the enthalpy values for such transitions, when compared with those of purified phospholipids, allow an estimate of the proportion of lipids undergoing a phase change. Using this technique thermotropic transitions have been demonstrated in a number of cells, isolated membranes and extracted lipids (Melchior and Steim, 1976). The transitions observed are often very broad due to the complex mixture of lipids present in most biological membranes (Table 2.2; Fig. 2.5). In some prokaryotes these broad phase transitions may encompass the growth temperature, suggesting that in these cell-types the membranes may be functional and yet in a state of phase separation.

Freeze-fracture electron microscopy

Following slow rates of cooling, an aggregation of intramembraneous particles appears on freeze-fracture electron micrographs. These particles are generally assumed to be membrane proteins, and this aggregation of intramembraneous particles has usually been interpreted as an exclusion of membrane proteins from the developing gel lipid matrix into regions which are still fluid. The aggregation of intramembraneous particles at low temperatures has been observed in a wide variety of cell-types (Table 2.3).

In many early studies protein clustering was scored simply as present or absent. Phase separations in biological membranes are, however, gradual events and protein aggregation is clearly not an all-or-nothing event. Some later studies have used numerical parameters such as an index of protein

Table 2.2 Evidence of thermotropic behaviour in biological membranes from differential scanning calorimetry

Cell-type	Preparation	Temperature of growth (°C)	Range of transition (°C)	Lipid undergoing transition (%)	Reference
Prokaryotes					
Acheloplasma laidlawii	Plasmalemma	37	10–40		Stein *et al*., (1969)
Escherichia coli	Plasmalemma	37	28–47		Heast *et al*., (1974)
E. coli	Extracted lipids	37	0–25		Jackson and Cronan (1978)
E. coli	Extracted lipids	25	0–8		Jackson and Cronan (1978)
E. coli (Auxotrophic mutant)	Extracted lipids	37	8–47		Jackson and Cronan (1978)
Bacillus subtilis	Plasmalemma	37	11–40		Haest *et al*., (1974)
Staphylococcus aureus	Plasmalemma	37	4–31		Haest *et al*., (1974)
Yersinia enterocolitica	Plasmalemma	37	−18–8		Abbas and Card (1980)
Y. enterocolitica	Plasmalemma	22	−24–−4		Abbas and Card (1980)
Y. enterocolitica	Plasmalemma	5	−29–−5		Abbas and Card (1980)
Higher plants					
Cauliflower floret	Total membranes		2–9	1	McMurchie *et al*., (1979)
(*Brassica oleracea*)	Extracted lipids		not detected		McMurchie *et al*., (1979)
Cucumber fruit	Total membranes		0–10	1	McMurchie *et al*., (1979)
(*Cucumis sativus*)	Extracted lipids		not detected		McMurchie *et al*., (1979)
Tomato fruit	Total membranes		not detected		McMurchie *et al*., (1979)
(*Lycopersicon esculentum*)	Extracted lipids		−6–13	1	McMurchie *et al*., (1979)
Tomato chloroplast (*L. esculentum*)	Polar lipids		−20–15	5	McMurchie *et al*., (1979)
Mammalian cells					
Myocardial membrane	Phospholipids	37	18–26	2	Charnock *et al*., (1980)
Small intestine (rat)	Plasmalemma	37	23–39		Brasitus *et al*., (1980)

Table 2.3 Freeze-fracture evidence for lateral phase separations

Cell-type	Membrane	Temperature of growth (°C)	Aggregation of IMP*	Temperature of aggregation† (°C)	Reference
Prokaryotes					
Acheloplasma laidlawii	Plasmalemma	37	+	5‡	Verkleij *et al.*, (1972)
A. laidlawii	Plasmalemma	37	+	37–5‡	James and Branton (1973)
A. laidlawii (Supplemented with lipids from *Staphylococcus aureus*)	Plasmalemma	37	–		Haest *et al.*, (1974)
A. laidlawii (Supplemented with lipids from *Staph. aureus*)	Plasmalemma	37	+ ('low' cholesterol) –4 ('high' cholesterol)	4	Rottem *et al.*, (1973)
Anabaena variabilis	Plasmalemma	38	–0		Ono and Murata (1982)
Anacystis nidulans	Photosynthetic membrane	30	+	5	Verwer *et al.*, (1978)
A. nidulans	Photosynthetic membrane	38	+	21	Armond and Staehelin (1979)
	Plasmalemma		+	21	
A. nidulans	Plasmalemma	39	+	0	Brand *et al.*, (1979)
		25	+	0	
A. nidulans	Photosynthetic membrane and plasmalemma	38	+	30–15 (range)	Furtudo *et al.*, (1979)
		28	+	25–5 (range)	
		18	+	15 to –5	
A. nidulans	Plasmalemma	38	+	16(midpoint)	Ono and Murata (1982)
		28	+	5(midpoint)	
Bacillus cereus	Plasmalemma	37	–(–10)		Haest *et al.*, (1974)
B. megaterium	Plasmalemma	37	–(–10)		Haest *et al.*, (1974)
B. subtilis	Plasmalemma	37	–(–10)		Haest *et al.*, (1974)
Escherichia coli	Plasmalemma	37	+	22–5‡	Haest *et al.*, (1974)
E. coli	Plasmalemma	13	+	–5	Haest *et al.*, (1974)
E. coli	Plasmalemma	13	+	20– –4	van Heerikhaizan *et al.*, (1975)
E. coli	Plasmalemma	37	+	37‡	Haest *et al.*, (1974)
E. coli	Plasmalemma	37	+	All temperatures below 37	Kleeman and McConnel (1974)
E. coli	Isolated plasmalemma vesicles	37	+	0	Shechter *et al.*, (1974)
Staphylococcus aureus	Plasmalemma	37	+	22–0	Haest *et al.*, (1974)
Streptococcus faecalis	Plasmalemma	37	–(–10)		Haest *et al.*, (1974)
Strep. faecalis	Plasmalemma	37	+	3	Tsien and Higgins (1974)
Synechococcus lividis	Plasmalemma	52	+	35	Golecki (1979)
(Thermophilic strain)	Thylakoids	52	+	35	Golecki (1979)
Synechococcus spp.	Plasmalemma	39	+	15	Golecki (1979)

(Continued over)

Table 2.3 (*Continued*)

Cell-type	Membrane	Temperature of growth (°C)	Aggregation of IMP*	Temperature of aggregation† (°C)	Reference
Eukaryotes					
Chlamydomonas reinhardii	Thylakoids	20	– (4)		Ojakian and Satir (1974)
Higher plants					
Solanum acaule (Potato)	Plasmalemma	27	+	2	Toivio-Kinnucan *et al.*, 1981)
S. tuberosum (Potato)	Plasmalemma	27	– (2)		Toivio-Kinnucan *et al.*, (1981)
Zea mays (Maize) root	E faces of cortical or endodermal cells	20	+ –	0	Robards *et al.*, (1981)
Zea mays (Maize) root	P faces of cortical or endodermal cells	20	– (0)		Robards *et al.*, (1981)
Mammalian cells and organelles					
Epithelial cells of prostate gland (rat)	Endoplasmic reticulum Golgi, nuclear envelope Plasmalemma	37	+	0	Kacher *et al.*, (1980)
Erythrocytes (human)	Plasmalemma	37	+	During freezing§	Fujikawa (1981)
Hepatocytes (rat)	Microsomes	37	+	4	Duppel and Dahl (1976)
Ehrlich ascites tumour cells	Nuclear membrane	37	+	4	Kim and Okada (1983)
Ehrlich ascites tumour cells	Plasmalemma	37	– (4)		Kim and Okada (1983)
Lymph node cells (pig) Lymphocytes (mouse) Lymphoma (mouse)	Inner and outer Nuclear membranes	37	+	17	Wunderluth *et al.*, 1974)
Lymphoma (mouse)	Plasmalemma	37	– (12)		Wunderluth *et al.*, (1974)
Leukaemic cell line (mouse)	Outer nuclear membrane	37	+	0	Feltkamp and Waerden (1982)
	Endoplasmic reticulum		+	0	Feltkamp and Waerden (1982)
	Plasmalemma		– (0)		Feltkamp and Waerden (1982)
Mitochondria (liver)	Inner and outer membrane	37	+	– 10	Hochli and Hackenbrock (1977)
Mitochondria (liver)	Outer membrane	37	– (10) +	 7	Hackenbrock *et al.*, (1976)
Mitochondria (liver)	Inner membrane	37	– (0) +	 – 8	Hackenbrock *et al.*, (1976)
Pancreatic cells	Endoplasmic reticulum	37	+	0	Kacher *et al.*, (1980)
	Plasmalemma		– (0)		Kacher *et al.*, (1980)

* IMP = Intramembraneous particles.
† The highest temperature at which aggregation of IMP was reported. Where examined the temperature range in which IMP aggregation was observed to increase is reported.
‡ Dependent upon the lipid composition of the organism (see Table 2.5 for example).
§ A freeze-induced aggregation was reported at rates of cooling of less than 700°C min^{-1}

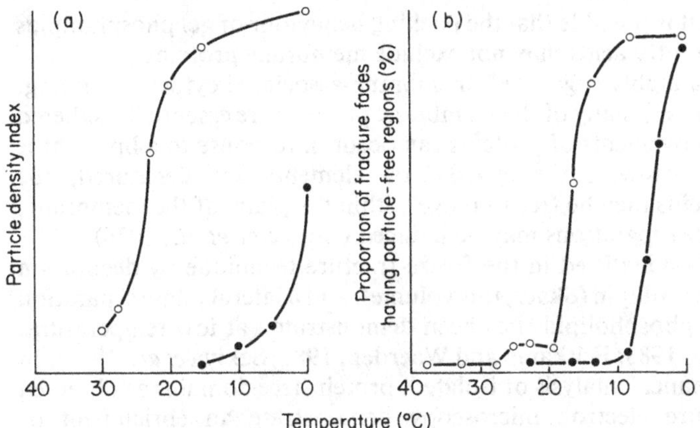

Fig. 2.6 (a) Shifts in the temperatures that induce phase separations in the outer alveolar membranes of *Tetrahymena pyriformis* as a result of altered growth temperature. Cells were grown at either 15°C (●) or 39°C (○) and incubated at various temperatures between 0°C and 33°C for 5 minutes before the cells were fixed with glutaraldehyde. The degree of aggregation of intramembranous particles observed by freeze-fracture electron microscopy was quantified as a particle density index. (Redrawn from Martin *et al.*, 1976.)

(b) The proportion of fracture faces having intramembranous particle-free regions in the cytoplasmic membrane of *Anacystis nidulans* showing dependence on fixation temperature. Cells were grown at either 28°C (●) or 38°C (○) (Redrawn from Ono and Murata, 1982.)

aggregation (James and Branton, 1973), a particle density index (Martin *et al* 1976) or the proportion of particle-free lipid (Ono and Murata, 1982). These analyses have demonstrated that protein aggregation may occur over a wide range of temperature, but this range usually has a discrete upper and lower limit (Fig. 2.6). It is thus possible to give the range and midpoints of such transitions. For a full description of methods of analysis of freeze-fracture electron micrographs see Robards *et al.*, 1981.

The morphology of membranes at low temperatures varies widely, and in eukaryotic cells the individual membrane types may respond differently (Table 2.3). For example, the endoplasmic reticulum, golgi and nuclear membranes of lymphoid and pancreatic cells undergo obvious phase separations at 0°C, whereas the plasmalemma exhibits no aggregation of proteins at this temperature. In isolated rat liver mitochondria, the inner mitochondrial membrane exhibits no clumping of protein at temperatures above −8°C, while aggregation is evident in the outer mitochondrial membrane at +7°C.

Although freeze-fracture electron microscopy can provide collaborative evidence of phase separations within membranes, the absence of protein aggregation does not necessarily indicate the absence of a phase separation.

Neither *Bacillus* sp. nor *Staphylococcus aureus* exhibit any aggregation of membrane proteins, even upon slow cooling to −10°C. This is in contrast to DCS which demonstrates broad lipid transitions between 11 and 40°C for *B. subtilis* and between 4 and 31°C for *Staph. aureus*. The membrane phospholipids of these bacteria contain relatively large amounts of branched-chain fatty acids, and it is possible that the packing behaviour of gel phospholipids containing such fatty acids may not exclude membrane proteins.

In cells with a highly organized, membrane-associated cytoskeleton (e.g. human erythrocytes) many of the membrane proteins are essentially tethered and no lateral movements of proteins can occur in response to a lipid phase transition. If, however, the cytoskeletal elements are denatured, the membrane proteins may be free to move within the plane of the membrane, and lateral phase separations may be induced (Elgsaeter *et al.*, 1976).

Sterols may be identified in the freeze-fracture technique by decoration with the antibiotic fillipin (Skaer, this volume), and a lateral phase separation of sterols from phospholipids has been demonstrated at low temperatures (Kim and Okada, 1983; Feltkamp and Waerden, 1982; Sekiya *et al.*, 1979). In addition, biochemical analysis of isolated, protein-free domains as observed by freeze-fracture electron microscopy demonstrate an enrichment of saturated phospholipids (Letellier *et al.*, 1977; Kameyama *et al.*, 1980).

Discontinuities in Arrhenius plots of membrane-sited reaction rates

In certain reactions a plot of the log of the reaction rate against the reciprocal of the absolute temperature yields a straight line, indicating compliance with the Arrhenius equation (Taylor; Clarke this volume). This relationship assumes the special case in which there is one rate-limiting step. If the nature of the rate-limiting step changes at a certain temperature, the linear relationship will be preserved, but the result will be two straight lines intersecting at a breakpoint coincident with the change in the rate-limiting step (Fig. 2.7). Discontinuities in Arrhenius plots have been used extensively as evidence for changes in the physical state of membranes. There are difficulties associated with this approach as indicated below.

The discontinuity in the Arrhenius plot of Ca^{2+} ATPase activity of rabbit sarcoplasmic reticulum at about 20°C (Fig. 2.7) has been attributed variously to a bulk phase change in the lipid bilayer, lipid cluster formation (Lee *et al.*, 1974) or an interaction between the ATPase and the surrounding lipid annulus (Hesketh *et al.*, 1976). The discontinuity is, however, still present in a delipidated preparation of the sarcoplasmic reticulum ATPase (Dean and Tanford, 1978). This suggests that the discontinuity is due to a direct temperature effect on the polypeptide itself rather than on the membrane lipids. Furthermore, DSC studies indicate no thermotropic events in sarcoplasmic reticulum above 10°C.

Temperature-induced changes in substrate binding kinetics have been demonstrated for a number of enzymes (Silvius *et al.*, 1978). Such alterations will produce discontinuities in the Arrhenius plot, unrelated to the physical state of the cellular membranes.

Comparison of the Arrhenius plots of three transport systems in human erythrocytes – the glucose carrier, the sodium pump and the L-system amino acid carrier – reveal breakpoints of 28, 21 and 16°C respectively (Ellory and Willis, 1981). Clearly, these three different 'break temperatures' cannot reflect phase behaviour of the bulk membrane lipids and furthermore no phase change can be detected in human erythrocytes by DSC. These discon-

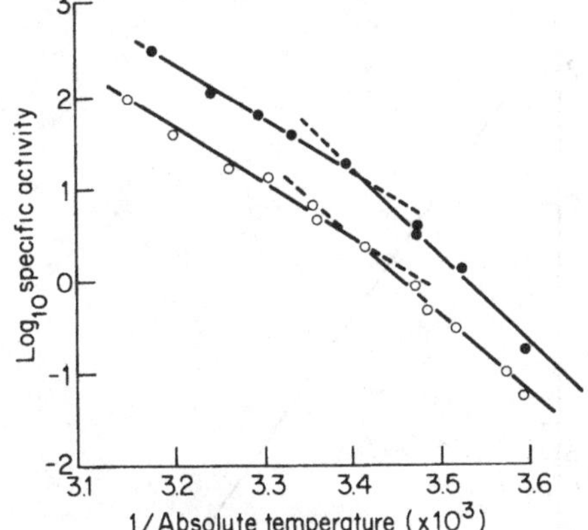

Fig. 2.7 Temperature dependence of the activity of sarcoplasmic reticulum ATPase (○) and of a delipidated enzyme preparation (●). (Redrawn from Dean and Tanford, 1978.)

tinuities may be due to the different temperature stability of the three carrier systems, changes in substrate binding kinetics or they may reflect the specific micro-environments of the pumps.

For a full discussion of the problems associated with the interpretation of this type of data, especially when used as evidence of a change in state of cellular membranes refer to Wilson in this volume. Although Arrhenius plots of membrane protein activity must obviously be interpreted with caution, the non-mediated uptake of solutes such as glycerol or erythritol (de Gier *et al.*, 1968) or osmotic dehydration may yield true breakpoints.

Rotational diffusion of membrane probes

A variety of techniques may be used to examine the rotational motion of a molecule (probe) sited within the membrane. These probes are either fluorescent molecules or labelled lipids, and the relative motion of these molecules has been employed in determining the physical state of the membrane. Two problems exist with this approach; the membrane structure may be altered by the insertion of a probe molecule, and the probe can only provide information about its immediate environment, which may not be typical of the membrane as a whole. Perturbance of the natural membrane organization is a particularly severe problem with the bulky modified fatty acids used in electron spin resonance (ESR).

The fluorescent probe most commonly employed to examine biological membranes is 1,6-diphenyl-1, 3, 5-hexatriene (DPH) (Cossins, 1981). DPH may reflect alterations in the fluidity of liposomal bilayers and distinguish bulk phase transitions in a pure phospholipid (Fig. 2.8a). In biological membranes, however, there is a gradual phase separation rather than a discrete phase transition. As DPH partitions approximately equally between gel and liquid-crystalline phases, alterations in the mobility of DPH within biological membranes as the temperature is lowered (Fig. 2.8b) may be due to

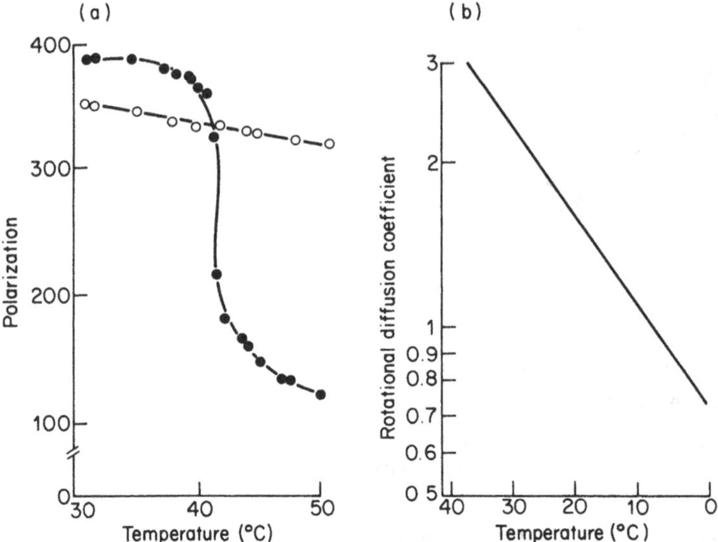

Fig. 2.8 (a) The polarization, at different temperatures, of diphenylhexatriene in liposomes of either dipalmitoyl phosphatidylcholine (●) or cholesterol and dipalmitoyl phosphatidylcholine (○) at equimolar ratios. (Redrawn from Johnson, 1981.)
(b) The rotational diffusion coefficient of diphenylhexatriene in a membrane preparation from goldfish. (Redrawn from Cossins, 1977.)

a decreased fluidity of the liquid-crystalline phase, a partial sequestering of DPH into gel phase lipid or a combination of these factors. The mean fluidity measured by DPH is therefore difficult to interpret in terms of the physical state of the membrane.

Effects of reduced temperatures on membrane lipid composition

Organisms that have evolved to live at low temperature have been shown to have membrane lipid compositions which are different from those of organisms living at warmer temperatures. Similar changes commonly occur in organisms acclimating to low temperature in the laboratory. The most frequently described alteration is an increase in the average degree of unsaturation of phospholipid fatty acyl chains. Other modifications may, however, occur, for example an increase in fatty acyl chain length (Russell, 1984), the proportion of branched-chain fatty acids (Kaneda, 1977), or the content of sterols (Wodtke, 1978; Sikorska and Farkas, 1982), alteration in lipid class composition (Lynch and Thompson, 1982; Fukushima *et al.*, 1976; Addink, 1980) and non-random modification in the positioning of fatty acids in phospholipids (Lynch and Thompson, 1984). The effect of all these modifications would be to decrease the upper temperature of membrane phase separations.

It has been suggested that in cell-types which have the potential to grow over a range of temperatures there is a homeostatic control of membrane fluidity. That is, as the temperature changes there are compensatory modifications to the membrane composition to maintain an optimal fluidity at all temperatures; this has been termed homeoviscous adaptation (Sinensky, 1974). This process necessarily involves some form of feedback control, the sensor for which would be expected to be some aspect of membrane function. The precise interaction between fluidity of the membrane and the biochemical processes resulting in the modification of lipid composition are, however, unclear.

It has been argued that lipid composition is modified at reduced temperatures to avoid the onset of lipid phase separations. This is a different hypothesis from the regulation of fluidity, but the two are in many ways difficult to separate by experiment as an increase in unsaturation (for example) at low temperatures will affect both fluidity and the tendency to undergo a phase separation.

An increase in the average degree of unsaturation of membrane phospholipid fatty acids following exposure to low temperatures has been reported in a wide variety of cell-types. The biochemistry of such modifications have, however, been investigated in only a few cell-types.

The membrane composition of *E. coli* is relatively simple. Only three phospholipids are present, having three fatty acids, 16:0 (palmitic), 16:17 (palmitoleic) and 18:1 7 (*cis* vaccenic). At 37°C, the optimum temperature for growth, most phospholipids have one saturated (*sn*–1) and one unsaturated (*sn*–2) fatty acid. At low temperatures there is an increase in the relative proportion of the 18:1 fatty acid with a corresponding reduction in 16:0, the amount of 16:1 being unaffected by temperature (Table 2.4). Subsequently, at reduced temperatures there is an increase in the proportion of phospholipids containing two unsaturated fatty acids.

The increased synthesis of phospholipids containing 18:1 occurs within 15 seconds of a temperature reduction from 42 to 24°C. This response is too rapid for transcription or translation to be involved. Furthermore, prolonged exposure to inhibitors of RNA and protein synthesis does not affect this response. A pre-existing enzyme is therefore responsible for the increased rate of synthesis of 18:1 at reduced temperatures (deMendoza and Cronan, 1983).

The rate-limiting enzyme in the synthesis of *cis* vaccenic acid in *E. coli* is β-ketoacyl-acyl carrier protein (ACP)-synthase. This enzyme elongates palmitoleoyl-ACP (the immediate precursor of 16:1 fatty acyl chains) to β-ketopalmitoleoyl-ACP which is then converted to *cis* vaccenoyl-ACP. Genetic analysis has demonstrated that this enzyme exists as two forms, synthases 1 and 2, which are different translation products. The elongation of palmitoleoyl-ACP is catalysed more efficiently by synthase 2 than by synthase 1, and this difference is increased at low temperatures. At 37°C the

Table 2.4 Effect of temperature on the phospholipid fatty acid composition of *Tetrahymena pyriformis* and *Escherichia coli*

Fatty acid*	*Tetrahymena pyriformis*†			*Escherichia coli*‡		
	Grown at 39°C	39–15°C (1 hour)	Grown at 15°C	Grown at 10°C	Grown at 25°C	Grown at 40°C
12:0	1.8	1.2	0.8			
14:0§	14.6	11.4	11.7	16.5	10.7	10.9
Ai 15:0	4.7	4.2	1.6			
15:0	2.0	2.2	tr.			
16:0	12.7	10.4	8.4	18.2	27.6	37.1
16:1	12.5	14.6	8.6	26.0	23.2	28.0
cyc 16:1¶				1.3	3.1	3.2
16:2	5.7	5.2	5.2			
18:0	0.6	0.7	tr			
18:1	6.5	5.4	8.1	37.9	35.5	20.8
18:2 $^{\Delta 6, 11}$	2.0	0.7	6.2			
18. $^{\Delta 9, 12}$	11.9	13.6	16.5			
18:3	21.2	25.7	28.5			
Minor components	1.2	1.1	2.1	3.51	2.13	1.31
Unsaturation Index**	1.26	1.41	1.62			

* Fatty acids are noted by two numbers, the first gives the number of carbon atoms and the second gives the number of double bonds.
† From Dickens and Thompson (1981).
‡ From Marr and Ingraham (1962).
§ Combined 14:0 and bhydroxymysistic.
¶ 16:1 containing a cyclopropane nmg.
** Average degree of unsaturation.

ratio of activity (synthase 1:2) was 12, this ratio was increased to 30 on cooling to 27°C. In addition the Michaelis constant of synthase 2 at 27°C was half that at 37°C. The increased level of *cis* -vaccenoyl-ACP competes for incorporation into position 1 of phospholipids giving di-unsaturated phospholipids.

A temperature-induced synthesis of a desaturase enzyme has been demonstrated in *B. megaterium* (Fulco and Fujii, 1980). This organism contains only one desaturase enzyme, which is membrane associated and desaturates only acyl chains attached to intact membrane phospholipids, but not those present as CoA or ACP thioesters. In bacteria grown at 35°C the enzyme is absent, but when the temperature is lowered, the enzyme is synthesized *de novo*; this can be prevented by blocking protein synthesis with cycloheximide before reducing the temperature.

The eukaryote *T. pyriformis* has been the subject of extensive studies of membrane alterations induced at low temperatures (Thompson, 1979a; Clarke, 1981). Following a reduction in temperature there is a rapid alteration in membrane fatty acid composition (Table 2.4), these changes may be mediated in at least two ways.

The activity of the enzyme which converts 16:0 to 16:1 (palmitoyl-CoA desaturase) increases following a reduction in temperature from 39.5 to 15°C, and this increase is prevented by cycloheximide, indicating that synthesis of new enzyme is involved (Fukushima *et al.*, 1979). The quantitative contribution of this enzyme to fatty acid unsaturation is, however, not clear, as unsaturation is normal in cells in which protein synthesis is inhibited by cycloheximide (Skriver and Thompson, 1979).

In *T. pyriformis*, membrane composition is controlled directly by the activity of the desaturases associated with the endoplasmic reticulum. Comparison of the rate of incorporation of ^{14}C-acetate into fatty acids by cells grown at 39.5 or 15°C showed that there was an increase in the rate of desaturase activity at the lower temperature compared with the rate of synthesis *de novo* (Martin *et al.*, 1976). This hypothesis has been supported by a number of studies in which the lipid composition of *Tetrahymena* is modified in response to the addition of membrane active compounds at a constant temperature. For example, following the application of 9- and 10-methoxystearic acid or anaesthetics, which increase the fluidity of membranes, the net desaturase activity was reduced (Kitajima and Thompson, 1977; Nandini-Kishore *et al.*, 1979). Supplementation with ergosterol, which reduces membrane fluidity, increases the activity of desaturases (Kasai *et al.*, 1976). For an extensive review of the effects on lipid composition induced by alterations in membrane fluidity see Nozawa (1980).

Although the control of membrane composition by the direct effect of fluidity on membrane associated desaturase enzymes offers an elegant mechanism for the cellular response to a rapid variation in temperature, the specific molecular rearrangements associated with enzyme activation or deactivation are not understood (Thompson, 1979).

In addition to the rapid response of fatty acid composition to a reduction in temperature, there is a longer term (10 hour) alteration in the phospholipid composition which accompanies cell division. The major feature is a switch in the relative amounts of ethanolamine phosphoglyceride and its phospholipid analogue, 2-aminoethylphospholipid. However, little is understood of the biochemistry underlying this process, except that it may be affected by the fluidity of the cellular membranes, the transfer of phospholipids by carrier molecules being governed by the fluidity of the donor and receptor membranes (Helmkamp, 1980).

Alterations in the lipid composition of membranes not directly attributable to low temperature

In the preceding section the increase in fatty acid unsaturation at reduced temperatures was considered as a direct cellular response to temperature. However, the lowering of temperature has a wide range of effects on the physical properties of solutions (Taylor, this volume) some or all of which will affect cellular metabolism. It has been suggested that some such factors may affect lipid composition directly rather than via temperature.

A simple scheme for the observed increase in unsaturation of plant phospholipid fatty acids at low temperatures is based on the effects of the increased availability of oxygen, which is twice as soluble in water at 0°C that at 40°C. Molecular oxygen is a co-substrate in fatty acid unsaturation, and it has been proposed that the increase in oxygen concentration at low temperatures is sufficient in itself to increase the rate of unsaturation (Harris and James, 1969). This direct effect of oxygen on lipid unsaturation has been confirmed in studies of suspension cultures of sycamore cells (Rebeille *et al.*, 1980). However, this effect appears to be species specific, for example *T. pyriformis* grown at different temperatures under various oxygen tensions has a fatty acid composition characteristic of the temperature and independent of the oxygen tension (Skriver and Thompson, 1976). This mechanism is unlikely to be relevant to higher animals which have subtle homeostatic controls for regulating extracellular and intracellular oxygen levels.

There have been several reports that the growth rate affects lipid composition in microorganisms. A survey of four mesophilic and three psychrophilic bacteria showed that these species react differently to a lowered temperature, but all showed variations in unsaturation with growth rate independent of the direct effect of temperature (Gill and Suisted, 1978). *Saccharomyces cerevisiae* showed little change in fatty acid composition in both batch and chemostat cultured cells when the growth temperature was altered. However, chemostat culture suggested that changes in lipid class composition were caused by both a lowering of the growth temperature and a decrease in growth rate (Hunter and Rose, 1972). Similarly, the phospholipid compositions of *Pseudomonas aeruginosa* (Gilbert and Brown, 1978) and *E. coli* (Calcott and Petty, 1980) were directly modulated by growth rate in chemostat culture.

When cultures of *T. pyriformis* were maintained at different temperatures (both sub- and supraoptimal) or at the optimum temperature for growth but in the presence of inhibitors and stimulators of growth there were variations

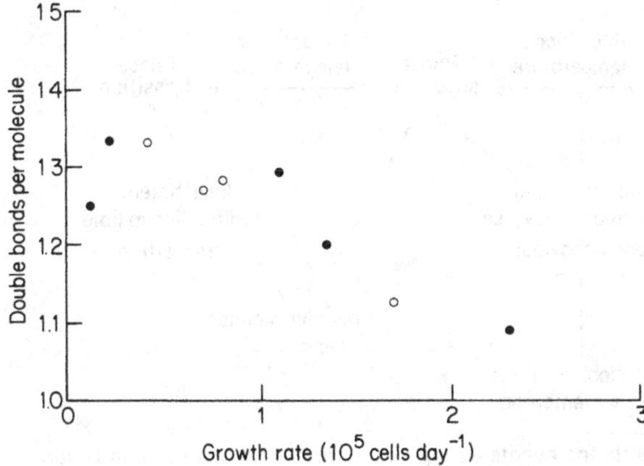

Fig. 2.9 Relationship between growth rate and phospholipid fatty acid unsaturation for *Tetrahymena pyriformis* grown under various conditions. Growth rates are modified by either alteration in temperature (○) or manipulation of the growth medium at a constant temperature (●). (Reproduced with permission from Morris *et al.*, 1981.)

in the fatty acid composition of total cell phospholipids (Morris *et al.*, 1981). Following the manipulation of growth rate at constant temperature there were alterations in mean phospholipid fatty acid unsaturation which, when plotted against growth rate, were indistinguishable from those obtained by varying temperature (Fig. 2.9). These data suggest the possibility that in *Tetrahymena* the relationship between temperature and lipid unsaturation may not be a direct one, but may be indirectly mediated through growth rate. The possibility that an increase in unsaturation may be a general response to a subtlethal stress should therefore be considered in studies both of thermal acclimation and of the effects of exogenous compounds on lipid composition.

A further factor associated with the analysis of lipid composition of cell cultures at reduced temperatures arises from possible cellular synchronization. Cells in which cell division is inhibited by reduced temperatures are commonly blocked at a specific stage of the cell cycle. Studies on synchronous cultures have demonstrated that phospholipid fatty acid composition is related to the stage in the cell cycle, suggesting that changes in lipid composition observed at low temperatures may in some cases be due in part to cell synchrony.

Effects of membrane thermotropic behaviour on membrane function

A potential sequence of events which may occur in cellular membranes in response to a reduction in temperature has been summarized as a flow diagram (Fig. 2.10.).

The initial effect of a reduction in temperature will be an increase in membrane viscosity. Depending on the cell-type and the extent of temperature

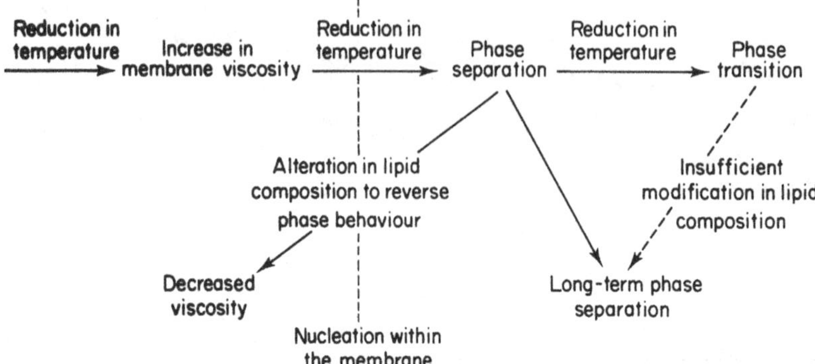

Fig. 2.10 Flow diagram of membrane events during a reduction in temperature, indicating potential long-term responses.

reduction, a phase separation may occur. These events will both have immediate effects on membrane function, and it is the ability of the cell to counter phase separations which will in part determine the long-term survival. Many cell-types can modify their lipid composition to offset phase changes and membrane function may be observed to 'recover'. These compensations are rarely perfect, however, and the membrane is generally more viscous at the reduced temperature, which will influence membrane physiology and function. If the cells lack the potential to compensate for changes in membrane phase behaviour at the reduced temperature, they will be trapped in a state of permanent phase change or transition.

Many biological properties of membranes are altered by the effects of a reduction in temperature (Cossins, 1983), and some of these processes are mediated by membrane-associated proteins (for a discussion of the direct effects of temperature on protein structure and function see below). However, the special case of membrane-associated proteins, with their unique environments, will be discussed here together with some other effects on cell biology which are directly attributable to lipid thermotropic behaviour.

An example of the effect of reduced temperature on the activity of a membrane protein is found in the uptake of cations by the unicellular green alga *Chlamydomonas reinhardii*. Cells must pump cations to remain viable, and it is usual for them to maintain a relatively high, internal potassium concentration. In *C. reinhardii* uptake of ^{86}Rb is a good tracer for potassium transport and is apparently mediated by an ATPase-driven carrier process (Clarke *et al.*, in preparation). With cells cultured at 20°C the uptake of ^{86}Rb decreased in a linear manner as the temperature was reduced to 5°C, and between 5°C and 0°C there was little change in activity (Fig. 2.11). At −5°C there was an increase in uptake, but the kinetics of this process suggested a loss of membrane permeability with a passive uptake of cations, rather than an active uptake. A similar increase in the accumulation of cations at low temperatures has been reported for other cell-types (Elford and Solomon,

1974). Cells of *C. reinhardii* acclimated at 8°C for 7 days, however, showed an increased in ^{86}Rb uptake at all temperatures and in these cells the kinetics of incorporation suggested active transport, even at −5°C.

Results of this type are usually interpreted in terms of thermotropic behaviour during cooling and a phase separation within the plasmalemma has been demonstrated by freeze-fracture electron microscopy upon slow cooling to 0°C. Nevertheless, cation uptake continues at this temperature. The increase in cation uptake at −5°C may be the consequence of a critical level of phase separation or even a complete phase transition within the membrane. Following acclimation at 8°C there is an increase in the average unsaturation of phospholipid fatty acids which should reduce the temperature at which a phase separation would occur. There are, however, other factors to consider. Following growth at 8°C an increase in ^{86}Rb uptake is observed over the whole temperature range of the experiment, which suggests that during acclimation there may be an increase in the total number of active transport proteins per cell. Such an increase has been demonstrated for ATPase activity in cold-hardened winter wheat (Jian *et al.*, 1982).

The energy source for cation transport is ATP (Poole, 1978), and therefore at low temperatures a reduction in metabolic rate may directly affect the rate of cation uptake. For active transport to continue at reduced temperatures, it is essential for metabolism to continue in an integrated manner. It is possible that the key alterations which occur during acclimation are those influencing cellular metabolism (e.g. processes in mitochondria or chloroplasts) rather than acclimation limited to the uptake system.

Following phase transitions, membrane-associated enzymes are concentrated into regions of lipid which are fluid. In such circumstances the rates of reactions may be increased as a consequence of the increased probability of collision with other membrane-bound molecules, and some enzymes may show maximum activity within membranes in a state of phase separation (i.e. phospholipase A2 hydrolyses saturated phosphatidylcholines only at the transition temperature, op den Kamp *et al.*, 1975). The most frequently observed pattern is the reduction in enzyme activity, which may be apparent in an Arrhenius plot as a discontinuity. Such breakpoints have been variously ascribed to the onset of a phase separation, the temperature at which a phase separation is complete or as some intrinsic property of the enzyme.

A feature of the fluid mosaic model of the lipid bilayer is that membrane components are free to diffuse unless bound to cytoskeletal elements or cross-linked into large patches. This concept has to be somewhat modified by the demonstration of lipid domains within certain membranes (Friend, 1982) and the suggestion that in some cases the diffusion of proteins may be directed (Kell, 1984). As the rates of many membrane processes are determined by the kinetics of diffusion-controlled reactions, phase separations would be predicted to have major effects on such processes. Direct effects of the physical state of lipids on diffusion processes have been obtained from studies on the mobility of proteins incorporated into liposomes. At temperatures above the phase transition, diffusion coefficients of 10^{-8}–10^{-10} cm^2 s^{-1} were obtained; below the phase transition, however, the value for the diffusion coefficient was reduced to 10^{-10}–10^{-12} cm^2 s^{-1}.

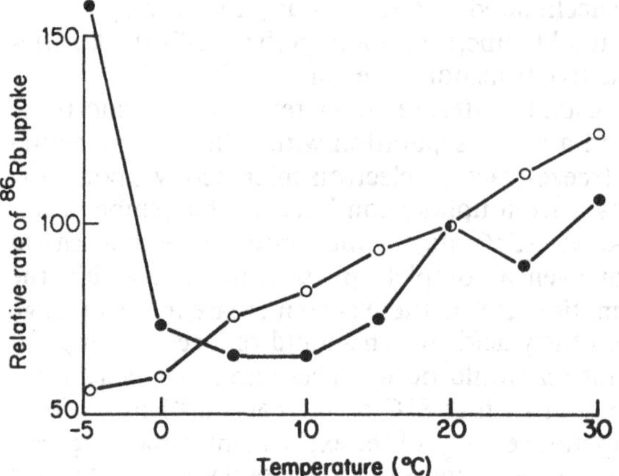

Fig. 2.11 Effect of assay temperature on the uptake of ^{86}Rb by *Chlamydomonas reinhardii* (●) cells cultured at 20°C, (○) cells acclimated at 8°C for 7 days. Uptake is expressed with respect to 20°C values.

In biological membranes, when a phase separation rather than a discrete transition occurs, the situation is more complex. For example, the diffusion coefficient for the intermixing of surface antigens in heterokaryons decreases as cells are cooled from 45°C to 22°C, further cooling to 15°C enhances diffusion, and cooling below 15°C again slows diffusion (Fig. 2.12). It has been suggested that 'islands of solid lipid' may 'canalize' diffusion at the phase separation causing more rapid intermixing (Petit and Edinin, 1974). The formation of pyrene dimers has also been examined to determine lateral diffusion rates and, using this method, lateral motion was found to be enhanced in the channels formed between gel domains (Galla and Sackmann, 1974), but long distance diffusion was hindered (Kapitza and Sackmann, 1980). This effect has been treated quantitatively (Saxton, 1982).

The lateral diffusion of membrane proteins may be restrained or directed by interactions with the cytoskeleton. A clear example of such interactions occurs with the phenomenon of capping of surface antigens. When a multivalent reagent reacts with a plasma membrane component it may become redistributed into a cap over one pole of the cell. Two stages of the process may be observed at low temperatures or in the presence of metabolic inhibitors. First, recognition and binding occur, the antigens being labelled in a random manner, and then, in a second energy requiring step, these patches are transported to the pole of the cell. Similar patching of surface receptors occurs as an early event in the many other cellular processes (e.g. receptor-mediated endocytosis). Capping in some cell-types is inhibited by compounds such as cytochalasin B and colchicine, which interact with the actin and tubulin components of the cytoskeleton respectively (Rittenhouse *et al.*, 1974) suggesting that the movement of membrane receptors to the pole of the cell is mediated by the cytoskeleton. The inhibition of capping at low temperatures may occur not simply because of membrane lipid phase separation

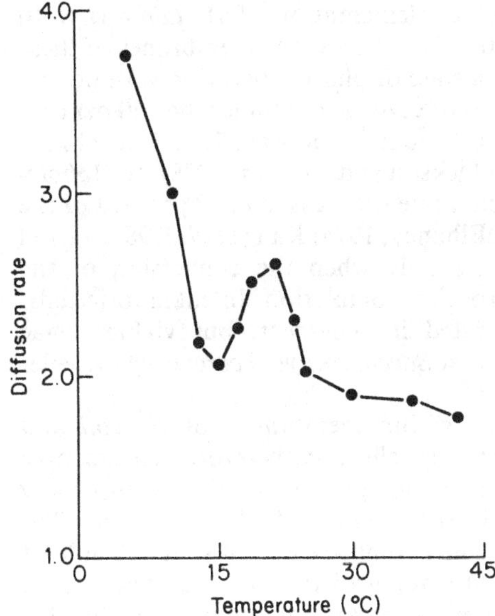

Fig. 2.12 Changes in the rate of mixing of fluorescent-labelled surface antigens. These were on the surfaces of two cells fused into a heterokaryon, and their movement results from diffusion of components within the surface membrane. (Redrawn from Petit and Edinin, 1974.)

but also as a consequence of the depolymerization of microtubules at low temperatures.

With the exception of fertilization, healthy cells rarely fuse with other cells in nature. Intracellular membrane fusion is, however, a very common event, especially in cells which are capable of endocytosis or secretion. The fusion of liposomes has been demonstrated to be inhibited at temperatures below the phase transition of the constituent lipids (Papahadjopoulos *et al.*, 1979). Consequently, thermotropic behaviour within cellular membranes would be expected to disrupt the cellular processes which are dependent upon fusion.

Lipid phase separations and cellular viability

The phase separations or transitions that are observed in many cell-types following a reduction in temperature have been correlated with alterations in various functions of the membranes and have been proposed as a cause of cellular injury at low temperatures (Lyons 1973). It is now evident, however, that some cell-types have the potential to metabolize and even divide when the cellular membranes are in a state of phase separation.

E. coli and *Acholeplasma laidlawii* at their optimum temperature for growth (37°C) have membranes which are entirely fluid. However, the use of fatty acid auxotrophs of *E. coli* or supplementation of *A. laidlawii* with saturated fatty acids has demonstrated that at 37°C the membranes of these organisms may be manipulated into a state of phase separation with no loss of cell viability. In *E. coli* normal rates of division continue when 20 per cent of the plasmalemma lipids are in the ordered state; only at 55 per cent gel state lipid is growth completely inhibited (Jackson and Cronan, 1978). In *Acholeplasma* growth and replication can continue with less than 10 per cent of the membrane lipid in the fluid state (McElhaney, 1974; Kang *et al.*, 1981; Jarrell *et al.*, 1982), and cell growth ceases only when the conversion of the membrane lipid to the gel state approaches completion. In eukaryotic cells, gel phase lipid has been demonstrated in senescent, but viable, algae (Thompson *et al.*, 1978) and in the yeast *Saccharomyces cerevisiae* (Moeller *et al.*, 1981).

The occurrence of ordered lipids within membranes of *E. coli* and *A. laidlawii* are obviously not lethal to the cells, and therefore thermotropic behaviour of other membranes during cooling is not proof in itself of any damaging processes. Furthermore, in the thermophilic bacterium *Bacillus staerothermophillus* the membrane is in a completely fluid state for at least 25°C below the minimum temperature for growth (Melchior and Steim, 1976). Rather than being detrimental, phase separations may be beneficial in certain specialized membranes (e.g. the microvilli and basal membranes of the rat intestine exist with ordered lipids at or near to the body temperature, (Brasitus *et al.*, 1980). The physiologial implications of this arrangement are as yet unknown, although the lateral compressibility of membranes lipids are enhanced at the phase transition temperature (Linden *et al.*, 1973; Marsh *et al.*, 1976; Marcelja and Wolfe, 1979). This increased compressibility could facilitate the insertion of compounds into the membrane and thereby enhance processes of transport or membrane assembly.

Proteins

Low temperatures may affect the structure and function of proteins in a way which modifies their biological activity. The effects may be categorized into two broad classes:

(a) decrease in the rate of enzyme activity – this may have important consequences for the integration of biochemical pathways.

(b) denaturation – when a protein spontaneously unfolds or when a structure with subunits dissociates into biologically inactive species which may or may not reassemble upon rewarming.

Considering the complex energy balance which is responsible for the conformational stability of proteins, such effects are not surprising. All the separate interactions, including hydrophobic forces, hydrogen-bonding and electrostatic contributions, vary in different ways with temperature; the weakening of hydrophobic interactions is of particular importance at low temperature. As it is not possible, however, to calculate the contribution of the individual forces, the result of the various temperature effects on protein

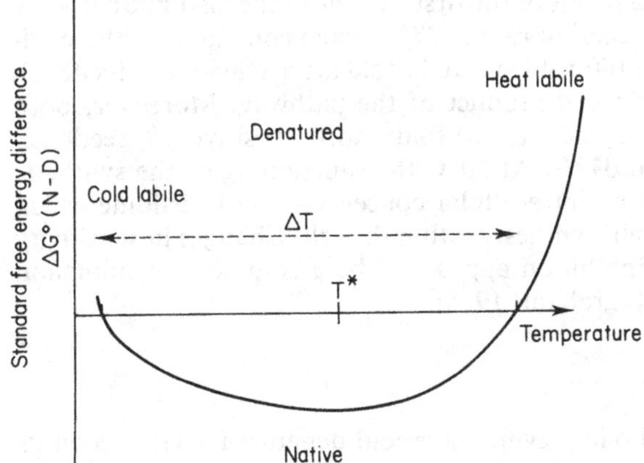

Fig. 2.13 Standard free energy difference of a hypothetical globular protein between the native (N) and a biologically denatured (D) state as a function of temperature. ΔT is the temperature region of protein stability and T* is the temperature of maximum stability. (Redrawn from Franks, 1982.)

stability may be examined as the standard free energy differences between the native and the denatured state (Fig. 2.13). From such considerations it has been suggested that all proteins should be stable only within a limited temperature range and cold denaturation should be as universal as heat denaturation (Franks, 1982).

In this section examples which illustrate basic principles will be discussed, but for extensive reviews of the effects of low temperatures on proteins see Fennema, 1979 and Jaenicke, 1981. (The special case of membrane-associated enzymes has been discussed above.)

Cytoplasmic enzymes

Even the simplest of enzyme-catalysed reactions involves several coupled steps, each of which is likely to have a different activation energy, so the overall effect of a temperature reduction on any individual enzyme is difficult to predict quantitatively. A reduction in temperature will also be expected to affect different enzyme reaction rates to a different extent. The effects of these alterations in enzyme activity on a series of linked reactions are complex and metabolic pathways are likely to become uncoupled (Clark, this volume).

In addition to altering the activation energies of enzyme catalysed reactions, low temperatures may modify enzyme in a number of other ways. Two examples of such processes will be presented, but no overall synthesis of the effects of low temperatures on metabolic pathways will be attempted.

(a) *Sensitivity to feedback inhibition*

A class of cold-sensitive mutants of *E. coli*, which cannot synthesize histidine at 20°C, have been isolated (O'Donovan and Ingraham, 1965). These mutants all contain altered forms of the first enzyme of the histidine pathway which have an unaltered reaction rate at 37°C when compared to the wild-type, but they are between 100-fold and 1000-fold more sensitive to feedback inhibition by histidine, the end-product of the pathway. Moreover, both wild-type and altered enzyme are 10-fold more sensitive to feedback inhibition at 20°C than at 37°C. At 20°C the functioning of the synthetic enzyme was prevented by an intracellular concentration of histidine which was too low to allow protein synthesis within the cell. Changes in sensitivity of enzymes to feedback inhibition appears to be a common phenomenon (Reviewed by Inniss and Ingraham, 1978).

(b) *Enzyme denaturation*

A clear example of a metabolic enzyme being cold-denatured *in vivo* accounts for the phenomenon of cold sweetening of potatoes. At temperatures below 10°C a key enzyme in glycolysis, phosphofruktokinase, is inactivated. Fructose is then diverted into sucrose synthesis instead of being phosphorylated. This enzyme inactivation is due to the dissociation of the tetrameric protein into two dimers, a process which is thermally reversible (Dixon *et al.*, 1981).

Cytoskeleton

The maintenance of cell shape, motility and chromosome movement in eukaryotic cells is generally considered to involve the cytoskeleton. This is composed of three dynamic, independently controlled, but interacting, arrays of protein filaments: microtubules, intermediate filaments and microfilaments. Each system is composed of distinct proteins and each has an array of regulatory proteins. The *in vitro* activity of these regulatory proteins is variously modified by physical and chemical factors, (e.g. the properties of some proteins from amoeba, which cross-link filamentous actin, have been demonstrated to alter with temperature). A 90 000-Da protein is Ca^{2+} sensitive at 28°C but not at 0°C, while two polypeptides exhibit a Ca^{2+}-sensitive cross-linking activity at 0°C but not at 28°C (Hellewell and Taylor, 1979).

Certain elements of the cytoskeleton, specifically microtubules, may depolymerize at low temperatures, and in this section the effects of reduced temperatures on the cytoskeleton will be reviewed. It must be stressed, however, that most such studies have been on the low temperature destabilization of the cytoskeleton of mammalian tissue culture cells. That many organisms grow and remain motile at low temperatures clearly demonstrates that the cytoskeleton is not intrinsically cold labile.

(1) *Microtubules*

Microtubules have a wide range of functions in cells, there is evidence that these may be affected differently by low temperatures (Behnke and Forer, 1967).

(a) *Cytoplasmic microtubules* The low temperature depolymerization of cytoplasmic microtubules has been demonstrated frequently, by electron microscopy (Behnke and Forer, 1967) and by immunofluorescence (Weber *et al.*, 1975), and it has been employed as a diagnostic feature to identify microtubules. However, this response is not universal, and a number of mammalian cell-types have been reported to have cold-stable cytoplasmic microtubules (Bershadsky *et al.*, 1979).

In spite of these observations, surprisingly few attempts have been made to link the effects of low temperatures on cells directly with the depolymerization of cytoplasmic microtubules. An exception to this is the recent study on the cell wall-less zoospores of the alga *Chlorosarcinopsis gelatinosa* which are ellipsoidal, upon cooling to 2°C they become spherical, and following rewarming they revert to their original shape. These changes in shape correlate with the disappearance and reappearance of peripheral cytoplasmic microtubules (Melkonian *et al.*, 1980).

(b) *Microtubules associated with the mitotic apparatus* The mitotic spindle is composed predominantly of microtubules, and in many cell-types this structure has been reported to disappear at low temperatures (Inoue, 1952; Roth, 1967) resulting in a complete inhibition of chromosome movements (Lambert and Bayer, 1977). However, in some cell-types exposure to low temperatures does not induce complete breakdown of the mitotic apparatus, but only causes a reduction in microtubule number (Brinkley and Cartwright, 1975; Lambert and Bayer, 1977). There is evidence that individual microtubules within the mitotic apparatus may have a differential sensitivity to low temperatures. For example, following exposure of mammalian tissue culture cells to 0°C for 1 hour, the pole to pole microtubules were disrupted while the pole to chromosome tubules remained intact (Welsh *et al.*, 1979).

(c) *Cortical microtubules of plant cells* In plant cells cortical microtubules lie close to and parallel with the plane of the plasmalemma. They may be dispersed over the entire wall of a cell or they may be in localized groups. It is generally assumed that these cortical microtubules are responsible for regulating the orientation in which cell wall microfibrils are deposited. Drugs such as colchicine, which depolymerize microtubules, result in cell wall abnormalities (Gunning and Hardman, 1982).

The microtubules associated with primary wall formation are either shortened (Hardman and Gunning, 1978) or destroyed (Juniper and Lawton, 1979) by reduction in temperature. In plant cells with secondarily thickened cell walls, cortical microtubules are stable at low temperatures (Nelmes *et al.*, 1973; Juniper and Lawton, 1979).

(d) *Microtubules of cilia and flagella* These structures are generally considered to be stable and they do not depolymerize at low temperatures. An extreme example is found in unicellular green algae, isolated from antarctic hypersaline lakes, which remain motile at −14°C (Burch and Marchant, 1983).

From the above it is evident that a variety of response is observed in microtubule-containing structures upon exposure to low temperatures. These responses have either been attributed to differences in polypeptide pattern between cold-labile and cold-resistant microtubules (Hesketh, 1984), or are thought to reflect different physical environments, for example, anchorage to non-microtubule structures (Lambert and Bayer, 1977).

(2) *Intermediate filaments*

The intermediate filaments of mammalian tissue culture cells persist after cold treatments which depolymerize the microtubules (Virtanen *et al.*, 1980; Maro *et al.*, 1983). By contrast, the intermediate filaments of epidermal cells of teleost fish (tonofilaments) dissociate at low temperatures, but the phenomenon is reversed within minutes when the cells are rewarmed. This is unrelated to cold-induced depolymerization of microtubules seen in the same cells, as neither colchicine nor vinblastin, which depolymerize microtubules, have any effect on tonofilaments (Schliwa and Euteneuer, 1979).

(3) *Microfilaments*

Microfilaments are composed primarily of actin, and are generally considered to be unaffected by exposure to low temperatures. The breakdown of transvacuolar strands of plant cells at reduced temperatures has, however, been associated with the dissociation of actin filaments (Patterson and Graham, 1977). Cytochalsin B, which promotes the disassembly of filamentous actin, also breaks down the transvacuolar strands of tomato tissue culture cells, mimiking the effects of cold (Woods *et al.*, 1984a). In parallel with the cytological changes which take place at low temperatures, differences in intracellular calcium levels, as assessed by chlorotetracycline fluorescence, occur. It has been suggested that such modifications in the intracellular calcium levels may induce changes in actin polymerization, not by acting directly upon microfilaments, but rather by calcium modulation of a regulatory protein (Woods *et al.*, 1984b).

Contractile proteins

Cytoplasmic streaming is mediated by actin microfilaments, and with reduction in temperature the rate of streaming may slow down (Das *et al.*, 1966; Patterson and Graham, 1977) or cease (Lewis, 1956). These changes occur within minutes of cooling, and it has been suggested that they are a direct effect of a disassembly of actin filaments. The rate of cytoplasmic streaming depends, however, on the synthesis of ATP, and is also very sensitive to Ca^{2+} concentration, being inhibited by micromolar amounts. Thus it is probable that the changes in the rate of cytoplasmic streaming observed at low temperatures also reflect alterations in cellular metabolism or the intracellular concentration or localization of calcium (Minorsky, 1985).

A particularly clear example of the compensation of the activity of contractile proteins, and the complexity of this response, is shown by fish white muscle (Clarke, this volume).

Ribosomal subunits

In *E. coli*, mutations in genes encoding ribosomal proteins frequently confer a cold-sensitive phenotype (Bayliss and Ingraham, 1974; Geyl *et al.*, 1977). Such mutants are unable to assemble ribosomal subunits at low temperatures but accumulate incomplete non-functional subunits within cells. In plants, a class of cold-sensitive mutants have been found to contain normal levels of cytoplasmic and chloroplast ribosomes at the permissive temperature, while at the restrictive temperature, cells are deficient in chloroplast ribosomes, but they have a normal level of cytoplasmic ribosomes (Hanson and Bogorad, 1978; Hoyer-Hansen and Casadoro, 1982).

There is no evidence that wild-type organisms have similar defects in ribosome assembly at low temperatures, but the formation of polysomes may be affected (see following discussion).

Protein synthesis

Alteration in temperature has been demonstrated to modify both the rate of protein synthesis and the types of proteins synthesized.

Rate of protein synthesis

At temperatures which are sub-optimal for cell division the rate of protein synthesis is decreased (Fig. 2.14a). In *E. coli* the initiation of protein

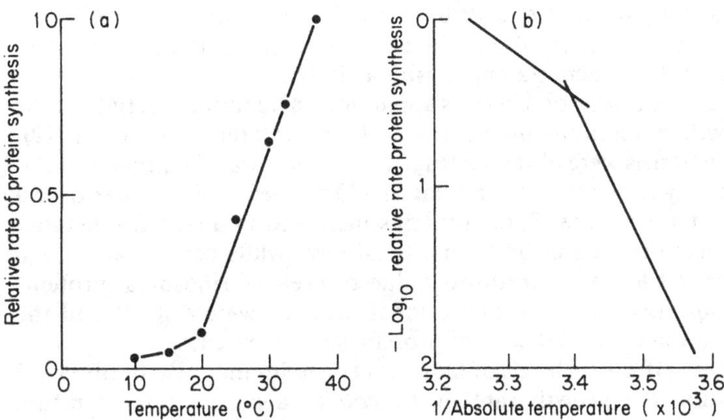

Fig. 2.14 Effect of temperature on the rate of protein synthesis in a mammalian tissue culture cell-line (mouse L cells). (a) Arithmetic plot with rates expressed relative to that of controls at 37°C. (b) Plotted as an Arrhenius treatment. (Redrawn from Craig, 1975.)

synthesis is blocked at temperatures below 8°C, while the elongation of initiated protein continues until they are completed. Upon rewarming, the cells initiate protein synthesis synchronously. The ribosomal subunits isolated from cold-treated cells seem to be identical to the subunits from exponentially growing cells. The block in protein synthesis appears to be at the joining of the 50s ribosomal subunit to the 30s initiation complex (Friedman *et al.*, 1971).

The process in eukaryotic cells appears to be more complex. For example, in a mammalian tissue culture cell-type the level of cytoplasmic polysomes decreases as the temperature is lowered to 10°C; below 10°C the level increases again reaching control levels at 0°C (Craig, 1975). In this case initiation of polysomes was considered to be the rate-limiting step for protein synthesis. Similar evidence to suggest that initiation is the rate-limiting step has been obtained for other mammalian cell-types (Oleinick, 1979). By contrast, in rabbit reticulocytes the level of polysomes remained constant for long periods of incubation between 0°C and 42°C, and in this cell-type it was suggested that as the temperature was reduced the rates of initiation and elongation-release were more equally affected (Craig, 1976).

An Arrhenius plot of the rate of protein synthesis *in vivo* reveals discrete breakpoints (Fig. 2.14b). These changes are not the result of a rate-limiting membrane phenomenon (see above) as they are also observed in a cell-free system (Craig and Fahrman, 1977), instead they are due to the temperature dependence of processes inherent in protein synthesis.

Effect of temperature on protein type

At temperatures near to the optimum for growth, metabolic co-ordination is achieved largely by modulation of enzyme activity rather than by variation in the amounts of enzyme synthesized. At reduced temperatures the relative amounts of different proteins within a cell-type may vary, as has been demonstrated for isoenzyme distribution of a number of cytoplasmic enzymes (Krasnuk *et al.*, 1975) and the electrophoretic patterns of plasmalemma polypeptides (Unemura and Yoshida, 1984).

The complexity of these processes is apparent in a quantitative study of the effect of growth temperature on proteins in *E. coli* (Herendeen *et al.*, 1979). A number of patterns were observed (Fig. 2.15), but it was clear that the relative level of many proteins (from a total of 133 studied) did not alter over a wide range of temperature. Some proteins increased in a stepwise manner upon temperature reduction below a critical level while others varied in a linear manner. At low temperatures, reduced levels of ribosomal proteins and other translational and transcriptional proteins were evident and the level of acetylation of one ribosomal protein was increased.

Many of these effects on levels of individual proteins may also be observed by manipulating the growth rate of *E. coli* at a constant temperature (Pedersen *et al.*, 1978). It is therefore possible that some effects of low temperatures on protein synthesis may be related to the temperature-induced reduction in growth rate rather than a direct effect of temperature.

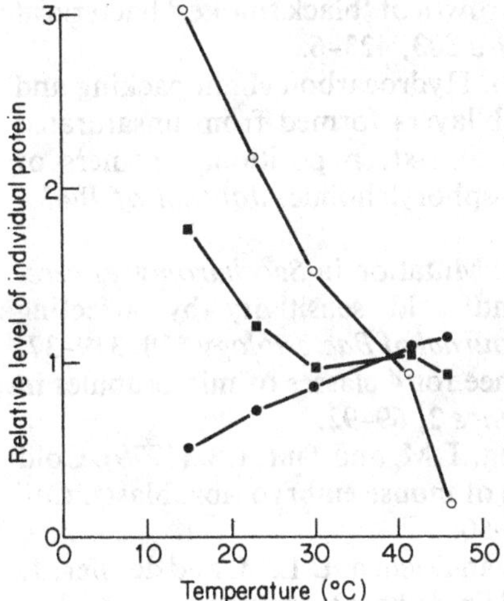

Fig. 2.15 Variation in the relative levels of three proteins of *Escherichia coli* as a function of growth temperature; the level of each protein is expressed with respect to its level at 37°C. (Redrawn from Herendeen *et al.*, 1979.)

Conclusions

From the many examples presented above it is evident that a reduction in temperature may affect the structure and activity of both cellular proteins and lipids. The resulting effects on cell physiology are bewildering in complexity. It has proved simple to induce and select mutants with a restricted range of growth temperatures, but no mutants capable of growth at temperatures significantly lower than that for the wild-type have been reported. This suggests that many genes determine the natural temperature range for growth, and attempts to identify individual biochemical sites responsible for limiting growth at low temperatures may be naïve.

References

Abbas, C.A. and Card, D.L. (1980). The relationships between growth temperature, fatty acid composition and the physical state and fluidity of membrane lipids in *Yersinia enterocolitica. Biochemica et Biophysica Acta* **602**, 469–76.

Addink, A.D.F. (1980). Activity of membrane bound enzymes of the respiratory chain during adaptation of fish to temperature changes. In *Membrane Fluidity: Biophysical Techniques and Cellular Regulation*, pp. 99–104. Edited by Kates, M. and Kukis, A. Humana Press, Clifton, New York.

Armond, P.A. and Staehelin, L.A. (1979). Lateral and vertical displacement of integral membrane proteins during lipid phase transition of *Anacystis nidulans. Proceedings of the National Academy of Sciences* **76**, 1901–5.

Baross, J.A. and Deming, J.W. (1983). Growth of 'black smoker' bacteria at temperatures of a least 250°C. *Nature* **303**, 423–6.

Barton, D.G. and Gunstone, F.D. (1975). Hydrocarbon chain packing and molecular motion in phospholipid bilayers formed from unsaturated lecithins. Synthesis and properties of sixteen positional isomers of 1,2-dioctodecenoyl-sn-glycero-3-phosphorylcholine. *Journal of Biological Chemistry* **250**, 4470–76.

Bayliss, F.A. and Ingraham, J.L. (1974). Mutation in *Saccharomyces cerevisiae* conferring streptomycin and cold sensitivity by affecting ribosome formation and function. *Journal of Bacteriology* **118**, 319–37.

Behnke, O. and Forer, A. (1967). Evidence for 4 classes of microtubules in individual cells. *Journal of Cell Science* **2**, 69–92.

Bershadsky, A.D., Gelfand, V.I., Suitking, T.M. and Tint, I.S. (1979). Cold stable microtubules in the cytoplasm of mouse embryo fibroblasts. *Cell Biology International Reports* **3**, 45–50.

Blok, M.C., van der Neut-Kok, E.C.M., van Deenen, L.L.M. and de Gier, J. (1975). The effect of chain length and lipid phase transitions on the selective permeability properties of liposomes. *Biochemica et Biophysica Acta* **406**, 187–96.

Braganza, L.F., Blott, B.H., Coe, T.J. and Melville, D. (1983). Dye permeability at phase transitions in single and binary component phospholipid bilayers. *Biochemica et Biophysica Acta* **731**, 137–44.

Brand, J.R., Kirchanski, S.J. and Ramirez-Mitchell, R. (1979). Chill-induced alterations in *Anacystis nidulans* as a function of growth temperature. *Planta* **145**, 63–8.

Brasitus, T.A., Tall, A.R. and Schachter, D. (1980). Thermotropic transitions in rat intestinal plasma membranes studied by differential scanning calorimetry and fluorescence depolarization. *Biochemistry* **19**, 1256–61.

Bretscher, M.S. (1972). Asymmetrical lipid bilayer structure for biological membranes. *Nature (New Biology)* **236**, 11–12.

Brinkley, B.R. and Cartwright, J. (1975). Cold-labile and cold-stable microtubules in the mitotic spindle of mammalian cells. *Annals of the New York Academy of Sciences* **253**, 428–39.

Brock, T.D. (1978). *Thermotropic Microorganism and Life at High Temperatures*. Springer, New York.

Burch, M.D. and Marchant, H.J. (1983). Motility and microtubule stability of Antarctic algae at sub-zero temperatures. *Protoplasma* **115**, 240–42.

Calcott, P.H. and Petty, R.S. (1980), Phenotypic variability of lipids of *Escherichia coli* grown in chemostat culture. *FEMS Microbiology Letters* **7**, 23–7.

Chapman, D., Williams, R.M. and Ladbrooke, B.D. (1967). *Chemistry and Physics of Lipids* **1**, 445–75.

Chapman, D., Cornell, B.A. and Quinn, P.J. (1977a). Phase transitions, phase aggregation and a new method of modulating membrane fluidity. In *Biochemistry of Membrane Transport*, pp. 72–85. Edited by Semenza, G. and Carofoli, E. Springer-Verlag, Berlin.

Charnock, J.S., Gibson, R.A., McMurchie, E.J. and Raison, J.K. (1980). Changes in the fluidity of myocardial membranes during hibernation:

Relationship to myocardial adenosinetriphosphatase activity. *Molecular Pharmacology* **18**, 476–82.

Clarke, A. (1981). Effects of temperature on the lipid composition of *Tetrahymena*. In *Effects of Low Temperature on Biological Membranes*, pp. 55–82. Edited by Morris, G.J. and Clarke, A. Academic Press, London.

Cossins, A.R. (1977). Adaptation of biological membranes to temperature. The effect of temperature acclimation of goldfish upon the viscosity of synaptosomal membranes. *Biochemica et Biophysica Acta* **470**, 395–411.

Cossins, A.R. (1981). The adaptation of membrane dynamic structure to temperature. In *Effects of Low Temperatures on Biological Membranes*, pp. 83–106. Edited by Morris, G.J. and Clarke, A. Academic Press, London.

Cossins, A.R. (1983). The adaptation of membrane structure and function to temperature. In *Cellular Acclimatisation to Environmental Change*, pp. 1–32. Edited by Cossins, A.R. and Sheterline, P. Cambridge University Press, Cambridge.

Craig, N. (1975). Effect of reduced temperatures on protein synthesis in mouse L cells. *Cell* **4**, 329–35.

Craig, N. (1976). Regulation of translation in rabbit reticulocytes and mouse L cells; comparison of the effects of temperature. *Journal of Cell Physiology* **87**, 157–66.

Craig, N. and Fahrman, C. (1977). Regulation of protein synthesis by temperature in mammalian cells: Non involvement of the plasma membranes. *Biochemica et Biophysica Acta* **474**, 478–90.

Cronan, J.E. and Gelman, E.P. (1975). Physical properties of membrane lipids: Biological relevance and regulation. *Bacteriological Reviews* **39**, 232–56.

Crowe, L.M., Mouradiam, R., Crowe, J.M., Jackson, S.A. and Wormersley, C. (1984a). Effects of carbohydrates on membrane stability at low water activities. *Biochemica et Biophysica Acta* **769**, 141–50.

Crowe, J.M., Whittam, M.A., Chapman, D. and Crowe, L.M. (1984b). Interaction of phospholipid monolayers with carbohydrates. *Biochemia et Biophysica Acta* **769**, 151–9.

Cullen, J., Phillips, M.C. and Shipley, G.G. (1971). The effects of temperature on the composition and physical properties of *Pseudomonas fluorescens*. *Biochemical Journal* **125**, 733–42.

Curatolo, E., Verma, S.P., Sakura, J.D., Small, D.M., Shipley, G.G. and Wallach, D.F.H. (1978). Structural effects of myelin proteolipid apoprotein in phospholipids: A Raman spectroscopic study. *Biochemistry* **17**, 1802–7.

Das, T.M., Hildebrandt, A.C. and Riker, A.J. (1966). Cinephotomicrography of low temperature effects on cytoplasmic streaming, nucleolar activity and mitosis in single tobacco cells in microculture. *American Journal of Botany* **53**, 253–9.

Dean, W.L. and Tanford, C. (1978). Properties of a delipidated, detergent-activated Ca^{2+}-ATPase. *Biochemistry* **17**, 1683–90.

Demel, R.A., Gearts van Kessel, W.S.M. and van Deenen, L.L.M. (1972). The properties of polyunsaturated lecithins in monolayers and liposomes and the intractions of these lecithins with cholesterol. *Biochemica et Biophysica Acta* **266** 26–40.

Dickens, B.F. and Thompson, G.A. (1981). Rapid membrane response during low temperature acclimation. Correlation of early changes in the physical properties and lipid composition of microsomal membranes. *Biochemica et Biophysica Acta* **644**, 211–18.

Dixon, W.L., Franks, F. and ap Rees, T. (1981). Cold-lability of phosphofructokinase from potato tubers. *Phytochemistry* **20**, 969–72.

Duppel, W. and Dahl, G. (1976). Effect of phase transition on the distribution of membrane associated particles in microsomes. *Biochemica et Biophysica Acta*, **426**, 408–17.

Elford, B.C. and Solomon, A.K. (1974). Temperature dependence of cation permeability of dog red cells. *Nature* **248** 522–4.

Elgsaeter, A., Shotton, D.M. and Branton, D. (1976). Intramembrane particle aggregation in erythrocyte ghosts. ii The influence of spectrin aggregation. *Biochemica et Biophysica Acta* **426**, 101–22.

Ellory, J.C. and Willis, J.S. (1981). Phasing out the sodium pump. In *Effects of Low Temperature on Biological Membranes*, pp. 107–19. Edited by Morris, G.J. and Clarke, A. Academic Press, London.

Feltkamp, C.A. and van der Waerden, A.W.M. (1982). Low temperature induced displacement of cholesterol and intramembrane particles in nuclear membranes of mouse leukemia cells. *Cell Biology International Reports* **6**, 137–46.

Fennema O. (Ed) (1979). Proteins at Low Temperatures *Advances in Chemistry Series* no. **180**, pp. 223 American Chemical Society, Washington, D.C.

Fettiplace, R. and Haydon, D.A. (1980). Water permeability of lipid membranes. *Physiological Reviews* **60**, 510–50.

Franks, F. (1982). Physiological water stress. In *Biophysics of Water*, pp. 279–94. Edited by Franks, F. and Matthias, S.F. John Wiley and Sons, Chicester.

Friedman, H., Lu, P. and Rich, A. (1971). Temperature control of initiation of protein synthesis in *Escherichia coli*. *Journal of Molecular Biology* **61**, 105–21.

Friend, D.S. (1982). Plasma-membrane diversity in a highly polarized cell. *Journal of Cell Biology* **93**, 243–9.

Fujikawa, S. (1981). The effect of various cooling rates on the membrane ultrastructure of frozen human erythrocytes and its relation to the extent of haemolysis upon thawing. *Journal of Cell Science* **49**, 369–82.

Fukushima, H., Martin, C.E., Iida, H., Kitajima, Y., Thompson, G.A. and Nozawaw, Y. (1976). Changes in membrane lipid composition during temperature adaptation by a thermotolerant strain of *Tetrahymena pyriformis*. *Biochemica et Biophysica Acta* **431**, 165–79.

Fukushima, H., Nagao, S. and Nozawa, Y. (1979). Further evidence for changes on the level of palmitoyl-CoA desaturase during thermal adaptation in *Tetrahymena pyriformis. Biochemica et Biophysica Acta* **572**, 178–82.

Fulco, A.J. and Fujii, D.K. (1980). Adaptive regulation of membrane lipid biosynthesis in Bacilli by environmental temperature. In *Membrane Fluidity. Biophysical Techniques and Cellular Regulation*, pp. 77–98. Edited by Kates, M. and Kukis, A. Humana Press, Clifton, New York.

Furtudo, D., Williams, Brian A.P.R. and Quinn, P.J. (1979). Phase separations in membranes of *Anacystis nidulans* grown at different temperatures. *Biochemica et Biophysica Acta* **555**, 352–7.

Galla, H.J. and Sackmann, E. (1974). Lateral diffusion in the hydrophobic region of membranes: Use of pyrene eximers as optical probes. *Biochemica et Biophysica Acta* **339**, 103–15.

Gavish, B. and Werber, M.M. (1979). Viscosity-dependent structural fluctuations in enzyme catalysis. *Biochemistry* **18**, 1269–75.

Geyl, P., Bock, A. and Wittman, H.G. (1977). Cold sensitive growth of a mutant of *Escherichia coli* with an altered ribosomal protein S8: Analysis of revertants. *Molecular and General Genetics* **152**, 331–6.

de Gier, J., Mandersloot, J.G. and van Deenen, L.L.M. (1968). Lipid composition and permeability of liposomes. *Biochemica et Biophysica Acta* **150**, 666–75.

Gilbert, P. and Brown, M.R.W. (1978). Influence of growth rate and nutrient limitations on the gross cellular composition of *Pseudomonas aeruginosa* and its resistance to 3- and 4-chlorophenol. *Journal of Bacteriology* **133**, 1066–72.

Gill, C.O. and Suisted, J.R. (1978). Effects of temperature and growth rate on proportion of fatty acids in bacterial lipids. *Journal of General Microbiology* **104**, 31–6.

Golecki, J.R. (1979). Ultrastructure of cell wall and thylakoid membranes of the thermophilic cyanobacterium *Synechococcus lividus* under the influence of temperature shifts. *Archives of Microbiology* **120**, 125–33.

Grisham, C.M. and Barnet, R.E. (1973). The role of lipid phase transitions in the regulation of the (sodium + potassium) adenosine triphosphatase. *Biochemistry* **12**, 2635–7.

Gunning, B.E.S. and Hardman, A.R. (1982). Microtubules. *Annual Review of Plant Physiology* **33**, 651–98.

Hackenbrock, C.R., Hochli, H. and Chau, R.M. (1976). Calorimetric and freeze-fracture analysis of lipid phase separations and lateral translational motion of intramembrane particles in mitochondrial membranes. *Biochemica et Biophysica Acta* **445**, 466–84.

Haest, C.W.M., Verklej, A.J., de Gier, J., Scheck, R., Ververgaert, P.H.J. and van Deenen (1974). The effect of lipid phase transitions on the architecture of bacterial membranes. *Biochemica et Biophysica Acta* **356**, 17–26.

Hanson, M.R. and Bogorad, L. (1978). Ery-H2 group of *Chlamydomonas reinhardii* – cold sensitive, erythromycin-resistant mutants deficient in chloroplast ribosomes. *Journal of General Microbiology* **105**, 253–62.

Hardman, A.R. and Gunning, B.E.S. (1978). Structure of cortical microtubule arrays in plant cells. *Journal of Cell Biology* **77**, 14–34.

Harris, P. and James, A.T. (1969). The effect of low temperature on fatty acid biosynthesis in plants. *Biochemical Journal* **112**, 325–30.

van Heerikhuizen, H., Kwak, E., van Bruggen, E.F.J. and Withott, B. (1975). Characterization of a low density cytoplasmic membrane subfraction isolated from *Escherichia coli. Biochemica et Biophysica Acta* **413**, 177–91.

Hellewell, S.B. and Taylor, D.L. (1979). The contractile basis of amoeboid movement. 6. The solation:construction coupling hypothesis. *Journal of Cell Biology* **83**, 633–48.

Helmkamp, G.M. (1980). Effects of phospholipid fatty acid composition and membrane fluidity on the activity of bovine brain phospholipid exchange protein. *Biochemistry* **19**, 2050–56.

Herendeen, S.L., van Bogelen, R.A. and Neidhardt, F.C. (1979). Levels of major proteins of *Escherichia coli* during growth at different temperatures. *Journal of Bacteriology* **139**, 185–94.

Hesketh, J.E. (1984). Differences in polpeptide composition and enzyme activity between cold-stable and cold-labile microtubules and the study of microtubule alkaline phosphatase activity. *FEBS Letters* **169**, 313–18.

Hesketh, K.R., Smith, G.A., Houslay, M.D., McGill, K.A., Birdsall, N.J.M., Metcalfe, J.C. and Warren, G.B. (1976). Annuluar lipids determine the ATPase activity of a calcium transport protein complexed with dipalmitoyllecithin. *Biochemistry* **15**, 4145–51.

Hochli, H. and Hackenbrock, C.R. (1977). Thermotropic lateral translational motion of intramembraneous particles in the inner mitochondrial membrane and its inhibition by artificial peripheral proteins. *Biochemica et Biophysica Acta* **72**, 278–92.

Hoyer-Hansen, G. and Casadoro, G. (1982). Unstable chloroplast ribosomes in the cold-sensitive barley mutant tigrina-034. *Carlsberg Research Communications* **47**, 103–19.

Hui, F.K. and Barton, P.G. (1973). Mesomorphic behaviour of some phospholipids with aliphatic alcohols and other nonionic substances. *Biochemica et Biophysica Acta* **296**, 510–17.

Hunter, K. and Rose, A.H. (1972). Lipid composition of *Saccharomyces* as influenced by growth temperature. *Biochemica et Biophysica Acta* **260**, 639–53.

Inniss, W.E. and Ingraham, J.L. (1978). Microbial life at low temperatures: Mechanisms and molecular aspects. In *Microbial Life in Extreme Environments*, pp. 73–104. Edited by Kushner, D.J. Academic Press, London.

Inoue, S. (1952). Effect of low temperature on the birefringence of the mitotic spindle. *Biological Bulletin* **103**, 316–24.

Jackson, M.B. and Cronan, J.E. (1978). An estimate of the minimum amount of fluid lipid required for the growth of *Escherichia coli. Biochemica et Biophysica Acta* **512**, 472–9.

Jaenicke, R. (1981). Enzymes under extremes of physical conditions. *Annual Review of Biophysics and Bioengineering* **10**, 1–67.

Jian, L.-G., Sun, L.-H. and Dong, H.-Z. (1982). Adaptive changes in ATPase activity in the cells of winter wheat seedlings during cold hardening. *Plant Physiology* **70**, 127–31.

James, R. and Branton, D. (1973). Lipid and temperature dependent structural changes in *Acholeplasma laidlawii*. *Biochemica et Biophysica Acta* **323**, 378-90.

Jarrell, H.C., Butler, K.W., Byrd, R.A., des Lauriers, I., Ekiel, I. and Smith, I.C.P. (1982). A 2H-NMR study of *Acholeplasma laidlawii* membranes highly enriched in myristic acid. *Biochemica et Biophysica Acta* **877**, 622-36.

Johnson, S.M. (1981). Steady state diphenyl hexatriene fluorescence polarization in the study of cells and cell membranes. In *Fluorescent Probes*, pp. 143-55. Edited by, Beddard, G.S. and West, M.A. Academic Press, London.

Juniper, B.T. and Lawton, J.R. (1979). The effect of caffeine, different fixation regimes and low temperature on microtubules of higher plants. *Planta* **145**, 411-16.

Kacher, B., Serrano, J.A. and Pinto de Silva, P. (1980). Particle displacement in epithelial cell membranes of rat prostate and pancreas induced by routine low temperature fixation. *Cell Biology International Reports* **4**, 347-56.

Kameyama, Y., Ohki, K. and Nozawa, Y. (1980). Thermally induced heterogenity in microsomal membranes of fatty acid-supplemented *Tetrahymena*: Lipid composition, fluidity and enzyme activity. *Journal of Biochemistry* **88**, 1291-303.

op den Kamp, J.A.F., de Gier, J. and van Dennen, L.L.M. (1975). Hydrolysis of phosphatidylcholine liposomes by pancreatic phospholipase A2 at the transition temperature. *Biochemica et Biophysica Acta* **345**, 253-6.

Kaneda, T. (1977). Fatty acids of the genus *Bacillus*: An example of branched chain preference. *Bacteriological Reviews* **41**, 391–418.

Kang, S., Gutowsky, H.S., Hshung, J.C., Jacobs, J.C., Jacobs, R., King, T.E., Rice, D. and Oldfield, E. (1979). Nuclear magnetic resonance investigation of the cytochrome oxidase-phospholipid interaction: A new model for boundary lipid. *Biochemistry* **18**, 3257-67.

Kang, S.Y., Kinsey, R.A., Rayan, S., Gutowsky, H.S., Gabridge, M.G. and Oldfield, E. (1981). Protein-lipid interactions in biological and model membrane systems. *Journal of Biological Chemistry* **256**, 1155-9.

Kapitza, H.G. and Sackmann, E. (1980). Local measurement of lateral motion in erythrocyte membranes by photobleaching techniques. *Biochemica et Biophysica Acta* **595**, 56-64.

Kasai, R., Kitajima, Y., Martin, C.E., Nozawa, Y., Skriver, L. and Thompson, G.A. (1976). Molecular control of membrane properties during temperature acclimation. Membrane fluidity regulation of fatty acid desaturase action. *Biochemistry* **15** 5228-33.

Kell, D.B. (1984). Diffusion of protein complexes in prokaryote membranes: Fast, free, random or directed. *Trends in Biochemical Sciences* **9**, 86-8.

Kim, J. and Okada, Y. (1983). Asymmetric distribution and temperature dependent clustering of fillipin-sterol complexes in the nuclear membrane of Ehrlich ascites tumour cells. *European Journal of Cell Biology* **29**, 244-52.

Kitajima, Y. and Thompson, G.A. (1977). *Tetrahymena* strives to maintain the fluidity interrelationships of all its membranes constant: Electron microscope evidence. *Journal of Cell Biology* **72**, 744–56.

Kleeman, W. and McConnell, H.M. (1974). Lateral phase separations in *Escherichia coli* membranes. *Biochemica et Biophysica Acta* **345**, 220–30.

Krasnuk, M., Jung, G.A. and Witham, F.H. (1975). Electrophoretic studies of the relationship of peroxidases, phenoloxidase and indolacetic acid oxidase to cold tolerance of alfalfa. *Cryobiology* **12**, 62–80.

de Kruijff, B., Demel, R.A. and van Dennen, L.L.M. (1972). The effect of cholesterol and epicholesterol incorporation on the permeability and on the phase transition of intact *Acholeplasma laidlawii* cell membranes and derived liposomes. *Biochemica et Biophysica Acta* **255**, 331–47.

de Kruijff, B., Cullis, P.R. and Verkleij, A.J. (1980). Non-bilayer lipid structures in model and biological membranes. *Trends in Biochemical Sciences* **5**, 79–81.

de Kruijff, B., Demel, R.A., Slotboom, A.J., van Deenen, L.L.M. and Rosenthal, R.F. (1973). The effect of polar headgroup on the lipid-cholesterol interaction: A monolayer and differential scanning calorimetry study. *Biochemica et Biophysica Acta* **307**, 1–19.

Ladbroke, B.D., Williams, R.M. and Chapman, D. (1968). Studies on lecithin-cholesterol-water interaction by differential scanning calorimetry and X-ray diffraction. *Biochemica et Biophysica Acta* **150**, 333–40.

Lambert, A.M. and Bayer, A.S. (1977). Microtubule distribution and reversible arrest of chromosome movements induced by low temperatures. *European Journal of Cell Biology* **15**, 1–23.

Lee, A.G. (1977a). Lipid phase transitions and phase diagrams. 1. Lipid phase transitions. *Biochemica et Biophysica Acta* **472**, 237–81.

Lee, A.G. (1977b). Analysis of the defect structure of gel phase lipid. *Biochemistry* **16**, 835–41.

Lee, A.G., Birdsall, N.J.M., Metcalfe, J.C., Troon, P.A. and Warren, G.B. (1974). Clusters in lipid bilayers and the interpretation of thermal events in biological membranes. *Biochemistry* **13**, 3699–704.

Lewis, D.A. (1956). Protoplasmic streaming in plants sensitive and insensitive to chilling temperatures. *Science* **124**, 75–6.

Letellier, L. Moudden, H. and Shechter, E. (1977). Lipid and protein segregation in *Escherichia coli* membrane. Morphological and structural study of different cytoplasmic membrane fractions. *Proceedings of the National Academy of Sciences* **74**, 452–6.

Linden, C.D., Wright, K.L., McConnell, H.M. and Fox, C.F. (1973). Lateral phase separations in membrane lipids and mechanism of sugar transport in *Escherichia coli. Proceedings of the National Academy of Sciences* **70**, 2271–5.

Lyman, G.H., Dreisler, H.D. and Papahadjopoulos, D. (1976). Membrane action of DMSO and other chemical inducers of Friend leukaemic cell differentiation. *Nature* **262**, 360–3.

Lynch, D.V. and Thompson, G.A. (1982). Low temperature induced alterations in the chloroplast and microsomal membranes of *Dunaliella salina*. *Plant Physiology* **69**, 1369–76.

Lynch, D.V. and Thompson, G.A. (1984). Microsomal phospholipid molecular species alterations during low temperature acclimation in *Dunaliella*. *Plant Physiology* **74**, 193–7.

Lyons, J.M. (1973). Chilling injury in plants. *Annual Review of Plant Physiology* **24**, 445–66.

McElhaney, R.N. (1974). The effect of alterations in the physical state of the membrane lipids on the ability of *Acholoplasma laidlawii* B to grow at various temperatures. *Journal of Molecular Biology* **84**, 145–57.

McMurchie, E.J. (1979). Temperature sensitivity of ion-stimulated ATPase associated with some plant membranes. In *Low Temperature Stress Crop Plants*, pp. 163–70. Edited by Lyons, J.M., Graham, D. and Raison, J.K. Academic Press, New York.

Marcelja, S. and Wolfe, J. (1979). Properties of bilayer membranes in the phase transition or phase separation region. *Biochemica et Biophysica Acta* **557**, 24–31.

Maro, B., Sauron, M.E., Paulin, D. and Bornens, M. (1983). Further evidence for interaction between microtubules and vimetin filaments: Taxol and cold effects. *Biology of the Cell* **47**, 243–6.

Marr, A.G. and Ingraham, J.L. (1962). Effect of temperature on the composition of fatty acids in *Escherichia coli*. *Journal of Bacteriology* **84**, 1260–67.

Marsh, D., Watts, A. and Knowles, D.F. (1976). Evidence for phase boundary lipid. Permeability of tempo-choline into dimyristroyl phosphatidylcholine vesicles at the phase transition. *Biochemistry* **15**, 3570–78.

Martin, C.E., Hiramitsu, K., Kitajima, Y., Nozawa, Y., Skriver, L. and Thompson, G.A. (1976). Molecular control of membrane properties during temperature acclimation. Fatty acid desaturase regulation of membrane fluidity in acclimating *Tetrahymena* cells. *Biochemistry* **15**, 5218–27.

Melchior, D.L. and Steim, J.M. (1976). Thermotropic transitions in biomembranes. *Annual Review of Biophysics and Bioengineering* **5**, 205–38.

Melkonian, M., Kroger, K.-H. and Marquardt, K.-G. (1980). Cell shape and microtubules in zoospores of the green alga *Chlorosarcinopsis gelatinosa*: Effects of low temperature. *Protoplasma* **104**, 283–93.

de Mendoza, D. and Cronan, J.E. (1983). Thermal regulation of membrane lipid fluidity in bacteria. *Trends in Biochemical Sciences* **8**, 49–52.

Minorsky, P.V. (1985) An heuristic hypothesis of chilling injury in plants: A role for calcium as the primary physiological transducer of injury. *Plant Cell and Environment* **8**, 75–94.

Moeller, C.H., Mudd, J.B. and Thompson, W.W. (1981). Lipid phase separations and intramembranous particle movements in the yeast tonoplast. *Biochemica et Biophysica Acta* **643**, 376–86.

Morita, R.Y. (1975). Psychrophilic bacteria. *Bacteriological Reviews* **39**, 144–67.

Morris, G.J., Coulson, G. and Clarke, A. (1981). Does growth rate rather than temperature regulate fatty acid composition in *Tetrahymena*? *CryoLetters* **2**, 111–16

Morris, G.J., Coulson, G.E. and Clarke, A. (1984). Cold shock injury in *Tetrahymena pyriformis. Cryobiology* **21**, 664–71.

Nandini-Kishore, S.G., Mattox, S.M., Martin, C.E. and Thompson, G.A. (1979). Membrane changes during growth of *Tetrahymena* in the presence of ethanol. *Biochemica et Biophysica Acta* **551**, 315–28.

Nelmes, B.J., Preston, R.D. and Ashworth, D. (1973). A possible function of microtubules suggested by their distribution in rubbery wood. *Journal of Cell Science* **13**, 741–51.

Nozawa, Y. (1980). Modification of lipid composition and membrane fluidity in *Tetrahymena*. In *Membrane Fluidity. Biophysical Techniques and Cellular Regulation*, pp. 399–418. Edited by Kates, M. and Kukis, A. Humana Press, Clifton, New York.

O'Donovan, G.A. and Ingraham, J.L. (1965). Cold sensitive mutants of *Escherichia coli* resulting from increased feedback inhibition. *Proceedings of the National Academy of Sciences* **54**, 451–7.

Ojakian, G.K. and Satir, P. (1974). Particle movements in chloroplast membranes: Quantitative measurements of membrane fluidity by the freeze-fracture technique. *Proceedings of the National Academy of Sciences* **71**, 2052–6.

Oldfield, E., Gilmore, R., Glaser, M., Gutovsky, H.S., Hshung, J.C., Kang, S.J., King, E., Meadows, M. and Rice, D. (1978). Deuterium nuclear magnetic resonance investigation of the effects of proteins and polypeptides on hydrocarbon chain order in model membrane systems. *Proceedings of the National Academy of Sciences* **75**, 4657–60.

Oleinick, N.L. (1979). The initiation and elongation steps in protein syntheses: Relative rates in Chinese hamster ovary cells during and after hyperthermic and hypothermic shock. *Journal of Cell Physiology* **98**, 185–92.

Ono, T.A. and Murata, N. (1982). Chilling susceptibility of the blue green alga *Anacystis nidulans* 3. Lipid phase of cytoplasmic membrane. *Plant Physiology* **69** 125–9.

Papahadjopoulos, D., Jacobsen, K., Nir, S. and Isac, T. (1973). Phase transitions in phospholipid vesicles: Fluorescence polarization and permeability measuremennts concerning the effect of temperature and cholesterol. *Biochemica et Biophysica Acta* **311**, 330–48.

Papahadjopoulos, D., Poste, G. and Vail, W. J. (1979). In *Methods in Membrane Biology*, **10**, pp. 1–21. Edited by Korn, E.D. Plenum Press, New York.

Patterson, B.D. and Graham, D. (1977). Effect of chilling temperatures on the protoplasmic streaming of plants from different climates. *Journal of Experimental Botany* **28**, 736–43.

Pedersen, S., Block, P.L., Rech, S. and Neidhardt, E.C. (1978). Pattern of protein synthesis in *Escherichia coli* – a catalog of the amount of 140 individual proteins at different growth rates. *Cell* **14**, 179–90.

Petit, V.A. and Edinin, M. (1974). Lateral phase separation of lipids in plasma membranes: Effects of temperature on the motility of membrane antigens. *Science* **184**, 1183–5.

Phillips, M.C. and Chapman, D. (1968). Monolayer characteristics of saturated 1,2,-diacyl phosphatidylcholines (lecithins) and phosphatidylethanolamines at the air-water interface. *Biochemica et Biophysica Acta* **163**, 301–13.

Phillips, M.C., Hauser, H. and Daltuuf, F. (1972). The inter- and intramolecular mixing of hydrocarbon chains in lecithin/water systems. *Chemistry and Physics of Lipids* **8**, 127–33.

Pirt, S.J. (1965). The maintenance energy of bacteria in growing cultures. *Proceedings of the Royal Society, Series B* **163**, 224–31.

Poole, R.J. (1978). Energy coupling for membrane transport. *Annual Review of Plant Physiology* **29**, 437–60.

Rebeille, F., Bligny, R. and Douce, R. (1980). Oxygen and temperature effects on the fatty acid composition of sycamore cells (*Acer pseudoplatanus* L.). In *Biogenesis and Function of Plant Lipids*, pp. 203–6. Edited by Mazliak, P., Benveniste, P., Costes, C. and Douce, R. Elsevier, Amsterdam.

Rittenhouse, H., Williams, R. and Fox, C.F. (1974). Effect of membrane lipid composition and microtubule structure on lectin interactions of mouse LM cells. *Journal of Supramolecular Structure* **2**, 629–45.

Robards, A.W., Bullock, G.R., Goodall, M.A. and Sibbons, P.D. (1981). Computer assisted analysis of freeze-fracture membranes following exposure to different temperatures. In *Effects of Low Temperatures on Biological Membranes*, pp. 219–38. Edited by Morris, G.J. and Clarke, A. Academic Press, London.

Roth, L.E. (1967). Electron microscopy of mitosis in Amoeba. 3. Cold and urea treatment: A basis for test of direct effect of mitotic inhibitors on microtubule formation. *Journal of Cell Biology* **34**, 47–59.

Rottem, S., Yashov, J., Neeman, Z and Rajin, S. (1973). Cholesterol in mycoplasma membranes. Composition, ultrastructure and biological properties of membranes from *Mycoplasma mycoides* var. capiri cells adapted to grow with low cholesterol concentrations. *Biochemica et Biophysica Acta* **323**, 495–508.

Russell, N.J. (1984). Mechanisms for thermal adaptation in bacteria: blueprints for survival. *Trends in Biochemical Sciences* **9**, 108–12.

Saxton, M.J. (1982). Lateral diffusion in an archipelago. Effects of impermeable patches on diffusion in a cell membrane. *Biophysical Journal* **39**, 165–73.

Schliwa, M. and Euteneuer, U. (1979). Structural transformations of epidermal tonofilaments upon cold treatment. *Experimental Cell Research* **122**, 93–101.

Seelig, J. (1981). Physical properties of model membranes and biological membranes. In *Membranes and Intracellular Communication*, pp. 15–78. Edited by Balian, P., Chabre, M. and Devaux, P.F. North Holland, Amsterdam.

Sekiya, T., Kitajima, K. and Nozawa, Y. (1979). Effects of lipid phase separation on the fillipin action on membranes of ergosterol replaced *Tetrahymena* cells, as studied by freeze-fracture electron microscopy. *Biochemica et Biophysica Acta* **550**, 269–78.

Shaw, M., Marr, A.G. and Ingraham, J.L. (1971). Determination of the minimal temperature of growth of *Escherichia coli. Journal of Bacteriology* **105**, 683–4.

Shechter, E., Letellier, E. and Gulik-Kryzwicki, T. (1974). Relations between structure and function in cytoplasmic membrane vesicles isolated from an *Eschericia coli* fatty acid auxotroph. High-angle X-ray diffraction, freeze-etch electron microscopy and transport studies. *European Journal of Biochemistry* **49**, 61–76.

Shimshick, E.J. and McConnell, H.M. (1973). Lateral phase separation in phospholipid membranes. *Biochemistry* **12**, 2351–60.

Sikorska, E. and Farkas, T. (1982). Sterols and frost hardening of winter rape. *Physiologia Plantarum* **56**, 349–52.

Sillerud, L.O. and Barnett, R.E. (1982). Lack of transbilayer coupling in phase transitions of phosphatidylcholine vesicles. *Biochemistry* **21**, 1756–60.

Silvius, J.R., Read, B.D. and McElhaney, R.N. (1978). Membrane enzymes: Artefacts in Arrhenius plots due to temperature dependence of substrate binding affinity. *Science* **199**, 902–4.

Sinensky, M. (1974). Homeoviscous adaptation – a homeostatic process that regulates the viscosity of membrane lipids in *Escherichia coli. Proceedings of the National Academy of Sciences* **71**, 522–5.

Simon, E.W. (1981). The low temperature limit for growth and germination. In *Effects of Low Temperatures on Biological Membranes*, pp. 173–88. Edited by Morris, G.J. and Clarke, A. Academic Press, London.

Skriver, L. and Thompson, G.A. (1976). Environmental effects on *Tetrahymena* membranes. Temperature-induced changes in membrane fatty acid unsaturation are independent of molecular oxygen concentration. *Biochemica et Biophysica Acta* **431**, 180–88.

Skriver, L. and Thompson, G.A. (1979). Temperature induced changes in fatty acid unsaturation of *Tetrahymena* membranes do not require induced fatty acid desaturase synthesis. *Biochemica et Biophysica Acta* **572**, 376.

Steim, J.M., Tourtellotte, M.E., Reinert, J.C., McElhaney, R.N. and Radar, R.L. (1969). Calorimetric evidence for the liquid crystalline state of lipids in a biomembrane. *Proceedings of the National Academy of Sciences* **63**, 104–9.

Thompson, G.A. (1979). Molecular control of membrane fluidity. In *Low Temperature Stress in Crop Plants*, pp 347–63. Edited by Lyons, K.J.M., Graham, D. and Raison, J.K. Academic Press, New York.

Thompson, J.E., Mayfield, C.I., Inniss, W.E., Butler, D.E. and Kruuv, J. (1978). Senescence related changes in the lipid transition temperature of microsomal membranes from algae. *Physiologia Plantarum* **43**, 114–20.

Toivio Kinnucan, M.A., Chen, H.H., Li, P.H. and Stushnoff, C. (1981). Plasma membrane alterations in callus tissues of tuber-bearing Solanum species during cold acclimatization. *Plant Physiology* **67**, 478–83.

Trent, J.A., Chastain, R.A. and Yajanos, A.A. (1984). Possible artefactual basis for apparent bacterial growth at 250°C. *Nature* **307**, 737–40.

Tsien, H.C. and Higgins, M.L. (1974). Effect of temperature on the distribution of membrane particles in *Streptococcus faecalis* as seen by the freeze fracture technique. *Journal of Bacteriology* **118**, 725–34.

Uemura, M. and Yoshida, S. (1984). Involvement of plasma membrane alterations in cold acclimation of winter rye seedlings (*Secale cereale* L. cv. Puma). *Plant Physiology* **75**, 818–26.

Verkley, A.J., Ververgaert, P.H.J., van Deenen, L.L.M. and Elbers, P.F. (1972). Phase transitions of phospholipid bilayers of *Acheoplasma laidlawii* visualized by freeze-fracturing electron microscopy. *Biochemica et Biophysica Acta* **288**, 326–32.

Verwer, W., Ververgaert, P.H.J.T., Leunissen-Bijvelt, J. and Verkley, A.J. (1978). Particle aggregation in photosynthetic membranes of the blue-green alga *Anacystis nidulans. Biochemica et Biophysica Acta* **504**, 231–4.

Virtanen, I., Lehto, V.-P., Lehtonen, E. and Brudleg, R.A. (1980). Organisation of intermediate filaments in cultured fibroblasts upon disruption of microtubules by cold treatment. *European Journal of Cell Biology* **23**, 80–84.

Weber, C., Pollack, R. and Bibring, T. (1975). Antibody against tubulin: The specific visualization of cytoplasmic microtubules in tissue culture cells. *Proceedings of the National Academy of Sciences* **72**, 459–563.

Welsh, M.J., Pedman, J.R., Brinkley, B.R. and Means, A.R. (1979). Tubulin and calmodulin. Effects of microtubule and microfilament inhibitors on localization of the mitotic apparatus. *Journal of Cell Biology* **81**, 624–34.

Wodtke, E. (1978). Lipid adaptation in liver microsomal membranes of carp acclimated to different environmental temperatures. Phospholipid composition, fatty acid pattern and cholesterol content. *Biochemica et Biophysica Acta* **529**, 191–280.

Woods, C.M., Reid, M.S. and Patterson, B.D. (1984a). Response to chilling stress in plant cells. 1. Changes in cyclosis and cytoplasmic structure. *Protoplasma* **121**, 8–16.

Wood, C.M., Reid, M.S. and Patterson, B.D. (1984b). Response to chilling stress of plant cells. 2. Redistribution of intracellular calcium. *Protoplasma* **121**, 17–24.

Wu, S.W. and McConnell, H.M. (1975). Phase separations in phospholipid membranes. *Biochemistry* **14**, 847–54.

Wunderlich, F., Wallach, D.F.H., Speth, V. and Fischer, H. (1974). Differential effects of temperature on the nuclear and plasma membranes of lymphoid cells. A study by freeze-etch microscopy. *Biochemica et Biophysica Acta* **373**, 34–43.

3

Direct chilling injury

G.J. Morris
Cell Systems Limited
Cambridge

Introduction

Many tissues and cell-types are damaged following exposure to low temperatures without freezing, and such injury can be classified into two distinct categories (Levitt, 1980):

(i) direct chilling injury – this is expressed quickly upon reduction in temperature and is dependent on the rate of cooling. Greater cellular injury is induced by 'rapid' than by 'slow' cooling. The values of 'rapid' and 'slow' cooling rates are not absolute, and they vary with cell-type;
(ii) indirect chilling injury – this is evident only following a long period (often days) at the reduced temperatures, and it is independent of the rate of cooling.

This chapter will concentrate on direct chilling injury; indirect chilling injury is discussed in detail elsewhere in this volume (Wilson; Fuller; Morris and Clarke). Direct chilling injury has been observed in a wide variety of cells (for a comprehensive literature survey see Morris and Watson, 1984) and has been reported following three types of experimental treatments. These are:

(i) **Cold shock** - rapid cooling of cells in isotonic medium;
(ii) **Cold osmotic shock** - cells are diluted into a large excess of cold, hypotonic solution;
(iii) **Cold shock haemolysis** or thermal shock - this phenomenon is restricted to erythrocytes, which are sensitized to a reduction in temperature following exposure to hypertonic solutions.

These three classes of direct chilling injury will be discussed independently. Emphasis will be placed on the phenomenon of cold shock and an analysis of the injury induced by this stress. Cold osmotic stress and thermal shock will then be considered as special cases of cold shock injury. The potential contribution of these stresses to freeze-thaw injury is discussed elsewhere in this volume (Grout and Morris).

Cold shock

Cellular injury following rapid cooling has been recognized in both mammalian spermatozoa (Watson, 1981) and Gram-negative bacteria (Strange, 1976), and it would appear that a wide range of cell-types and tissues can be lethally damaged by the stress of rapid cooling (Table 3.1). In addition, under certain conditions cellular viability may be retained following rapid cooling when physiological processes are altered, for example cold-induced contraction of mammalian striated muscle (Sakai and Kurihara, 1974), induction of polyploidy in fish eggs (Purdom, 1972; Valenti, 1975; Lemoine and Smith, 1980), insects (Kawamura, 1979) and *Xenopus* (Kawahara, 1978).

Cell viability

Many variables determine the viability of cells following rapid cooling and rewarming.

Rate of cooling

Cell viability following a reduction in temperature is determined by the rate of cooling (Fig. 3.1). Results such as these for *Amoeba* sp. demonstrate two important features of cold shock injury.

Firstly, cell injury is not simply the result of exposure to a critical temperature, as all cells in the experiment described in Fig. 3.1 were exposed to the same final temperature; it is, however, the rapidity of temperature change which is the critical factor. In addition, the terms 'chilling-sensitive' and 'chilling-resistant' are not absolute, and they are valid only if the rate of cooling and the final temperature attained are defined. In this example (Fig. 3.1), at a cooling rate of 0.1°C min^{-1} both strains of amoeba

Table 3.1 Cell-types lethally damaged by a rapid reduction in temperature

Cell-type	Reference
Bacteria	
Many Gram-negative species	Reviewed in Strange, 1976; Farrell & Rose, 1967
Gram-positive species	Ring, 1965; Smeaton & Elliot, 1967; Truci & Duncan, 1974
Yeasts	
Saccharomyces cerevisiae	Calcott & Rose, 1982
Protozoa	
Tetrahymena pyriformis	Morris *et al.*, 1983; 1984
Blepharisma japonicum	Giese, 1973
Amoeba spp.	McLellan *et al.*, 1984
Algae	
Anacystis nidulans	Siva *et al.*, 1977; Forrest *et al.*, 1957; Jansz & Maclean, 1973; Ono & Murata, 1981
Chlamydomonas reinhardii	Morris *et al.*, 1983
Higher plants	
Sycamore suspension culture	Morris *et al.*, 1983
Plants of tropical origin	Reviewed in Levitt, 1980; Wilson & McMurdo, 1981
Mammalian cells	
Spermatozoa of many species	Reviewed by Watson, 1981
Embryos	Wilmut *et al.*, 1975
Isolated hepatocytes	Morris *et al.*, 1983
Lymphoid cell line	Morris *et al.*, 1983
Granulocytes	Knight *et al.*, 1976; Takahashi *et al.*, 1981

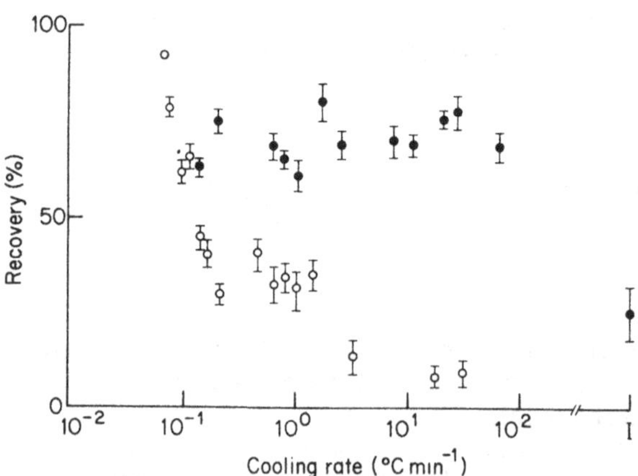

Fig. 3.1 Recovery (%) of *Amoeba* sp. strain Bor (o) and *A. proteus* (●) following cooling at different rates to − 10°C. All cell suspensions were undercooled (i.e. in the absence of ice) and maintained at − 10°C for 5 minutes before rewarming.
I = Instantaneous cooling achieved by direct injection of 0.1 ml of a concentrated cell suspension into 10 ml of precooled medium. (Reproduced with permission from McLellan *et al.*, 1984.)

demonstrate significant resistance; following 'instantaneous cooling', however, both are very markedly sensitive. It is only within the range of intermediate cooling rates that a distinction between resistant and sensitive may be made.

Secondly, although cell injury following exposure to low temperature is determined by cooling rate, it is virtually independent of warming rate. Only when very slow rates of warming are employed is there a further reduction in viability (McLellan *et al.*, 1984). It can be assumed, therefore, that the primary damage is induced during the cooling phase and may be compounded by significantly longer exposure to low temperature as will occur during very slow warming.

Length of incubation at reduced temperature

Following cooling, cell recovery is dependent on the period of incubation at the reduced temperature before rewarming. At any rate of cooling a greater loss of viability occurs with increasing time at the reduced temperature (Fig. 3.2).

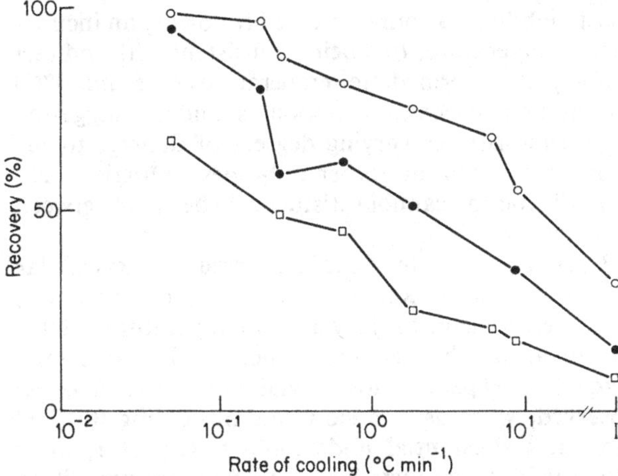

Fig. 3.2 Recovery (%) of *Chlamydomonas reinhardii* CW15+ following cooling at different rates to −7.5°C. All cell suspensions were undercooled (i.e. in the absence of ice) and maintained at −7.5°C for 1 hour (o), 2 hours (●) or 4 hours (□) before rewarming.

Final temperature attained

Cell viability following cold shock is also dependent on the final temperature attained (Fig. 3.3). In these experiments with amoeba the cellular reaction following rapid cooling to −10°C was typically that of cold shock, with maximum lethal injury being sustained by the population after only 5 minutes incubation at the final temperature. In further experiments

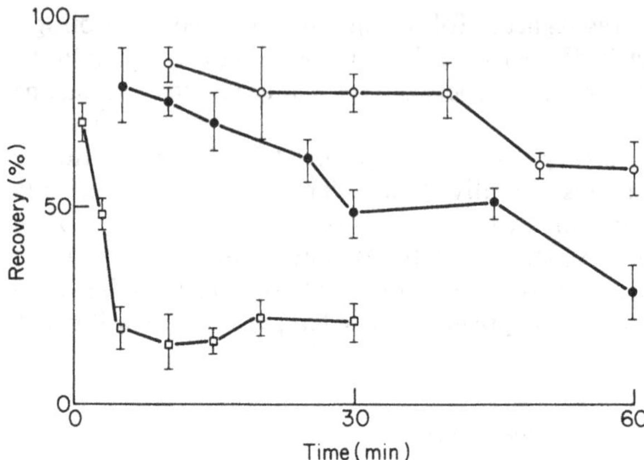

Fig. 3.3 Recovery (%) of *Amoeba* sp. strain Bor following exposure to −10°C (□), −5°C (●) or 0°C (o) for different times before rewarming to 20°C. (Reproduced, with permission from McLellan *et al.*, 1984.)

(Morris, unpublished observation), when cells were cooled as rapidly as possible to +10°C, loss of viability was only observed following an incubation period of days at this temperature, this being consistent with indirect chilling injury (Levitt, 1980). At intermediate temperatures (−5 and 0°C) there appears to be a continuum between these responses, and it is suggested that this graded response demonstrates varying degrees of damage to the same site of injury. Thus, in amoeba and other cell-types, (Morris *et al.*, 1983; Morris *et al.*, 1984) a distinction cannot satisfactorily be made between direct and indirect chilling injury.

It can be shown (Fig. 3.1) that as the rate of cooling decreased, less cellular injury was observed at any temperature. Reduction in cooling rate does not, however, protect cells completely against injury at low temperatures, but it may alter the temperature at which such injury occurs. For example, following rapid cooling to 0°C a 99 per cent loss of viability of *Tetrahymena pyriformis* has been observed, whereas at a slow rate of cooling (0.25°C min^{-1}) the organism tolerates substantial undercooling. However, after cooling at such a rate to −10°C a complete loss of viability occurs within 15 minutes. This rapid expression of injury indicates that this is still typical cold shock.

It can thus be argued that all cell-types are sensitive to cold shock if they are cooled rapidly enough and to a sufficiently low temperature (Morris *et al.*, 1983). This has important consequences, for cells sensitive to cold shock were previously considered as atypical and 'normal' biological material was considered to be resistant to such effects. Indeed, much research into the basis of cold shock injury has consisted of studies of the comparative biochemistry of cell-types considered to be resistant or sensitive. The widespread occurrence of cold shock injury now observed (Table 3.1) also provokes the thought that the primary site of injury may be common to both prokaryotic and eukaryotic cells.

Age of culture

For bacteria grown in batch culture the response of cells to the stress of rapid cooling varies with the phase of culture; cells from the exponential phase are the most sensitive (Strange, 1976). Exceptions may be demonstrated, for example, *Pseudomonas* cells appear to be resistant to cold shock in the early stages of culture but become sensitive with increasing age (MacKelvie *et al.*, 1968). In contrast, for *Tetrahymena* there is no difference in the response to rapid cooling from the early exponential phase to the late stationary phase of culture.

When the growth rate of microorganisms has been controlled using continuous culture methods, the sensitivity of cells to cold shock appears to be proportional to growth rate (Kenward and Brown, 1978; Green, 1978, Gilbert *et al.*, 1979) and absolute survival depends on the specific nature of the limiting nutrient (Fig. 3.4).

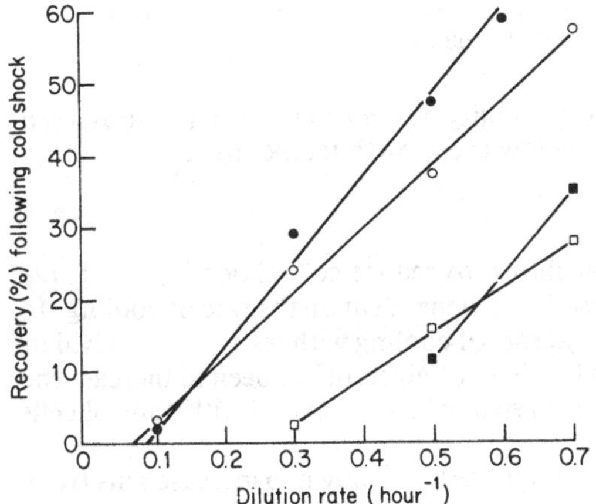

Fig. 3.4 Effect of growth rate on the sensitivity of *Vibrio cholerae* to cold shock. Growth limiting nutrients: magnesium (●), glucose (o), phosphate (□) or nitrogen (■). (Reproduced with permission from Green, 1978.)

Temperature of growth

Maintenance of cells at reduced temperatures may modify the response to subsequent rapid cooling (Farrell and Rose, 1967; Paton *et al.*, 1978a; Ono and Murata, 1982). This treatment does not, however, induce resistance to cold shock, but alters the temperature at which such injury is induced. For example, when cells of *T. pyriformis* grown at 20°C were cooled rapidly to +4°C there was a 50 per cent reduction in viability, whereas for cultures maintained at 10°C the equivalent loss of viability occurred on cooling to −5.8°C (Fig. 3.5). It can be seen that for both culture types a 50 per cent reduction in viability occurred at approximately 15°C below the temperature

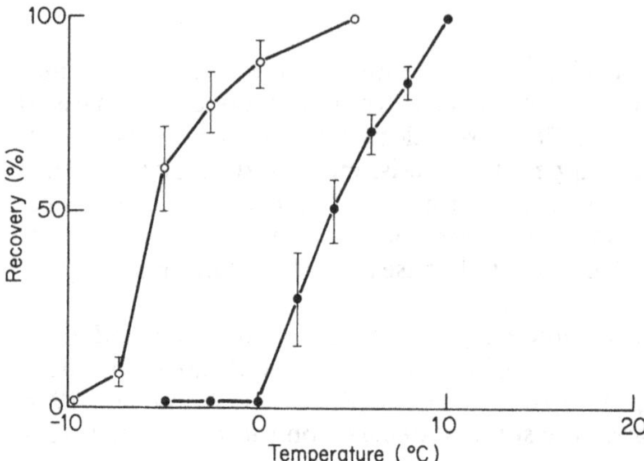

Fig. 3.5 Recovery (%) of *Tetrahymena pyriformis* following rapid cooling to different temperatures. Cells were from cultures which had been either maintained at 20°C for 3 days (●), or following 3 days at 20°C had been transferred to 10°C for a further 3 days (○). (Reproduced with permission from Morris *et al.*, 1984.)

of growth and complete loss of viability (> 99 per cent) was observed following rapid cooling to 20°C below the growth temperature.

Additives

A variety of additives have been shown to reduce cold shock injury (Table 3.2). The degree of protection also being dependent on the rate of cooling. In the presence of glycerol, an optimal rate of cooling with respect to survival of amoeba is observed (Fig. 3.6). The effect of glycerol has been to increase the cooling rate that affords optimal survival when compared with control cells (Fig. 3.1).

In addition, a number of compounds has been shown to increase sensitivity to cold shock. These include dimethylsulphoxide (DMSO) for lymphocytes (Morris *et al.*, 1983) and both colchicine and vinblastine for amoeba (McLellan *et al.*, 1984). The special case of the sensitization of mammalian erythrocytes to rapid cooling is discussed below.

Table 3.2 Additives which reduce cold shock injury

Additive	Cell-type	Reference
Mg^{2+}, Ca^{2+}, Mn^{2+}	*Aerobacter aerogenes*	Strange & Dark, 1962
Tween 80	*Bacillus amyloliquefaciens*	Paton *et al.*, 1978a
Sucrose (0.3M)	*Escherichia coli*	Meynell, 1958
Sucrose (20% w/v) NaCl (1%w/v)	*E. coli*	Bosl & Birck, 1981
Ca^{2+}, glycerol (0.07M) glucose (0.1M)	*Amoeba* sp.	McLellan *et al.*, 1984
EDTA, Mg^{2+}	Ram and Bull spermatozoa	Quinn & White, 1968
Egg yolk, lecithin	Mammalian spermatozoa (Variation between species)	Reviewed in Watson, 1981

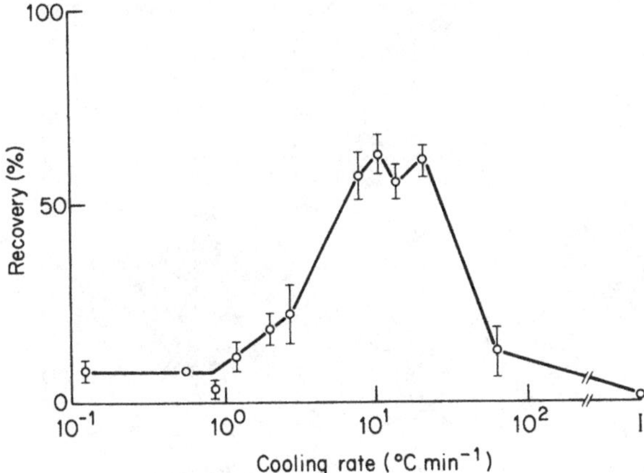

Fig. 3.6 Effect of pregrowth in glycerol (0.07 M for 17 hours) on the recovery (%) of *Amoeba* sp. strain Bor following cooling at different rates to −10°C. All cells were maintained at −10°C for 5 minutes before rewarming. (Reproduced with permission from McLellan *et al.*, 1984.)

Morphology of cold shock injury

Following rapid cooling a variety of morphologies have been observed. In some cell-types a loss of membrane selective permeability may occur, as demonstrated by the loss of pigment from the ciliated protozoan *Blepharisma* (Giese, 1973). In mammalian spermatozoa alterations to the outer acrosomal membrane and the plasmalemma overlying the acrosomes, and a loss of acrosomal contents are apparent. Mitochondrial changes may occur in some cold shocked sperm but are variable; nuclear membranes are, however, apparently unaltered (Watson, 1981).

Cryomicroscopy (McGrath, this volume) has been employed to study cold shock in amoeba (Fig. 3.7). As the temperature is lowered cytoplasmic streaming is reduced and ceases at approximately 10°C. There were no further morphological changes during undercooling to −10°C. Upon rewarming, a contraction of the cytoplasm occured together with gross deformation of the plasmalemma. Following cold shock, cellular shrinkage in response to hypertonic conditions will still occur, clearly demonstrating that the plasmalemma is osmotically intact. Any alterations which may occur in the physical state of the plasmalemma during cooling (Morris and Clarke, this volume) are, by implication, reversed upon rewarming.

Thin section electron microscopy of amoeba following such cold shock demonstrates the aggregation of actin, a compound known to be a significant component of the cytoskeleton (McLellan *et al.*, 1984). Such a response to cooling may be analogous to the cold-induced contraction of mammalian striated muscle (Sakai and Kurihara, 1974) and involve elements of the cyto-skeleton. Although these may be cold labile, and thus responsible for the observed cell shrinkage (see Morris and Clarke, this volume), the trigger for

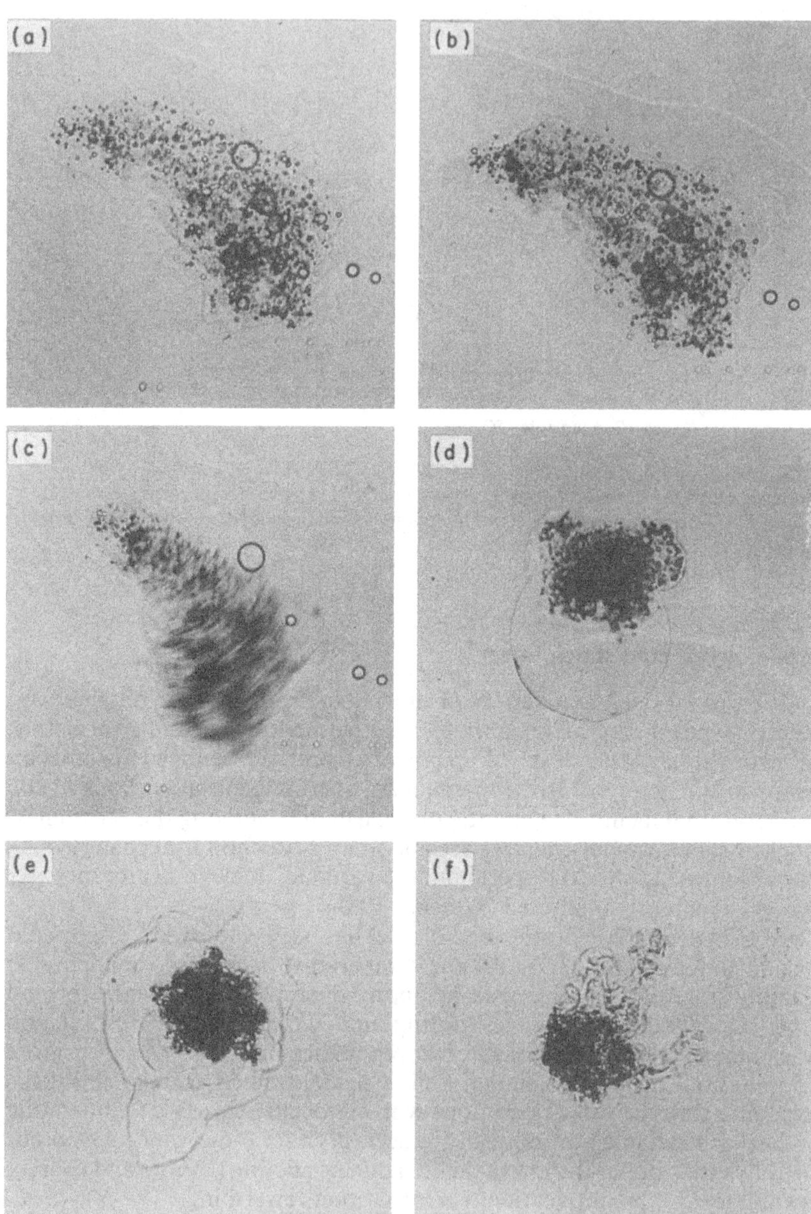

Fig. 3.7 Direct observation of *Amoeba* sp. strain Bor during cooling and rewarming. All photomicrographs are at the same magnification. (a) Cell at 20°C. (b) Cell following cooling at 10°C min^{-1} to −10°C Cell during rewarming at 30°C min^{-1}, (c) at +7°C and (d) at +15°C. Osmotic response of a cell following cold shock, prior to (e) and following (f) the addition of hypertonic sodium (0.6 M). (Reproduced with permission from McLellan *et al*., 1984.)

cellular contraction following cold shock might also be due to alterations in intracellular ion concentrations, especially Ca^{2+}, as a result of a loss of membrane selective permeability (see below).

Biochemical and physiological alterations induced by cold shock

Many changes in the biochemistry of cells have been reported following cold shock but it is often difficult to distinguish between the primary effects, which are responsible for the loss of cellular viability, and secondary, pathological events.

Plasmalemma selective permeability

It is generally agreed that changes in the selective permeability properties of the plasmalemma occur as an early cold shock injury. Such changes have been demonstrated by a variety of methods.

(i) *Cation redistribution* In spermatozoa sodium and calcium are gained by cells following cold shock while potassium and magnesium are lost (Watson, 1981). Although the movements of sodium, potassium and magnesium are by passive diffusion, accumulation of calcium is against a concentration gradient (Quinn and White, 1966). This suggests that, following changes in the selective permeability of the plasma membrane, calcium is actively taken up into the cells. The cold-induced contraction of certain mammalian striated muscle cells is initiated also by alterations in the free calcium level rather than by changes in the calcium affinity of the contractile apparatus (Jeacocke, 1982).

Following cold shock of plant cells, increased permeability to potassium (Zsoldos and Karvaly, 1979; Ono and Murata, 1981) and calcium (Zocchi and Hanson, 1982) have also been reported. In addition, intracellular cations are released from *Escherichia coli* following rapid cooling (Haest *et al.*, 1972).

(ii) *Release of intracellular constituents* Following cold shock a wide range of cytoplasmic compounds can be released. These have been reported to include cytoplasmic enzymes from spermatozoa (Watson, 1981), adenosine triphosphate (ATP), UV-absorbing compounds, amino acids and small molecular weight peptides from *Aerobacter aerogenes* (Strange, 1976), rhodanase from *Pseudomonas aeruginosa* (Ryan *et al.*, 1979) and sugars from *E. coli* (Novotny and Englesberg, 1965; Haest *et al.*, 1972).

(iii) *Penetration of compounds into cells* Concomitant with the release of intracellular constituents *Aerobacter aerogenes* becomes permeable to extracellular ribonuclease (MW 12 000) and anilino-naphthalene-8-sulphonate (Strange and Postgate, 1964). The permeability of *Streptomyces hydrogenans* to amino acids and thiourea also increases following rapid cooling (Ring, 1965). Similarly, rapid cooling of *Clostridium perfringens* prevents subsequent cell growth if plated on a medium containing the antibiotic neomycin, a compound that does not normally permeate these cells (Traci and Duncan, 1974). Cold shock has also been used experimentally to increase permeability of cells to phospholipase to examine membrane lipid asymmetry (Paton *et al.*, 1978b).

(iv) *Osmotic response* Bacteria susceptible to cold shock do not shrink in response to a hypertonic stress following rapid cooling to 0°C, while those cooled slowly to this temperature remain osmotically sensitive (Strange, 1964). In contrast, amoebae damaged by cold shock are still osmotically sensitive following rewarming (Fig. 3.7).

Respiration

Evidence of respiratory damage following cold shock can be found in a number of systems. Anaerobic glycolysis and respiratory activity are both diminished in ram and bull spermatozoa following rapid cooling and rewarming, and cellular ATP levels are reduced, with no resynthesis occuring (Watson, 1981). Cold shock of cereal seedlings will result in a decrease in respiratory control and an uncoupling of oxidation and phosphorylation upon rewarming. This has been correlated with the release of free fatty acids within the mitochondria (Vojnikov *et al.*, 1983).

Photosynthesis

Genuine cold shock studies in this area are limited, but in *Anacystis nidulans* photosynthesis is reduced following rapid cooling which has been linked with the release of cytoplasmic ions (Jansz and MacLean, 1973; Siva *et al.*, 1977; Ono and Murata, 1981). Furthermore, the activity of the Hill reaction, diminished by cold shock, could be restored by a second cold shock in a medium containing elevated levels of K^+ and Mg^{2+} (Ono and Murata, 1981). This implicates both the role of the ions in the process, and also altered plastid permeability resulting from cold shock.

Other biochemical changes

A wide variety of other biochemical changes have been demonstrated in cold-shocked cells including the inability to metabolize certain carbohydrates, an increase in ammonia formation (indicating a breakdown of intracellular protein) and breakdown of DNA (Watson, 1981). Many of these reactions, together with some of the metabolic changes outlined above, may be secondary reactions occurring as a consequence of loss of cytoplasmic ions.

Mechanisms of cold shock injury

The essential features of cold shock that must be accounted for by any hypothesis are that:

(i) all cell-types may be considered to be sensitive to cold shock, provided that they are cooled rapidly enough to a sufficiently low temperature;

(ii) cellular viability is dependent on the rate of cooling, with greater mortality being observed following 'rapid', rather than 'slow', cooling;
(iii) cold shock injury is independent of the rate of warming, when this does not significantly increase the length of exposure to low temperature (see iv);
(iv) injury is increased as the period of isothermal incubation at the reduced temperature is extended;
(v) loss of membrane permeability occurs following rapid cooling, and, in some instances, may be reversed upon rewarming;
(vi) the response of any cell-type may be modified by the culture conditions before cooling or by certain additives.

Injury following rapid cooling cannot be due to metabolic imbalances resulting from different temperature coefficients of enzyme reactions or accumulation of toxic by-products, because maximum damage is observed following the most rapid cooling, during which cells are exposed to low temperatures for the shortest period. Increased exposure times, such as those which occur at slower rates of cooling, may even be protective (Fig. 3.1).

It may be proposed that the thermotropic behaviour of membrane lipids determines cold shock injury (see Morris and Clarke, this volume). This is well illustrated in the ciliated protozoan *T. pyriformis* for which there is an extensive literature on the effects of a reduction in temperature on the structure and function of membranes. Freeze-fracture studies of the intramembraneous particle distribution (see Skaer, this volume) of cells grown at 39°C, 24°C or 15°C and then cooled 'slowly' to different temperatures for 5 minutes demonstrated that a phase separation was initiated within the plane of the membrane at approximately 10°C below the growth temperature (Martin *et al.*, 1976). This was complete, as judged by freeze-fracture electron microscopy, at 20°C below the growth temperature. For cells grown at 39°C or 15°C the temperatures at which a 50 per cent phase separation had occurred were separated by 24°C; this demonstrates a perfect temperature compensation in membrane structure. Parallel studies have revealed an increase in the membrane phospolipid fatty acid unsaturation following growth at reduced temperatures. These observations, suggesting alterations in membrane structure, have been confirmed qualitatively in wider studies. For example, using wide-angle X-ray diffraction, the transition temperature (T_c) for pellicle lipids of cells grown at 39°C was found to be 26°C, while for cells grown at 15°C the transition was first evident at 6°C (Nakajama *et al.*, 1983). In addition, Arrhenius plots of the ATPase activity of the pellicle membrane from cells grown at 39°C showed a discontinuity at 28°C, while for cells grown at 15°C the discontinuity shifts to lower temperatures (Nozawa, 1980). The injuries sustained by whole cells of *Tetrahymena* subjected to cold shock (Fig. 3.5) or isothermal incubation at low temperatures show an obvious parallel to this type of membrane phase behaviour. The initial loss of viability becomes evident at 12–15°C below the growth temperature and increases as the temperature is further reduced, suggesting a direct correlation with altered membrane structure. A similar correlation between loss of viability following rapid cooling and lipid transitions in membranes has been reported for other cell-types (Farrell and Rose, 1967; Haest *et al.*, 1972; Paton *et al.*, 1978a; Ono and Murata, 1981).

To a first approximation, however, at any temperature below the nucleation point of the membrane lipids the same amount of gel phase lipid will be formed following either rapid or slow cooling (Morris and Clarke, this volume). As cellular viability is evidently a function of cooling rate, it may not simply be the presence of gel phase lipid within the membrane which determines cold shock injury, but rather the more specific aspects of the nature and pattern of crystal growth. The effects of rapid cooling on nucleation events within membranes will be discussed and demonstrated here in terms of an accepted model using the packing behaviour of a monolayer of

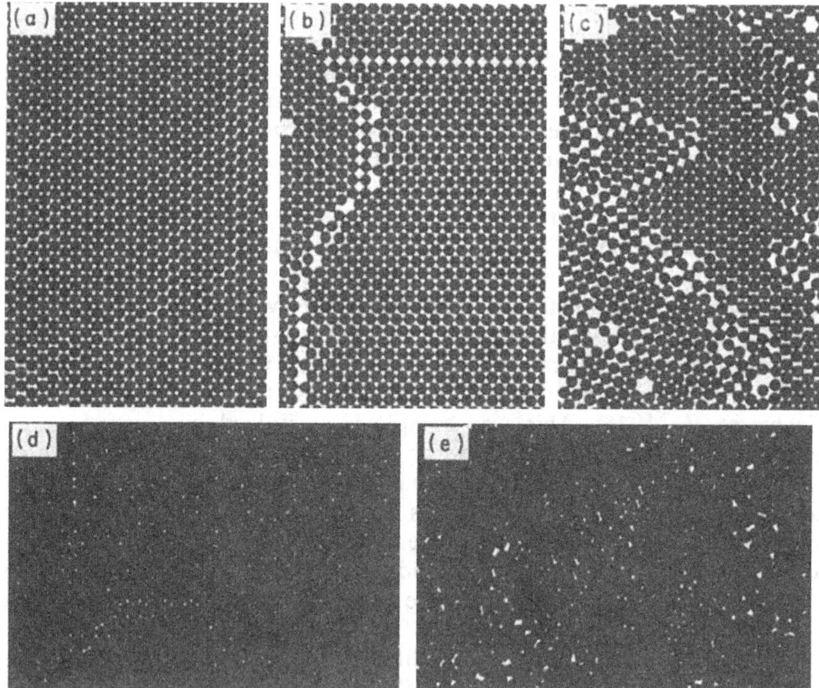

Fig. 3.8 Potential nucleation patterns within a membrane as modelled by the packing characteristics of ball bearings.
(a) Ideal hexagonal packing within a monolayer.
(b) Junction of three hexagonally packed lattices, with grain boundaries and packing faults in the boundary structures.
(c) Numerous small hexagonally packed domains with extensive grain boundaries and packing faults between the adjacent regular lattices.
(d) Superimposition of the monolayer structures Fig. 3.8a onto Fig. 3.8b, to represent a bilayer. No large imperfections occur in the bilayer structure.
(e) Superimposition of Fig. 3.8c onto an area of similar packing characteristics. Imperfections coincident in the two layers result in structural defects, equivalent to pores, which are 'set' into the bilayer structure.

ball bearings (Fig. 3.8 and see Morris and Clarke, this volume). Although this model is obviously simplistic, it does serve to illustrate the basic processes which may occur, and similar models have been used effectively by other workers (Hui *et al.*, 1974; Chapman *et al.*, 1979).

During rapid cooling, events equivalent to undercooling are observed within biological membranes and in model systems of lipid bilayers. This undercooling is clearly demonstrated by differential scanning calorimetry (DSC), and it is shown to be reduced at slower rates of cooling (Black and Dixon, 1981). As the critical nucleation size is related to the extent of undercooling (see Taylor, this volume), many nucleation sites would be expected within membranes following rapid cooling. Crystal growth would be rapid, and there would be insufficient time for significant redistribution of components within the plane of the membrane. This proposal is supported by freeze-fracture electron microscopy following rapid cooling, which shows no gross diffusion of proteins, Lateral separation of membrane proteins is commonly observed following slow cooling (see Morris and Clarke, this volume). Studies using DSC combined with X-ray diffraction indicate that, following rapid cooling, bilayers are crystalline, but domains of hexagonally-packed, crystalline phospholipid are small or otherwise disordered (Melchior *et al.*, 1982).

As rapid cooling will result in multiple nucleation sites and, consequently, small gel phase domains, extensive grain boundaries and numerous packing faults would be expected to occur between adjacent regions of hexagonally-packed phospholipids (Fig. 3.8c), and such structures have been demonstrated in phospholipid bilayers by electron diffraction (Hui, *et al.*, 1974). The occurrence of lipids with different packing characteristics, or the presence of proteins, would increase the incidence of imperfections in gel lipid structure (Chapman *et al.*, 1979). The two halves of the bilayer will nucleate independently (Sillerud and Barnett, 1982) and, if packing defects within the two monolayers are coincident, leakage of small molecules across the membrane might occur (Fig. 3.8e).

As the rate of cooling is reduced, the probability of undercooling and, consequently, the number of nucleation sites within the membrane will decrease. At slower rates of cooling there is more time for gel phase growth, lateral lipid phase separations and diffusion of proteins; this can be clearly demonstrated by freeze-fracture electron microscopy. The areas of gel lipid formed will be larger than those nucleated following rapid cooling and, consequently, the incidence of grain boundaries, packing faults and dislocations will be reduced. Ideally, the structure would approach that shown in Fig. 3.8a, and superimposition of such crystal structures (to represent both layers of the membrane) would not result in imperfections in the bilayer structure of gel phospholipids. At slow rates of cooling the loss of intracellular constituents from packing faults within gel membrane structure would therefore be reduced, with an increase in the probability of the cells remaining viable.

At very slow rates of temperature reduction a further loss of cellular viability may occur (Fig. 3.6) which cannot be ascribed to the occurrence of packing faults as described above, However, at such slow rates of cooling, membranes will be in a state of phase separation (i.e. an equilibrium coexistence of gel and liquid phase lipids) for relatively long times. At the interfaces between the gel and liquid lipid there will be regions of molecular mismatch, which are sites of increased permeability (see Morris and Clarke, this volume). The duration of such a state of phase separation will be directly related to the rate of cooling and thus may provide a clue to a loss of viability observed at such slow rates of cooling. It must be stressed, however, that the occurrence of a phase separation within membranes is not necessarily lethal to all cell-types (see Morris and Clarke, this volume). Additionally, if cells are exposed to low temperatures for long periods during slow cooling, other factors, such as depletion of metabolites or the accumulation of toxic by-products may lead to a loss of membrane selective permeability and hence viability.

In summary it may be said that membranes show a biphasic response to cooling rate. At rapid rates of cooling, cellular injury is determined by the pattern of nucleation within the membranes, whereas at slow cooling rates injury may be related to the prolonged time during which phase separations will be present in the membrane. At intermediate rates of cooling and at higher temperatures, survival may be determined by the relative importance of these two processes. This pattern of response accounts for many of the features of cellular injury at low temperatures and may explain the continuum between the cellular responses defined as direct and indirect chilling injury. The processes of nucleation and crystal growth will depend on the composition of the membranes, and the occurence (if any) of structures within the membrane which promote or depress nucleation. In addition, membrane thermotropic behaviour will be influenced by compounds which interact with membrane components. This hypothesis also explains the relative independence of cell survival to the rate of warming and the reversibility of membrane leakage upon rewarming observed in some cell-types.

Cold osmotic shock

This experimental treatment has been used extensively to study transport processes in bacteria and plant cells. The protocol as originally described for *E. coli* (Neu and Heppel, 1965; Heppel, 1967) uses cells from the exponential phase of growth suspended in a solution containing ethylenediamine tetra-acetic acid (EDTA) (1×10^{-4} M) and sucrose (0.5 M) and then rapidly dispersed into an excess (80 volumes) of cold dilute $MgCl_2$ (10^{-4} M).

Immediately following this treatment viability is retained and cells are osmotically responsive. If, however, the cells are maintained at the reduced temperature for longer period, a progressive reduction in viability occurs. Upon resuspension into nutrient medium the ability of cells to take up substances by specific transport mechanisms, is drastically reduced. Approximately 5 per cent of the total cellular protein is released by this treatment, and it is assumed that these released proteins are involved in transport, possibly being peripheral proteins which serve as recognition sites (Pardee *et al.*, 1966; Piperno and Oxender, 1966; Anraku, 1967; Wilson and Holden, 1969;

Wiley, 1970). In addition, the cells become sensitive to a range of compounds to which they are normally resistant, suggesting further alterations in cellular permeability. Following an extended lag period, the normal transport processes and selective permeability of the cell envelope are regained and cell division resumes (Anraku and Heppel, 1967). This series of experimental procedures and their effects on cellular physiology obviously have much in common with cold shock as discussed above.

Cold shock haemolysis

Human erythrocytes suspended in plasma or isotonic salt solutions are not lysed by the stress of cold shock. However, pre-treatments, such as exposure to phospholipases, calmodulin antagonists (Takahashi, 1983) or hypertonic solutions sensitize the cells to a reduction in temperature. The following discussion will concentrate on the sensitization of erythrocytes to cold shock by exposure to hypertonic solutions, a phenomenon originally termed thermal shock (Lovelock, 1953b). It is, however, evidently a special case of cold shock injury, and this relationship can be emphasized by redefining this stress as cold shock haemolysis.

Haemolysis

Many variables determine the extent of haemolysis following hypertonic-induced cold shock. A large number of these factors are interrelated, and all contribute to lysis, but they will be considered separately for ease of explanation. It must be stressed, however, that exposure of erythrocytes to hypertonic solutions renders the cells sensitive to stresses other than a reduction in temperature (Lovelock, 1953b). Most studies on cold shock in erythrocytes have measured lysis in the hypertonic solution and not upon subsequent resuspension. This has resulted in problems of interpretation, especially following short-term exposure to solutions of greater than 2.5 Osm kg^{-1} or prolonged incubation in solutions of 1–2 Osm kg^{-1} (see below). In this discussion therefore only the sensitization of erythrocytes to cold shock induced by short-term exposure to solutions of less than 2 Osm kg^{-1} will be examined in detail.

Hypertonic solutions

(i) *Hypertonicity* There is a critical concentration necessary to sensitize erythrocytes to a reduction in temperature. No lysis is observed on cooling of cells suspended in solutions of sodium chloride less than 1.4 Osm kg^{-1}, but at osmolalities above this haemolysis occurs (Fig. 3.9) and is maximal following exposure to sodium chloride solutions of 1.8 Osm kg^{-1}.

At concentrations of sodium chloride or sucrose greater thaan 2 Osm kg^{-1} haemolysis induced by cold shock is reduced and at 4 Osm kg^{-1} no additional leakage of haemoglobin occurs upon rapid cooling (Fig. 3.9) (Lovelock, 1955; Takahashi and Williams, 1983). It must be emphasized, however, that

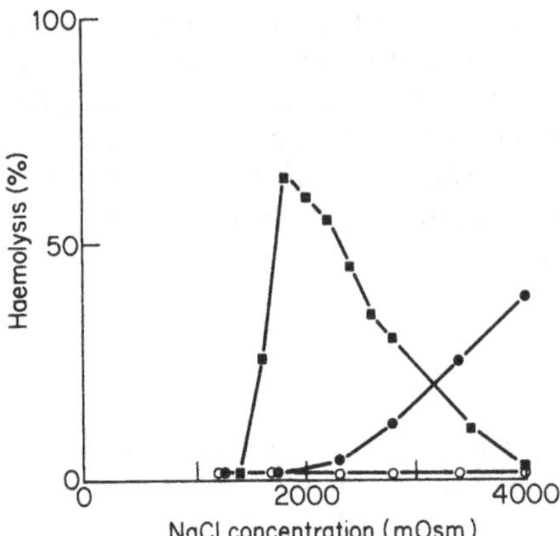

Fig. 3.9 Haemolysis (%) of human erythrocytes following exposure to hypertonic sodium chloride for 5 minutes at 37°C, (○), upon resuspension into isotonic NaCl (●) and following rapid cooling to 0°C (■). (Redrawn from Lovelock, 1953b and Takahashi and Williams, 1983.)

haemolysis was determined in the supernatant of cells suspended in hypertonic solutions. At these high osmolalities a loss of membrane selective permeability occurs and cell volume increases from its apparent minimum (Farrant and Woolgar, 1972; Zade-Oppen, 1968).

(ii) *Composition* Cold shock haemolysis is induced by a variety of hypertonic solutions of ionic and non-ionic non-permeating compounds (Table 3.3). Additives which are freely permeable to erythrocytes do not sensitize cells to cold shock lysis. The extent of cold shock haemolysis induced following exposure to solutions of equal osmolality varies with the composition of the solution. The effects of different anions and cations in hypertonic solutions of approximately equal osmolality (1.8 Osm kg^{-1}) have been compared with that of sodium chloride (Morris, 1975). The replacement of sodium by lithum, caesium, potassium or ammonium cations had no significant effect, but the substitution of sodium by choline increased survival following cooling. In contrast, the substitution of chloride by iodide, sulphate or nitrate increased haemolysis induced by cold shock, whereas acetate increased survival.

Table 3.3 Hypertonic solutions which sensitize human erythrocytes to cold shock injury

Additive	Concentration	Reference
NaCl	> 1.5 Osm kg^{-1}	Lovelock, 1953
Sucrose	> 1.5 Osm kg^{-1}	Woolgar & Morris, 1973
LiCl, KCl, CsCl, NH_4Cl	> 1.7 Osm kg^{-1}	Morris, 1975
NaI, Na_2SO_4, $NaNO_3$, Na acetate		
$CaCl_2$	> 1.0 Osm kg^{-1}	Green & Jung, 1977
Mannitol	Not stated	Jung & Green, 1978

With cells suspended in hypertonic sucrose the extent of haemolysis following rapid cooling was modified by the presence of ions, maximum haemolysis being observed with sucrose in distilled water, the addition of phosphate buffer (20mM) or phosphate buffered saline significantly reducing cold shock haemolysis (Green and Jung, 1977).

(iii) *Time and temperature of exposure* The effects of length of exposure to hypertonic solutions before cooling are complex and depend on both the temperature of exposure and composition of the hypertonic solution. For example, following exposure of erythrocytes at 25°C to solutions of sucrose and sodium sulphate the sensitivity to a subsequent reduction in temperature increases continuously with increasing time of exposure (Fig. 3.10). In contrast, cells exposed to hypertonic solutions of sodium chloride and sodium iodide are initially at their most sensitive and increasing time of exposure reduces cold shock haemolysis.

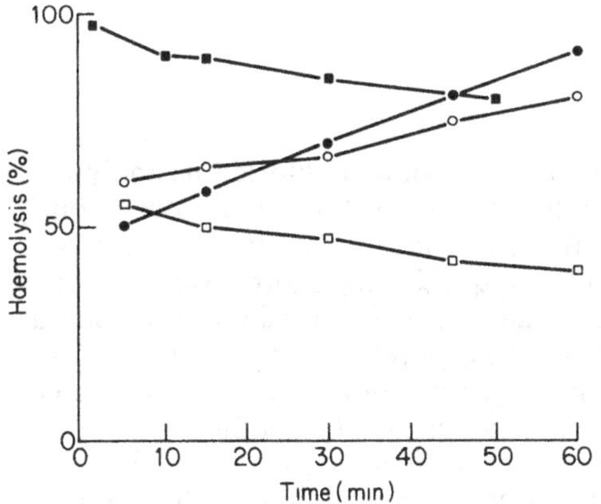

Fig. 3.10 Haemolysis (%) of human erythrocytes cooled rapidly from 37°C to 0°C in hypertonic solutions (approximately 1.8 Osm kg^{-1}) of sodium chloride (●), sodium iodide (○), sodium sulphate (□) and sucrose (■). The erythrocytes were exposed to the hypertonic solutions for different periods before cooling. (Redrawn from Morris and Farrant, 1973; Morris, 1975.)

The combined effects of temperature and period of exposure on the sensitization of erythrocytes to a reduction in temperature has been examined at a single concentration of sodium chloride (Takahashi and Williams, 1983). During exposure of erythrocytes to hypertonic solutions at 37°C and 30°C cells are maximally sensitive following short periods of exposure. In contrast, at 15°C cells are initially resistant to cold shock and during increasing periods of exposure sensitization to cold shock develops. No haemolysis upon cooling to 0°C was observed following exposure to hypertonic sodium chloride at 13°C for periods of up to 90 minutes.

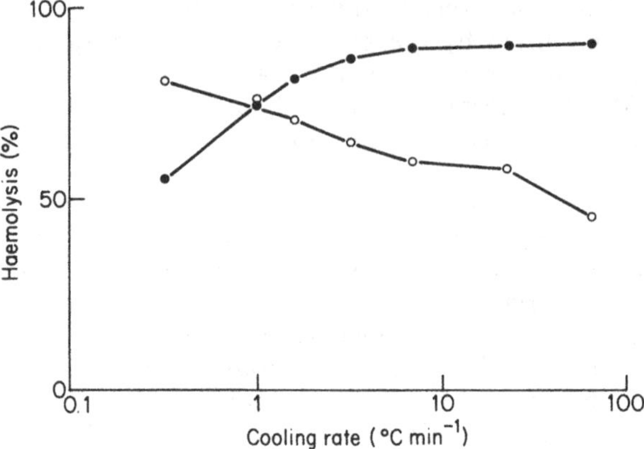

Fig. 3.11 Haemolysis (%) of human erythrocytes suspended for 5 minutes at 25°C in either 1.2 M NaCl (○) or 1.17 M sucrose (●) and then cooled at different rates to 0°C. (Redrawn from Morris and Farrant, 1973.)

Rate of cooling

After exposure to hypertonic solutions of sodium chloride haemolysis is greater following rapid cooling to 0°C than following slow cooling (Fig. 3.11) (Lovelock, 1955; Morris and Farrant, 1973; Takahashi and Williams, 1983). The opposite is observed following short periods of exposure to hypertonic sucrose (i.e. maximum injury is induced following slow as compared to rapid cooling). It has been suggested that these differences are simply a consequence of the kinetics of sensitization observed in these solutions (Morris and Farrant, 1973). For example, the longer the time of exposure to sucrose either isothermally before rapid cooling or during slow rates of cooling, the more cold shock haemolysis occurs. Although there is qualitative support for this relationship, this hypothesis has been criticized because of the lack of quantitative support (Takahashi and Williams, 1983).

Final temperature attained

When erythrocytes are exposed to hypertonic sodium chloride (1.8 Osm kg^{-1}) at 37°C no haemolysis is induced upon rapid cooling to temperatures above 12.5°C; cooling the cells from 37°C to 10°C induces lysis, which is increased at lower temperatures. In addition, at any final temperature, haemolysis is further increased as the period of isothermal incubation is extended (Takahashi and Williams, 1983).

Protective compounds

The effects of a large variety of compounds on cold shock haemolysis have been reported (Morris, 1975; Green and Jung, 1977; Jung and Green, 1978; Dubbleman *et al.*, 1979). Some additives which decrease the extent of lysis are

Table 3.4 Secondary additives which reduce cold shock haemolysis induced by hypertonic solutions

Hypertonic solution	Secondary additive	Reference
NaCl (1.2 M)	Egg yolk lecithin	Lovelock, 1955
NaCl (1.2 M)	Glycerol (5–30% w/v)	Morris, 1975
	Dimethylsulphoxide (5–30% w/v)	Morris, 1975
NaCl (1.37 M), KCl (1.03 M)	Amphotericin B	Jung & Green, 1978
KCl (1.03 M)	Valinomycin	Jung & Green, 1978
NaCl (1.0 M), Sucrose (0.86 M)	Hexanol (15 mM)	Dubbleman *et al.*, 1979
	Chlorpromazine (0.12 mM)	
	Ethanol (2.5 M)	

listed in Table 3.4. The effects of ionophores are particularly significant. Amphotericin B protects cells from cold shock lysis induced by hypertonic solutions of sodium and potassium chloride but has no effect on haemolysis induced by sodium sulphate, mannitol or sucrose. Amphotericin B induces membrane channels which are permeable to sodium and potassium ions but are virtually impermeable to sulphate, mannitol or sucrose. Valinomycin inhibits cold shock haemolysis induced by potassium chloride but has no effect on haemolysis in sodium chloride or sucrose. This ionophore specifically increases potassium permeability of erythrocytes; its effect on sodium permeability is negligible by comparison.

Cold shock haemolysis is also modified by the pH of the hypertonic solution. Using hypertonic sucrose, haemolysis following rapid cooling was minimal at pH 5 and increased with increasing pH to a maximum at pH 8–8.5. The effect of modifying pH is more complex in hypertonic sodium chloride solutions, a maximum sensitivity to cooling being observed at pH 5.7, whereas at both higher and lower pH values haemolysis following cooling is reduced.

Other factors

The physiological state of the erythrocytes also determines the response to cold shock haemolysis. Using freshly drawn erythrocytes stratified according to *in vivo* age, it has been demonstrated that old cells are more susceptible to cold shock than younger cells (Green and Jung, 1977). Depletion of ATP by incubating erythrocytes for 24 hours at 37°C also increases the resistance to cold shock haemolysis (Dubbleman *et al.*, 1979).

Morphology of cold shock haemolysis

Following suspension of erythrocytes in hypertonic sodium chloride the normal biconcave disc is converted to a flattened or irregular discocyte. Upon cooling of such discocytes to below 10°C on a cryomicroscope (see McGrath, this volume) the flattened discs, about 8 μm diameter, are observed to change rapidly (1–2 seconds) into 5 μm diameter spheres with an accompanying loss of some 40 per cent of membrane surface area and release of haemoglobin (Williams and Takahashi, 1984).

Biochemical alterations in cold shock haemolysis

(i) *Loss of membrane selective permeability* By definition cold shock haemolysis demonstrates breakdown of the fundamental properties of the erythrocyte membrane. Further evidence of a loss of membrane selective permeability has been obtained following cold shock in hypertonic sodium chloride or sucrose where a loss of potassium and a gain of sodium, water and glucose on cooling to 0°C has been recorded (Daw *et al.*, 1973).

(ii) *Loss of membrane lipids* Lipids are present in cell-free supernatants following exposure of erythrocytes to hypertonic solutions (Lovelock, 1955; Souzu *et al.*, 1971; Araki *et al.*, 1982). It has been suggested that a differential loss of membrane phospholipid from intact erythrocytes, compared to cholesterol, is the basis of the sensitization to cold shock (Lovelock, 1955). However, this supposition ignores the small amount of haemolysis that occurs on exposure to hypertonic solutions and the correlation between the loss of membrane lipid and cellular lysis at a constant temperature (Morris, 1975; Green and Jung, 1977). The lipids released are probably derived from erythrocyte ghosts and membrane fragments produced by haemolysis rather than from intact cells.

Mechanisms of injury in cold shock haemolysis

Unlike most other cell types, erythrocytes in isotonic solutions are not sensitive to a reduction in temperature. For cold shock haemolysis to occur, cellular shrinkage is essential at all stages before and during cooling. Erythrocytes exposed to a hypertonic solution at 37°C and then resuspended into an isotonic solution at 37°C are not susceptible to a subsequent reduction in temperature. Any changes occurring in response to hypertonic solutions which sensitize the cells to the stress of rapid cooling are fully reversible. In addition, there is hysteresis in the system. If cells are exposed to a hypertonic solution at 0°C and then rewarmed to 37°C before rapid cooling to 0°C, haemolysis is reduced in comparison to addition of cells to hypertonic medium at 37°C (Dubbleman *et al.*, 1979).

The extent of cold shock haemolysis is directly proportional to the reduction in cell volume before cooling rather than to the osmolality of the solution (Takahashi and Williams, 1979). Additives which are freely permeable to erythrocytes, such as glycerol and dimethyl sulphoxide (DMSO) do not sensitize erythrocytes to the stress of rapid cooling. The storage of cells in hypertonic solutions can modify the sensitivity to cold shock. This is dependent on the composition of the hypertonic solution (Fig. 3.10) and this may be related to the effects of different hypertonic solutions on erythrocyte volume. For example, prolonged exposure to hypertonic sodium chloride would be expected to result in a loss of membrane selective permeability, after which cell volume would increase and the sensitivity to cold shock would decrease.

In contrast, in hypertonic sucrose solutions potassium would leak from the erythrocyte, the cell volume would decrease and the sensitivity to rapid cooling would increase with increasing exposure. With both sodium chloride and sucrose at osmolalities above 2 Osm kg^{-1} reduction in cold shock lysis is observed, which again may be related to loss of membrane selective permeability and subsequent increase in cell volume. In addition, ionophores reduce hypertonic induced cold shock lysis by dissipating ionic gradients (see above). There is one reported exception to these observations, namely cold shock lysis that occurs following valinomycin-induced shrinkage in isotonic sodium chloride (Jung and Green, 1978).

The processes by which a cell resistant to the stress of rapid cooling is converted, by cellular shrinkage, to a cell sensitive to cold shock are the key steps for an understanding of cold shock haemolysis. The membrane of the human erythrocyte has been extensively studied, and the extrinsic protein spectrin in association with actin, comprises a cytoskeletal network (Shotton, 1984). This cytoskeleton is closely associated with the lipid bilayer and reduces the lateral mobility of lipids and proteins within the membrane, for example, the lateral diffusion coefficient of the red cell intrinsic protein, band 3, at 37 °C is two orders of magnitude lower than that of rhodopsin in retinal rod outer segments (Pringle and Chapman, 1981).

Erythrocytes in isotonic solutions do not exhibit thermotropism on cooling to 0 °C as detected by differential scanning calorimetry (DSC) (Ladbrooke *et al.*, 1968) or freeze fracture electron microscopy; this may be a consequence of the high cholesterol:phospholipid ratio of erythrocyte membranes (see Morris and Clarke, this volume). Alternatively, physical association with the cytoskeleton may modify lipid thermotropic behaviour. Phase separations have been demonstrated to occur following exposure to hypertonic solutions at low temperatures (Araki *et al.*, 1982). As a working hypothesis it has been suggested that cellular shrinkage either directly disrupts the cytoskeleton:membrane interactions or renders them susceptible to cooling (Jung and Green, 1978; Green *et al.*, 1981). With the removal of this physical restriction the membrane may be free to undergo thermotropic events during cooling and the physical processes that occur during classic cold shock (see above) may then account for loss of membrane permeability. However, there is, as yet no direct evidence for the destabilization of the erythrocyte cytoskeleton during cellular shrinkage.

References

Anraku, Y. (1967). The reduction and restoration of galactose transport in osmotically shocked cells of *Escherichia coli. Journal of Biological Chemistry* **242**, 793–900.

Anraku, Y. and Heppel, L.A. (1967). On the nature of the changes induced in *Escherichia coli* by cold shock. *Journal of Biological Chemistry* **242**, 2561–9.

Araki, T., Roelofsen, B., op den Kamp, J.A.F. and van Deenen, L.L.M. (1982). Temperature-dependent vesiculation of human erythrocytes caused by hypertonic salt: A phenomenon involving lipid separation. *Cryobiology* **19**, 353-61.

Black, S.G. and Dixon, G.S. (1981). A.C. calorimetry of dimyristoyl phosphatidylcholine multilayers; Hysteresis and annealing near the gel to liquid-crystal temperature. *Biochemistry* **20**, 6740–44.

Bosl, A. and Birck, A. (1981). Ribosomal mutation in *Escherichia coli* affecting membrane stability. *Molecular and General Genetics* **182**, 358–60.

Calcott, P.H. and Rose, A.H. (1982). Freeze-thaw and cold-resistance of *Saccharomyces cerevisiae* as effected by plasma membrane lipid composition. *Journal of General Microbiology* **128**, 549–56.

Chapman, D., Gomez-Fernandez, J.P. and Goni, F.M. (1979). Intrinsic protein-lipid interactions: Physical and biochemical evidence. *FEBS Letters* **98**, 211–28.

Daw, A., Farrant, J. and Morris, G.J. (1973). Membrane leakage of solutes after thermal shock or freezing. *Cryobiology* **10**, 126–33.

Dubbleman, T.M., De Bruijne, A.W., Christianne, K. and Van Stevenick, J. (1979) Hypertonic cryohemolysis in human red blood cells. *Journal of Membrane Biology* **50**, 225–240.

Farrant, J. and Woolgar, A.E. (1972). Human red cells under hypertonic conditions; A model system for investigating freezing damage. 1. Sodium chloride. *Cryobiology* **9**, 9–15.

Farrell, J. and Rose, A.H. (1967). Temperature effects on micro-organisms. In *Thermobiology*, pp. 142–69. Edited by Rose, A.H. Academic Press, London.

Forrest, H.S., van Baalen, C. and Myers, J. (1957). Occurrence of pteridines in a blue-green alga. *Science* **125**, 699–700.

Giese, A.C. (1973). *Blepharisma: The biology of a light-sensitive protozoan*, pp. 163–4, Stanford University Press, Stanford, Connecticut.

Gilbert, P., Dickinson, N.A. and Brown, M.R.W. (1979). Interrelation of DNA replication, specific growth rate and growth temperature on the sensitivity of *Escherichia coli* to cold shock. *Journal of General Microbiology* **115**, 89–94.

Green, F.A. and Jung, C.J. (1977). Cold induced haemolysis in a hypertonic mileu. *Journal of Membrane Biology* **33**, 249–63.

Green, F.A., Jung, C.J., Cuppoletti, J. and Owens, N. (1981). Hypertonic cryo-haemolysis and the cytoskeleton system. *Biochemica et Biophysica Acta* **648**, 225–30.

Green, J.A. (1978). The effects of nutrient limitation and growth rate in the chemostat on the sensitivity of *Vibro cholerae* to cold shock. *FEMS Microbiology Letters* **4**, 217–19.

Heast, C.W.M., de Gier, J., van Es, G.A., Verkleij, A.J. and van Deenan, L.L.M. (1972). Fragility of the permeability barrier of *Escherichia coli. Biochemica et Biophysica Acta* **288**, 43–53.

Heppel, L.A. (1967). Selective release of enzymes from bacteria. *Science* **156**, 1451–5.

Hui, S.W., Parsons, D.F. and Cowden, M. (1974). Electron diffraction of wet phospholipid bilayers. *Proceedings of the National Academy of Sciences* **71**, 5068–72.

Jansz, E.R. and MacLean, F.I. (1973). The effect of cold shock on the blue-green alga *Anacystis nidulans. Canadian Journal of Microbiology* **19** 381–7.

Jeacocke, R.E. (1982). Calcium efflux during cold-induced contraction of mammalian striated muscle fibres. *Biochemica et Biophysica Acta* **682**, 238–44.

Jung, C.J. and Green, F.A. (1978). Hypertonic haemolysis: Ionophore and pH effects. *Journal of Membrane Biology* **39**, 273–84.

Kawahara, H. (1978). Production of triploid and gynogenetic diploid *Xenopus* by cold treatent. *Development, Growth and Differentiation* **20**, 227–36.

Kawamura, N. (1979). Cytological studies on mosaic silkworms induced by low temperature treatment. *Chromosoma* **74**, 179–88.

Kenward, M.A. and Brown, M.R.W. (1978). Relation between growth rate and sensitivity to cold shock of *Pseudomonas aeruginosa. FEBS Microbiology Letters* **3**, 17–19.

Knight, S.C., O'Brien, J. and Farrant, J. (1976). Cryopreservation of granulocytes. *Inserm* **62**, 139–50.

Ladbrooke, B.D., Williams, R.H. and Chapman, D. (1968). Studies on lecithin cholesterol water interactions by differential scanning calorimetry and X-ray diffraction. *Biochemica et Biophysica Acta* **150**, 333–40.

Lemoine, M.L. and Smith, L.T. (1980). Polyploidy induced in brook trout by cold shock. *Transactions of the American Fish Society* **109**, 626–31.

Levitt, J. (1980). Chilling, freezing and high temperature stress. In *Responses of Plants to Environmental Stress, 2nd edition Volume 1*, pp. 23–64. Academic Press, New York.

Lovelock, J.E. (1953b). The haemolysis of human red blood cells by freezing and thawing. *Biochemica et Biophysica Acta* **10**, 414–26.

Lovelock, J.E. (1955). Haemolysis by thermal shock. *British Journal of Haemotology* **1**, 117–29.

MacKelvie, R.H., Gronland, A.F. and Campbell, J.J.R (1968). Influence of cold shock on the endogenous metabolism of *Pseudomonas aeruginosa. Canadian Journal of Microbiology* **14**, 633–8.

McLellan, M.R., Morris, G.J., Coulson, G.E., James, E.R. and Kalinina, L.V. (1984). Role of cytoplasmic proteins in cold shock injury in *Amoeba. Cryobiology* **21**, 44–59.

Martin, C.E., Hiramitsuj, K., Nozawa, Y., Skriver, L. and Thompson, G.A. (1976). Molecular control of membrane properties during temperature acclimation. Fatty acid desaturase regulation of membrane fluidity in acclimating *Tetrahymena* cells. *Biochemistry* **15** 5218–27.

Melchior, P.L., Bruggeman, E.P. and Stein, J.M. (1982). The physical state of quick frozen membranes and lipids. *Biochemica et Biophysica Acta* **690** 81–8.

Meynell, G.G. (1958). The effect of sudden chilling on *Escherichia coli.*

Journal of General Microbiology **19**, 380–9.
Morris, G.J. (1975). Lipid loss and haemolysis by thermal shock: Lack of correlation. *Cryobiology* **12**, 192–201.
Morris, G.J. and Farrant, J. (1973). Effects of cooling rate on thermal shock haemolysis. *Cryobiology* **10**, 119–125.
Morris, G.J. and Watson, P.F. (1984). Cold shock bibliography. *Cryoletters* **5**, 352–72.
Morris, G.J., Coulson, G.E. and Clarke, A. (1984). Cold shock injury in *Tetrahymena pyriformis. Cryobiology* **21**, 664–71.
Morris, G.J., Coulson, G.E., Meyer, M.A., McLellan, M.R., Fuller, B.J., Grout, B.W.W., Pritchard, H.W. and Knight, S.C. (1983). Cold shock: A widespread cellular reaction. *CryoLetters* **4**, 179–92.
Nakajama, H., Goto, M., Ohki, K., Hitsui, T. and Nozawa, Y. (1983). An X-ray diffraction study of phase transition temperatures of various membranes isolated from *Tetrahymena pyriformis* cells grown at different temperatures. *Biochemica et Biophysica Acta* **730** 17–24.
Neu, H.C. and Heppel, L.A. (1965). The release of enzymes from *Escherichia coli* by osmotic shock and during the formation of spheroplasts. *Journal of Biological Chemistry* **240**, 3685–92.
Novotny, C.P. and Englesberg, E. (1965). The L-arabinose permease system in *Escherichia coli* B/r. *Biochemica et Biophysica Acta* **117**, 217–30.
Nozawa, Y. (1980). Modification of membrane lipid composition and membrane fluidity in *Tetrahymena*. In *Membrane Fluidity: Biophysical Techniques and Cellular Regulation*, pp. 399–418. Edited by Kates, M. and Kuksis, A. The Humana Press, Clifton, New York.
Ono, T.A. and Murata, N. (1981). Chilling susceptibility of the blue-green alga *Anacystis nidulans* 1. Effect of growth temperature. *Plant Physiology* **67**, 176–81.
Ono, T.A. and Murata, N. (1982). Chilling susceptibility of the blue-green alga *Anacystis nidulans* 3. Lipid phase of the cytoplasmic membrane. *Plant Physiology* **69**, 125–9.
Pardee, A.B., Prestidge, L.S., Whipple, M.B. and Dreyfuss, J. (1966). A binding site for sulphate and its relation to sulphate transport into *Salmonella typhimurium. Journal of Biological Chemistry* **241**, 3962–9.
Paton, S.C. McMurchie, E.J., May, B.K. and Elliot, W.H. (1978a). Effect of growth temperature on membrane fatty acid composition and susceptibility to cold shock of *Bacillus amyloliquefaciens. Journal of Bacteriology* **135**, 754–9.
Paton, S.C., May, B.K. and Elliot, W.H. (1978b). Membrane phospholipid asymmetry in *Bacillus amyloliquefaciens. Journal of Bacteriology* **135**, 393–401.
Piperno, J.R. and Oxender, D.L. (1966). Amino acid binding protein released from *Escherichia coli* by osmotic shock. *Journal of Biological Chemistry* **241**, 5732–94.
Pringle, M.J. and Chapman, D. (1981). Biomembrane structure and effects of temperature. In *Effects of Low Temperatures on Biological Membranes*, pp. 21–37. Edited by Morris, G.J. and Clarke, A.

Academic Press, London.

Purdom, C.E. (1972). Induced polyploidy in plaice (*Pleuroncetes platessa*) and its hybrid with the flounder. (*Platichthys flesus*). *Heredity* **29**, 11–24.

Quinn, P.G. and White, I.G. (1966). The effect of cold shock and deep freezing on the concentration of major cations in spermatozoa. *Journal of Reproduction and Fertility* **12**, 263–70.

Quinn, P.J. and White, I.G. (1968). The effect of pH, cations and protective agents on the susceptibility of ram spermatozoa to cold shock. *Experimental Cell Research* **49**, 31–9.

Ring, K. (1965). The effect of low temperatures on permeability in *Streptomyces hydrogenans*. *Biochemical and Biophysical Research Communications* **19**, 576–81.

Ryan, R.W., Gourlie, M.P. and Tilton, R.C. (1979). Release of rhodanase from *Pseudomonas aeroginosa* by cold shock and its localization within the cell. *Canadian Journal of Microbiology* **25**, 340–51.

Sakai, T. and Kurihara, S. (1974). A study of rapid cooling contracture from the viewpoint of excitation–contraction coupling. *Jikei Medical Journal* **21**, 47–88.

Shotton, D.M. (1984). The proteins of the erythrocyte membrane. In *The Electron Microscopy of Proteins*, Vol. 3. Edited by Harris, J.R. Academic Press, London.

Sillerud, L.O. and Barnett, R.E. (1982). Lack of transbilayer coupling in phase transitions of phosphatidylcholine vesicles. *Biochemistry* **21**, 1756–60.

Siva, V., Rao, K., Brand, J.J. and Myers, J. (1977). Cold-shock syndrome in *Anacystis nidulans*. *Plant Physiology* **59**, 965–9.

Smeaton, J.R. and Elliot, W.H. (1967). Selective release of ribonuclease from *Bacillus subtilis* cells by cold-shock treatment. *Biochemical and Biophysical Research Communications* **26**, 75–81.

Souzou, H., Nei, T. and Sato, T. (1971). Haemolysis and release of phospholipids of human erythrocytes caused by freeze-thawing or exposure to hypertonic salt solutions. *Low Temperature Science, Series B* **29**, 75–82.

Strange, R.E. (1964). Effect of magnesium on permeability control in chilled bacteria. *Nature* **203**, 1304–5 .

Strange, R.E. (1976). *Microbial response to mild stress. Patterns in Progress* **6**, pp. 44–61. Meadowfield Press, Shildon, UK.

Strange, R.E. and Dark, F.A. (1962). Effect of chilling on *Aerobacter aerogenes* in aqueous suspension. *Journal of General Microbiology* **29**, 719–30.

Strange, R.E. and Postgate, J.R. (1964). Penetration of substances into cold shocked bacteria. *Journal of General Microbiology* **36**, 393–403.

Takahashi, T. (1983). Calmodulin antagonists induce isotonic thermal shock haemolysis in human erythrocytes. *Cryobiology* **20**, 726.

Takahashi, T. and Williams, R.J. (1979). Thermal shock in red cells. *Cryobiology* **16**, 588–9.

Takahashi, T. and Williams, R.J. (1983). Thermal shock haemolysis in

human red cells. 1. The effects of temperature, time and osmotic stress. *Cryobiology* **20**, 507–20.
Takahashi, T., Hammett, M.F. and Cho, M.S. (1981). Mechanisms of extracellular freezing injury at high sub-zero temperatures in human polymorphonuclear leukocytes. *Cryobiology* **18**, 622–3.
Traci, P.A. and Duncan, C.L. (1974). Cold shock lethality and injury in *Clostridium perfringens. Applied Microbiology* **28**, 815–21.
Valenti, P.J. (1975). Induced polyploidy in *Tilapia anrea* (Steindachner) by means of temperature shock treatment. *Journal of Fish Biology* **7**, 519–28.
Vojnikov, V.K., Luzova, G.B. and Kerzun, A.M. (1983). The composition of free-fatty acids and mitochondrial activity in seedlings of winter cereals under cold shock. *Planta* **158**, 194–8.
Watson, P.F. (1981). The effects of cold shock on sperm cell membranes. In *The Effects of Low Temperatures on Biological Membranes* pp. 189–218. Edited by Morris, G.J. and Clarke, A. Academic Press, London.
Wiley, W.R. (1970). Tryptophan transport in *Neurospora crassa*: a tryptophan-binding protein released by cold osmotic shock. *Journal of Bacteriology* **103**, 656–62.
Williams, R.J. and Takahashi, T. (1984). Erythrocyte metastability: Microscopic observation of thermal shock haemolysis. *CryoLetters* **5**, 111–16.
Wilmut, I., Polge, C. and Rowson, L.E.A. (1975). The effect on cow embryos of cooling to 20, 0 and −196°C. *Journal of Reproduction and Fertility* **45**, 409–11.
Wilson, J.M. and McMurdo, A.C. (1981). Chilling injury in plants. In *The Effects of Low Temperatures on Biological Membranes*, pp. 145–72. Edited by Morris, G.J. and Clarke, A. Academic Press, London.
Wilson, O.H. and Holden, J.T. (1969). Arginine transport and metabolism in osmotically shocked and unshocked cells of *Escherichia coli* W. *Journal of Biological Chemistry* **244**, 2737–42.
Woolgar, A.E. and Morris, G.J. (1973). Some combined effects of hypertonic solutions and changes in temperatures on posthypertonic haemolysis of human red blood cells. *Cryobiology* **16**, 82–6.
Zade-Oppen, A.M.M. (1968). Posthypertonic haemolysis in sodium chloride systems. *Acta Physiologica Scandinavica* **73**, 341–64.
Zocchi, G. and Hanson, J.B. (1982). Calcium influx into corn roots as a result of cold shock. *Plant Physiology* **70**, 318–19.
Zsoldos, F. and Karvaly, B. (1979). Cold shock injury and its relation to ion transport by roots. In *Low Temperature Stress in Crop Plants: The Role of the Membranes*, pp. 123–39. Edited by Lyons, J.M., Graham, D. and Raison, J.K. Academic Press, New York.

4

Freezing and cellular organization

B.W.W. Grout
Department of Biological Sciences
Plymouth Polytechnic
G.J. Morris
Cell Systems Limited
Cambridge

Introduction

During freezing and thawing cells are exposed to a wide range of stresses and the nature and extent of cellular response to such stresses determines whether or not injury is sustained. During freezing the stresses will result from:

(1) the exposure of cellular components to low temperatures;
(2) the mechanical effects of extracellular ice crystals at cell surfaces. In tissues, cellular interconnections may be of particular significance;

(3) alterations in the physical properties of solutions outside the cell (Taylor, this volume), including the concentration of solutes which results when a proportion of the extracellular water undergoes the transition to ice. When the residual solution becomes hypertonic to those of the intracellular compartment, an osmotic stress will be imposed on the cell. Stresses due to changes in concentration of soluble ionic and molecular species will also develop. In addition, other physical properties such as pH and viscosity will change, often in a non-linear manner.

Further stress will result if intracellular freezing occurs, with the delicate structures of the cytoplasm experiencing stresses comparable to those outlined above. If the water within individual organelles does not nucleate at the same time as the cytoplasm, they will be stressed in the same way as a non-frozen cell entrapped in partially frozen, extracellular medium.

Cellular responses to freezing are the direct result of physical laws and are not under any sort of biological control. The immediate responses of a cell, or organelle, are governed by physical and chemical properties of both the biological and physical systems involved, and they will tend to establish a new set of equilibrium conditions, those of normal physiology having been disrupted. The increase in freezing tolerance following exposure of some plants, insects and microorganisms to low, non-freezing, temperatures (cold hardening – see below) may bring about changes in cellular components that subsequently effect the injuries resulting from response to stress, but it cannot affect the nature of the stresses involved.

A further series of stresses are imposed upon the cells by rewarming and thawing; particularly as the ice melts the biological systems must respond again to radical change of temperature, solute concentration and water availability.

If the essential structural and metabolic components of the cell remain unaltered, despite their inevitable responses to freezing and thawing, they may be described as intrinsically 'freeze-resistant'. Any alteration of cellular components that is detrimental to function and results directly from freezing can be described as primary freezing injury, and may be either repairable or permanent. For practical reasons injury is usually assessed immediately following thawing and so any observed injury described as primary may actually be a result of either freezing or thawing stress. Only investigations undertaken in the frozen state (e.g. freeze-fracture, freeze-substitution, frozen-tissue histochemistry) can make the necessary distinctions. Primary injury may lead to secondary injury (e.g. a disruption of lysosomal membrane integrity following freezing and thawing is a primary injury; the damage caused by leakage of lytic enzyme is a secondary injury). Persistent injury that is expressed in cells following freezing and thawing results, therefore, from interactions of irreversible primary and secondary injuries.

This discussion will focus on stresses imposed on cells by the presence of ice, and the responses to such stresses. The additional stress of reduced temperature *per se* must be acknowledged as a perpetual, contributory factor during freezing (see Morris, Morris and Clarke, this volume) but will not be dealt with specifically here. For clarity, the arguments will centre on single cell systems, which in many ways reflects the attitudes and literature of laboratory-based cryobiology. It must also be emphasized that this is a general discussion and does not concentrate on any one cell-type for specific detail. The plasmalemma is taken to be the limiting barrier of the cell defining the protoplast. Unless otherwise stated, in cells with a wall, the cell wall will be considered to be inert and extracellular, and cells from which the wall has been removed will be referred to as isolated protoplasts.

Freezing of cells in suspension

Following the nucleation of ice in dilute aqueous solutions many complex changes occur in the physical properties of the system (Taylor, this volume). All these changes may modify cell structure and function, and since the studies of Lovelock (1953) particular emphasis has been placed on the effects of the increase in solute concentration that accompanies freezing. When freezing is initiated in an aqueous solution only a proportion of the water that contributes to the solution undergoes the transition to ice. The effective molarity of the residual solution, when in equilibrium with the ice, is determined by the relationship:

$$\Delta T°C/1.86 \text{ (osmoles litre}^{-1}\text{)}$$

where $\Delta T°C$ is the depression of freezing point.

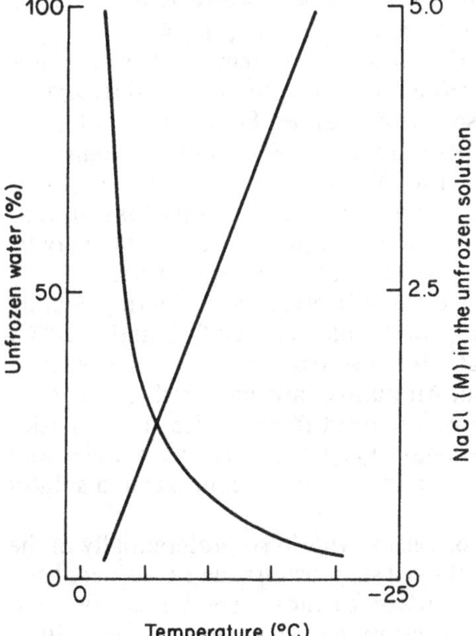

Fig. 4.1 Percentage of unfrozen water and the molarity of solute in unfrozen solutions of sodium chloride at different sub-zero temperatures. The initial molarity of sodium chloride was 0 15 M. (Redrawn from Lovelock, 1954.)

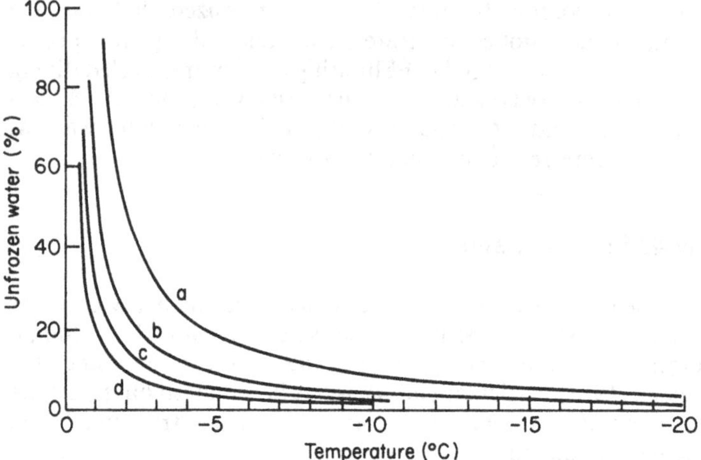

Fig. 4.2 The relationship between the initial osmolality before freezing and the equilibrium percentage of unfrozen water at various sub-zero temperatures. The letters on the graphs refer to the osmolality before freezing (a) 0.50, (b) 0.25, (c) 0.15, (d) 0.10 Osmol Kg^{-1}. (Redrawn from Morris and McGrath, 1981.)

As temperature decreases, osmolality is seen to increase and the amount of water that remains liquid is seen to decrease (Fig. 4.1). A corollary of this is that the proportion of unfrozen solution at any sub-zero temperature is directly related to the initial osmolality before freezing (Fig. 4.2).

The residual liquid volume will reduce in size as the temperature decreases and ice will coexist with the concentrating solution until the eutectic temperature is reached, at which point the solutes crystallize. For aqueous solutions of sodium chloride the eutectic temperature is −21.8°C, and at all temperatures above this, ice and concentrated solutions of sodium chloride coexist. At temperatures below −21.8°C the system solidifies. For aqueous solutions containing more than one dissolved solute multiple eutectic points may be observed, with some solutes coming out of solution before others.

Ice formation will also have an effect on dissolved gases. During cooling the solubility of gases increases and will double between 0°C and −20°C. However, when a portion of the water is converted to ice, the gases are then concentrated in the residual solution. Air bubbles are nucleated ahead of the advancing ice/water interface and are excluded from the ice crystal lattice (Hobbs, 1974). Therefore, a three phase system of ice, gas bubbles and aqueous solution saturated with gases and containing concentrated solutes will exist.

When ice forms in a suspension of cells it will do so preferentially in the extracellular medium. The medium of a cell suspension can be considered as a single, large compartment when compared to those represented by each, individual cell and the probability of nucleation in an aqueous system will be related to the compartment size (Hobbs, 1974). Once nucleation occurs, ice will propagate throughout the extracellular solution to re-establish thermodynamic stability. At this stage, the suspended cells are excluded from the ice

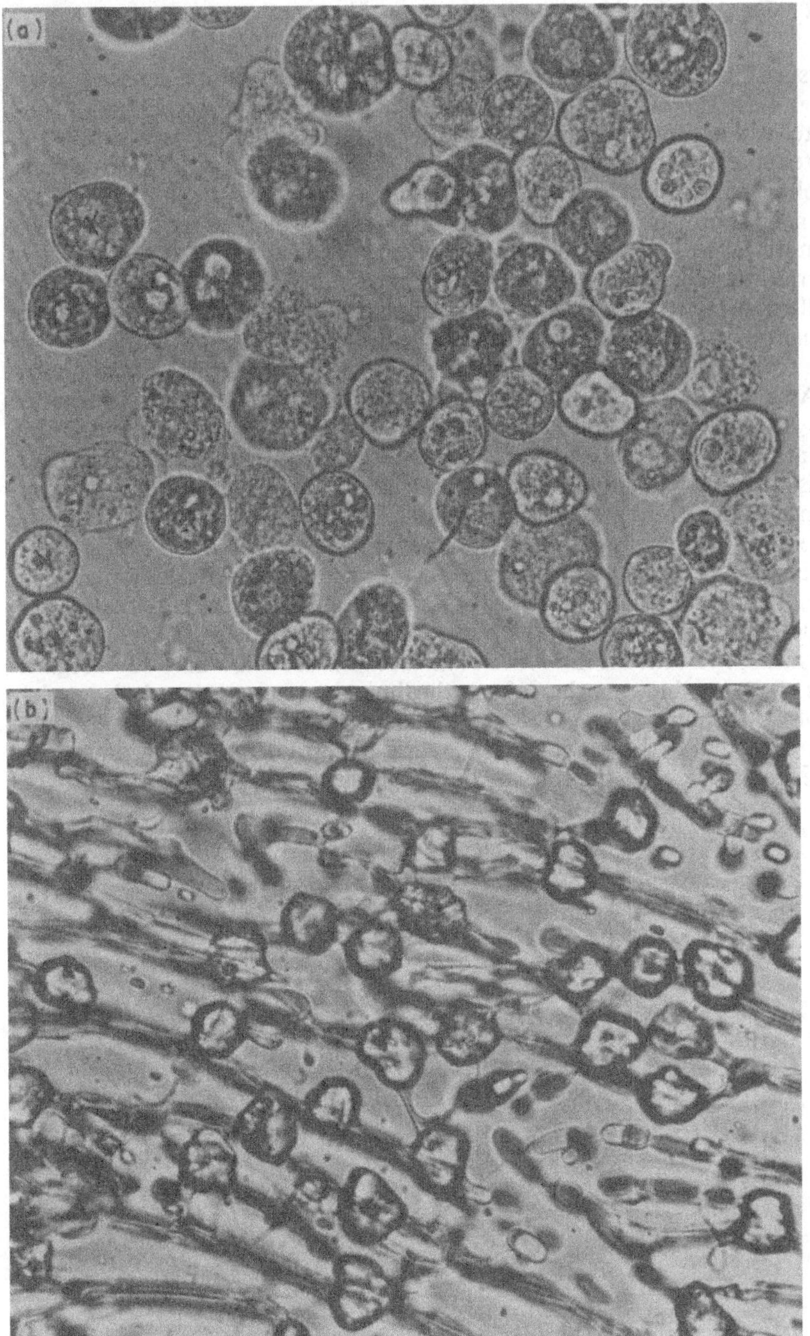

Fig. 4.3 (a) A population of isolated rat hepatocytes suspended in physiological medium at room temperature. (× 638). (b) Cells from the same population, following freezing to − 10°C on a cryomicroscope stage. The cells are entrapped between ice crystals and shrinkage is obvious. (× 615).

crystals and are confined in channels of residual solution between them (Fig. 4.3). This residual solution will be hypertonic to the cell (Fig. 4.1) and water will move from the protoplast in response to this imposed gradient of water potential, resulting in a reduction of cell volume (Fig. 4.3). Progressive cooling will therefore result in extensive cellular dehydration. The semi-permeable properties of the plasmalemma must be maintained to sustain this osmotic activity, and such a functional membrane appears to be an effective barrier to the propagation of ice crystals into the cell (Mazur, 1977). Following slow cooling to very low temperatures the residual cell water will either nucleate, or, because of the concentrated intracellular solutes, a glass will form (see Taylor, this volume).

An alternative way in which thermodynamic equilibrium could be attained is by the bulk freezing of the intracellular solutions. The equilibrium freezing point of many cell cytoplasms is between −0.5°C and −2°C, although most cells are capable of undercooling by at least 10°C under appropriate circumstances. The probability of intracellular ice formation depends upon the extent of undercooling of the cytoplasm and rate of cooling (McGrath, this volume). At relatively rapid rates of cooling there is insufficient time for the osmotic loss of water and the cytoplasm will remain dilute and increasingly undercooled, raising the probability of intracellular nucleation. For isolated rye mesophyll cells the incidence of intracellular ice is low at −10°C regardless of the cooling rate; this reflects the limited extent of intracellular undercooling (Steponkus *et al.*, 1982). Between −10°C and −40°C a significant degree of undercooling may occur and the probability of intracellular freezing is related to the rate of cooling and temperature (Fig. 4.4).

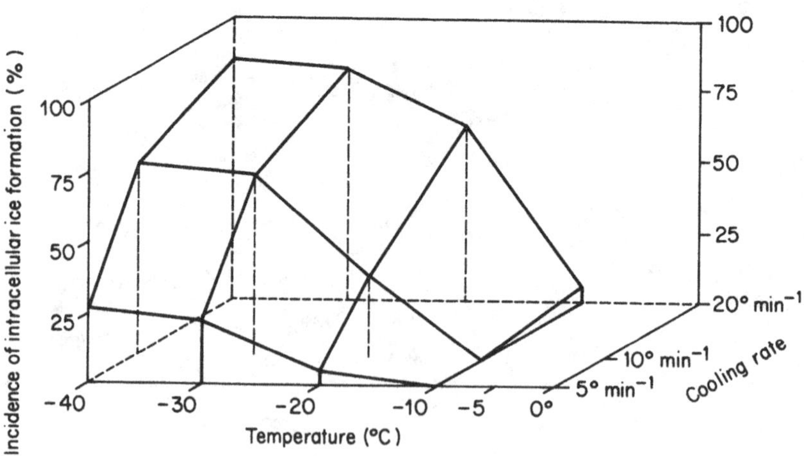

Fig. 4.4 The incidence of intracellular ice formation in isolated plant protoplasts as a function of rate of cooling. (Redrawn from Steponkus *et al.*, 1982.)

Freezing injury and viability

From the preceding discussion it is clear that cooling can impose either of two conditions of stress on cells. Firstly, during 'slow' cooling the major stresses are likely to result from extracellular freezing, whereas during 'rapid' cooling the stresses are likely to result from intracellular ice formation. This situation is recognized in the 'two-factor' hypothesis of freezing injury (Mazur *et al.*, 1972) which essentially ascribes cellular injury at sub-optimal cooling rates to the damaging effects of prolonged exposure to hypertonic solutions ('solution effects') whilst at super-optimal rates of cooling the nucleation of intracellular ice results in cell death. It must be stressed that 'rapid' and 'slow' rates of cooling have no absolute values and vary with cell-type.

Optimal rates of cooling that maintain cell survival have been observed for a range of cell-types (Fig. 4.5) which may be described as intrinsically freeze-hardy. Similarly, optimal rates of cooling may be found for cells that have been modified to resist freezing injury by either chemical cryoprotection or environmental hardening (see below). The rate of cooling at which optimum survival is observed is related to a number of factors including the permeability of the cell to water, hydraulic conductivity, cell surface area to volume ratio and the type and concentration of any additives (Mazur *et al.*, 1972; Leibo, 1981). A cell with a high permeability to water and a high surface area to volume ratio will lose water rapidly and so may tolerate a relatively rapid rate of cooling before the probability of intracellular nucleation becomes high. Conversely, a cell with a low permeability to water and a low surface area to volume ratio cannot lose water as rapidly, and is therefore likely to suffer intracellular freezing at lower rates of cooling. This means that the optimal cooling rate can vary considerably depending on the natural properties of the cells involved and how these may have been modified by laboratory or environmental treatments.

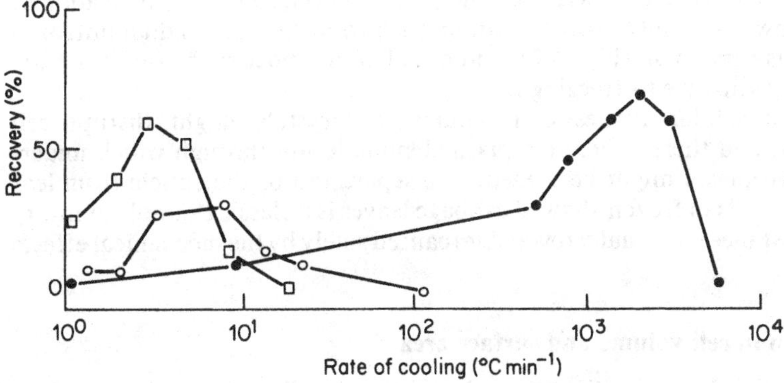

Fig. 4.5 The effect of cooling rate on the post-thaw survival of a range of cell-types, no cryoprotectants were added. Red blood cells (●), the yeast *Saccharomyces cerevisiae* (○), a unicellular green alga *Chlamydomonas nivalis* (□).

Table 4.1 The categories of stress associated with extracellular freezing of a cell suspension medium

(a) Reduction in temperature
(b) Mechanical effects of extracellular ice crystals
(c) Altered physical properties of the residual, extracellular solution (e.g. viscosity, pH)
(d) Generation of gas bubbles and electrical fields at the ice–solution interface
(e) Concentration of solutes in the extracellular solution and the related decrease in the water potential
(f) Volume reduction of the protoplast and possible surface area reduction of the plasmalemma
(g) Increase in concentration of cytoplasmic solutes
(h) Effects of an increasingly concentrated cytoplasm on the contained organelles

Slow rates of cooling

The presence of extracellular ice will impose a series of stresses upon a cell; these are summarized in Table 4.1 and discussed below. It must be emphasized that during freezing and thawing a cell will be exposed to a sequence of potentially damaging stresses; not all of these will be of equal importance to each cell-type.

Mechanical effects of ice

There is little evidence for direct mechanical injury of cells in suspension by the ice surrounding them. However, lysis of human erythrocytes during slow freezing does appear to be more dependent on the fraction of water that remains unfrozen than on the ionic concentration in the unfrozen solution (Mazur *et al.*, 1981). The decrease in volume of the aqueous compartment increases the density of cells within it, and therefore increases the possibility of cellular interactions. A further indication of injury under such circumstances is given by the increase in haemolysis of human erythrocytes when frozen at high cell densities (Pegg, 1981). In addition, following nucleation of ice in a low osmolality growth medium the proportion of residual unfrozen solution is very small (Fig. 4.2), and cells frozen under such conditions are often very sensitive to freezing injury.

In multicellular tissues extracellular ice crystals might disrupt cell junctions, and thereby create a plasmalemma lesion through which undercooled cytoplasm might be seeded. The separation of the cuticle from leaf epidermal cells in frozen-thawed cabbage leaves is a classic example of injury (in this instance it is usually reversible) caused solely by the mechanical effects of ice.

Reduction in cell volume and surface area

Once freezing begins in the external medium, a cell in suspension will lose water as the result of the imposed gradient of water potential. At slow rates of cooling extensive cell shrinkage will occur; this can be readily observed by cryomicroscopy (McGrath, this volume) and electron microscopy (Skaer, this volume). The inevitable changes in cell volume, surface area and water content of the cell are likely to affect plasmalemma structure and function, and it is this membrane that has been widely investigated as a primary site of freezing injury. As a result of the many practical problems associated with the direct examination of freeze-induced dehydration many workers have examined osmotic shrinkage and rehydration at a constant temperature, often in the region of 20°C, as a system intended to simulate aspects of freezing injury.

An explanation of the injury sustained by erythrocytes has been attempted in terms of volume reduction and the damaging effects of solute movement from the high concentrations of the external medium into the cell (Lovelock, 1953; Meryman, 1968, 1971; Zade-Oppen, 1968). This theory, broadly described as the 'minimum-volume hypothesis', suggested that when the erythrocyte volume diminished to a certain value there was a resistance to further shrinkage (Meryman, 1968), associated with a hydrostatic pressure across the plasmalemma. This pressure was responsible for altered osmotic behaviour and allowed the influx of solute to intracellular compartments. Alterations in the properties of the plasmalemma in this system have been taken to include the loss of membrane components (Meryman *et al.*, 1977; Williams and Hope, 1981). When restored to isotonic conditions the increased solute content of these cells will result in an excessive water uptake sufficient to rupture the plasmalemma (i.e. haemolysis).

Exposure of erythrocytes to hypertonic solutions results in a flattening of the erythrocyte, allowing large volume changes to occur with only minor alterations in surface area. By contrast, protoplasts isolated from higher plant cells behave as osmometers and remain spherical within a wide range of external osmolalities implying relatively large changes in surface area

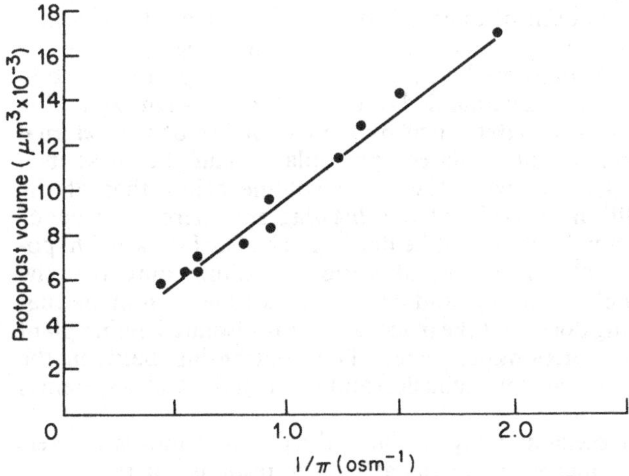

Fig. 4.6 Boyle van't Hoff plot of volume of isolated protoplasts from rye cells after a 10-minute exposure to a solution of $CaCl_2$ + NaCl. (Redrawn from Steponkus and Weist, 1979.)

(Fig. 4.6). Biological membranes are not significantly elastic, (i.e. they cannot increase or decrease in thickness - Kwok and Evans, 1981) and so large changes in surface area can occur only if material is added to or removed from the plasmalemma (Steponkus, 1984). From direct measurements of the mechanical properties of the plasmalemma of isolated plant protoplasts undergoing shrinkage and rehydration it has been argued that lateral intramembrane pressures of the magnitude invoked in the 'minimum-volume' hypothesis are unlikely (Wolfe and Steponkus, 1983a) and that the critical factor is a small reduction in membrane surface tension. As the surface area of the protoplast decreases the tension within the plasmalemma is reduced, and conversely the surface area may increase under large imposed tensions (Wolfe and Steponkus, 1981). This implies the existence of a reservoir of membrane material which can act as a source, and as a sink, for structural membrane components (Wolfe and Steponkus, 1981, 1983b). Microscopic observations have been invoked to demonstrate that such a membrane reservoir exists (Steponkus *et al.*, 1983).

Injury to isolated plant protoplasts in the presence of extracellular ice can therefore be described in the following terms. Upon exposure to hypertonic conditions the protoplast decreases in volume and tension within the plasmalemma is relaxed to zero. Membrane material is taken into a cytoplasmic reservoir until the protoplast regains a stable resting tension, which represents an equilibrium between plasmalemma and reservoir. When the newly-adjusted protoplast is exposed to the original isotonic conditions (e.g. during thawing), the protoplast will increase in volume, which is determined by the intracellular solute concentration. Membrane tension would increase altering the equilibrium between plasmalemma and reservoir such that material would tend to be reincorporated into the plasmalemma. If insufficient material is taken back into the membrane, lysis will result as the protoplast volume and hence surface area is increased. The avoidance of lysis will, therefore depend upon the amount of material that is deleted from the plasmalemma during shrinkage and retained in the reservoir to be available for reincorporation during expansion. The critical factor is therefore the incremental increase in surface area demanded of the protoplast when expanding after shrinkage. This will be determined by the osmolality of the external solution. This behaviour of isolated protoplasts can be described diagramatically (Fig. 4.7), showing lysis at a volume below that of the original isotonic conditions. It is clear from this diagram where a concept of an apparent 'minimum volume' might be derived. The 'surface area' hypothesis, however, does not require any alteration in solute content of the isolated protoplast to effect injury, and observations have been made that suggest that such loading does not take place, at least for isolated higher plant protoplasts (Weist and Steponkus, 1978). The relationship between the extent of contraction, subsequent expansion and survival for such a system is described in Fig. 4.8.

The structural and molecular composition of the plasmalemma will determine its mechanical properties and thus affect the response of the cell to hypertonic stress. This has been inferred in studies of protoplasts isolated from yeast (Hossack *et al.*, 1977) and mycoplasma (Razin *et al.*, 1966) where experimental manipulation of the lipid composition of the cellular

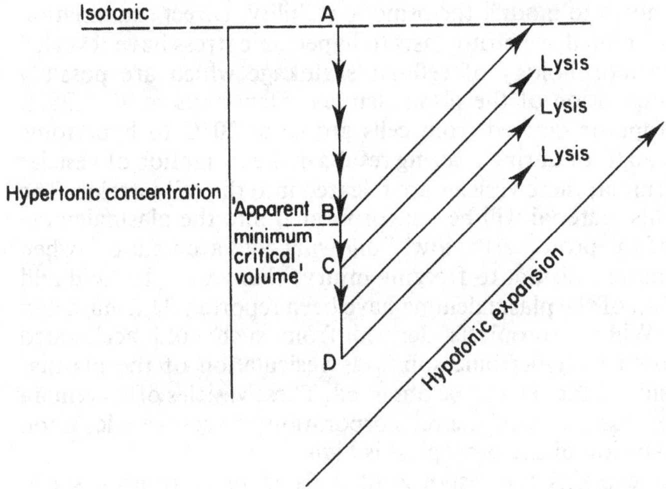

Fig. 4.7 A diagrammatic description of protoplast contraction and lysis. If a protoplast reduces in surface area by an amount represented by A to B, it is able to reincorporate sufficient material upon rehydration to avoid lysis. If, however, the reduction in surface area is as great as A to C, the necessary reincorporation is not possible and the protoplast will lyse. The volume that corresponds to the surface area at D could be interpreted as the apparent 'minimum volume'. (Redrawn from Steponkus and Weist, 1979.)

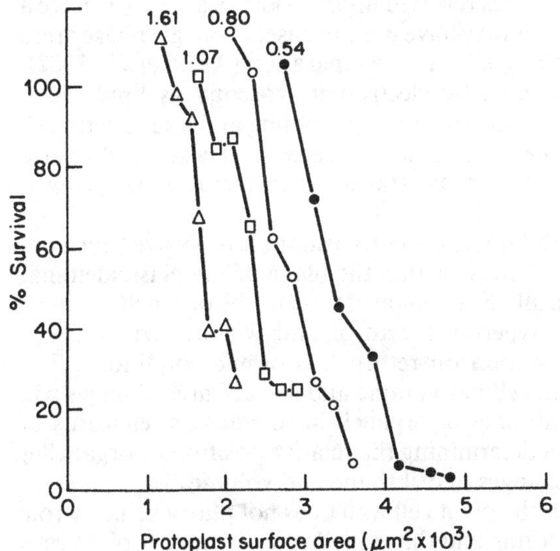

Fig. 4.8 Protoplast survival plotted in relation to expansion and contraction. The number above each curve indicates the osmolality to which the sample was exposed. Each point on the curve represents a subsequent expansion (achieved by dilution of the osmoticum) and the corresponding protoplast viability. (Reproduced from Steponkus and Weist, 1979.)

membranes was shown to modify the osmotic stability. Direct examination of the response of isolated rye protoplasts to hypertonic stress have revealed differences in the morphology of cellular shrinkage which are possibly related to the composition of the plasmalemma (Steponkus *et al.*, 1983). Exposure of protoplasts derived from cells grown at 20°C to hypertonic solutions either at 20°C or during freezing results in the formation of vesicles from the plasmalemma; these vesicles are released into the cell interior. The probability that this material will be reincorporated into the plasmalemma upon expansion of the protoplast is low. Following cold acclimation, when the cells become more resistant to freezing injury, changes in the lipid and protein composition of the plasmalemma have been reported (Unemura and Yoshida, 1984). When protoplasts derived from such cold acclimated material are exposed to hypertonic solutions vesiculation of the plasmalemma to the outside of the cell can be observed. These vesicles often remain attached, and the probability of the reincorporation of such vesicles upon subsequent re-expansion of the protoplast is high.

This hypothesis explains the response of isolated plant protoplasts to osmotic stress at both constant temperature and during freezing, but in other cell-types additional stresses may also be important. Specifically, at low temperatures cell membranes may undergo a lipid phase separation (Morris and Clarke, this volume) and the mechanics of the release of material from the plasmalemma under such conditions may be significantly different from those occurring at higher temperatures. For example, exposure of erythrocytes to hypertonic solutions of sodium chloride at 37°C results in the release of vesicles with a lipid and protein content similar to that of the red cell membrane. By contrast the vesicles released upon exposure at −10°C have a composition which suggests that they have been released from gel phase areas of a membrane which has undergone a phase separation (Araki *et al.*, 1982). These findings can be confirmed by electron microscopy as lipid phase separations are observed following slow rates of cooling and a subduction of plates of gel phase membrane occurs during freezing. These would form vesicles when warmed above their phase transition temperature (Fujikawa, 1981).

In the above discussion cellular responses to osmotic stress have been considered in terms of mechanical stress within the plane of the plasmalemma. However, during osmotic shrinkage of eukaryotic cells the organelles will be exposed to an increasingly hypertonic cytosol and will in turn undergo shrinkage followed by rehydration on return to isotonic conditions. The types of stress associated with cellular volume and surface area changes will be operating within the membranes of organelles. In addition, elements of the cytoskeleton important in determining the relative position of organelles may be destabilized by the changes in cell shape and volume.

It is generally assumed that the plant cell wall does not play a primary role in the response of a cell to freezing and thawing. Typically, higher plant cells exposed to hypertonic solutions at 20°C show a separation of the protoplast from the cell wall (plasmolysis); the cell wall retains its shape and dimensions. By contrast, cryomicroscopy of isolated rye cells at slow rates of cooling, demonstrates that the cell wall collapses to a greater or lesser extent

(Steponkus *et al.*, 1982). Unfortunately, there was no correlation between the spatial relation of the cell wall to the protoplast and survival or injury upon thawing in this study. An investigation of potato cells and their derived protoplasts indicates that collapsed cells suffer more injury than derived protoplasts when both are subjected to comparable dehydration and volume reduction (Tao *et al.*, 1983).

The cell wall of higher plant cells is composed predominantly of rigid polymers and under normal turgor the cell wall is only slightly strained. By contrast, the walls of Gram-negative bacteria and yeasts are composed of more elastic polymers and are under considerable tension at isotonic equilibrium (Koch, 1984). As the volume of the protoplast reduces as a result of osmotic water loss the wall remains adpressed to the plasmalemma, as it is under tension, and there is no plasmolysis. The cell wall will continue to shrink with the plasmalemma until the tension in the wall is zero, when eventual separation of the plasmalemma from the wall may occur. In yeasts the reduction in surface area of the cell wall results in changes in thickness and ultrastructure which may be important in cellular response to osmotic stress, both at constant temperature and during freezing (Morris *et al.*, In prep.).

Effects of concentrated solutions on protein structure and activity

Freeze-induced dehydration will expose outer-plasmalemma proteins to the effects of concentrated extracellular solutions and cytoplasmic proteins to concentrated intracellular solutes. At physiological temperatures the biological activity of such proteins will be affected by both the water activity of the suspending solution and the specific effects of various solutes. These effects are also temperature-dependent, being reduced at low temperatures. For clarity the case of soluble enzymes and membrane-associated proteins will be discussed, but it must be remembered that other proteins, including those of the cytoskeleton and ribosomal sub-units, will be exposed to similar stresses.

The possible effects of an increase in solute concentration on soluble enzymes are complex. Initially, the enzyme activity may be inhibited, in a reversible manner, with normal reaction rates being observed on return to isotonic conditions. The sensitivity of enzymes varies with both the specific enzyme and the source of the enzyme. Different solutes at the same concentration have different effects upon activity (Table 4.2, Fig. 4.9) and the addition of a second compound, for example glycerol or proline, may reverse the inhibitory effects of added solutes such as NaCl (Fig. 4.9).

At higher electrolyte levels denaturation of enzymes may occur by modification of the weak molecular interactions responsible for maintaining the native protein structure (Jaenicke, 1981). Depending on the solvation properties of the respective ions, intramolecular hydrogen bonds may be formed or broken; hydrophobic interactions are strengthened, and ion pairs destabilized because of charge shielding. Although protein denaturation has frequently been observed following freezing and thawing, especially in frozen foods such as fish muscle (Fennema, 1979; Reid this volume); there

Table 4.2 Effect of various solutes on the activity (as % of control) of barley leaf malate dehydrogenase. (From Pollard and Wyn-Jones, 1979.)

Compound	Concentration		
	100 mM	200 mM	500 mM
Glycine betaine		100	99
Glycerol		103	98
Proline		96	96
Sucrose		98	91
Dimethylglycine		93	88
Urea		88	77
Glycine		85	62
Monomethylglycine		78	52
Potassium chloride	70		
Sodium chloride	59	35	10
Lithium chloride	40		
Sodium nitrate	26		
Sodium iodide	22		
Calcium chloride	0		

have been few systematic studies of the effects of freezing and thawing on isolated enzymes.

Following cellular shrinkage a vertical displacement of membrane proteins has been reported (Niedermeyer *et al.*, 1976) possibly as a result of a decrease in membrane tension. Such vertical movements of conical-shaped proteins into or away from the plane of the membrane may be a factor which allows minor changes in surface area (see above). Exposure of isolated membranes to hypertonic solutions may also induce the release of proteins away from the membrane matrix. This release of peripheral membrane proteins is not the result of osmotic stress as it occurs from membrane fragments as well as from osmotically responsive structures and is a direct effect of solute ions suppressing intramembrane, ionic interactions. This results in protein molecules dissociating from the membrane matrix (e.g. protein release from thylakoid membranes following hypertonic exposure either at a constant temperature or during freezing and thawing – Heber *et al.*, 1981). The amount and type of protein released depends on the nature of the solute. It has also been suggested that freezing injury to onion epidermal cells is related to damage of a membrane cation transport system (Palta and Li, 1978).

Isolated chloroplasts suspended in hypertonic solutions initially shrink and then re-expand, and this has been demonstrated to occur by a process of reverse loading of cations following the release of regulatory proteins of membrane pumps. A similar phenomenon has been observed during the freezing of a protoplast mutant of *Chlamydomonas reinhardii*, which when entrapped in extracellular ice will swell and increase in surface area by approximately 60 per cent during 3 minutes at −2.5°C (Grout, unpublished observations). This implies influx of both solutes and water into the cell at sub-zero temperatures and possibly alteration in the properties of cytoskeletal elements. These cells respond similarly to a wide range of ionic and non-ionic solutes, at hypertonic concentrations, in simulations of freezing stress conducted at room temperature.

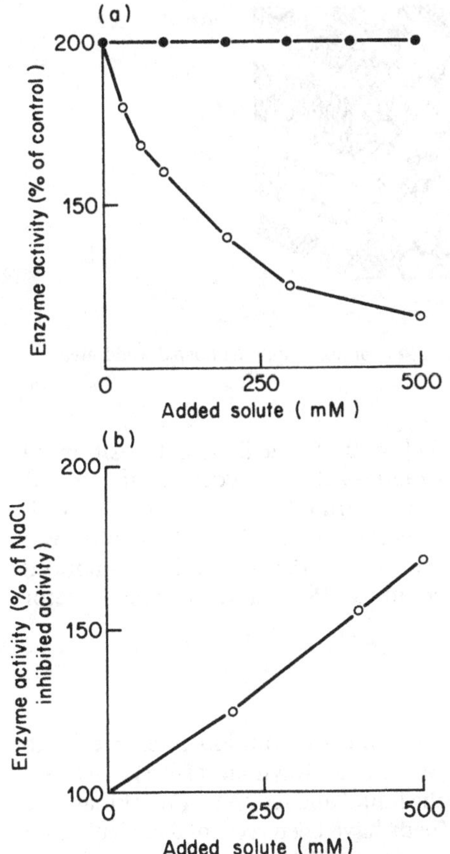

Fig. 4.9 (a) The activity of barley leaf malate dehydrogenase in the presence of different concentrations of sodium chloride (○) and glycine betaine (●). (b) The effects of different concentrations of glycine betaine on the inhibition of barley leaf malate dehydrogenase by 300 mM NaCl. (Redrawn from Pollard and Wyn-Jones, 1979.)

Alterations in pH

During cooling and freezing complex alterations occur to the pH of solutions (Taylor, this volume) and it has been inferred that such changes may affect the stability of proteins in suspension. The recovery of mammalian smooth muscle after storage in the undercooled state at −13°C was markedly pH* dependent with a clearly defined optimum at $pH^*_{-13} = 9.2$ (Taylor, 1982). Unfortunately as a result of practical difficulties there have not yet been any studies of the role of pH* and buffer capacity in the recovery of biological material following freezing and thawing.

Removal of water of hydration

In an excess of water phospholipids are predominantly in the lamellar phase.

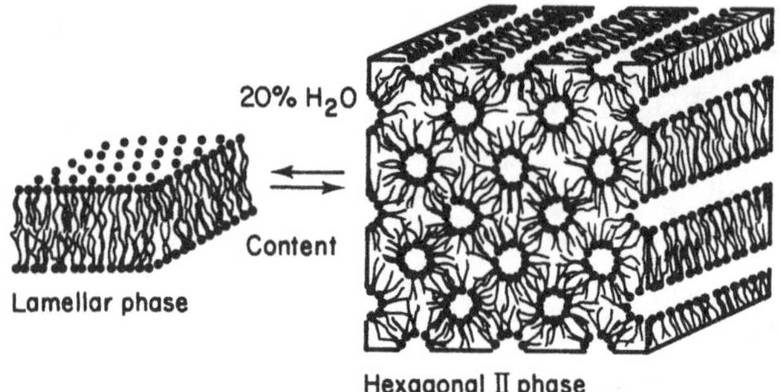

Fig. 4.10 Diagram of hydration-dependent phase changes in a phospholipid. (Redrawn from Crowe *et al.*, 1982.)

Following the removal of water to below 0.25 g/g lipid a transition to hexagonal II phase may occur. The hexagonal II phase consists of tubes of lipid with polar ends orientated into a hydrated core (Fig. 4.10). Such hydration-dependent changes in natural membranes have been observed during isothermal dehydration (Crowe *et al.*, 1982) and freeze-induced dehydration (Gordon-Kamm and Steponkus, 1984) and may be a factor which contributes to cellular injury.

Electrical effects

Electrical fields may be generated across the ice–solution interface in an aqueous solution that is being frozen (Workman–Reynolds effect) as a result of the differential solubility of ions in the liquid and ice phase (Workman and Reynolds, 1950; Hobbs, 1974). Such fields have been recorded at significant levels in dilute aqueous solutions (e.g. as high as 50 V during freezing of NaCl solutions between 10^{-3} and 10^{-4} molar - Hobbs, 1974).

Biological membranes can be perturbed by the application of electrical fields and may subsequently show altered properties (e.g. with respect to solute permeability and mechanical integrity - Zimmerman, 1981). This response has been usefully exploited in techniques for controlled plasmalemma fusions of isolated plant protoplasts, but it is clear that uncontrolled perturbations could contribute significantly to cellular injury.

It has been suggested that during freezing of aqueous cell suspensions the Workman–Reynolds effect may perturb the plasmalemma of individual cells and may contribute to observed cellular injury (Steponkus *et al.*, 1984). This suggestion will undoubtedly lead to further systematic investigations.

Rapid rates of cooling

In the natural environment rates of cooling are slow (e.g. in a temperate climate a cooling rate in the order of 1 °C $hour^{-1}$ to sub-zero temperatures on a night with frost would be typical). Initial ice formation is invariably restricted to the extraprotoplast compartment, which may include the cell wall. In the laboratory rapid rates of cooling are attainable leading to an increased probability of intracellular ice formation (McGrath, this volume). A reduction in cellular viability following rapid cooling is typical and has been correlated with intracellular ice formation (Mazur, 1977). The specific mechanism of injury and the interactions with ice recrystallization during rewarming are unclear (Mazur, 1977; Fujikawa, 1981). There are also examples of cell survival following the formation of intracellular ice (Morris *et al.*, 1977; Rall *et al.*, 1980). In addition, the formation of intracellular ice is a fundamental problem associated with ultrastructural techniques which employ rapid freezing as a method of fixation (Skaer, this volume).

There are two processes by which intracellular ice formation may occur.

(1) Ice may be seeded across the plasmalemma from extracellular ice. It is generally assumed that the intact plasmalemma is a barrier to seeding from extracellular ice. However, if alterations to the plasmalemma properties occur, as a consequence of extracellular ice formation (see above) or the effects of reduced temperatures on membrane structure (Morris, this volume), ice may propagate across the membrane.

(2) Nucleation may occur within the cell independently of extracellular ice. This nucleation may either be homogeneous (i.e. due to a clustering of water molecules) or heterogeneous with a non-aqueous nucleus acting as the catalyst for ice formation.

Evidence of each of these processes has been obtained, but they are not mutually exclusive and both may occur within one cell-type under different experimental conditions. For example, during rapid cooling of human erythrocytes in the presence of extracellular ice, intracellular freezing is observed, by cryomicroscopy at −12°C (Diller, 1979). By contrast, when placed in an emulsion (where the cells are surrounded only by a thin layer of extracellular fluid which does not nucleate during cooling) intracellular nucleation occurred at −29°C (Franks *et al.*, 1983). These data suggest that in the presence of extracellular ice, alterations in the membrane occur at −12°C allowing ice to be propagated across the altered membrane structure. If no such extracellular ice is present, erythrocytes may undercool to −29°C before intracellular ice nucleates. It has been argued from studies of emulsions of haemoglobin solutions, which do not nucleate until −38°C that erythrocytes possess ice-nucleating sites, possibly on the inside of the cell membrane (Mathias *et al.*, 1984).

Events occurring in the frozen state

Assuming that no gross damage, such as physical rupture of the plasmalemma or internal membranes, is expressed during freezing, further stresses may be imposed upon the cell in the frozen state. Following freezing to subzero temperatures above the appropriate eutectic point, cells will be suspended in concentrated solutions and encased in ice. Under these conditions the effects of low temperatures on the structure and function of cell components, together with the potential interactions with concentrated solutions, will be important. For example, enzyme reactions may still occur, albeit in an unregulated manner with an increased solute concentration, gaseous exchange will be altered, and the activity of membrane-bound pumps will change. The effects of such stresses will be a progresive increase in cell damage as the period of storage in the frozen state is extended.

At temperatures below the glass transition temperature (−139°C for water, and higher for aqueous solutions) no recrystallization of ice will occur (see Taylor, this volume) and rates of chemical reactions and biophysical processes will be too slow to affect cell survival. For example, if a simple chemical reaction could occur at −200°C, it would be approximately eight million times slower than at 0°C. Consequently, below the glass transition temperature cell viability should be independent of the period of storage, which provides the basis for cryopreservation of biological material in cryogens such as liquid nitrogen.

At these temperatures photochemical events may still occur and macromolecules, including the genetic material of cells, will still be accessible to irradiation. Since no chemical repair mechanism is operational at such temperatures, any genetic damage that is induced will be cumulative and expressed upon thawing. It has been calculated that with background irradiation at the current level, the median lethal dose for mammalian tissue-culture cells stored in liquid nitrogen will accumulate within 10 000 years (Ashwood-Smith and Grant, 1977).

Stresses associated with thawing

There have been few systematic studies of the effect of warming conditions on frozen cells but it is believed that the survival rate following rapid rewarming is usually higher than that following slow rewarming. There are, however, exceptions, for example following slow cooling (0.5°C min^{-1}) of mammalian embryos in dimethylsulphoxide (1.5 M), a higher recovery rate was obtained following slow rather than rapid warming (Leibo, 1977). The stresses that are imposed upon cells during thawing are, in part, determined by the manner in which the cells were frozen (Fig. 4.11) and two generalized cases are discussed below.

Following 'slow' cooling

When cells in suspension are cooled relatively slowly they become embedded in a frozen matrix, are more or less shrunken, and may contain a quantity of intracellular ice. The amount of ice they contain will depend upon the precise details of cooling and cellular changes that might have occurred as as consequence of cooling, or freezing, that enhance the probability of intracellular nucleation.

Changes in cell structure and function that occur during slow cooling (see above) may not be evident in the frozen state and will only be expressed following thawing when metabolic demands are again imposed upon the cell. If the cell is to resume normal, integrated metabolism injury must be rapidly

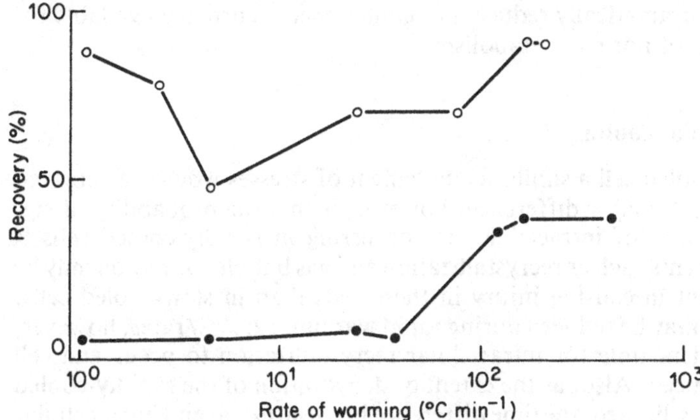

Fig. 4.11 Recovery of *Saccharomyces cerevisiae* as a function of rate of warming. Cells were frozen either at 212°C min^{-1} (●) or at 1 4°C min^{-1} (○). (Winters, unpublished data)

repaired, otherwise it will be expressed as continuing, and possibly lethal, damage. Further, reactions that occur in a metabolically imbalanced, injured cell may themselves contribute to secondary imbalance or injury.

The first stress that thawing will impose upon the cell is the increase in temperature, which will result in the recrystallization of ice (see Taylor, this volume). Such an event in the extracellular medium is likely to be of little consequence to embedded cells. Following the rise in temperature and the melting of extracellular ice crystals, water will be available to rehydrate the cell, and much of thawing injury is believed to be associated with the osmotic events that result. Provided that the plasmalemma has retained semipermeability (and if it has not, the cell is effectively dead), osmotic rehydration will begin with an associated increase in cellular volume. If, however, plasmalemma components have been irreversibly removed during contraction then, at some point below the original isotonic protoplast volume, expansion-induced lysis will be observed. This sequence of events also applies to dehydrated organelles contained within the rehydrating cytoplasm.

If the thawed cell is not to be metabolically imbalanced, necessary concentrations and compartmentalization of metabolites must be restored, and maintained, following the osmotic disturbances associated with freezing and thawing. This relies on preservation of selective permeability and pumping mechanisms associated with the cellular membranes, which are dependent upon membrane-associated proteins. If altered physical conditions during freezing, or thawing have significantly altered the structure or location of such proteins, faulty metabolism can be expected. As a specific example of membrane injury one might consider the electron transport chains of mitochondria in which the precise spatial relationship of cytochromes within the membrane is vital to the synthesis of adequate quantities of ATP. Small changes in the structural configuration of these membranes, or in the effective structures of cytochrome molecules, would severely restrict electron transport and dramatically reduce the cellular energy currency available for the restoration of normal metabolism.

Following 'rapid' cooling

In a rapidly-cooled cell a similar complement of stresses would be expected upon thawing. A major difference, however, is that the probability of significant quantities of intracellular ice occurring in rapidly-cooled cells is high, and so events such as recrystallization and gas bubble formation may be more important in causing injury in these cells than in slow-cooled cells. These stresses may be reduced during rapid warming; at slow rates, however, there is sufficient time for intracellular recrystallization to occur and cell viability is reduced. Also, as the extent of dehydration of the rapidly-cooled cell is likely to be limited, the time difference between extra- and intra-cellular nucleation being small, the significance of osmotic effects following thawing may be somewhat reduced.

Sites of freezing injury

Many studies have been carried out to determine the site(s) of cellular freezing injury but it is difficult, if not impossible to distinguish between the primary causes of freezing injury and secondary pathological changes. There have been few systematic studies of cellular damage induced by rapid cooling, where intracellular ice nucleation is likely to occur. Following slow cooling, however, a number of cellular sites have been implicated in primary freezing injury. Information concerning the most important of these sites is considered below.

Plasmalemma

The plasmalemma is the first barrier of the cell to encounter the effects of extracellular freezing and is therefore an obvious site for the primary freezing injury. Transport processes occur across this membrane and changes in cell shape and volume are closely associated wtih it. Alterations in the properties of the plasmalemma are observed following freezing and thawing and are utilized in tests that use dye exclusion and measurement of intracellular solute leakage to evaluate cellular viability. Specific damage to the plasmalemma may occur by the loss of lipid or protein components or by the denaturation of membrane proteins (see above). Alternatively, this membrane may be mechanically ruptured following an excess influx of solutes or water.

In many cell types, alterations in the phospholipid composition of membranes are observed following freezing and thawing (Yoshida and Sakai, 1974). These changes are consistent with the activation of intracellular phospholipases, and it has been proposed that activation of phospholipase D and its action on the plasmalemma is the specific mechanism of freezing injury in woody plants. However, it appears that such alterations in phospholipid composition are a secondary, pathological event and not the primary cause of freezing injury (Clarke *et al.*, 1982).

Mitochondria

Isolated mitochondria are sensitive to freeze–thaw damage (Araki, 1977; Singh *et al.*, 1977). However, completely functional mitochondria may be isolated from non-viable, frozen-thawed rye coleoptile cells (Singh *et al.*, 1977). The maintenance of mitochondrial function was dependent on the rapid isolation from thawed material.

Chloroplast

In vitro studies using purified chloroplasts have been used to investigate both the biochemistry and biophysics of freezing injury. However, many differences are apparent in the pattern of damage to chloroplasts frozen *in vitro* and *in vivo* and it appears unlikely that *in situ* damage to chloroplasts is a primary cause of freezing injury in leaf cells. For example, chloroplasts isolated rapidly from frozen leaves were not inactivated until they are cooled to 5 °C below the threshold temperature for leaf survival (Heber *et al.*, 1981).

Vacuoles

In the unicellular alga *Chlorella* there is a direct correlation between the presence of a large vacuole and the sensitivity to freezing and thawing (Morris and Clarke, 1978). With many other plant cell-types, pretreatments which reduce the size of the vacuole, such as growth in hypertonic solutions (Pritchard *et al.*, 1982) and selection of exponential phase cells (Withers and Street, 1977) result in an increase in freezing tolerance. Enzymes may be released from the vacuoles of plant cells (Pitt, 1978) and lysosomes of mammalian granulocytes (Rowe and Lenny, 1980) following freezing injury, although this may be a pathological symptom of injury.

DNA and the nucleus

The DNA double helix has been shown to be remarkably stable to freezing and thawing both *in vitro* and *in vivo*, and freezing and thawing is generally not considered to be mutagenic (Ashwood-Smith and Grant, 1977). This can be supported by the lack of evidence of genetic aberrations following the use of cryopreserved bull semen. These genetic studies had to be carried out on viable cells and do not provide information on the physiological state of the nucleus in non-functional cells after thawing or on the contribution of nuclear damage to cellular injury.

Despite apparent genetic stability, in terms of heredity, following freezing there is some evidence for altered physiological activity. This may indicate an alteration of gene expression rather than mutation. Evidence can be presented from the altered growth regulation requirements of frozen-thawed tomato meristem cultures (Grout *et al.*, 1978) and the worsened storage properties of broccoli seeds produced from cryopreserved pollen (Crisp and Grout, 1984). With *Amoeba proteus* it is possible, by micromanipulation, to transfer a nucleus from one cell to another previously enucleated cell. Following freezing to and thawing from −196°C less than 0.1% of cells were recovered. However, all nuclei isolated from such cells were active in stimulating cytoplasmic streaming, and in 16% of the cells into which nuclei were transferred cell division subsequently occurred (Kalinina and Morris, 1982). Damage to the nucleus cannot be the sole primary cause of lethal freezing injury.

Cytoskeleton

Certain elements of the cytoskeleton are known to be destabilized upon exposure to low temperatures (Morris and Clarke, this volume); and this may be a factor in determining the cellular response during freezing. In addition, changes in cell shape and volume would be expected to modify the structure and function of this network. It is therefore surprising that there is only one report of the specific effects of freezing on the cytoskeleton, where lethal freezing of onion root-tip cells results in an irreversible depolymerization of microtubules (Carter and Wick, 1984).

Although the above summary is not comprehensive, it does clearly indicate that there are many potential sites of injury in cells, and that injury will occur when the weakest component is altered beyond a critical level. It is probable that this 'weakest link' varies from cell to cell, and within an individual cell-type under different freezing conditions. Non-lethal, repairable injury may typically occur following freezing and thawing, and the uninjured cell occurs only exceptionally. Once irreversible damage occurs within one cellular component or organelle a progressive sequence of secondary injuries will result.

It is also apparent that following freezing and thawing functional chloroplasts, mitochondria and nuclei may be 'rescued' from non-viable cells but if these organelles are frozen in a cell-free suspension, they may be irreparably damaged. The intracellular compartment appears therefore to buffer organelles against some of the stresses associated with freezing and thawing. Similar observations have been made following freezing of malarial parasites, which are more resistant to freezing injury when contained within a host cell than when isolated (James, this volume).

Cryoprotection

This term describes the protection of cell structure and metabolism against injury associated with freezing events either within or around the cell.

Natural cryoprotection can result from adaptive metabolism of the organism, with changes in cellular structure, composition and metabolic balance giving an enhanced tolerance of freezing. Characteristically, adaptive cryoprotection is genetically defined (i.e. some organisms are able to respond in this fashion whilst others cannot). It requires specific direction of metabolic effort to achieve the necessary cellular changes, and the necessary redistribution of normal metabolism is commonly triggered by changes in environmental parameters (e.g. reduction in temperature, shortening daylength) that signal the impending low temperature conditions of winter (Clarke, Grout, Wang, this volume).

Associated with natural cryoprotection there are several changes that may have significant functional consequences. The most commonly discussed of these is a change in the composition of membranes during adaptation to lower temperatures, but the relative levels of many other cellular components will also change (Morris and Clarke, this volume). Changes in membrane composition may help to maintain a high permeability to water and thereby minimize osmotic stresses during freeze-induced dehydration and subsequent thawing. Chemical changes are also associated with natural cryoprotection including the accumulation of compounds such as glycerol, proline, betaine, sugars and alcohols (Grout, Clarke, this volume). Only a limited range of compounds are synthesized as natural cryoprotectants and are observed in both prokaryotic and eukaryotic cells. In addition to their colligative effects during freezing (Taylor, this volume) it has been suggested that some of these compounds may have specific interactions with membrane components (Crowe *et al.*, 1984).

In laboratory experiments when cell viability or ultrastructure are to be preserved following freezing a different approach to cryoprotection may have to be taken. If the tissues involved have an ability for natural cryoprotection, a period of acclimation can be employed before the freezing treatment, either by using low temperature *per se*, or by imposing osmotic stresses to condition the tissues to the osmotic stresses associated with freezing. This latter approach is particularly successful with plant tissues (Withers, this volume).

If such acclimation is inadequate, or if the tissue has no acclimative response (e.g. human tissues), chemical additives must be incorporated to confer a degree of cryoprotection. In addition to the range of naturally occurring cryoprotective additives a wide range of non-physiological compounds have been demonstrated to be cryoprotective. These range from organic solvents such as dimethylsulphoxide to high molecular weight polymers such as polyvinylpyrollidone (many authors this volume).

References

Araki, T. (1977). Freezing injury in mitochondrial membranes. 1. Susceptible components in the oxidative systems of frozen and thawed rabbit liver mitochondria. *Cryobiology* **14**, 144–50.

Araki, T., Roelofsen, B., op den Kamp, J.A.F. and van Deenen, L.L.M. (1982). Temperature-dependent vesiculation of human erythrocytes caused by hypertonic salt: A phenomenon involving lipid separation. *Cryobiology* **19**, 353–61.

Ashwood-Smith, M.J. and Grant, E. (1977). Genetic stability in cellular systems stored in the frozen state. In *The Freezing of Mammalian Embryos*, pp. 251–68. Edited by Elliot, K. and Whelan, J. Ciba Foundation Symposium 52 (New Series) Elsevier, Amsterdam.

Carter, J.V. and Wick, S.M. (1984). Irreversible microtubule depolymerization associated with freezing injury in *Allium cepa* root tip cells. *Cryo-Letters* **5**, 372–82.

Clarke, A., Coulson, G. and Morris, G.J. (1982). Relationship between phospholipid breakdown and freezing injury in a cell wall-less mutant of *Chlamydomonas reinhardii. Plant Physiology* **70**, 97–103.

Crisp, P.C. and Grout, B.W.W. (1984). Storage of broccoli pollen in liquid nitrogen. *Euphytica* **33**, 819–23.

Crowe, J.H., Crowe, L.M. and Deamer, D.W. (1982). Hydration dependent phase changes in a biological membrane. In *Biophysics of Water*, pp. 295–99. Edited by Franks, F. and Mathias, S.F. John Wiley and Sons, Chichester.

Crowe, J.H., Whittam, M.A., Chapman, D. and Crowe, L.M. (1984). Interactions of phospholipid monolayers with carbohydrates. *Biochemica Biophysica Acta* **769**, 151–9.

Diller, K.R. (1979). Intracellular freezing of glycerolized red cells. *Cryobiology* **16**, 125–31.

Fennema, O. (Ed). (1979). Proteins at Low Temperatures. In *Advances in Chemistry Series. no. 180*, pp. 233. American Chemical Society, Washington, D.C.

Franks, F., Mathias, S.F., Galfre, P., Webster, S.D. and Brown, D. (1983). Ice nucleation and freezing in undercooled cells. *Cryobiology* **20**, 298–309.

Fujikawa, S. (1981). The effect of various cooling rates on the membrane ultrastructure of frozen human erythrocytes and its relation to the extent of haemolysis after thawing. *Journal of Cell Science* **49**, 369–82.

Gordon-Kamm, W.J. and Steponkus, P.L. (1984). A freeze-fracture study of lamellar to hexogonal phase transitions of the plasma membrane during freeze-induced dehydration of isolated protoplasts. *Plant Physiology* **75**, 115 (Abstract).

Grout, B.W.W., Westcott, R.J. and Henshaw, G.G. (1978). Survival of shoot meristems of tomato seedlings frozen in liquid nitrogen. *Cryobiology* **15**, 478–83.

Heber, U., Schmitt, J.M., Krause, G.H., Klosson, R.J. and Santarius, K.A. (1981). Freezing damage to thylakoid membranes *in vitro* and *in vivo*. In *Effects of Low Temperatures on Biological Membranes*, pp. 263–284. Edited by Morris, G.J. and Clarke, A. Academic Press, London.

Hobbs, P.V. (1974). *Ice Physics*, pp. 837. Clarendon Press, Oxford.

Hossack, J.A., Sharpe, V.J. and Rose, A.H. (1977). Stability of the plasma membrane in *Saccharomyces cerevisiae* enriched with phosphatidylcholine or phosphatidylethanolamine. *Journal of Bacteriology* **129**, 1144–7.

Jaenicke, R. (1981). Enzymes under extremes of physical conditions. *Annual Review of Biophysics and Bioengineering* **10**, 1–67.

Kalinina, L.V. and Morris, G.J. (1982). Nuclear transplantation as a means of preservation in non-viable frozen-thawed *Amoeba proteus. CryoLetters* **3**, 239–44.

Koch, A.L. (1984). Shrinkage of growing *Escherichia coli* cells by osmotic challenge. *Journal of Bacteriology* **159**, 919–24.

Kwok, R. and Evans, E. (1981). Thermoelasticity of large lecithin bilayer vesicles. *Biophysics Journal* **35**, 637–52.

Leibo, S.P. (1981). Fundamental cryobiology of mouse ova and embryos. In *The Freezing of Mammalian Embryos*, pp. 69–92. Edited by Elliot, K. and Whelan, J. Ciba Foundation Symposium 52 (New Series), Elsevier, Amsterdam.

Lovelock, J.E. (1953). The haemolysis of human red blood cells by freezing and thawing. *Biochemica Biophysica Acta* **10**, 414–26.

Mathias, S.F., Franks, F. and Trafford, K. (1984). Nucleation and growth of ice in deeply undercooled erythrocytes. *Cryobiology* **21**, 123–32.

Mazur, P. (1977). The role of intracellular freezing in the death of cells cooled at supraoptimal rates. *Cryobiology* **14**, 251–72.

Mazur, P., Leibo, S.P. and Chu, E.H.Y. (1972). A two-factor hypothesis of freezing injury. Evidence from Chinese hamster tissue culture cells. *Experimental Cell Research* **71**, 345–55.

Mazur, P., Rall, W.F. and Rigopoulos, N. (1981). Relative contributions of the fraction of unfrozen water and of salt concentration to the survival of slowly frozen human erythrocytes. *Biophysics Journal* **36**, 653–75.

Meryman, H.T. (1968). Modified model for the mechanism of freezing injury in erythrocytes. *Nature* **218**, 333–6.

Meryman, H.T. (1971). Osmotic stress as a mechanism of freezing injury. *Cryobiology* **8**, 489–500.

Meryman, H.T., Williams, R.J. and st J. Douglas, M. (1977). Freezing injury from "solution" effects and its prevention by natural or artificial cryoprotection. *Cryobiology* **14**, 287–302.

Morris, G.J. and McGrath, J.J. (1981). Intracellular ice nucleation and gas bubble formation in *Spirogyra*. *Cryoletters* **2**, 341–52.

Morris, G.J. and Clarke, A. (1978). The cryopreservation of *Chlorella*. 4. Accumulation of lipid as a protective factor. *Archives of Microbiology* **119**, 153–6.

Morris, G.J., Clarke, K.J. and Clarke, A. (1977). The cryopreservation of *Chlorella*. 3. Effect of heterotrophic nutrition on freezing tolerance. *Archives of Microbiology* **114**, 249–54.

Morris, G.J., Winters, L.C., Coulson, G.E., Jackson, K., Stewart, H.S. and Clarke K.J. The effect of osmotic stress on the ultrastructure and viability of the yeast *Saccharomyces cerevisiae*. *Journal of General Microbiology*. (In press).

Niedermeyer, W., Parish, G.R. and Moor, R. (1976). The elasticity of the yeast cell tonoplast related to its ultrastructure and chemical composition. 1. Induced swelling and shrinkage: a freeze-etch membrane study. *Cytobiologie* **13**, 364–74.

Palta, J.P. and Li, P.H. (1978). Cell membrane properties in relation to freezing injury. In *Plant Cold Hardiness and Freezing Stress. Mechanisms and Crop Implications*, pp. 93–115. Edited by Li, P.H. and Sakai, A. Academic Press, New York.

Pegg, D.E. (1981). The effect of cell concentration on the recovery of human erythrocytes after freezing and thawing in the presence of glycerol. *Cryobiology* **18**, 221–8.

Pitt, D. (1978). Effect of freezing and thawing on the distribution of lysomal hydrolases in leaves of *Solanum tuberosum*. *Planta* **138**, 79–86.

Pollard, A. and Wyn-Jones, R.G. (1979). Enzyme activities in concentrated solutions of glycinebetaine and other solutes. *Planta* **144**, 291–98.

Pritchard, H.W., Grout, B.W.W., Short, K.C. and Reid, D.S. (1982). The effects of growth under water stress on the structure, metabolism and cryopreservation of cultured sycamore cells. In *Biophysics of Water*, pp. 315–18. Edited by Franks, F. and Mathias, S.F. John Wiley and Sons, Chicester.

Rall. W.F., Reid, D.S. and Farrant, J. (1980). Innocuous biological freezing during warming. *Nature* **286**, 511.

Razin, S., Tourtelotte, M.E., McElhaney, R.N. and Pollock, J.D. (1966). Influence of lipid components of mycoplasma membranes on osmotic fragility of cells. *Journal of Bacteriology* **91**, 609–16.

Rowe, A.W. and Lenny, L.L. (1980). Cryopreservation of granulocytes for transfusion: Studies on human granulocyte isolation, the effect of glycerol on lysosomes, kinetics of glycerol uptake, and cryopreservation with dimethylsulphoxide and glycerol. *Cryobiology* **17**, 198–212.

Singh, J. de la Roche, A.I. and Siminovitch, D. (1977). Relative insensitivity of mitochondria in hardened and nonhardened rye coleoptile cells to freezing *in situ*. *Plant Physiology*. **60**, 713–15.

Steponkus, P.L. (1984). Role of the plasma membrane in freezing injury and cold acclimation. *Annual Review of Plant Physiology* **35**, 543–84.

Steponkus, P.L. and Dowgert, M.F. (1981). Gas bubble formation during intracellular ice formation. *Cryoletters* **2**, 42–47.

Steponkus, P.L. and Weist, S.C. (1979). Freeze-thaw induced lesions in the plasma membrane. In *Low Temperature Stress in Crop Plants*, pp. 231–54. Edited by Lyons, J.M., Graham, D. and Raison, J.K. Academic Press, New York.

Steponkus, P.L., Dowgert, M.F. and Gordon-Kamm, W.J. (1983). Destabilization of the plasma membrane of isolated plant protoplasts during a freeze–thaw cycle: The influence of cold acclimation. *Cryobiology* **20**, 448–65.

Steponkus, P.L., Evans, R.Y. and Singh, J. (1982). Cryomicroscopy of isolated rye mesophyll cells. *Cryoletters* **3**, 101–14.

Steponkus, P.L., Stout, D.G., Wolfe, J. and Lovelace, R.V.E. (1984). Freeze-induced electrical transients and cryoinjury. *Cryoletters* **5**, 343–8.

Tao, D., Li, P.H. and Carter, J.V. (1983). Role of cell walls in freezing tolerance of cultured plant cells and their protoplasts. *Physiologia Plantarum* **59**, 527–32.

Taylor, M.J. (1982). The role of pH* and buffer capacity in the recovery of function of smooth muscle cooled to −13°C in unfrozen media. *Cryobiology* **19**, 585–601.

Unemura, M. and Yoshida, S. (1984). Involvement of plasma membrane alterations in cold acclimation of winter rye seedlings (*Secale cereale* L. cv. Puma). *Plant Physiology* **75**, 818–26.

Weist, S.C. and Steponkus, P.L. (1978). Freeze-thaw injury to isolated spinach protoplasts and its simulation at above freezing temperatures. *Plant Physiology* **62**, 699–705.

Williams, R.J. and Hope, H.J. (1981). The relationship between cell injury and osmotic volume reduction. III Freezing injury and frost resistance in winter wheat. *Cryobiology* **18**, 133–45.

Withers, L.A. and Street, H.E. (1977). Freeze-preservation of cultured plant cells. 3. The pregrowth phase. *Physiologia Plantarum* **39**, 171–8.

Wolfe, J. and Steponkus, P.L. (1981). The stress:strain relation of the plasma membrane of isolated protoplasts. *Biochemica Biophysica Acta* **643**, 663–8.

Wolfe, J. and Steponkus, P.L. (1983a). Tension in the plasma membrane during osmotic contraction. *Cryoletters* **4**, 315–322.

Wolfe, J. and Steponkus, P.L. (1983b). Mechanical properties of the plasma membrane of isolated plant protoplasts. *Plant Physiology* **71**, 276–85.

Workmann, E.J. and Reynolds, S.E. (1950). Electrical phenomena occurring during the freezing of dilute aqueous solutions and their possible relationship to thunderstorm electricity. *Physics Review* **78**, 254–59.

Yoshida, S. and Sakai, A. (1974). Phospholipid degradation in frozen plant cells associated with freezing injury. *Plant Physiology* **53**, 509–11.

Zade-Oppen, A.M.M. (1968). Post hypertonic haemolysis in sodium chloride systems. *Acta Physiol Scandanavia* **73**, 341–64.

Zimmerman, U. and Scheurich, P. (1981). High frequency fusion of plant protoplasts by electric fields. *Planta* **151**, 26–32.

Section II

Techniques

5

Low temperature and biological electron microscopy

H. le B. Skaer
ARC Unit of Insect Neurophysiology and Pharmacology
Department of Zoology
University of Cambridge

Introduction

The following discussion covers two major areas of current interest in low temperature biology and electron microscopy

(1) the use of freezing as a method of physical fixation (cryofixation) in the preparation of biological tissues for electron microscopy;

(2) analysis and observation of ultrastructural changes in cells brought about by low temperatures, including freezing. Adaptive responses to low temperatures are also considered.

Fundamental differences between these areas can be demonstrated by considering a combined study of ultrastructure and viability after low temperature treatment of chinese hamster fibroblasts (Farrant *et al.*, 1977). A

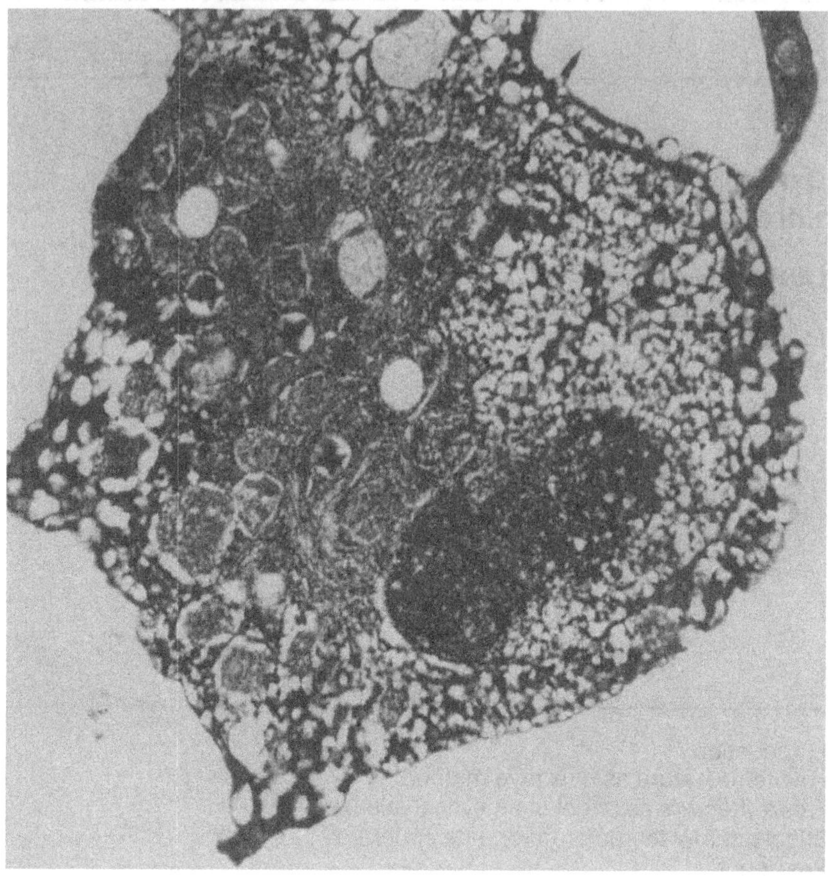

Fig. 5.1 Electron micrograph of Chinese hamster fibroblast freeze-substituted at 193K after rapid cooling (473K min^{-1}) in DMSO (5 %v/v) to 77K. The cell is not shrunken in appearance but contains numerous intracellular ice cavities. (From Farrant *et al.*, 1977.)

rapidly-cooled cell showing 'accepted' ultrastructure (i.e. reasonably fixed with respect to conventional criteria) proved to be non-viable when rewarmed to physiological temperatures (Fig. 5.1). A slowly-cooled cell which had totally abnormal ultrastructural characteristics and was apparently extensively dehydrated (Fig. 5.2) was viable upon rewarming. This demonstrates that although fast-cooling may be an acceptable fixation technique, as it leaves the ultrastructure of the cell unaltered, the lack of ultrastructural change is not necessarily conducive to preservation of viability. The potentially viable cells, however, reveal very little useful information about the normal structure of the cell.

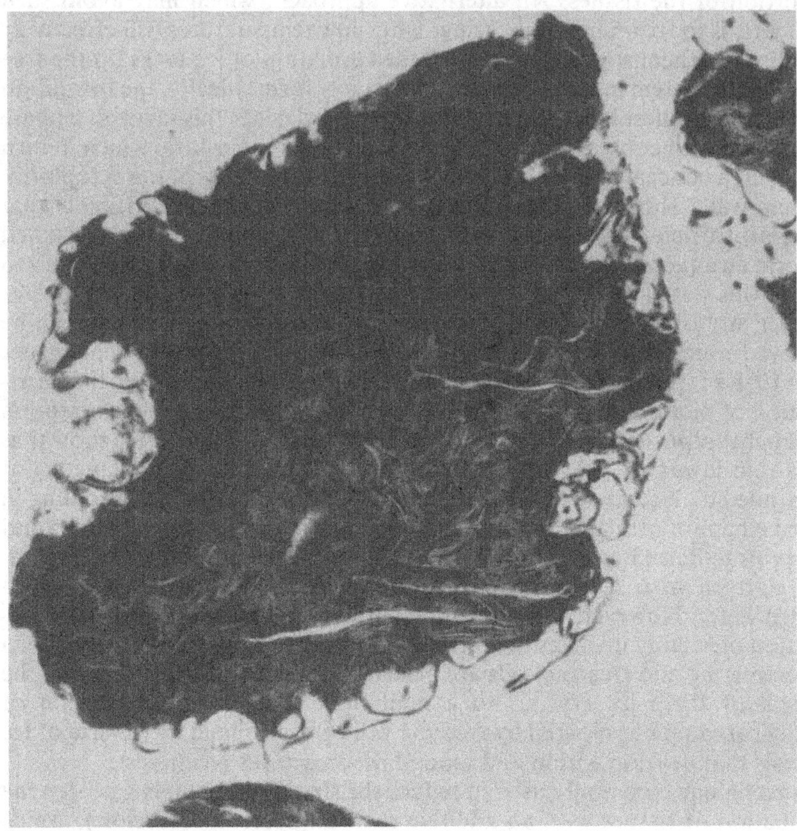

Fig. 5.2 Electron micrograph of Chinese hamster fibroblast freeze-substituted at 193K after two-step cooling. This potentially functional cell shows a typical shrunken appearance without any evidence of intracellular ice cavities. (From Farrant *et al.*, 1977.)

Low temperature as a fixative method

The preparation of biological tissues for transmission electron microscopy involves stabilization of the tissues in such a way that they survive the rigorous environment of the electron microscope, which is essentially a high vacuum in which the tissues are bombarded with highly accelerated electrons. Traditionally, this has been achieved by fixation with chemical agents such as aldehydes to stabilize the cellular components by molecular crosslinking, and then dehydration of the tissue by substitution of water with nonaqueous solvents such as alcohols. Although a vast amount of valuable information has been gained from thin-sectioned material after treatment in this way, inevitably both chemical fixation and dehydration lead to alteration and distortion of the tissues. An alternative approach, which may avoid such artefacts, is to fix tissues by freezing. Ultra-low temperatures will effectively stabilize the chemistry of the cell and the formation of ice locks up the free water in the system in an essentially unavailable form. Ideally, the formation of even the smallest ice crystals,should be avoided, since this involves a phase separation with consequent redistribution of water molecules, which in turn results in a concentration of solutes in still fluid domains in the cytoplasm. Instead water should be immobilized by cooling the tissues so rapidly that there is insufficient time for ice to nucleate and the whole sample is subcooled to form an aqueous solid or glass. However, the rates of cooling necessary to ensure this state (10^7–10^{10} Ks^{-1} for *c.* 1 μm diameter droplets of pure water [Mayer and Bruggeller, 1982]) are orders of magnitude higher than can be achieved even with the most rapid cooling methods for biological tissues (10^5–10^6 Ks^{-1} [Heuser *et al.*, 1979]). It therefore appeared unlikely that vitrification of biological structures would be possible. However, recent reports (Dubochet *et al.*, 1982a,b; Lepault *et al.*, 1983; Adrian *et al.*, 1984) show that ultra-thin layers (< 300 nm) or fine droplets (up to 1 μm in diameter) of suspended biological material can be completely vitrified by rapid cooling in liquid ethane or propane. In this way, preparations such as virus particles and cell subfractions (Stewart and Vigers, 1986) can be examined by low temperature electron microscopy, unfrozen and fully hydrated, suspended in supercooled water. However, for any samples other than fine suspensions, freeze-fixation inevitably involves the formation of ice, and the methods employed in pretreating and freezing cells are primarily concerned with restricting the growth of these ice crystals to a minimum. Clearly, the resolution of biological samples prepared by freezing will be determined by the size of ice crystals that develop within and around the structures of interest.

A technique commonly used to reduce the size of ice crystals has been the incubation of tissues with an additive such as glycerol (see below), which extends the subcooling before ice nucleation is initiated and, when the cooling rate is high, results in many, small ice crystals rather than fewer, larger crystals (Taylor, this volume).

Unfortunately, such additives may alter the cells both structurally and chemically, and so chemical prefixation may be employed in an attempt to stabilize cells. Much of the information that has been produced by freeze-fracture, for example, has concerned chemically prefixed and additive-treated samples. The expected benefits of cryofixation may therefore have been diminished or lost as a result of artefacts caused by chemical prefixation.

Recently, ultrarapid methods of freezing (see below) have been devised that allow such rapid cooling in the sample surface regions that only tiny ice crystals (of nanometer dimensions) develop in the peripheral 10–15 μm. These very rapidly frozen samples represent the ultimate in the preservation of unaltered biological structure by cryofixation alone.

The principal methods that are discussed in this section are summarized in Fig. 5.3.

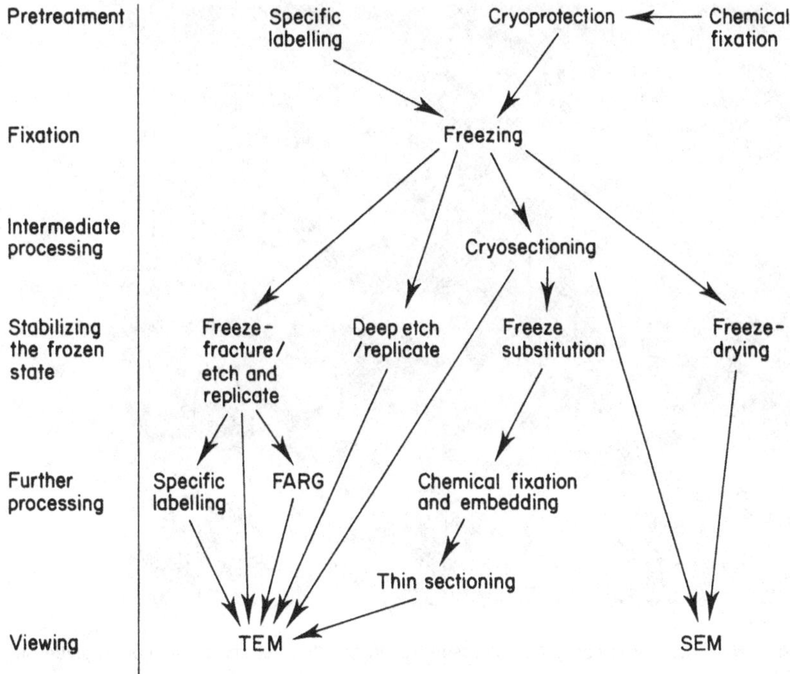

Fig. 5.3 Flow diagram to summarize the most commonly used preparative methods for low temperature electron microscopy. Although by no means an exhaustive list of available methods, this diagram shows the ways in which techniques described in this chapter can be integrated.

Use of compounds to reduce ice crystal size (cryoprotection)

Although ultrarapid cooling is the method of choice for freeze-fixation there are obvious practical difficulties when bulky samples are investigated. If internal structures of tissue pieces larger than 30–50 μm in diameter are to be studied, samples must be incubated with an additive to prevent the formation of undesirably large ice crystals.

In practice additives can be divided into two groups: those that must penetrate all compartments of the cell to be effective and those that remain outside the cell and exert a protective effect from the external medium (Skaer, 1982).

The most commonly used penetrating additive is glycerol, although dimethyl sulphoxide (Stolinski and Breathnach, 1975), methanol and ethanol (Boyde and Wood, 1969; Humphreys *et al.*, 1974, 1975; Schiller and Taugner, 1980) have been used. Dimethyl sulphoxide, methanol and ethanol have not been widely adopted as they produce unacceptable artefacts in the tissues (McIntyre *et al.*, 1974). Glycerol is also known to produce a range of structural artefacts and does not permeate readily into the intracellular compartments of plant cells. Nevertheless, it is the most popular additive for

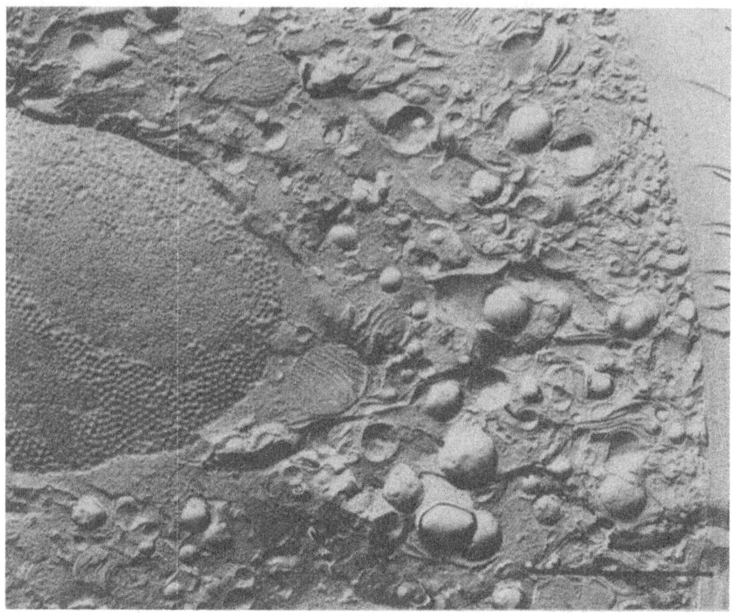

Fig. 5.4 Freeze-fracture micrograph of tissue frozen rapidly after cryoprotection in glycerol. Scale bar = 3μm.

freeze-fracture studies and seems likely to remain so as it is non-toxic, simple to use and results obtained with it can be compared with a wealth of published data from similarly treated material (e.g. Fig. 5.4). The artefacts it causes are easily recognized and have been documented (Böhler, 1975; Rash and Hudson, 1979). These include particle clustering, membrane blebbing, vesiculation of internal membranes, fusion of intracellular secretory granules with each other and with the surface membrane and alterations in the size and fracturing characteristics of intramembranous particles (IMPs). Furthermore, glycerol treatment has been shown to result in the breakdown of intercellular structures (Filshie and Flower, 1977). It is common to find that material has been fixed chemically and then incubated in glycerol in an attempt to stabilize the tissues before cryoprotection, and, in the case of plants, to increase the permeability of the cells to glycerol. This may not be totally effective in stabilizing structures and it has been shown to produce further artefacts, for example particle-free blisters in the membrane and changes in the fracturing characteristics of IMPs (Hasty and Hay, 1978; Cartaud *et al.*, 1978; Hay and Hasty, 1979; van Deurs and Luft, 1979; Willison and Brown, 1979; Green, 1981; Arancia *et al.*, 1980).

Additives that remain exclusively in the external medium exert their cryoprotective effect by enhancing subcooling and reducing the rate of ice crystal growth. This latter effect is a result of an increase in viscosity that occurs sharply as the temperature drops. These compounds are less likely to interfere with the internal structure and chemistry of the cells. A disadvantage of such non-penetrating additives is that the cells respond to the increase in external osmotic potential by shrinking (*see* Figs. 2,5 in Skaer, 1982) and, although this will reduce the probability of intracellular freezing, it will also preclude accurate morphological preservation. To avoid this problem the use of sucrose as a cryoprotectant may be combined with chemical fixation to enhance additive penetration into cells.

Higher molecular weight compounds, for example polyvinyl pyrrolidone (PVP), hydroxyethyl starch (HES) and dextran, weight for weight exert a lower osmotic pressure than sucrose and therefore theoretically should be less liable to shrink the cells. These hydrophilic polymers are used as cryoprotectant additives and as a mechanical support for specimens to be cryosectioned or cryofractured, especially in the preparation of material for microanalysis (Echlin, 1984). Excellent cryofixation has been achieved with both HES, PVP and dextran as demonstrated by frozen sections and freeze-fracture (Fig. 5.5) (Skaer, 1982). There are, however, certain limitations to their use that appear to be inseparable from their cryofixative capacity. It is not yet understood how they cryoprotect but they are believed to reduce the mobility of water and so restrict the ability of water to form into ice crystals. The viscosity of the solution is also high and increases as the temperature is lowered so that crystallization of water into ice is inhibited (Franks, 1980). It has, however, become clear that the tissues suspended in a polymer solution are cryoprotected only if the cooling rate is high enough to vitrify the polymer

Fig. 5.5 Insect nerve cells prepared for freeze-fracture after cryoprotection in 50% PVP. There is no evidence of intracellular ice. N = nerve cell. G = glial cells. Scale bar = 1.0µm.

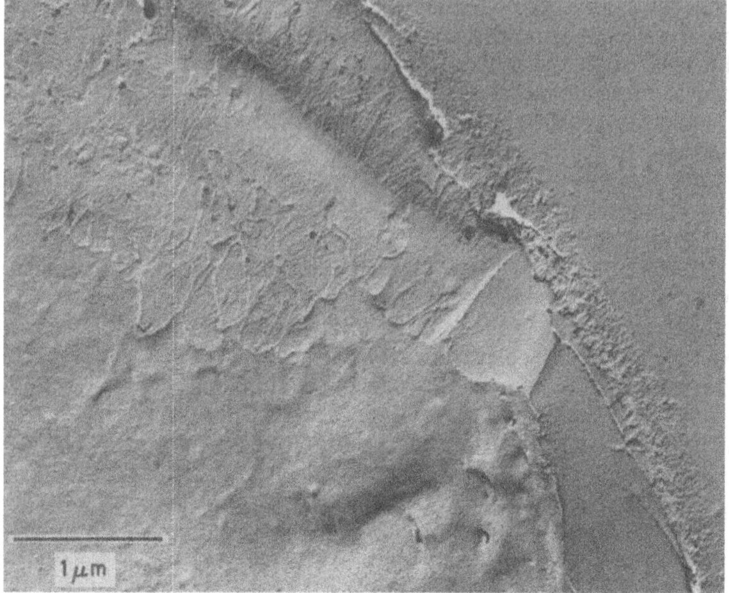

Fig. 5.6 Evidence of cell shrinkage after 30 minutes incubation in 50% PVP. The membrane of a plasmolysed carrot cell has retracted from its cell wall. Scale bar = 1.0μm.

solution. This suggests that prevention of freezing outside the cells delays nucleation inside so that the tissue freezes at the lowest possible nucleation temperature of cytoplasm (close to −40°C). Two problems arise from the nature of such cryoprotection. Firstly any intra- or intercellular reservoirs of tissue water that cannot equilibrate with the polymer solution (e.g. vacoules or luminal contents in gut or blood vessel) initiate freezing at higher sub-zero temperatures. This freezing spreads throughout the tissues, resulting in poor cryofixation. Secondly, although high molecular weight polymers exhibit a low colligative osmotic pressure, their ability to make water unavailable results in an effective osmotic gradient and cell shrinkage (Fig. 5.6). In practice tissues show a great variation in their tendency to shrink (compare Wilson and Robards, 1980 with Pihakaski and Seveus, 1980 and Echlin *et al.*, 1980a) and this problem can be mitigated by reducing to a minimum the concentration of the cryoprotective solution employed and the time of incubation in it. Although these cryoprotectants are of use for freeze-fracture (Skaer *et al.*, 1977, 1978; Schiller *et al.*, 1978, 1979; Swales and Lane, 1983), the area in which they show most promise is in preparation for microanalysis; there is evidence to suggest that normal electrolyte concentrations and gradients are preserved in the tissues during cryoprotection and further processing (e.g. Echlin *et al.*, 1980b, 1982; Gupta *et al.*, 1976, 1978a,b).

Table 5.1 Average cooling rates for thermocouples to cool from 273 K to 173 K when plunged into liquid coolants. The same thermocouple (25 μm diameter wire, 70 μm diameter bead) was used for all measurements (Reproduced with permission from Costello and Corless, 1978)

Coolant	Temperature* (K)	Rate × $10^{-3\dagger}$ (K/s)
Freon 22	118	66
Freon 12	121	47
Freon 13	88	78
Freon 22 : 12 (4 : 1)	110	64
Freon 22 : 13 (3 : 1)	110	64
Propane	83	98
Isopentane	113	45
Propane : Isopentane : Methylcyclohexane (20 : 5 : 1)	82	96
Nitrogen, liquid	77	16
Nitrogen, solid/liquid slush	66§	21

* Temperature of the coolant surface at the time of the drop. A single temperature measurement is accurate to ± 1 K. The values reported represent average from two to six runs, and maximum variation in reported temperatures is ± 3 K.

† Rate calculated from time required to cool from 273 K to 173 K. Average variation is about 5%.

§ The surface temperature was 66–64 K. The temperature at lower levels in the bath was 63 K.

Freezing techniques

The simplest way to freeze tissue rapidly is to plunge it into a liquid coolant. The rate of cooling depends on the mass of the sample, the geometry and conductivity of the sample holder, the coolant used and the speed with which the specimen moves through the coolant (Costello and Corless, 1978 and Table 5.1). Cryoprotected material is routinely frozen on copper or gold planchettes in the fluorocarbon Freon 22, cooled in a jacket of liquid nitrogen to its melting point at −163°C (110 K); in general this produces adequate freezing (see Fig. 5.4). However, for optimal fixation without additives the rate of cooling must be sufficiently rapid to produce ice crystals of dimensions below that of the resolution of the electron microscope employed. In practice this has been achieved by minimizing the sample size (spray freezing, thin layer freezing), improving specimen coolant contact by using a cold polished metal surface (slam freezing), increasing the rate of coolant movement over the specimen surface (jet and plunge freezing) and freezing samples under high pressure, which effectively inhibits the growth of ice crystals (Plattner and Bachmann, 1982; Elder *et al.*, 1982; Rash, 1983).

Spray freezing (Bachmann and Schmitt, 1971; Plattner *et al.*, 1972) involves forcing small droplets through a fine nozzle into the coolant and produces excellent results for cell suspensions, (Fig. 5.7). Very thin layers (120–300 nm) of suspended material can be prepared on electron microscope grids and, by allowing them to fall free into liquid propane or ethane, cooling rates high enough to vitrify the entire sample can be achieved (Adrian *et al.*, 1984). Slam freezing produces good specimen coolant contact by firing or

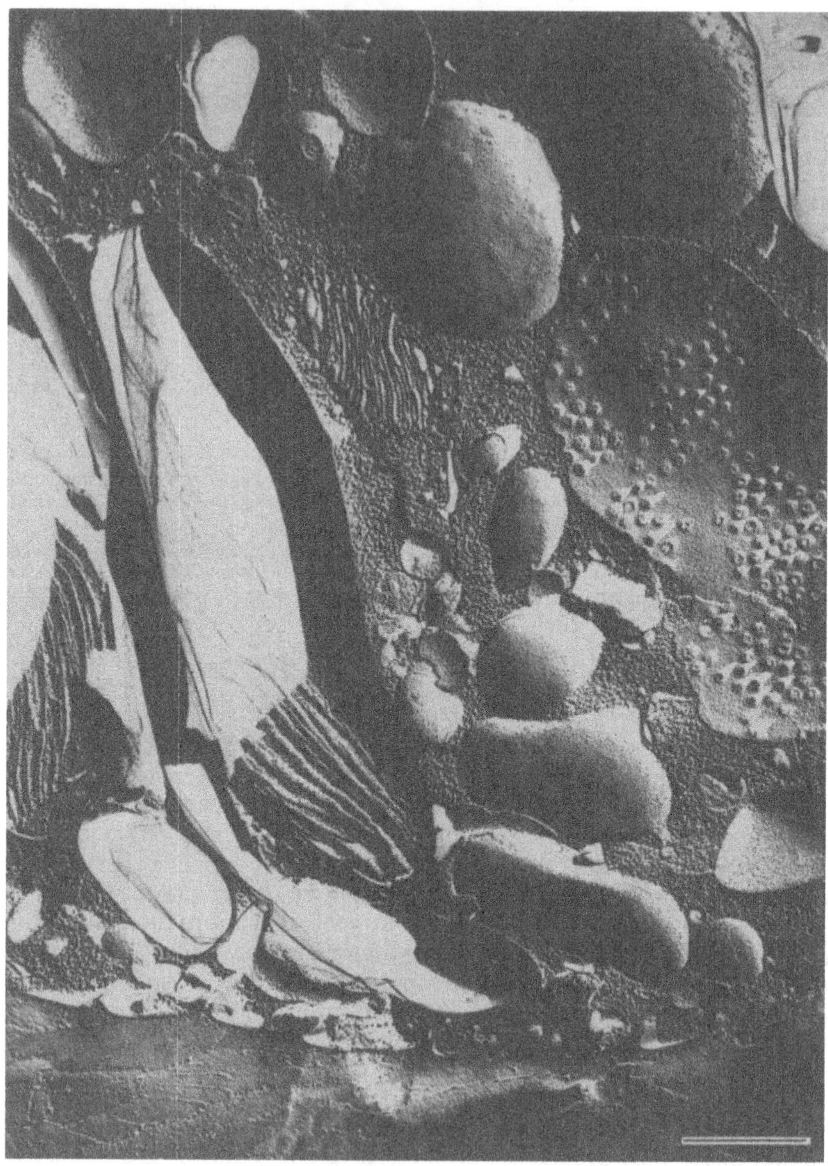

Fig. 5.7 Spray frozen preparation of an uncryoprotected *Euglena* cell. Scale bar = 3μm. (From Plattner *et al.*, 1972. Reproduced from *The Journal of Cell Biology*, 1972, **53**., 116–26, by copyright permission of the Rockefeller University Press.)

Fig. 5.8 Slam frozen preparation of frog neuromuscular junction. (From Heuser *et al.*, 1975.)

dropping the sample onto the metal block, cooled to −196°C with liquid nitrogen or −269°C with liquid helium (van Harrevald and Crowell, 1964; Kretzchner and Wilkie, 1969; Christenson, 1971; van Harrevald *et al.*, 1974; Dempsey and Bullivant, 1976 a,b; Heuser *et al.*, 1975, 1979). Excellent freezing can thus be achieved in the superficial layers of the sample (Fig. 5.8), and the method is suitable for pieces of tissue as well as for cell monolayers or suspensions. A particular advantage of the slamming technique is that tissues can be physiologically stimulated milliseconds before freezing, so that very short-lived events can be preserved (Heuser *et al.*, 1979). The shock waves produced on impact can, however, distort membrane structure in the split second before freezing (Pinto da Silva and Kachar, 1980). This can be avoided by jet freezing; firing the coolant (propane) over the sample (Moor *et al.*, 1976; Burstein and Maurice, 1978; Müller *et al.*, 1980 a,b; Giddings and Staehelin, 1980; Espevik and Elgsaeter, 1981; Pschied *et al.*, 1981; Knoll *et al.*, 1982), but there have been problems with reproducibility of cooling rates (Robards and Severs, 1981; Swalés and Lane, 1983). Jet freezing has been used more often for ultra-thin samples (suspensions of cells or cell fragments and cell monolayers) than for tissue pieces (*see*, however, Swales and Lane, 1983) largely because of the difficulty of obtaining a fracture that passes through the frozen tissue block. By moving the sample rapidly through the coolant (plunge freezing) it is possible to freeze more reproducibly. The advantages of this method are that it is very simple, it is suitable for all types of sample and it should theoretically be capable of freezing short-lived physiological events within milliseconds of stimulation (Bald and Robards, 1978; Costello and Corless, 1978; Handley *et al.*, 1981; Severs and Green, 1983).

All the rapid cooling methods described above suffer from the limitation that acceptable patterns of freezing are restricted to the surface layer of *c.* 20 μm. Deeper within the tissue the size of ice crystals increases as a result of the poor thermal conductivity of biological tissue when frozen (Dempsey and Bullivant, 1976a,b; Heuser *et al.*, 1975). Furthermore, the latent heat of fusion produced when ice is formed causes a rise in temperature, resulting in yet slower freezing, or even recrystallization, deeper in the tissue where heat extraction is poor (van Venrooij *et al.*, 1975).

High pressure mimics the effects of cryoprotectants in that solutions under hyperbaric conditions subcool and exhibit lowered nucleation temperatures and lower crystallization rates (Kanno *et al.*, 1975). The application of a pressure of 2100 bars to biological material is claimed to have the same effect on freezing patterns as a cryoprotectant in a 20 per cent solution (Riehle, 1968; Moor, 1973) but of course has the advantage that no chemical interference with the tissue is involved. Pressure freezing (Moor and Riehle, 1968; Moor, 1971; Bald and Robards, 1978) has produced good results in terms of fixation (Riehle and Hoechli, 1973; Moor *et al.*, 1980); structures deep within the tissue are well preserved. However, the application of high pressures for longer than extremely short periods of time (< 0.1 second) is known to kill cells (Moor and Hoechli, 1970). Recently, a method has been devised in which the coolant (liquid nitrogen) acts as the pressurizing medium and in this way reduces to a minimum the application of high pressures before freezing (Moor *et al.*, 1980). The use of this technique has so far been limited by the complexity of the apparatus required and uncertainty over the deleterious effects of high pressure on the tissue. This is, however, the only method available which enables chemically untreated tissue to be acceptably cryofixed throughout a 0.5 mm cube.

An entirely different technique, which capitalizes on freezing under pressure, is described by Buchheim and Welsch (1977). Tissue fragments up to 30 μm in diameter were taken into emulsion with paraffin oil, using glycerol mono-oleate as an emulsifier. Reasonable cryofixation was achieved in the uncryoprotected tissue pieces by subsequent immersion in cooled Freon 22. The restriction of ice crystal growth results both from the dispersion of the sample into very small droplets (improving heat extraction and reducing the probability of nuclei being formed) and from the differential thermal contraction of the two phases, which results in an increased pressure on the aqueous droplets.

Stabilizing the frozen state

A sample cooled sufficiently rapidly to give good cryofixation is in a metastable state with respect to ice and temperature, and unless it is viewed at extremely low temperatures, it cannot be used for electron microscopy without further stabilizing treatment. The major stabilizing techniques are concerned either with making a replica of the frozen sample (e.g. freeze-fracture) or with replacing or removing the ice either by substitution (freeze-substitution) or by sublimation (e.g. preparation of material for scanning electron microscopy).

If the frozen sample is rewarmed before the structures within it are stabilized, the quality of the original cryofixation will be obliterated by recrystallization. This results in the redistribution of ice such that some crystals grow larger at the expense of others and eventually the sample becomes indistinguishable from a poorly frozen specimen. Pure water recrystallizes at −130°C and the recrystallization temperature is raised by dissolved solutes just as the freezing point is lowered. Cytoplasm and cryo-protectant solutions thus have higher temperatures of recrystallization than pure water. As a rule, wet tissues should never be rewarmed above −80°C and, to be completely sure of preventing recrystallization, the temperature should be maintained nearer to −100°C (Skaer *et al.*, 1979).

The above requirement and the dangers of contamination by the condensation of atmospheric water vapour onto cold materials pose considerable problems for the handling and examination of frozen specimens. Tissue can be examined either as an intact piece by scanning electron microscopy (SEM) or after cutting frozen sections (Frederik *et al.*, 1982; Barnard, 1982) by transmission electron microscopy (TEM) or SEM. In both cases the microscope is specially modified with a low temperature stage. Interpretation of the images produced is far from straightforward as the electron contrast of wet tissues is very low. However, for the analysis of the distribution of elements in tissues by X-ray microanalysis, the frozen hydrated state is closest to the *in vivo* condition and is therefore the preparation of choice. There is a large volume of literature concerned with details of such preparation and with interpretation (e.g. Goldstein *et al.*, 1981, Echlin, 1984).

Freeze-drying

The simplest way to stabilize frozen tissues for examination in the electron microscope under vacuum at ambient temperature is to air dry them at the appropriate low temperature. A compromise must be found with regard to the rate of sublimation of ice required, for the higher the temperature, the greater the danger of recrystallization but also the faster the drying process (Umrath, 1983). The rate of sublimation can be increased by drying under vacuum and various methods involving intermediate substitution can be used, for example critical point drying (Echlin, 1984). A recent development of preparative methods for SEM by freeze-drying involves cryoprotection, chemical fixation and replication as well as freezing and drying. Specimens are infiltrated with cryoprotectant (normally DMSO), frozen (against a metal block or in Freon 22) and fractured. They are then thawed and washed in a fixative solution to leach out soluble components of the cytoplasm. Specimens are subsequently dehydrated and critical point dried (Haggis *et al.*, 1976; Haggis and Phipps-Todd, 1977; Tanaka, 1980). These methods of preparation, combined with improved resolution in SEM (Komoda and

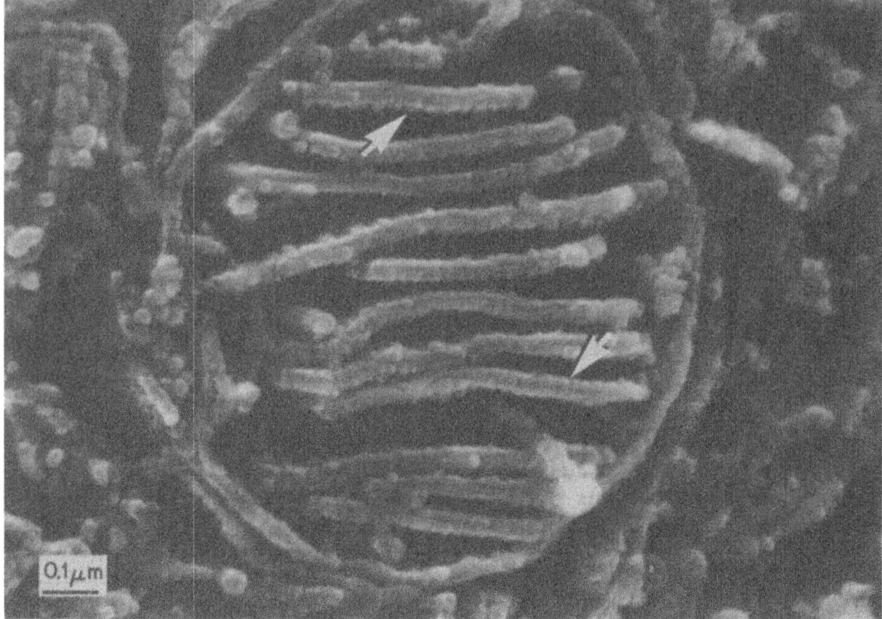

Fig. 5.9 Scanning electron micrograph of a mitochondrion from a rat epididymal cell, prepared by the osmium-DMSO-osmium method. Numerous small granules (arrows) are seen on the membranes of the cristae. Scale bar = 0.1μm. (From Tanaka, 1980.)

Saito, 1972) have resulted in pictures showing details of the internal structure of cell organelles (Fig. 5.9).

Tissue drying poses considerable problems of interpretation. Tissue collapse is relatively readily assessed but the precipitation of cellular solutes and the decoration of existing structures during drying to produce novel features require very careful evaluation. For example, a new cytoskeletal structure, the microtrabecular lattice, seen in thick sections by high voltage EM and in freeze-dried preparations has been described (Wolosewick and Porter, 1979) but claimed as a drying artefact by others (Hirokawa, 1982; Schnapp and Reese, 1982). Certainly, it can be shown that filaments and vesicles can be produced by freeze-drying non-biological samples such as salt, sugar or buffer solutions (Miller *et al.*, 1983). Despite such difficulties in interpretation, the increased electron contrast produced by tissue drying may be necessary to reveal details that are not visible in frozen hydrated preparations (Fig. 5.10).

Freeze-substitution

An alternative to the sublimation of ice is to substitute it with non-aqueous solutes that remain liquid at low temperatures, such as acetone, ethanol or methanol. The tissue is fixed during substitution and can then be embedded

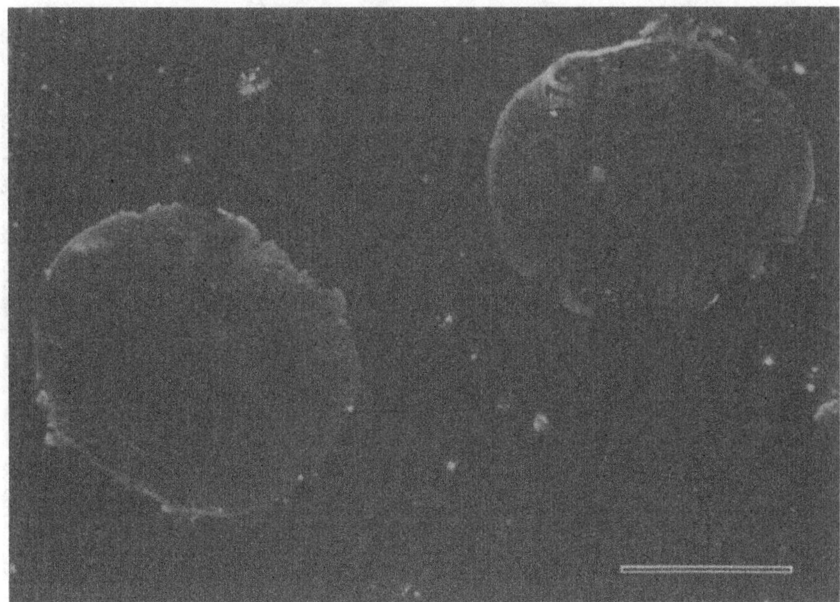

Fig. 5.10 Frozen dried (right) and frozen-hydrated (left) tissue sections of rat renal papillae as seen in the SEM. Note the detail revealed by drying the section. Scale bar = 0.3 μm. (From Saubermann *et al.*, 1981. Reproduced from *The Journal of Cell Biology*, 1981, **88**., 257–67, by copyright permission of the Rockefeller University Press.)

in plastic and sectioned in the normal way. Freeze-substitution allows a comparison of cryofixation with standard chemical fixation and, especially when samples are small so that cooling can be very rapid, freezing appears to be the better method (Fig. 5.11, and Steinbrecht, 1980). For the majority of samples, however, adequate freezing beneath a superficial 20 μm is impossible without cryoprotection, and ice artefacts obscure the normal structure of the cytoplasm (Fig. 5.12).

Freeze-fracture

Freeze drying and substitution involve a change in the state of the frozen specimen before the image is formed. By making a replica of the frozen specimen, this alteration is avoided. The frozen sample is fractured and a metal mould is made of the exposed and still frozen surface. Freeze-fracture was developed by Steere (1957) and by Moor *et al.* (1961) and is now widely used as it provides a unique view of cytoplasmic and, more especially, membrane structure; the planes of fracture pass through the frozen tissue in regions that cannot be explored readily by thin sectioning, for example the intramembrane domain.

The basic essentials of freeze-fracture are:

(1) cleaving the sample;

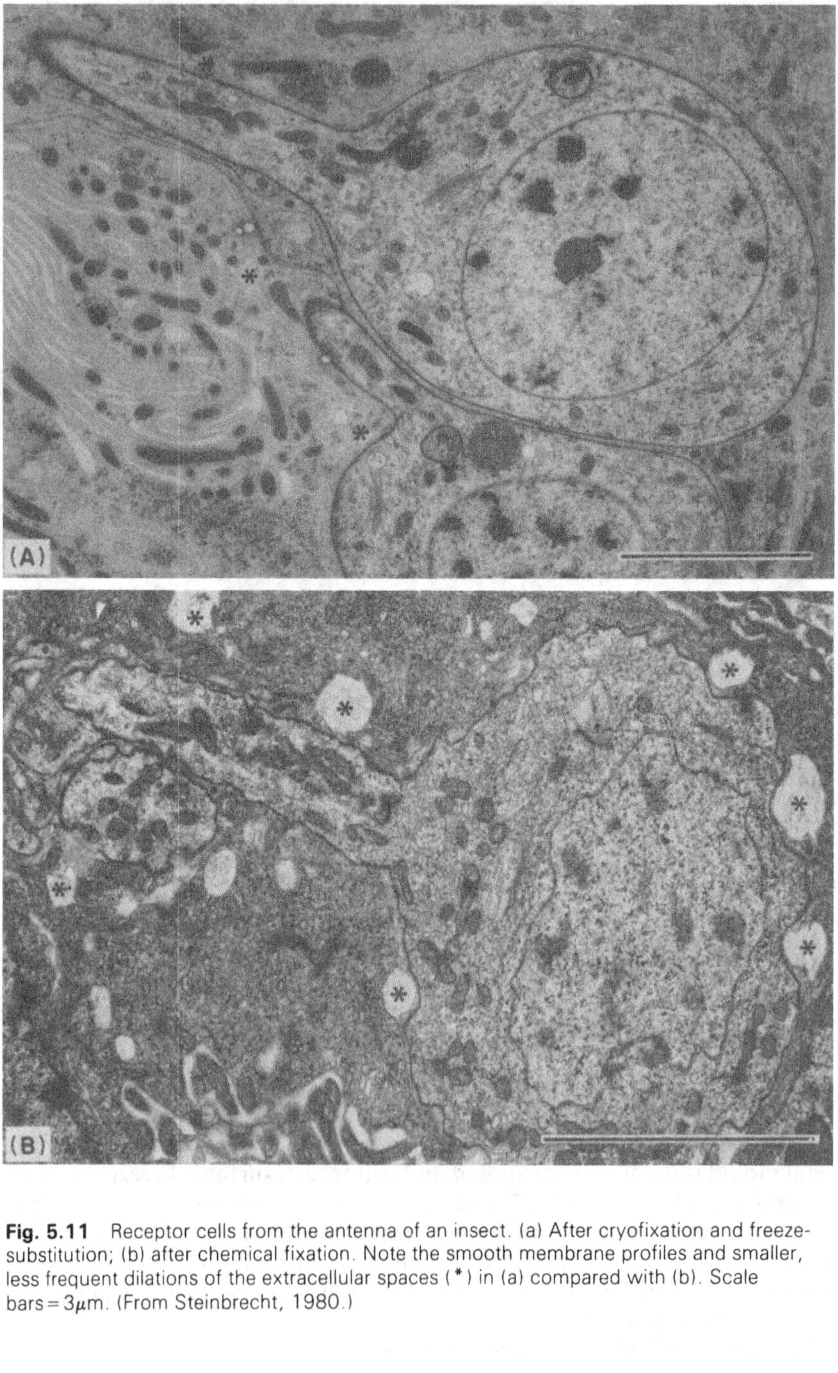

Fig. 5.11 Receptor cells from the antenna of an insect. (a) After cryofixation and freeze-substitution; (b) after chemical fixation. Note the smooth membrane profiles and smaller, less frequent dilations of the extracellular spaces (*) in (a) compared with (b). Scale bars = 3 μm. (From Steinbrecht, 1980.)

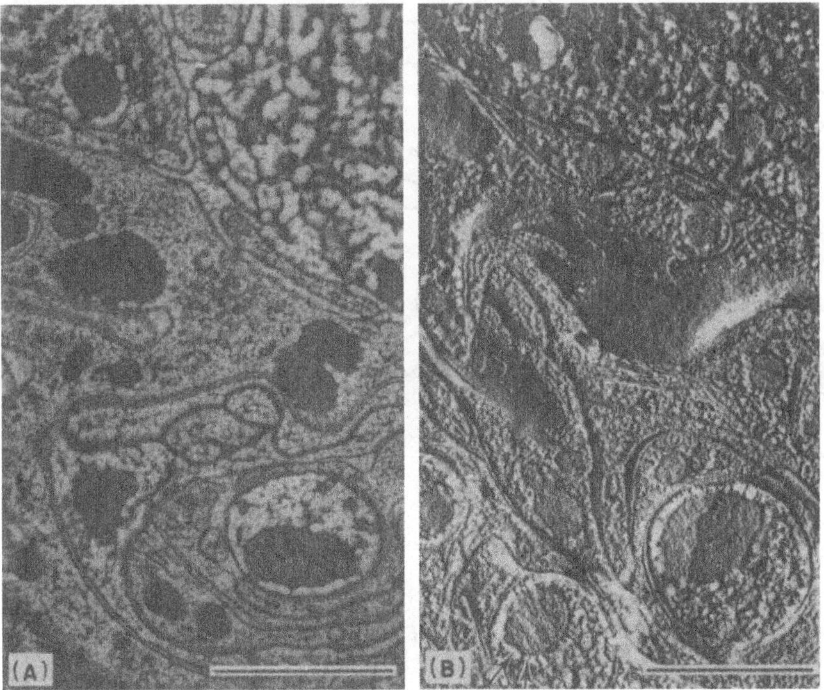

Fig. 5.12 Deeper tissues of the insect antenna, frozen without cryoprotection, showing by freeze-substitution (a) and also by freeze-fracture (b) that the cells are damaged by ice crystals. Scale bars = 1 μm. (From Steinbrecht, 1980.)

(2) shadowing of the exposed surface with heavy metal from an angle (to create electron contrast);

(3) deposition of a film of carbon over the whole surface (to make a coherent replica); and

(4) cleaning away all organic material from the replica after thawing (Fig.5.13).

A range of apparatus has been developed to carry out the first three stages, all of which operate on the same basic principles (Bullivant, 1973, 1974, Sleytr and Robards, 1977b).

(1) Fracturing can be carried out under liquid nitrogen followed by transfer to a vacuum evaporator for metal shadowing. Alternatively, the specimen is mounted on a cold stage (−100°C to −150°C) and the vacuum is established before fracturing. Fracturing of the sample can be achieved by a single blow to the specimen, with the possibility of producing complementary replicas from the fractured halves, or by a series of sweeps with a microtome, which allows the exposure of chosen structures. Cleaving must be carried out at low temperature to ensure against recrystallization and to dissipate as rapidly as possible the local heat produced by the fracturing process, which

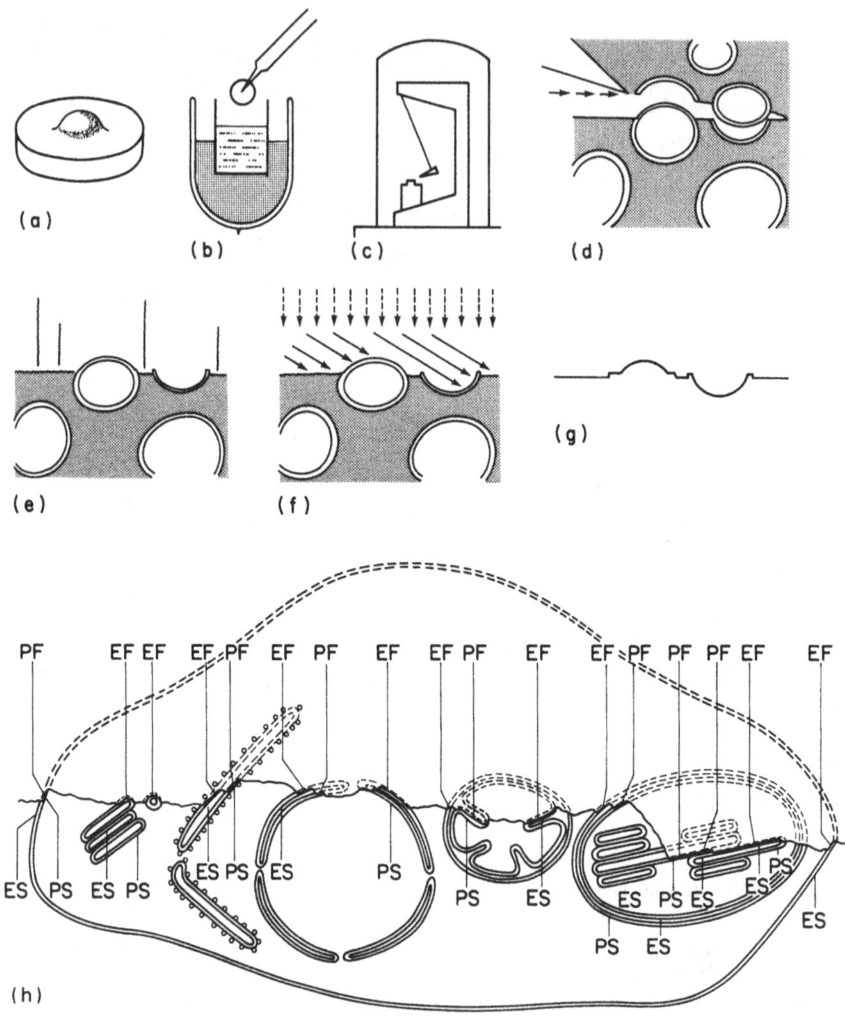

Fig. 5.13 (a)–(g) The basic steps in conventional freeze-fracture procedure, as described in the text (adapted from Branton, 1966). (h) Labels for some frequently studied membrane fracture faces and surfaces. The dark line through the cell and some of its organelles traces the course of a hypothetical fracture prior to etching. PF and EF labels are used here to distinguish the faces of the P and E halves of each membrane produced by fracture, while PS and ES indicate the true hydrophilic surfaces that could be exposed by etching. However, simple P and E labels would suffice when a distinction between fracture faces and surfaces is unnecessary. (From Branton *et al.*, 1975.)

tends to result in the deformation of non-elastic structures (i.e. most macromolecules at low temperature). With some specimens this deformation appears to be unavoidable even at temperatures of liquid helium (Sleytr, 1974; Sleytr and Robards, 1977a).

(2) Typically fractured surfaces are shadowed with a mixture of platinum and carbon (Fig.5.14), although tungsten/tantalum mixtures can also be used (Fig. 5.15) and theoretically produce replicas of higher resolution. The metals are vapourized from an arc or electron beam source at an angle of *c.* 30–50° to the specimen and then condense on the bumps and ridges in the fractured surface, leaving clear the dips and hollows. The resolution of some structures can be enhanced by spinning the specimen table during evaporation (rotary shadowing, Fig. 5.16, Margaritis *et al.*, 1977).

(3) A backing film of carbon is evaporated uniformally over the whole fractured surface from an overhead source to provide mechanical strength. Sufficient carbon must be deposited to form a strong replica but, if the coating is too thick, it will obscure the fine detail of the fractured structures. In practice a thickness of 15–20 nm is normally used.

(4) The replica cannot be examined by TEM until the tissue to which it is moulded is thawed and completely cleaned away. This is normally done by digesting the organic material with bleach, acid, enzymes or detergents, or commonly a combination of such reagents. Moving replicas through cleaning solutions tends to damage them and various methods have been devised to increase their stability during this stage, for example casting a film of collodion over the replica surface, so that it is held together during cleaning and mounting on a grid. The film is then washed away in amyl acetate before the replica is examined microscopically.

Freeze-fracture produces an inorganic replica of the fractured tissue. This means that direct cytochemical, immunological or X-ray analysis is impossible and special methods have been devised to allow the combination of specific labelling with freeze-fracture. However, high resolution microscopy of fractured specimens (compared with standard SEM images) has revealed details of structures previously inaccessible to ultrastructural study. This is because the fracture splits the membrane bilayer, producing two membrane halves. Controversy over the ubiquity of this fracture plane (Sjöstrand, 1979) has been resolved by evidence that membrane splitting is confined to the bilayer continuum. The fracture plane may be deflected at the site of membrane particles and thus, in certain specialized membranes, such as intercellular junctions or mitochondria and chloroplast membranes, the extent of bilayer splitting may be low; this increases the range of observed fracturing patterns (Pinto da Silva *et al.*, 1981a). As a result of bilayer splitting, the hydrophobic interior of membranes is revealed and structural aspects of membrane differentiation can be studied (Fig. 5.17). This has proved of particular importance in the study of areas such as secretion (Satir *et al.*, 1973), synaptic transmission (Landis, 1982), intercellular junctions (Lane and Skaer, 1980) and organelle membranes, for example of chloroplasts (Staehelin, 1976; Staehelin and Arnatzen, 1983), mitochondria, (Sjostrand, 1979) and rough endoplasmic reticulum (Giddings and Staehelin, 1980).

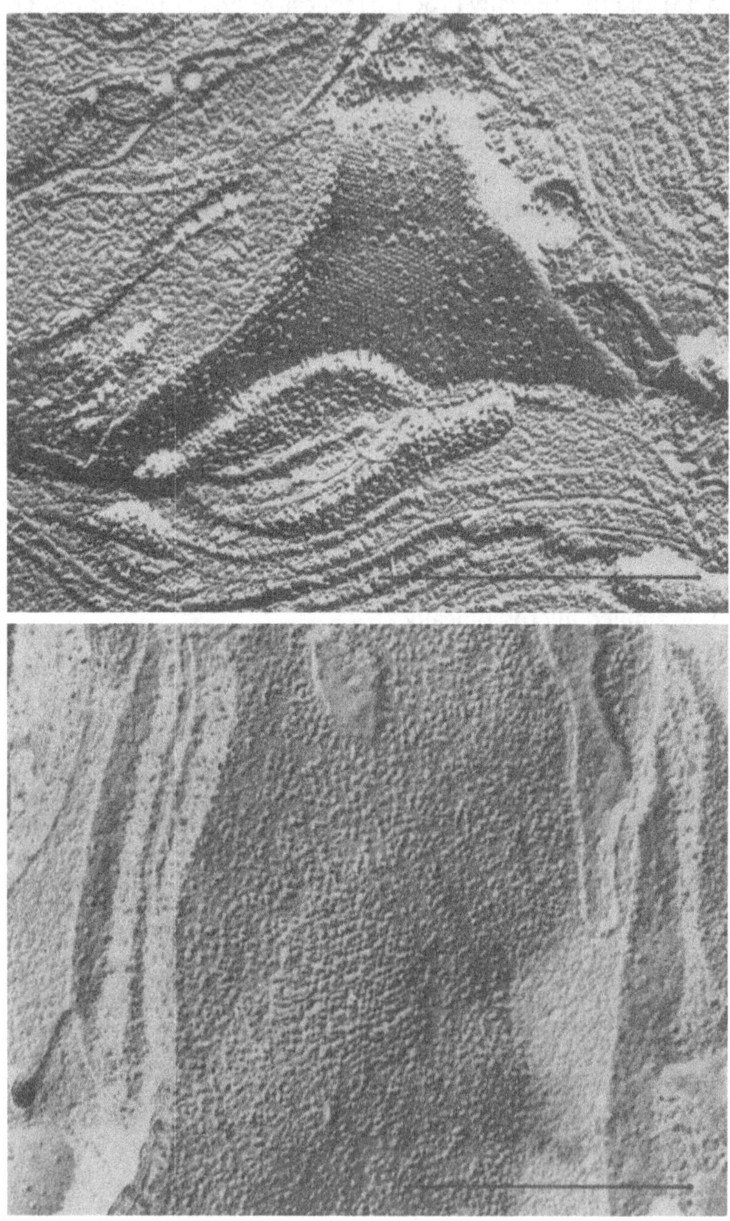

Fig. 5.14, 15 Replicas made by shadowing Pt-C(14) and W-Ta(15) before backing with carbon. Typically, Pt-C produces a more grainy replica with sharp contrast while W-Ta produces a softer, greyer image with a higher theoretical resolution. Scale bars = 0.5μm.

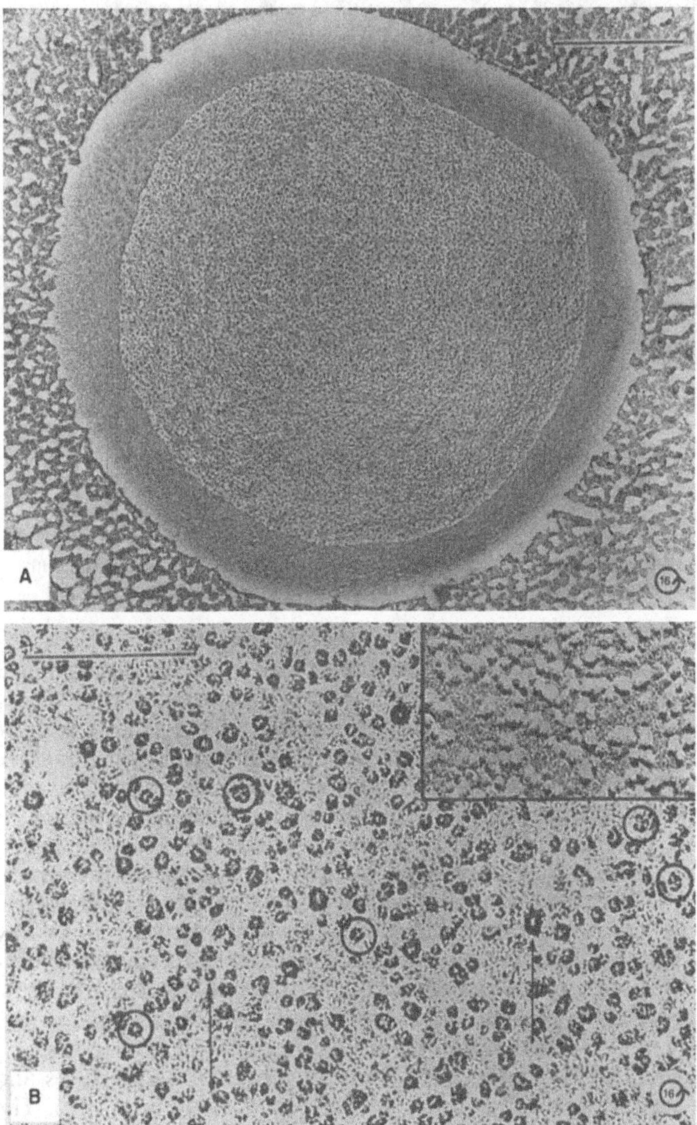

Fig. 5.16 Freeze-etched erythrocyte ghost after rotary replication. (a) The overall contrast is high compared to that of conventional unidirectionally shadowed membranes. Scale bar = 0.5μm. (b) At high magnification many of the intramembrane particles exhibit tetrameric subunit structure (circles) normally not visible after conventional unidirectional shadowing (insert). However, heterogeneity in substructure is also evident (arrows). Scale bar = 0.1μm. (From Margaritis *et al.*, 1977. Reproduced from *The Journal of Cell Biology*, 1977, **72**., 47–56, by copyright permission of The Rockefeller University Press.)

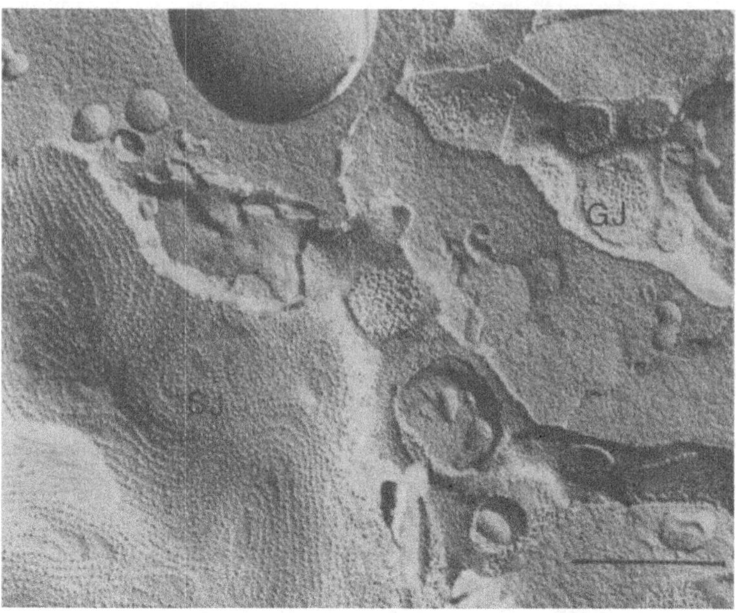

Fig. 5.17 Freeze-fracture replica revealing the intramembrane structure associated with two kinds of intercellular junction, the gap junction(GJ) and septate junction(SJ). Scale bar = 0.5μm.

Ice may be removed from the superficial layers of the fractured specimen if it is left for a short time in the vacuum before the replica is made. This etching requires a precise balance between the temperature of the sample and the pressure in the vacuum chamber to ensure sublimation (water being removed from the specimen) rather than condensation (which would lead to contamination of the fractured surface). The colder the specimen, the higher the vacuum must be (Robards, 1974; Umrath, 1983). Replicas of etched specimens may reveal the surfaces of membranes together with typical fracture-faces (Fig. 5.18). Cytoplasm will etch only to a very limited degree, but by treating the tissue before freezing with chemical fixatives followed by washing with distilled water the soluble phase can be etched away to a significant extent. This will expose something of the internal architecture of the cells. This deep etching technique, in combination with rotary shadowing, (Heuser, 1981) is a powerful tool in the study of cellular ultrastructure.

The nomenclature ascribed to the various membrane fracture faces and surfaces has been standardized (Branton *et al.*, 1975, and Fig. 5.13h). The membrane fracture faces are the PF, referring to that half of the membrane remaining attached to the cytoplasm (protoplasmic), and the EF, referring to that half leaflet attached to the extracellular medium or, in the case of organelles, to the internal domain of the organelle. Where compartments in the cytoplasm are surrounded by two membranes, the situation is more complex, the EFs of the two membranes being those adjacent to the intermembrane space. The genuine surfaces of the membrane, revealed by

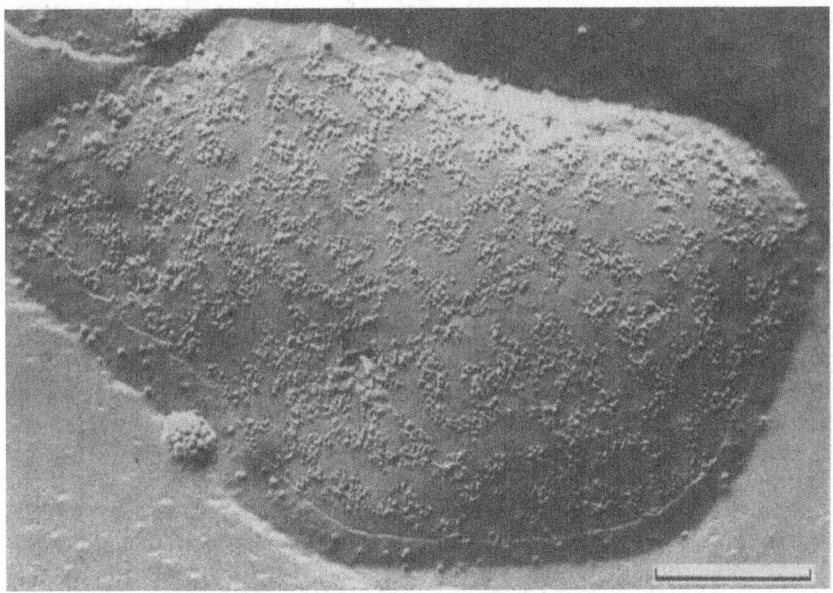

Fig. 5.18 Freeze-etch preparation of a red blood cell. A ridge separates the fracture face from the etched surface. The preparation was labelled with ferritin prior to freezing. Ferritin molecules(F) are associated with the etched surface but not with the fracture face. Scale bar = 0.5μm. (From Branton, 1971.)

etching, are the P and E surfaces, P being the surface exposed by etching away the cytoplasm (PS) and the E the surface revealed by sublimation of the extracellular medium (ES).

The initial hopes, that freeze-fracture would be a technique free of artefacts, (Steere, 1957; Moor, 1964) have proved to be over-optimistic. However, the artefacts that do arise have been described very fully (Stolinski and Breathnach, 1975; Böhler, 1975; Bullivant, 1977; Rash and Hudson, 1979; Sleytr and Robards, 1982) and they can, to a great extent, be avoided. Where this is not possible (e.g. with plastic deformation [Sleytr and Robards, 1977b] and lipid collapse during fracturing and replication [Bullivant, 1977]) they can be taken into account when interpreting the replicas.

Freeze-fracture as a tool in analytical microscopy

Specific labelling

The descriptive role of electron microscopy can be extended by combination with specific labelling techniques such as autoradiography, immunochemistry or cytochemistry, to allow biochemical and spatial analysis of cellular structures. The product of freeze-fracture, a metal replica, is biologically inert so that standard cytochemical probes are inappropriate.

However, the ways in which cytochemical information can be incorporated into, and extracted from, such replicas by specialized techniques is well illustrated by the study of red blood cell membranes (Shotton, 1984). The techniques that have been applied to these erythrocytes are equally useful for other cell membranes.

Ferritin labelling

Biological membranes may be labelled by incorporation of a chemical probe that has a characteristic 3-dimensional structure which will show up in a freeze-fracture replica. This method has been used to demonstrate membrane bilayer separation during fracturing (Pinto da Silva and Branton, 1970). Red blood cells were labelled before freezing with the surface marker ferritin, and, on membranes in etched preparations, characteristic globules of ferritin could be found only where the true outer membrane surface was revealed (Fig. 5.18).

Cationized ferritin binds preferentially to the anionic, sialic acid rich N-terminal ends of glycophorin molecules (a major protein of the red cell membrane), and this property has been exploited to demonstrate, indirectly, the probable continuity of the surface marker with the IMPs found on the PF of erythrocyte membranes. Clustering of the IMPs induced by incubation at pH 5.0 and 35°C was mirrored by clustering of the ferritin labelled surface marker (Pinto da Silva *et al.*, 1973). Cationized ferritin binding has also allowed an analysis of the progressive removal of sialic acid residues from erythrocyte membranes with age. The cationized ferritin binding sites decrease in older erythrocytes (Danon *et al.*, 1972). As well as anionic sites, particular antigens can be labelled by conjugating ferritin with antibody IgG fractions (Pinto da Silva *et al.*, 1971).

Surface labelling of membranes can be specific for particular proteins if the ferritin is conjugated with a specific cytochemical or immunochemical marker or if a specific label is used that itself has a characteristic shape in etched preparations. In this way it has been possible to label Band 3 components with concanavalin A-ferritin conjugates (Pinto da Silva *et al.*, 1971; Pinto da Silva and Nicolson, 1974), and glycophorin with plant lectins and influenza virus (Grant and McConnell, 1974), and to establish an association between these two molecules to form oligomeric structures (Pinto da Silva and Nicolson, 1974). Furthermore, immunoferritin labelling has revealed an association between these oligomeric structures and the peripheral protein spectrin (Shotton *et al.*, 1978).

Membrane reconstruction

Modification of membranes before freeze-fracture can contribute significantly to an understanding of the organization of normal, undisturbed membranes. Specific digestion of red blood cell membranes with the proteolytic enzyme pronase indicated that the intramembrane particles were predominantly proteinaceous (Engstrom, 1970; Branton, 1971; Branton and Deamer, 1972), and reconstitution experiments have shown that intramembrane particles reappear when anion channel protein, Band 3, is added to lipid vesicles, but not when glycophorin alone is added (Yu and Branton, 1976; Grant and McConnell, 1974; Branton and Kirchanski, 1977) (Fig. 5.19). Furthermore, when a spectrin–actin extract from fresh red cell ghosts was added to vesicles containing Band 3 components, freeze-fracture revealed that pH-dependent aggregation patterning, known to be mediated by spectrin–actin activity in intact ghost membranes (Elgsaeter *et al.*, 1976), was restored (Yu and Branton, 1976).

Specific pre- and post-fracture labelling

The associations that proteins make within and across the membrane can be investigated by their behaviour when the frozen membrane fractures. This was exploited indirectly when Fisher (1976a) devised a method for harvesting

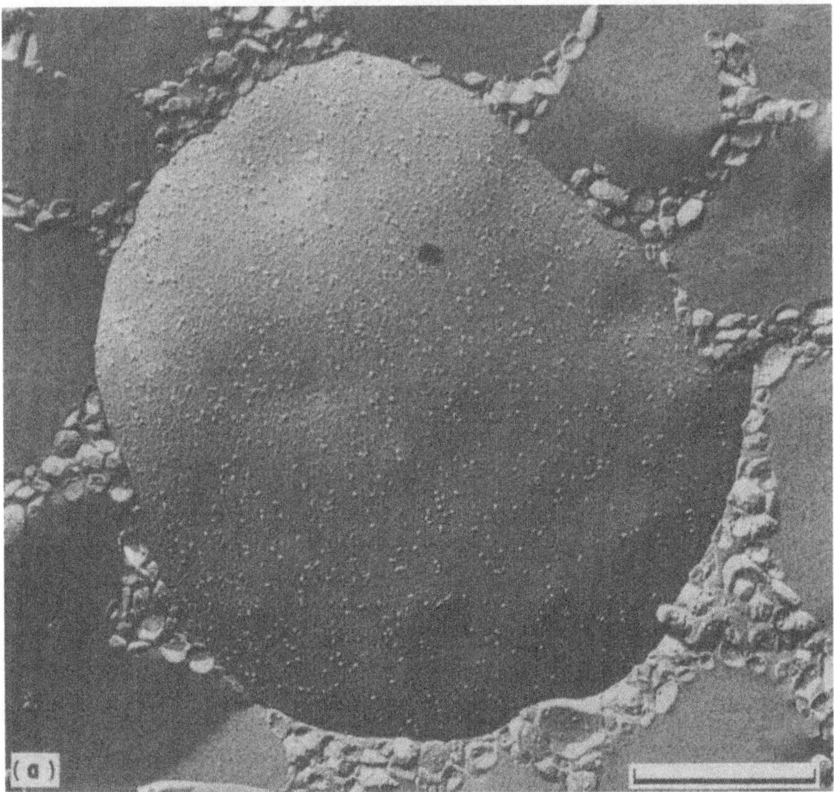

Fig. 5.19 (a), (b and c see p. 202) Freeze-etched phosphatidyl choline vesicles. (a) Recombinant vesicles containing a crude band 3 preparation. Note the particles which are similar in appearance to those of freeze-etched erythrocyte membranes. (b) Similarly prepared vesicle containing lipid only. The fracture face is smooth and devoid of particles. (c) Recombinant vesicle containing glycophorin. The protein produces very small particles easily distinguished from those produced by Band 3. Scale bar = 0.5μm. (From Branton and Kirchanski, 1977.)

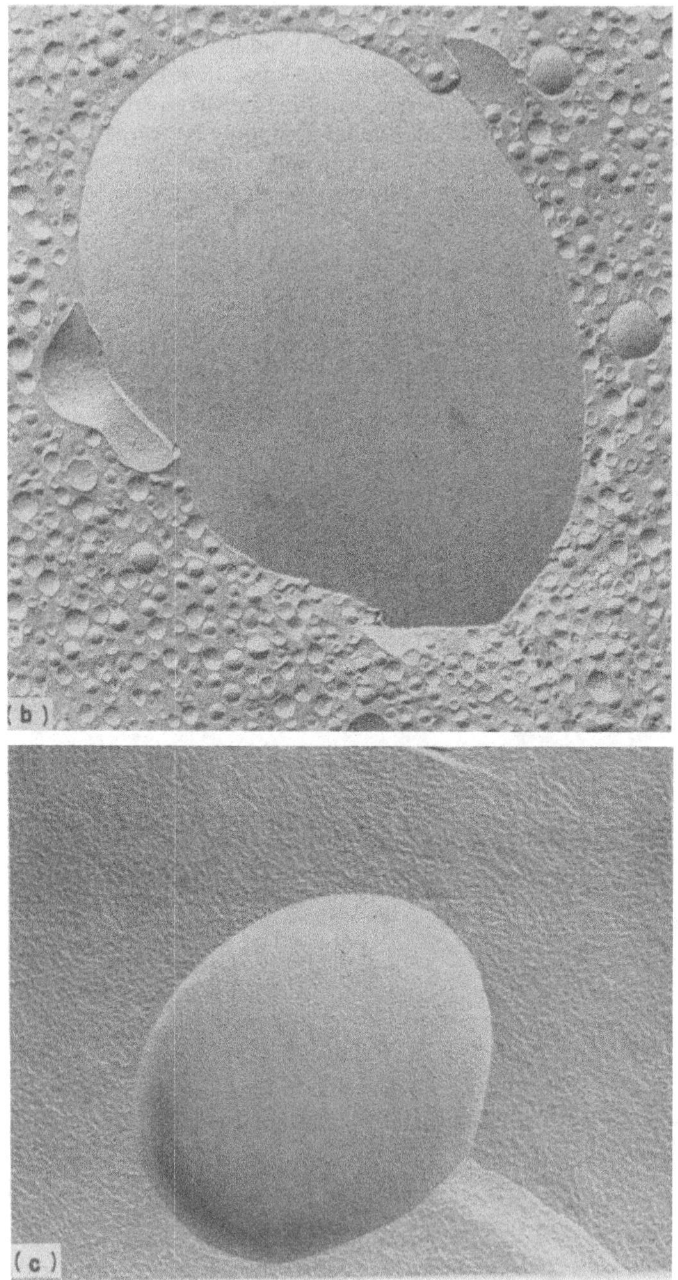
(b)
(c)

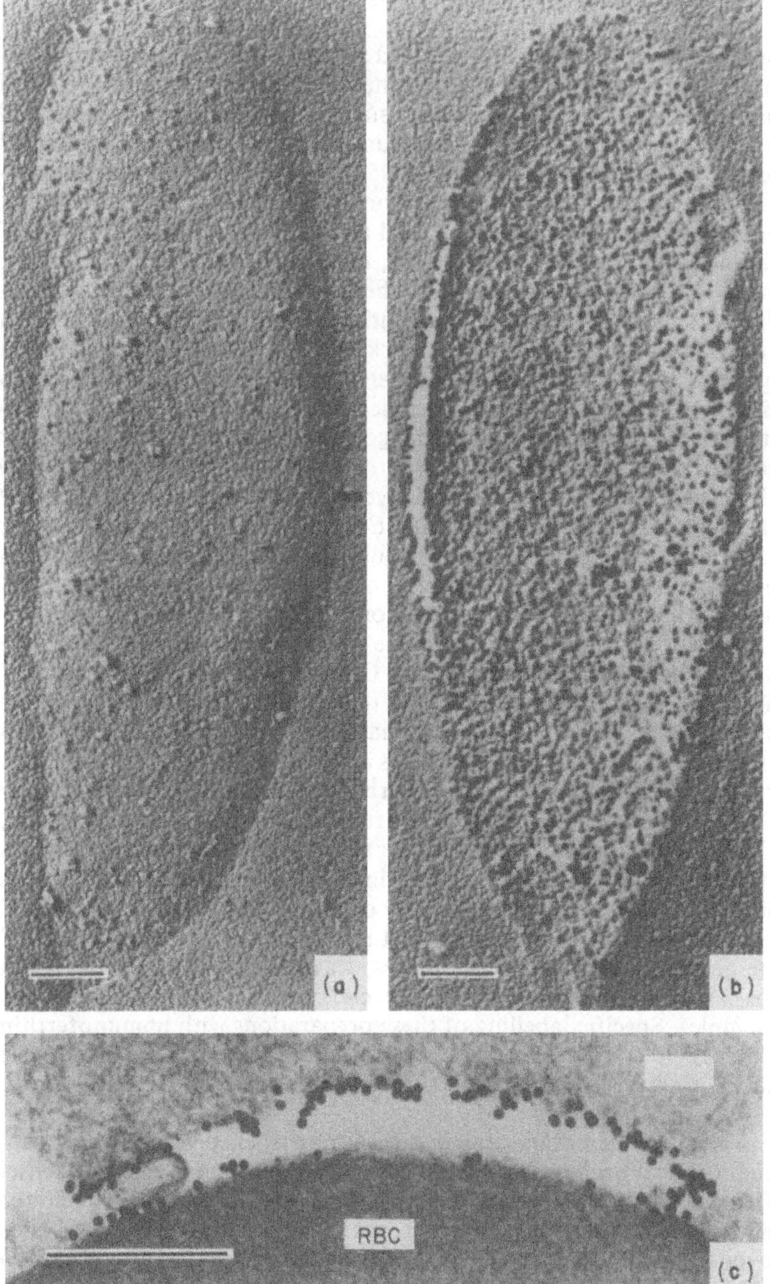

Fig. 5.20 Partition of WGA binding sites between P and E faces as observed in (c) thin section and (a,b) replicated red blood cells after post fracture labelling. Both preparations show that label is strong over the EF and weaker over the PF. Scale bars = 0.5 μm(a,b); 0.1 μm(c). (From Pinto da Silva and Torrisi, 1982. Reproduced from *The Journal of Cell Biology* 1982, **93**., 463–69, by copyright permission of the Rockefeller University Press.)

fractured membrane halves. This allowed an analysis of the asymmetric distribution of lipids in erythrocyte membranes (Fisher, 1976a) and also of Band 3 components and sialoglycoproteins such as glycophorin. Biochemical analysis of these freeze-fractured membrane halves indicated that the Band 3 component associated with the inner half (PF) and the glycoproteins associated with the outer half (EF) of the membrane (Edwards *et al.*, 1979). An ingenious post-fracture labelling method allows structural confirmation of this finding (Pinto da Silva *et al.*, 1981a; Pinto da Silva and Torrisi, 1982). Specimens are frozen and fractured and then thawed out before cytochemical labelling with markers (wheat germ agglutinin [WGA] for glycophorin, Con A for the Band 3 component). These markers can be made electron dense by conjugating them with colloidal gold or ferritin. The labelled membranes can then either be critical point dried and replicated followed by cleaning in hypochlorite (leaving the unaltered colloidal gold incorporated in the replica) or they can be freeze-substituted and prepared for thin-sectioning. Figure 5.20 provides evidence that WGA-labelled glycophorin fractures predominantly with the outer leaflet of the membrane (EF) but also serves to illustrate the potential of these techniques. The distribution and fracturing characteristics of specified components of any cellular membrane can be investigated with replicas of critical point dried preparations, and thin sections of fracture-label preparations are of particular value when it is necessary to determine which cell or organelle has been labelled (Pinto da Silva *et al.*, 1981b).

Rash (1979) has developed a method that combines freeze-fracture and thin sectioning to allow a direct correlation of sectioned and replicated images. Fixed and cryoprotected samples are frozen and fractured and then rather lightly shadowed. The backing carbon layer may be omitted to allow post-thaw cytochemical labelling. After replication, specimens are thawed, post-fixed, dehydrated and embedded for subsequent sectioning. The blocks are cut approximately parallel to the original replica surface, so that sections revealing both fixed cells and replica are obtained (Fig. 5.21). This method has been used to identify membrane features associated with particular cell types (e.g. fast versus slow muscle fibres, Rash *et al.*, 1981) and, as serial sections can be cut, maximal information can be extracted from rare or very small samples. Specific labelling of these preparations with immunoferritin or colloidal gold conjugates after fracture and partial replication is possible and has allowed the identification of acetylcholine receptors (Rash *et al.*, 1982).

The proteins of membranes can be specifically labelled with cytochemical or immunological probes that, by conjugation, can be made electron dense. Specific markers for the identification of membrane lipids are less exploited and, as yet, are less specific. However, molecules that complex with lipids and so modify their appearance in freeze-fracture replicas have been used to investigate the distribution of lipid types in cell membranes. The polyene antibiotic, filipin, complexes with 3-hydroxysterols, for example cholesterol (de Kruijff and Demel, 1974) to produce distinct circular deformations *c.* 25 nm in diameter (Fig. 5.22). This molecule has been used widely to localize cholesterol-rich regions of cell membranes (Friend, 1982; Feltkamp and van der Waerden, 1982a). Other probes for 3-hydroxysterols include digitonin

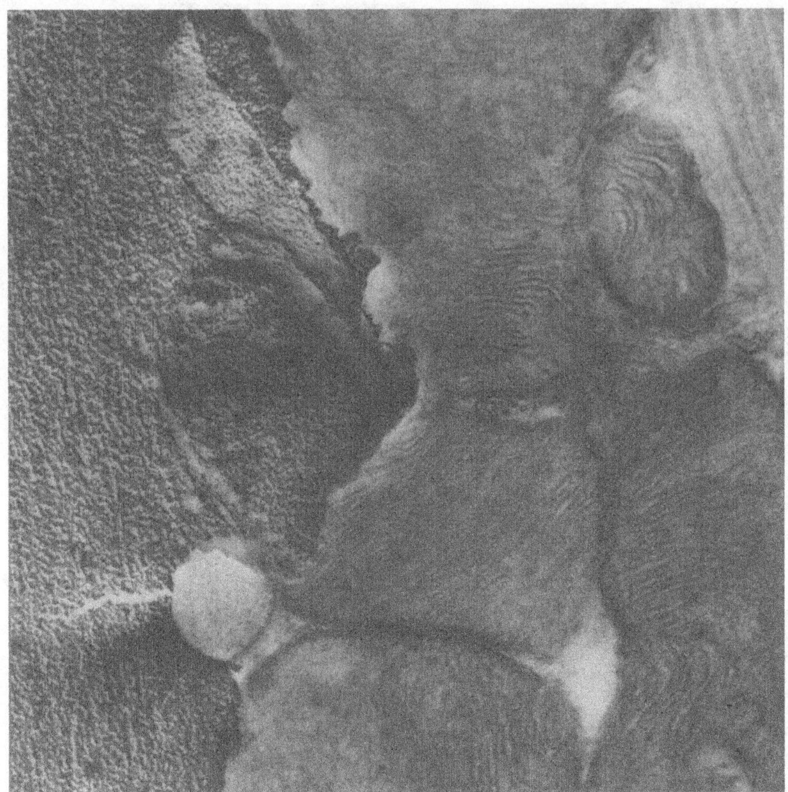

Fig. 5.21 Section-replica of rabbit myocardium. A conventional thin-section view of the myofilaments and mitochondria are visible to the right of the field, and replicated myofilaments and a mitochondrion to the left. (Kindly provided by Severs.)

(Elias *et al.*, 1978) and the saponin tomatin (Elias *et al.*, 1979) which complex to produce ridge-like deformations in freeze-fracture replicas. Filipin and tomatin have different sensitivities to cholesterol (Elias *et al.*, 1979) and have recently been shown to lack equivalence in their labelling of cholesterol in smooth muscle membrane (Severs and Simons, 1983). These experiments call into question the reliability of both probe molecules. Indeed filipin binding is affected not only by closely-packed IMPs in the membrane (Severs *et al.*, 1981), but also by the association of peripheral protein with the membrane (Feltkamp and van der Waerden, 1982a; Gotow and Hashimoto, 1983) and by restricted access to cholesterol as a result of the packing of phospholipids or binding of cholesterol to certain proteins (Pal *et al.*, 1980; Henderson *et al.*, 1979; Feltkamp and van der Waerden, 1982a). Any arrangement of components that renders the membrane too rigid to be readily deformed may reduce the efficacy of filipin labelling (Severs and Simons, 1983).

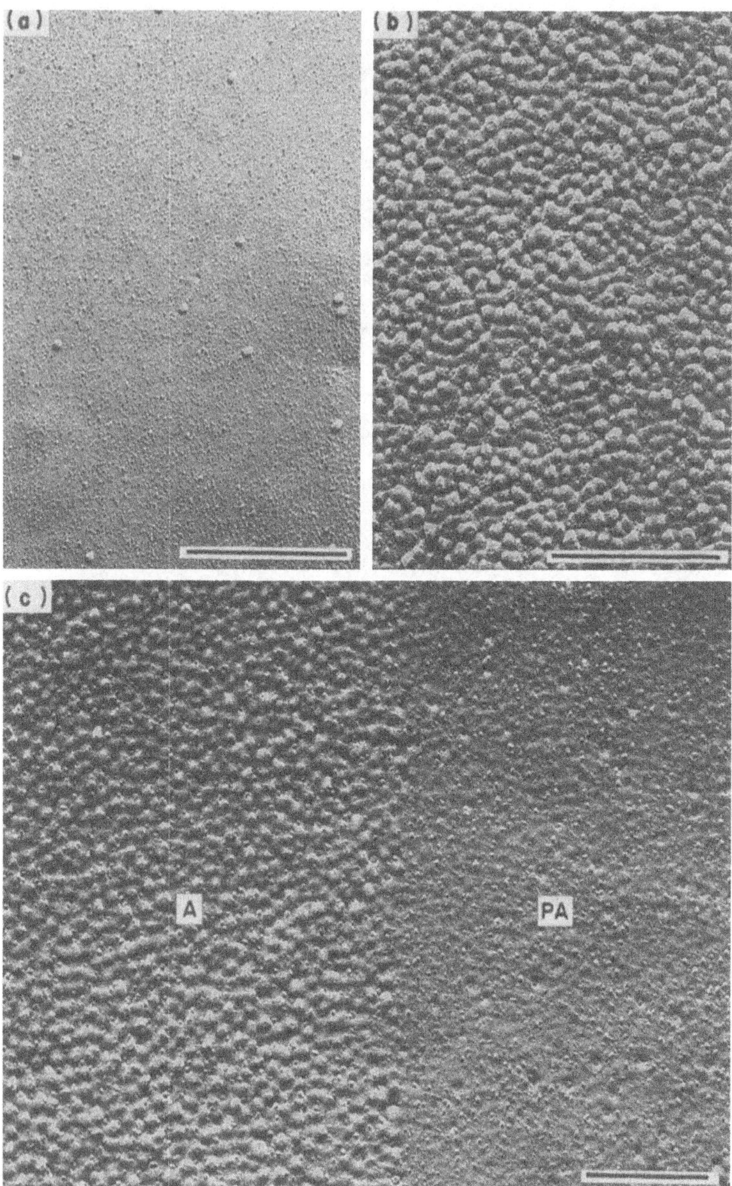

Fig. 5.22 (a) Filipin/sterol complexes on the PF of a *Drosophila* larval cell from a tissue-culture population of cells containing 1.5×10^{-3}µmol sterol/g protein. (b) Filipin/sterol complexes in the plasma membrane of a larval cell population which contained 18.4×10^{-3}µmol sterol/g protein. (c) Acrosomal(A) :post-acrosomal (PA) junction of a capacitated sperm plasma membrane. The fusigenic acrosomal portion contains four times the number of sterol/filipin complexes than the stable post-acrosomal segment. Scale bars = 0.5µm. (From Friend, 1982. Reproduced from *The Journal of Cell Biology*, 1982, **93**., 243–9, by copyright permission of The Rockefeller University Press.)

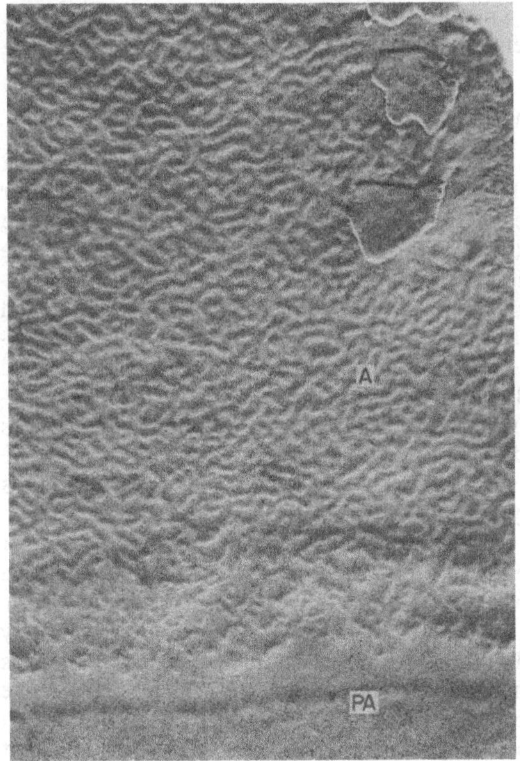

Fig. 5.23 Polymixin B labelling of the acrosomal(A)/post-acrosomal(PA) area of sperm plasma membrane. The anionic phospholipid-rich acrosome is crenelated, leaving the post-acrosomal region unaltered. (Kindly provided by Friend.)

Another class of lipid can be localized by the antibiotic polymixin B, which interacts with membrane rich in anionic phospholipids to produce characteristic crenelations of the membrane in freeze-fracture replicas (Bearer and Friend, 1980, 1982; Fig. 5.23).

The potential value of these labels in cell biology is demonstrated by a review on the membrane domains of vertebrate sperm (Friend, 1982). The labels are also of great potential in low temperature biology, allowing an analysis of the lipid components of membranes during chill and cold adaptation when changes in composition have been implicated (Clarke, 1981; Cossins, 1981). The way in which different lipid types are sorted and excluded from the membrane during acclimation (Gordon-Kamm, 1983) is also an intriguing problem. Furthermore, the use of these labels might reveal the processes of lipid:lipid phase separation that may occur when membranes are cooled, paralleling the more familiar protein:lipid phase separation (Feltkamp and van der Waerden, 1982b).

Further details of the use and pitfalls of specific membrane probes are discussed in a review by Severs and Robenek (1983).

Freeze-fracture autoradiography (FARG)

The attraction of FARG, as with conventional autoradiography, is that it allows the possibility of analysis in space and time of identified molecular species. In addition, the cleavage of membranes produced by fracturing frozen specimens allows autoradiographic analysis within and across the membrane; this is not possible with thin sectioning techniques. The possibility of laying down photographic emulsion on fractured and shadowed specimens, either *ex vacuo* after thawing or preferably *in vacuo* at low temperature immediately after shadowing has been explored (Fisher and Branton, 1976). Using red blood cell ghosts labelled with ^{125}I-lactoperoxidase-glucose oxidase, which tags both sides of the membrane, it was possible to produce autoradiographs with a half distance (HD) (i.e. the distance from the membrane within which 50 per cent of the developed grains are found) of approximately 250 nm (Fig. 5.24). However, this could be done only by the *ex vacuo* method; emulsions were insufficiently flexible at low temperature to allow deposition on frozen specimens. A considerable problem was the very low level of efficiency (< 1 per cent). Specimens are exposed to the emulsion at − 80°C and a drop in efficiency of approximately 30 per cent might be expected to result from this. A further factor involved is the thickness of the parlodian–gelatin strengthening film which separates the replica from the emulsion (50–90 nm). A similar FARG method in which films remained flexible at low temperature by incorporation of 4 per cent

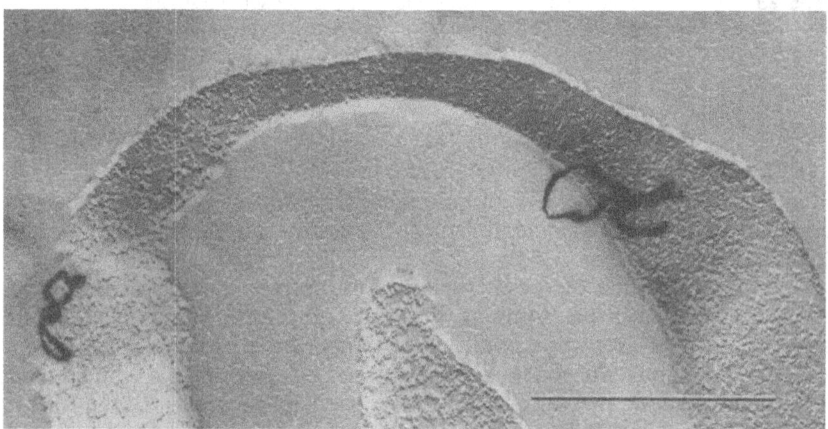

Fig. 5.24 FARG preparation of lactoperoxidase-glucose oxidase ^{125}I-labelled erythrocyte ghost membrane. 8.5 day exposure at − 80°C. Silver grains overlying the EF indicate the presence of radioisotope on the exterior half of the membrane. Scale bar = 1.0μm. (From Fisher and Branton, 1976. Reproduced from *The Journal of Cell Biology*, 1976, **70**., 453–8 by copyright permission of The Rockefeller University Press.)

glycerol into the emulsions has been reported (Rix *et al.*, 1976). Tissue-emulsion separation was reduced to the thickness of the replica (2–3 nm) by applying the strengthening film of gelatin over the emulsion and, by developing new spreading techniques, films were deposited on fractured specimens at − 100°C under vacuum (Schiller *et al.*, 1978a). Using these techniques, autoradiography of whole tissue pieces has been possible (Rix *et al.*, 1976; Schiller *et al.*, 1978b), but precise localization of cell-specific radioactive markers has not yet been achieved.

A simple technique of FARG was developed to investigate surface-related events such as exocytosis or membrane insertion (Skaer and Skaer, 1979). This method reverses the order of fracture and photographic exposure, so that the tissue with its developed, adherent autoradiographic emulsion is freeze-fractured and an analysis can then be made of those areas where the fracture passes through the cell surface with adjacent developed grains (Fig. 5.25).

FARG for tissue pieces introduces two further problems in technique. Cryoprotection by permeant molecules is not possible, since soluble radioactive molecules might be displaced. The use of non-penetrating protectants, such as polyvinyl pyrrolidone, may be appropriate in this circumstance (Rix *et al.*, 1976). Replica cleaning also represents a problem as silver grains are dissolved by acid and bleach. Strengthening the replica-emulsion-gelatin

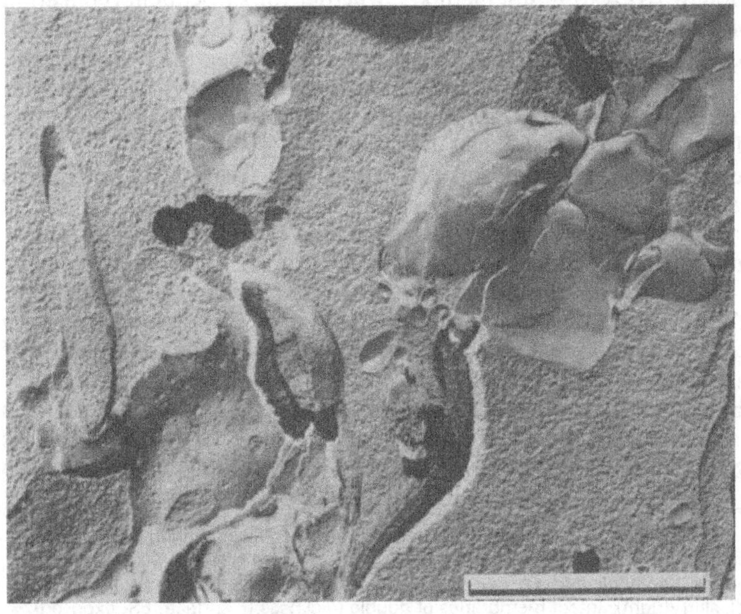

Fig. 5.25 Freeze-fracture autoradiograph of human platelets labelled with ^{3}H-arachidonic acid which, on platelet aggregation, becomes localized in lipid-rich surface structures protruding from the cell surface. Silver grains have developed in the emulsion adjacent to these structures. Scale bar = 1.0μm.

sandwich with an extra carbon layer before gentle cleaning in 1N KOH has been used to resolve this difficulty (Rix *et al.*, 1976). Alternatively, a solvent such as biological detergent can be used, without damaging the silver grains (Skaer and Skaer, 1979).

More recently FARG has been exploited in conjunction with the splitting of membranes and the harvesting of the two separate halves (monoFARG) (Fisher, 1976b, 1978, 1982a,b). Membranes can be prepared intact, by cell lysis, or fractured after a monolayer of cells is stuck down on a polycationized surface (Fig. 5.26). This method has proved useful for red blood cell membranes (Fisher, 1982a; Nermut and Williams, 1980) with an efficiency of 25–45 per cent and an HD of 150 nm (Fisher, 1981). The advantage of monoFARG, as with post-fracture labelling techniques, lies in the possibility of analysis across the membrane. The resolution of standard autoradiography is in the region of 100 nm, and therefore of no use in transmembrane resolution. By splitting the membrane or by using preparations of the membrane surface, the distribution of membrane components and the sidedness of intercalated proteins can be analysed. Furthermore, pulse-chase experiments allow an analysis of events with time, although this has not yet been exploited with either FARG or monoFARG.

Deep etching techniques

A limitation of freeze-fracture methods, whether for subsequent replication and TEM or as a preparative procedure for SEM, has been the difficulty of etching cryoprotected cytoplasm to reveal inner structures of the cell. Recently, methods have been devised to wash out soluble cytoplasmic components and deep etch, or critical point dry, the fractured specimen to reveal details of the cytoskeleton and cell organelles.

Deep etching of the cytoplasm, after ultrarapid freezing, is possible as cryoprotectants are not necessary to ensure acceptable fixation, at least in superficial layers of the sample. By combining deep etching with rotary shadowing, intricate 3-dimensional details of cytoplasmic structures can be revealed (Heuser and Saltpeter, 1979; Hirokawa and Heuser, 1981). (Fig. 5.27). However, deep etching of untreated cytoplasm results in a heavy deposition of salts and other soluble components. Glutaraldehyde fixation of tissue fragments followed by washing in distilled water before freezing solves this problem but inevitably introduces the possibility of other artefacts (Heuser and Saltpeter, 1979; Heuser, 1981). This method produces replicas revealing information about cytoplasmic structures at high resolution

Fig. 5.26 Summary of the monoFARG method. (a) sequence for preparing intact membranes(left) and fractured 'half' membranes(right). Cells are labelled with radioisotope, column-purified, and applied to a planar cationic surface, PL-glass, either after dilution(left) or as a thick slurry(right). Intact membranes of double thickness (E surface, ES, exposed) and single thickness (cytoplasmic surface, PS, exposed) are prepared for lysing and freeze-drying bound cells(left). Split membranes (E fracture face, EF, exposed) are prepared by monolayer freeze-fracture(right). Intact and split samples are processed together for heavy metal shadowing and autoradiography. After exposure and photographic processing, emulsion-coated replicas are stripped from the glass and examined by TEM. (From Fisher, 1981.)

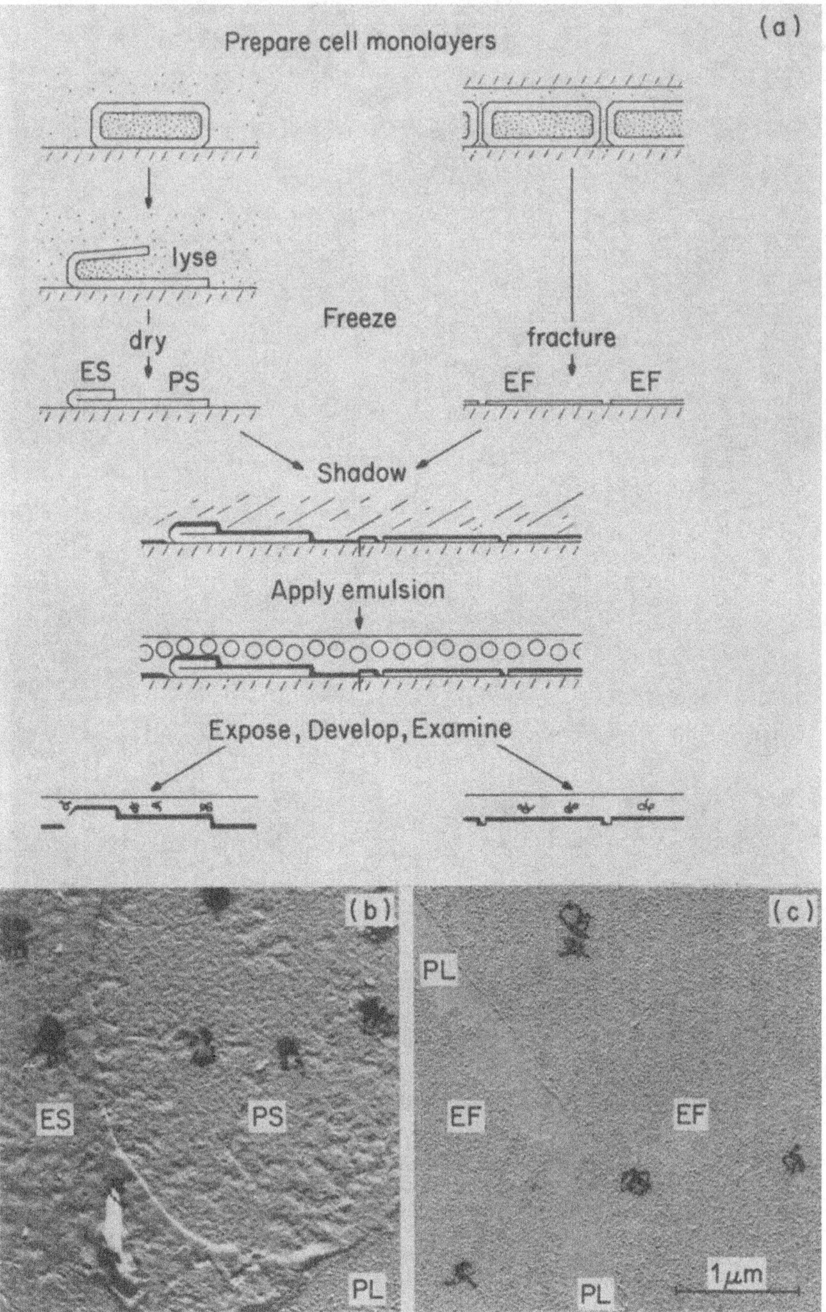
(a)
Prepare cell monolayers
lyse
Freeze
dry
fracture
ES
PS
EF
EF
Shadow
Apply emulsion
Expose, Develop, Examine
(b)
ES
PS
PL
(c)
PL
EF
EF
PL
1 μm

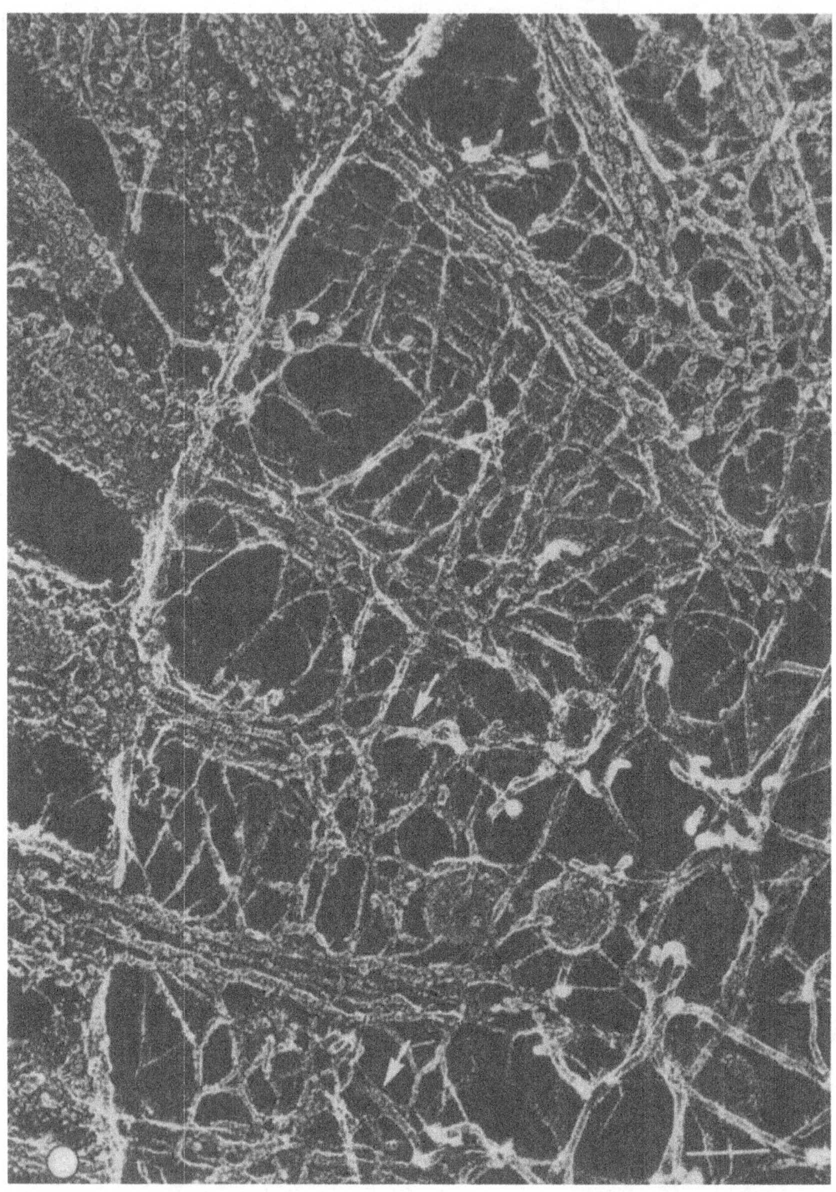

Fig. 5.27 Deep-etch rotary shadowed preparation of the terminal web of a fractured intestinal cell, to show (1) the internal extensions of the bundles of filaments that form microvillar cores, (2) the underlying foundation of intermediate filaments, which are clearly thicker and smoother than the actin filaments in the bundles, and (3) the anastomosing network of thin wisps that are found in between the above elements. These wisps appear to be neither actin nor intermediate filaments and may be the critical structural links in the terminal web. Intermediate filaments sometimes loop up and tangle with core bundles (arrows). Scale bar = 0.1 μm. (From Hirokawa and Heuser, 1981. Reproduced from *The Journal of Cell Biology*, 1981, **91**., 399–409 by copyright permission of The Rockefeller University Press.)

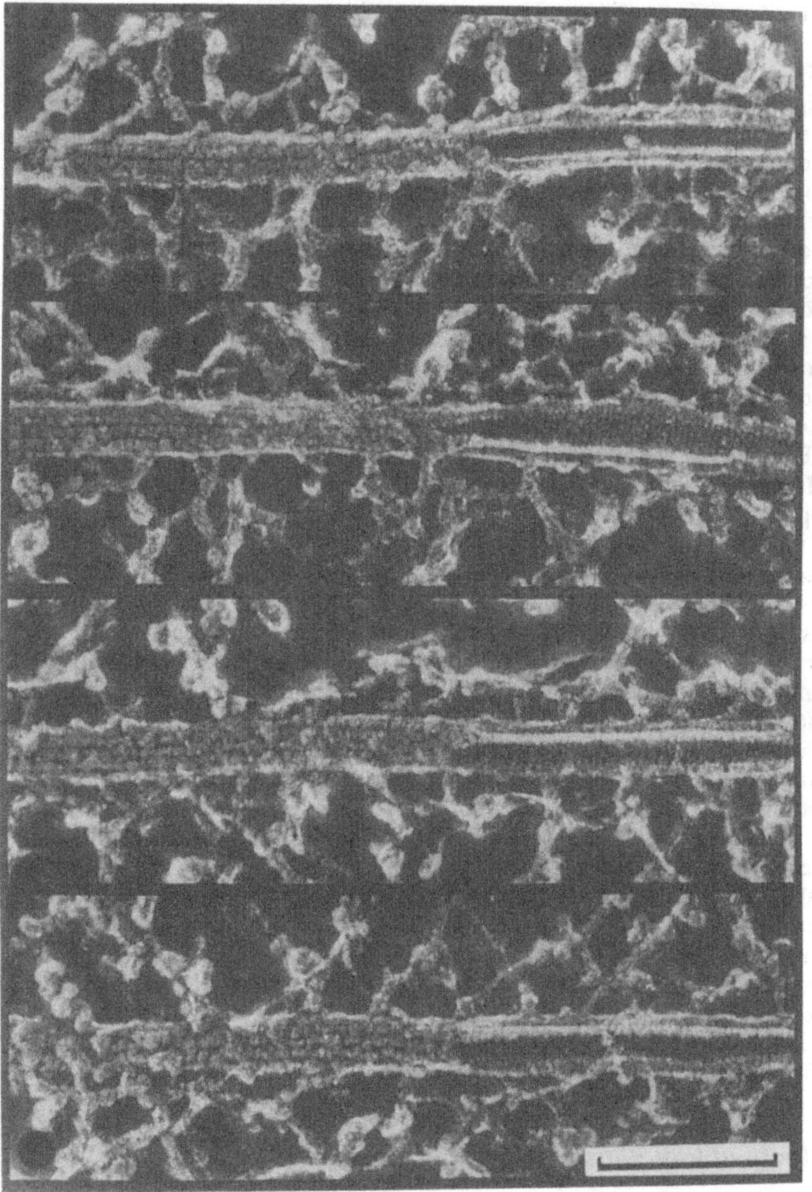

Fig. 5.28 High magnification of purified microtubules that were fractured and deep etched after quick-freezing, to illustrate the resolution of the rotary replication technique. In the example shown, the left half of the field illustrates the outer surface of the microtubule, which displays longitudinal bands of bumps spaced 5.5 nm apart, which may represent the microtubule's protofilaments. To the right of each figure, the microtubule is fractured open to reveal the inner luminal wall, which displays characteristic oblique striations separated by 4 nm. This 3-strand helix has been seen before only in optically filtered electron micrographs. Scale bar = 0.1 μm. (From Heuser and Kirchner, 1980. Reproduced from *The Journal of Cell Biology*, 1980, **86**., 212–34 by copyright permission of The Rockefeller University Press.)

(Fig. 5.28) and can be used in combination with specific labelling techniques (Fig. 5.29) (Heuser and Kirchner, 1980, Hirokawa and Heuser, 1981; Hirokawa *et al.*, 1982) and membrane reconstruction (Roof *et al.*, 1982). In addition to the study of true membrane surfaces (Heuser and Saltpeter, 1979; Roof and Heuser, 1982), deep etching is suited to the elucidation of dynamic membrane events such as vesicle release and coated pit formation (Heuser and Evans, 1980) and has proved of particular importance in the study of the cytoskeleton (Heuser and Kirchner, 1980; Gulley and Reese, 1981). The interactions of different components of the cytoskeleton with cell membranes can be analyzed by, for example, antibody labelling and myosin S1 decoration of actin components before deep etching and rotary shadowing (Heuser and Kirchner, 1980; Hirokawa and Heuser, 1981; Hirokawa and Tilney, 1982; Hirokawa *et al.*, 1982, 1983).

Elements of the cytoskeleton are known to be unstable at low temperature (Morris and Clarke, this volume) and so these techniques have an obvious application in the elucidation of low temperature injury and in the assessment of modifications of the cytoskeleton in those species that can tolerate cold. The microtrabecular lattice (Wolosewick and Porter, 1976, 1979; Porter and Tucker, 1981; Ellisman and Porter, 1980; Schliwa and van Blerkom, 1981) has been described as a system of fine strands 3–6 nm in diameter, which form a 3-dimensional network connecting the cell membrane, organelles and components of the cytoskeleton such as microtubules and filaments. The evidence for such a network is largely from high voltage electron microscopy of whole cells and is the subject of some controversy (Gray, 1975; Small and Langange, 1981; Hirokawa, 1982). Deep etching, combined with rotary shadowing, can reveal such filaments, and fine cross-linker filaments between neurofilaments, microtubules and membranous organelles have been demonstrated in frog axons (Hirokawa, 1982). However, the microtrabecular lattice has not been observed in rapidly frozen preparations of fibroblasts and axons (Heuser and Kirchner, 1980; Schnapp and Reese, 1982; Hirokawa, 1982). Whether or not the lattice is an artefact is of some importance to cryobiologists as the existence of a high density of fine filaments in the cytoplasm may well reduce the mobility of water molecules, thus influencing both their solvent characteristics and their readiness to form ice (Clegg, 1982).

Effects of low temperature on fine structure

Membranes

All cells are surrounded by a plasma membrane composed of lipid and protein. Furthermore, eukaryotic cells' contain membrane bounded

Fig. 5.29 A terminal web decorated with myosin S_1. The actin filaments assume the appearance of a two-stranded, twisted rope, while the intermediate filaments to the right have not been decorated. Scale bar = 0.1 μm. (From Hirokawa *et al.*, 1982. Reproduced from *The Journal of Cell Biology*, 1982, **94**., 425–43 by copyright permission of The Rockefeller University Press.)

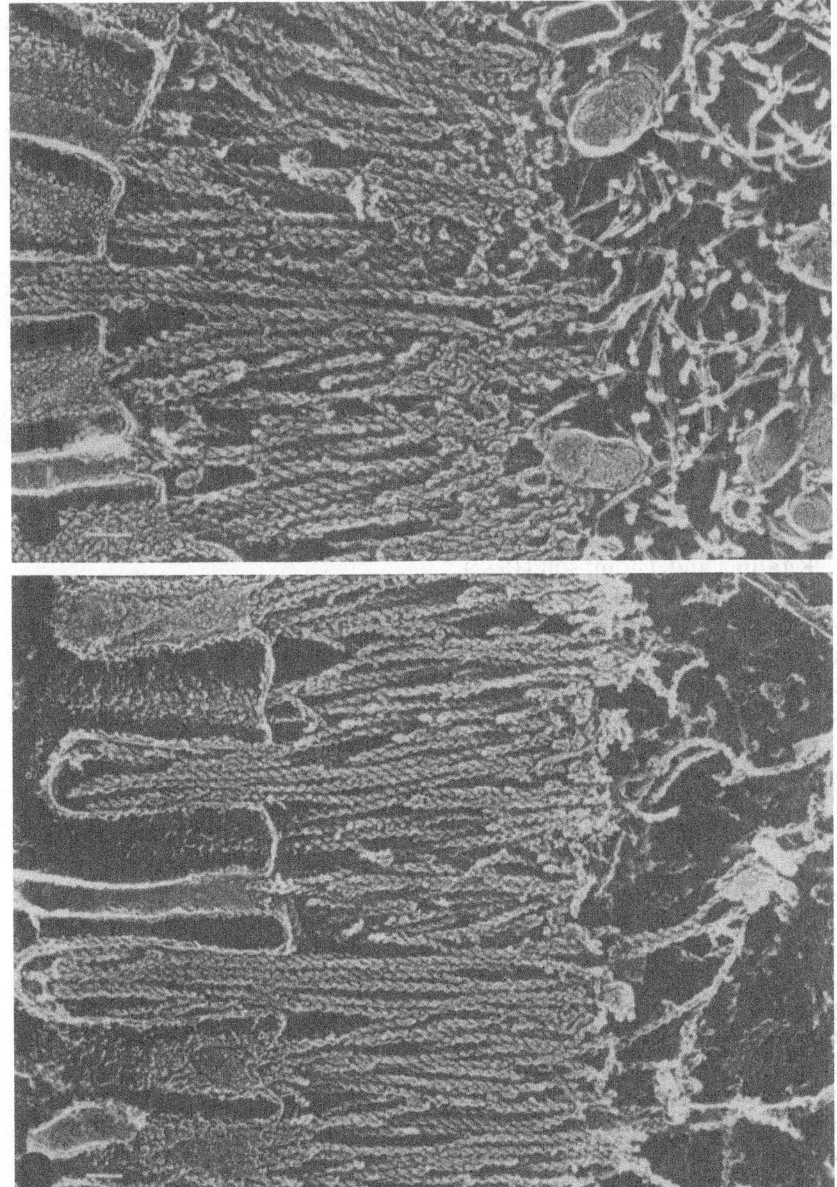

organelles and cytoplasm which is subdivided into compartments by internal membranes. The physical and biochemical changes induced in such membranes by the lowering of the temperature is considered in detail elsewhere in this volume (Morris and Clarke). As the temperature is reduced, the phospholipids of the membrane become less fluid and eventually gel into a crystalline array. The proteins embedded in the liquid bilayer tend to be excluded from this crystal lattice and forced into pockets of still fluid membrane. This can be seen in freeze-fracture replicas of slowly cooled membranes where IMPs, generally assumed to be the integral membrane proteins, have become clustered. However, if the membrane is cooled rapidly enough, there is insufficient time for the proteins to migrate and the particles are revealed dispersed in the membrane as they are thought to occur *in vivo* at the normal temperature (Fig. 5.30). This change in pattern with temperature has been shown in a wide variety of cell and organelle membranes, for example, in prokaryotes (Verkleij *et al.*, 1972; James and Branton, 1973), eukaryotes (Speth and Wunderlich, 1973; Fujikawa, 1981a), mitochondria (Hoechli and Hackenbrock, 1979), nuclear membrane (Wunderlich *et al.*, 1974), and endoplasmic reticulum (Wunderlich *et al.*, 1975). For a useful comparison of the behaviour of different membrane types in *Tetrahymena* see Kitajima and Thompson (1977).

In addition to protein/lipid phase separation, lipid/lipid separation can be induced by low temperatures; for example, cholesterol displacement has been demonstrated by special labelling techniques in cultured mouse cells equilibrated at 0°C (Feltkamp and van der Waerden, 1982b). The crystallization of lipids can be seen clearly in freeze fracture (Crowe *et al.*, 1982; Gulik and Costello, 1978; Verkleij *et al.*, 1980). Known bilayer configurations of lipid have been demonstrated in *Solanum* species during cold acclimation (Fig. 7, Toivio-Kinnucan *et al.*, 1981 and Toivio-Kinnucan, personal communication to Morris). Such partitioning of membrane lipids could have a profound influence on the thermotropic behaviour of the remaining, fluid membrane.

The relationship of the temperature of phase separation to growth and adaptive mechanisms has been discussed previously (Morris and Clarke, this volume). It has been assumed that thermotropic behaviour of the membrane lipids, associated with IMP migration and clustering may result in impaired membrane function, with packing faults and reduced mobility of proteins in the membrane being implicated (Morris and Clarke, this volume). Although there is evidence to support this assumption (Cossins, 1981), it is also clear that some species can function at their normal growth temperature with membranes exhibiting localized phase separations (Morris and Clarke, this volume). Freeze-fracture analysis has a clear role to play in resolving these apparent contradictions.

Shrinkage and intracellular ice

If, during cooling, an unfrozen cell is surrounded by ice, a gradient of water potential will have become established resulting in water loss from the cell. As a consequence, the cell volume contained within the plasma membrane will

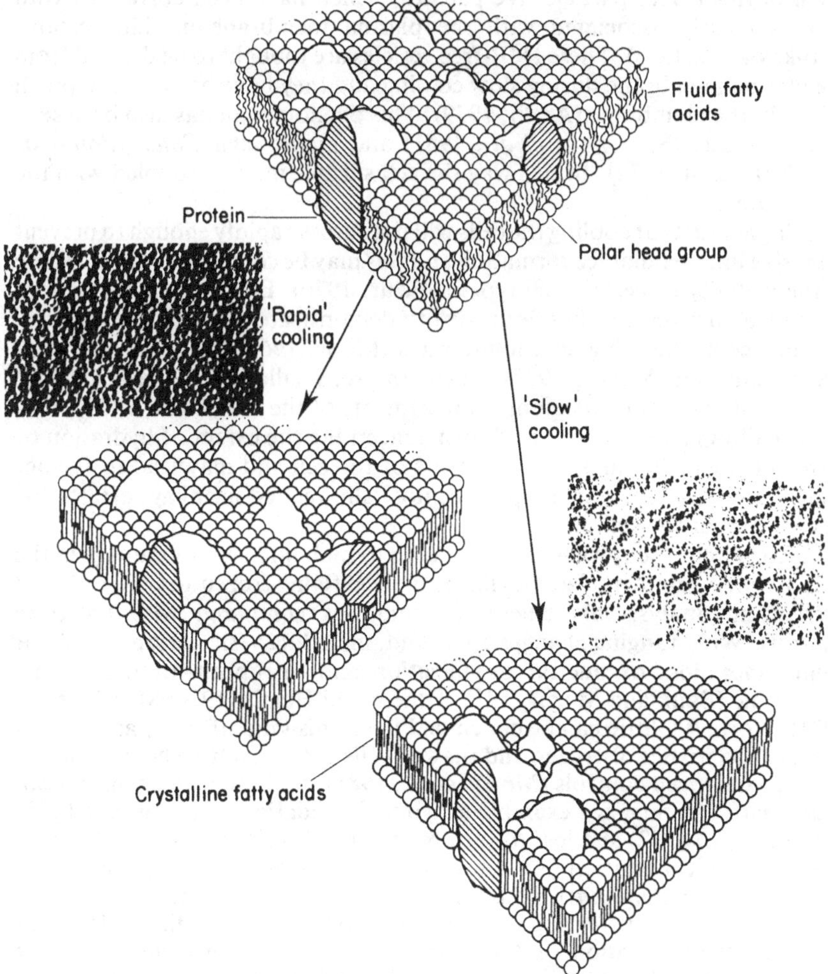

Fig. 5.30 The fluid mosaic model of the structure of cell membranes and the effects of slow and rapid cooling on membrane protein distribution. (Adapted from Morris, 1981.)

be reduced, and accompanying changes will reduce the effective surface area of the plasma membrane. Where the volume, and hence the surface area, reduction is small, the necessary changes may be accommodated by lateral compression of the physiological bilayer, but beyond a certain limit surface area reduction is achieved by microscopic buckling of the membrane, vesiculation and pinching off of small blebs of membrane (Grout and Morris, this volume). The relative importance of these mechanisms and the effects upon them of cold acclimation has been studied by light microscopy (McGrath, this volume) and also by electron microscopy both of thin sections and of freeze-fracture replicas. Replicas of the membranes of slowly frozen erythrocytes reveal particle-free patches, which have been correlated with vesicles closely associated with the plasma membrane in thin sections (Fujikawa, 1981a,b). These IMP-free vesicles are thought to be derived from the plasma membrane during slow cooling and they have been seen to pinch off from the membrane (Araki, 1979). This phenomenon has also been seen in protoplasts (Steponkus *et al.*, 1981) and in the alga *Chlamydomonas nivalis* (Leeson, 1983). In this way the cell surface may be coupled with the cell volume.

At higher rates of cooling the cells will not shrink rapidly enough to prevent extensive intracellular ice formation and this may be correlated with a decline in the viability of certain cell types (Mazur, 1970). Electron microscopy of freeze-fracture replicas has been used to demonstrate the presence of intracellular ice at these higher cooling rates (Moor, 1964; Sherman and Kim, 1967; Bank and Mazur, 1973). Furthermore, it allows an analysis of ice crystal distribution in different compartments of the cell (Bank, 1974). This in turn allows inferences to be drawn concerning the extent of hydration of different organelles and compartments of the cell and may lead to further understanding of the ways in which osmotically-related shrinkage affects different parts of the cell.

Electron microscopy has also been used to highlight sites of damage in the frozen cell. A number of studies have involved fixing frozen-thawed cells and present an analysis of structural alterations compared with unfrozen controls. Wide-ranging changes are found, including vesiculation of cellular membranes, loss of nuclear organization and disruption of the plasma membrane (Bank and Mazur, 1972; Beadle and Harris, 1974; Morris *et al.*, 1981). However, little can be concluded from this kind of study as the fixation characteristics of frozen and thawed cells are unlikely to be the same as those of unfrozen controls (Morris *et al.*, 1981). Pathological changes associated with freezing, for example enzymic degradation of membrane lipids (Clarke *et al.*, 1982) and loss of membrane proteins (Heber *et al.*, 1981)may modify chemical fixation (e.g. the observed reversal of staining contrast in the alga *Chlamydomonas*). Thus it is impossible to tell whether the observed changes result from freezing damage *per se* or from the indirect effects of freezing on the fixation of the cells. Freeze-fixation provides a way of avoiding chemical fixation and so methods such as freeze-fracture or freeze-substitution would seem more suitable to investigate freezing damage.

Changes in the appearance of membranes in freeze-fracture replicas have been reported, ranging from holes in the membrane (Nei, 1976) to 'worm-eaten' spots (areas of depressed and wrinkled membrane [Fujikawa, 1980]) and deposits of material scattered over the fractured face of the membrane (Allen *et al.*, 1980). It is likely that ice does not punch holes in cell membranes (Nei, 1976) and the images of membrane holes are the artefactual result of deep etching 'worm-eaten' spots, which themselves result from a change in the path of membrane fracturing where extracellular and intracellular ice crystals compress the plasma membrane. Such compression might result in molecular disorganization in the membrane and be related to the altered membrane permeability that has been implicated in reduced post-thaw viability (Fujikawa, 1980, 1981). However, a direct causal relationship has not been established and some other basic biochemical lesion may be responsible for the loss of membrane integrity, for example, intracellular phospholipases might be released leading to a loss of membrane components; this phenomenon has been shown to occur (Morris *et al.*, 1979, 1981).

Analysis of the detailed structure of freeze-fracture replicas of cellular membranes can reveal changes in the populations of molecules making up the membrane (Robards *et al.*, 1981). This technique has been used to follow the loss of chloroplast coupling factor 1 (CF_1) from the surface of thylakoid membranes. During freezing CF_1 is lost from the membrane, resulting in a decrease in the particle population on the thylakoid membranes (Steponkus *et al.*, 1977) and a loss of Ca-dependent ATPase activity associated with the membrane fraction (Garber and Steponkus, 1976). This combination of biochemical analysis and freeze-fracture morphology is a powerful tool in elucidating and localizing sites of structural and physiological change by freezing.

The drawbacks of attempting to analyse the structural lesions in freeze–thawed cells by chemical fixation are circumvented by freeze-fracture, but these studies also have limitations. Cells that are to be preserved by freezing must also be thawed, whereas freeze-fracture replicas are made at low temperatures. Attempts have been made to mimic thawing by rewarming tissues to high sub-zero temperatures before recooling and replicating (Bank, 1974). The fact remains, however, that replicas are made of deeply frozen tissues that have never actually thawed completely. During rewarming to physiological temperatures, ice in and around the cells will recrystallize, resulting in a redistribution of water molecules to form fewer larger ice crystals, followed by a sudden release of pure water at the melting temperature. Both these effects are potential traumas which threaten cell survival.

Postscript

New developments in electron microscopy allow an analytical approach to the study of cellular ultrastructure to be taken. Specialized labelling methods make it possible to identify, chemically, different structures in the cell and therefore provide a clearer idea of their functional significance. Apparently uniform structures such as the membrane matrix or IMPs can now be dissected into chemically distinct categories and domains. Furthermore, increases in analytical resolution allow a more accurate assessment of the distribution of both macromolecules and small diffusible moieties such as ions both in the cytoplasm and within the membrane. As a result, electron microscopy has become a much more powerful tool in the search for an understanding of chill and freezing injury and in the changes that occur at the cellular level during acclimation.

References

Adrian, M., Dubochet, J., Lepault, J. and McDowall, A.W. (1984). Cryo-electron microscopy of viruses. *Nature* **308**, 32–6.

Allen, E.D., Weatherbee, L. and Permoad, P.A. (1980). Unusual freeze-structure pattern of unprotected red cells. *Cryobiology* **17**, 610.

Araki, T. (1979). Release of cholesterol-enriched microvesicles from human erythrocytes caused by hypertonic saline at low temperatures. *FEBS Letters* **97**, 237–40.

Arancia, G., Rosati Valente, F. and Trovalusci Crateri, P. (1980). Effects of glutaraldehyde and glycerol on freeze-fractured *Escherichia coli. Journal of Microscopy* **118**, 161–76.

Bachmann, L. and Schmitt, W.W. (1971). Improved cryofixation applicable to freeze etching. *Proceedings of the National Academy of Sciences* **68**, 2149–52.

Bald, W.B. and Robards, A.W. (1978). A device for the rapid freezing of biological specimens under precisely controlled and reproducible conditions. *Journal Microscopy* **112**, 3–15.

Bank, H. (1974). Freezing injury in tissue cultured cells as visualized by freeze-etching. *Experimental Cell Research* **85**, 367–76.

Bank, H. and Mazur, P. (1972). Relation between ultrastructure and viability of frozen-thawed chinese hamster tissue-culture cells. *Experimental Cell Research* **71**, 441–54.

Bank, H. and Mazur, P. (1973). Visualization of freezing damage. *Journal of Cell Biology* **57**, 729–42.

Barnard, T. (1982). Thin frozen-dried cryosections and biological X-ray microanalysis. *Journal of Microscopy* **126**, 317–31.

Beadle, D.J. and Harris, L.W. (1974). Relationship between freezing rate, ultrastructure and recovery in a human diploid line. *Journal Cell Science* **15**, 419–27.

Bearer, E.L. and Friend, D.S. (1980). Anionic lipid domains:correlation with functional topography in a mammalian cell membrane. *Proceedings of the National Academy of Sciences* **77**, 6601–6605.

Bearer, E.L. and Friend, D.S. (1982). Modifications of anionic-lipid domains preceding fusion in guinea pig sperm. *Journal of Cell Biology* **92**, 604–15.

Böhler, S. (1975). *Artefacts and Specimen Preparation Faults in Freeze Etch Technology*. Balzers, A.G., Liechtenstein.

Boyde, A. and Wood, C. (1969). Preparation of animal tissue for surface-scanning electron microscopy. *Journal of Microscopy* **90**, 221–49.

Branton, D. (1966). Fracture faces of frozen membranes. *Proceedings of the National Academy of Sciences* **55**, 1048–55.

Branton, D. (1971). Freeze-etching studies of membrane structure. *Philosophical Transactions of the Royal Society Series B* **261**, 133–8.

Branton, D., Bullivant, S., Gilula, N.B., Karnovsky, M.J., Moor, H., Muhlethaler, K., Northcote, D.H., Packer, L., Satir, B., Satir, P., Speth, V., Staehelin, L.A., Steere, R.L. and Weinsten, R.S. (1975). Freeze-fracture nomenclature. *Science* **190**, 54–6.

Branton, D. and Deamer, D.W. (1972). Membrane Structure. *Protoplasmatologia II E.1.* Edited by Alfert, M., Bauer, H., Sandritter, W. and Sitte, P. Springer-Verlag, Wien, N.Y.

Branton, D. and Kirchanski, S. (1977). Interpreting the results of freeze-etching. *Journal of Microscopy* **111**, 117–24.

Bridgman, P.C. and Reese, T.S. (1982). Structure of cytoplasm in whole mounts of directly frozen cells. *Journal of Cell Biology* **95**, 464a.

Buchheim, W. and Welsch, V. (1977). Freeze-etching of unglycerinated tissue dispersions by application of the oil emulsion technique. *Journal of Microscopy* **111**, 339–49.

Bullivant, S. (1973). Freeze-etching and freeze-fracturing. In *Advanced Techniques in Biological Electron Microscopy*, pp. 67–112. Edited by Koehler J.K. Springer-Verlag, Berlin.

Bullivant, S. (1974). Freeze-etching techniques applied to biological membranes. *Philosophical Transactions of the Royal Society London B* **268**, 5–14.

Bullivant, S. (1977). Evaluation of membrane structure facts and artefacts produced during freeze-fracturing. *Journal of Microscopy* **111**, 101–16.

Burstein, N.L. and Maurice, D.M. (1978). Cryofixation of tissue surfaces by a propane jet for electron microscopy. *Micron* **9**, 191–8.

Cartaud, J., Benedetti, E.L., Sobel, A. and Changeux, J-P. (1978). A morphological study of the cholinergic receptor protein from *Torpedo marmorata* in its membrane environment and in its detergent-extracted purified form. *Journal of Cell Science* **29**, 313–37.

Christensen, A.K. (1971). Frozen thin sections of fresh tissue for electron microscopy, with a description of pancreas and liver. *Journal of Cell Biology* **51**, 772–804.

Clarke, A. (1981). Effects of temperature on the lipid composition of *Tetrahymena*. In *Effects of Low Temperatures on Biological Membranes*, pp. 55–82. Edited by Morris, G.J. and Clarke, A. Academic Press, New York, London.

Clegg, S. (1982). Alternative views on the role of water in cell function. In *Biophysics of Water*, pp. 365–83. Edited by Franks, F. and Mathias, S.F. John Wiley, Chichester.

Cossins, A.R. (1981). The adaptation of membrane dynamic structure to temperature. In *Effects of Low Temperature on Biological Membranes*, pp. 83–106. Edited by Morris, G.J. and Clarke, A. Academic Press, London.

Costello, M.J. and Corless, J.M. (1978). The direct measurement of temperature changes within freeze-fracture specimens during rapid quenching in liquid coolants. *Journal of Microscopy* **112**, 17–37.

Crowe, J.H., Crowe, L.M. and Deamer, D.W. (1982). Hydration dependent phase changes in a biological membrane. In *Biophysics of Water*, pp. 295–9. Edited by Franks, F. and Mathias, S.F. John Wiley, Chichester.

Danon, D., Goldstein, L., Marikovsky, Y. and Skutelsky, E. (1972). Use of cationized ferritin as a label of negative charges on cell surfaces. *Journal of Ultrastructure Research* **38**, 500–510.

Dempsey, G.P. and Bullivant, S. (1976a). A copper block method for freezing non-cryoprotected tissue to produce ice-crystal free regions for electron microscopy. I Evaluation using freeze-substitution. *Journal of*

Microscopy **106**, 251–60.

Dempsey, G.P. and Bullivant, S. (1976b). A copper block method for freezing non-cryoprotected tissue to produce ice-crystal free regions for electron microscopy. II Evaluation using freeze-fracturing with a cryo-ultramicrotome. *Journal of Microscopy* **106**, 261–72.

Dubochet, J., Chang, J-J., Freeman, R., Lepault, J. and McDowall, A.W. (1982a). Frozen aqueous suspensions. *Ultramicroscopy* **10**, 55–62.

Dubochet, J., Lepault, J., Freeman, R., Berriman, J.A. and Homo, J-C. (1982b). Electron microscopy of frozen water and aqueous solutions. *Journal of Microscopy* **128**, 219–37.

Echlin, P.E. (1984). Cryomicroscopy. In *Practical Scanning Electron Microscopy*. Vol. II. Edited by Goldstein, J.I. Newbury, D.E. Echlin, P. Joy, D.C., Fiori, C. and Lifshin, E. in press.

Echlin, P., Lai, C.E. and Hayes, T.L. (1982). Low temperature X-ray microanalysis of the differentiating vascular tissue in root tips of *Lemma minor* L. *Journal of Microscopy* **126**, 285–306.

Echlin, P., Lai, C., Hayes, T. and Saubermann, A. (1980a). Cryofixation of *Lemma minor* roots for morphological and analytical studies. *Cryo-Letters* **1**, 289–98.

Echlin, P., Lai, C., Hayes, T. and Hook, G. (1980b). Elemental analysis of frozen-hydrated differentiating phloem parenchyma in roots of *Lemma minor* L. SEM. 1980 Vol. 2. 383–394.

Edwards, H.H., Mueller, T.J. and Morrison, M. (1979). Distribution of transmembrane polypeptide in freeze-fracture. *Science* **203**, 1343–5.

Elder, H.Y., Gray, C.C., Jardine, A.G. and Chapman, J.N. (1982). Optimum conditions for cryoquenching of small tissue blocks in liquid coolants. *Journal of Microscopy* **126**, 45–61.

Elgsaeter, A., Shotton, D.M. and Branton, D. (1976). Intramembrane particle aggregation in erythrocyte ghosts. II The influence of spectrin aggregation. *Biochemica et Biophysica Acta* **426**, 101–122.

Elias, P.M., Friend, D.S. and Goerke, J. (1979). Membrane sterol heterogeneity. Freeze-fracture detection with saponins and filipin. *Journal of Histochemistry and Cytochemistry* **27**, 1247–60.

Elias, P.M., Goerke, J., Friend, D.S. and Brown, B.E. (1978). Freeze-fracture identification of cholesterol-digitonin complexes in cell and liposome membrane. *Journal of Cell Biology* **78**, 577–96.

Ellisman, M.H. and Porter, K.R. (1980). Microtrabecular structure of the axoplasmic matrix: visualization of cross-linking structures and their distribution. *Journal of Cell Biology* **87**, 464–79.

Engstrom, L.H. (1970). *Structure in the erythrocyte membrane*. Ph.D. dissertation, University of California, Berkeley.

Espevik, T. and Elgsaeter, A. (1981). *In situ* liquid propane jet-freezing and freeze-etching of monolayer cell cultures. *Journal of Microscopy* **123**, 105–110.

Farrant, J., Walter, C.A., Lee, H., Morris, G.J. and Clarke, K.J. (1977). Structural and functional aspects of biological freezing techniques. *Journal of Microscopy* **111**, 17–34.

Feltkamp, C.A. and van der Waerden, A.W.M. (1982a). Membrane associated proteins affect the formation of filipin–cholesterol complexes in viral membranes. *Experimental Cell Research* **140**, 289–97.

Feltkamp, C.A. and van der Waerden, A.W.M. (1982b). Low-temperature induced displacement of cholesterol and intramembrane particles in nuclear membranes of mouse leukemia cells. *Cell Biology International Reports* **6**, 137–45.

Filshie, B.K. and Flower, N.E. (1977). Junctional structures in *Hydra*. *Journal of Cell Science* **23**, 151–72.

Fisher, K.A. (1976a). Analysis of membrane halves. *Proceedings of the National Academy of Sciences USA* **73**, 173–7.

Fisher, K.A. (1976b). Autoradiography of membrane 'halves': ^{3}H-cholesterol labeled erythrocytes. *Journal of Cell Biology* **70**, 218a.

Fisher, K.A. (1978). Split membrane lipids and polypeptides. In *Proceeding of the 9th International Congress of Electron Microscopy Toronto* **Vol. III**, pp. 521–32. Edited by Sturgess, J.M. Imperial Press, Toronto.

Fisher, K.A. (1981). Monolayer freeze-fracture autoradiography: preliminary estimates of 125-I-resolution. *Journal of Cell Biology* **91**, 249a.

Fisher, K.A. (1982a). Monolayer freeze-fracture autoradiography: quantitative analysis of the transmembrane distribution of radioiodinated concanavalin A. *Journal of Cell Biology* **93**, 155–63.

Fisher, K.A. (1982b). Monolayer freeze-fracture autoradiography: origins and directions. *Journal of Microscopy* **126**, 1–8.

Fisher, K.A. and Branton, D. (1976). Freeze-fracture autoradiography: feasibility. *Journal of Cell Biology* **70**, 453–8.

Franks, F. (1980). Physical, biochemical and physiological effects of low temperatures and freezing – their modification by water soluble polymers. In *Scanning Electron Microscopy 1980 II*, pp. 349–60. Edited by O'Hare, A.M.F. Chicago.

Frederik, P.M., Busing, W.M. and Persson, A. (1982). Concerning the nature of the cryosectioning process. *Journal of Microscopy* **125**, 167–76.

Friend, D.S. (1982). Plasma-membrane diversity in a highly polarized cell. *Journal of Cell Biology* **93**, 243–9.

Fujikawa, S. (1980). Freeze-fracture and etching studies on membrane damage on human erythrocytes caused by formation of intracellular ice. *Cryobiology* **17**, 351–62.

Fujikawa, S. (1981a). The effect of different cooling rates on the membrane of frozen human erythrocytes. In *Effects of Low Temperatures on Biological Membranes*, pp. 323–34. Edited by Morris, G.J. and Clarke, A. Academic Press, London.

Fujikawa, S. (1981b). The effect of various cooling rates on the membrane ultrastructure of frozen human erythrocytes and its relation to the extent of haemolysis after thawing. *Journal of Cell Science* **49**, 369–82.

Fukushima, H., Martin, C.E., Iida, H., Kitajima, Y., Thompson, G.A. and Nozawa, Y. (1976). Changes in membrane lipid composition during temperature adaptations by a thermotolerant strain *Tetrahymena pyriformis*. *Biophysica et Biochemica Acta* **431**, 165–179.

Garber, M.P. and Steponkus, P.L. (1976). Alterations in chloroplast thylakoids during an *in vitro* freeze-thaw cycle. *Plant Physiology* **45**, 1343–50.

Giddings, T.H. and Staehelin, L.A. (1980). Ribosome binding sites visualized on freeze-fractured membranes of rough endoplasmic reticulum. *Journal of Cell Biology* **85**, 147–52.

Goldstein, J.I., Newbury, D.E., Echlin, P., Joy, D.C., Fiori, C. and Lifshin, E. (1981). *Scanning Electron Microscopy & X-ray Microanalysis.* Plenum, New York.

Gotow, T. and Hashimoto, P.H. (1983). Filipin resistance in intermediate junction membranes of guinea pig ependyma: possible relationship to filamentous underlying. *Journal of Ultrastructure Research* **84**, 83–93.

Grant, C.W.M. and McConnell, H.M. (1974). Glycophorin in lipid bilayers. *Proceedings of the National Academy of Sciences USA* **71**, 4653–7.

Gray, E.G. (1975). Synaptic fine structure and nuclear, cytoplasmic and extracellular networks. The stereoframework concept. *Journal of Neurocytology* **4**, 315–39.

Green, G.R. (1981). Fixation-induced intramembrane particle movement demonstrated in freeze-fracture replicas of a new type of septate junction in Echinoderm epithelia. *Journal of Ultrastructure Research* **75**, 11–22.

Gulik-Krzywicki, T. and Costello, M.J. (1978). The use of low temperature X-ray diffraction to evaluate freezing methods used in freeze-fracture electron microscopy. *Journal of Microscopy* **112**, 103–13.

Gulley, R.L. and Reese, T.S. (1981). Cytoskeletal organization at the post-synaptic complex. *Journal of Cell Biology* **91**, 298–302.

Gupta, B.L., Berridge, M.J., Hall, T.A. and Moreton, R.B. (1978a). Electron microprobe and ion-selective microelectrode studies of fluid secretion in the salivary glands of *Calliphora. Journal of Experimental Biology* **72**, 261–84.

Gupta, B.L., Hall, T.A., Maddrell, S.H.P. and Moreton, R.B. (1976). Distribution of ions in a fluid-transporting epithelium determined by electron-probe X-ray microanalysis. *Nature* **264**, 284–7.

Gupta, B.L., Hall, T.A. and Naftalin, R.J. (1978b). Microprobe measurement of Na, K and Cl concentration profiles in epithelial cells and intercellular spaces of rabbit ileum. *Nature, London* **272**, 70–73.

Haggis, G.H., Bond, E.F. and Phipps. B. (1976). Visualization of mitochondrial cristae and nuclear chromatin by SEM. In *9th 11TR1 SEM symp*, p. 281. Chicago.

Haggis, G.H. and Phipps-Todd, B. (1977). Freeze-fracture for scanning electron microscopy. *Journal of Microscopy* **111**, 193–201.

Handley, D.A., Alexander, J.T. and Chiens, S. (1981). The design and use of a simple device for rapid quench-freezing biological samples. *Journal of Microscopy* **121**, 273–82.

Hasty, D.L. and Hay, E.D. (1978). Freeze-fracture studies of the developing cell surface. II. Particle-free membrane blisters on glutaraldehyde fixed corneal fibroblasts are artifacts. *Journal of Cell Biology* **78**, 756–68.

Hay, E.D. and Hasty, D.L. (1979). Extrusion of particle-free membrane blisters during glutaraldehyde fixation. In *Freeze Fracture: Methods,*

Artifacts, and Interpretations pp. 59–66. Edited by Rash, J.E. and Hudson, C.S. Raven Press, New York.

Heber, V., Schmitt, J.M., Krause, G.H., Klosson, R.J. and Santarius, K.A. (1981). Freezing damage to thylakoid membranes *in vitro* and *in vivo*. In *Effects of Low Temperatures on Biological Membranes*, pp. 263–283. Edited by Morris, G.J. and Clarke, A. Academic Press, London.

Henderson, D., Eibl, H. and Weber, K. (1979). Structure and biochemistry of mouse heptatic gap junctions. *Journal of Molecular Biology* **132**, 193–218.

Heuser, J.E. (1981). Quick freeze deep-etch preparation of samples for 3-D electron microscopy. *Trends in Biochemical Sciences* **6**, 64–8.

Heuser, J.E. and Evans, L. (1980). Three-dimensional visualization of coated vesicle formation in fibroblasts. *Journal of Cell Biology* **84**, 560–83.

Heuser, J.E. and Kirschner, M.W. (1980). Filament organization revealed in platinum replicas of freeze-dried cytoskeletons. *Journal of Cell Biology* **86**, 212–34.

Heuser, J.E., Reese, T.S. and Landis, D.M.D. (1975). Preservation of synaptic structure by rapid freezing. *Cold Spring Harbor Symposium on Quantitative Biology* **40**, 17–24.

Heuser, J.E., Reese, T.S., Dennis, M.J., Jan, Y., Jan, L. and Evans, L. (1979). Synaptic vesicle exocytosis captured by quick freezing and correlated with quantal transmitter release. *Journal of Cell Biology* **81**, 275–300.

Heuser, J.E. and Saltpeter, S.R. (1979). Organization of acetylcholine receptors in quick frozen, deep etched and rotary replicated *Torpedo* postsynaptic membrane. *Journal of Cell Biology* **82**, 150–73.

Hirokawa, N. (1982). Cross-linker system between neurofilaments, microtubules, and membrane organelles in frog axons revealed by the quick-freeze, deep-etching method. *Journal of Cell Biology* **94**, 129–42.

Hirokawa, N. and Heuser, J.E. (1981). Quick-freeze deep-etch visualization of the cytoskeleton beneath surface differentiations of intestinal epithelial cells. *Journal of Cell Biology* **91**, 399–409.

Hirokawa, N., Keller, T.C.S., Chasan, R. and Mooseker, M.S. (1983). Mechanism of brush border contractility studied by the quick-freeze, deep-etch method. *Journal of Cell Biology* **96**, 1325–36.

Hirokawa, N. and Tilney, L.G. (1982). Interactions between actin filaments and membranes in quick-frozen and deeply etched hair cells of the chick ear. *Journal of Cell Biology* **95**, 249–61.

Hirokawa, N., Tilney, L.G. and Heuser, J.E. (1982). Interaction of actin, myosin and intermediate filaments in the brush border of intestinal epithelial cells. *Journal of Cell Biology* **94**, 425–43.

Hoechli, M. and Hackenbrock, C. (1979). Lateral translational diffusion of cytochrome C oxidase in the mitochondrial energy-transducing membrane. *Proceedings of the National Academy of Sciences USA* **76**, 1236–40.

Humphreys, W.J., Spurlock, B.O. and Johnson, J.S. (1974). Critical point drying of ethanol-infiltrated cryofractured biological specimens for scanning electron microscopy. In *Proc. 7th Ann. 11TR1 SEM Symp.* p. 275. Edited by Johari, O. and Corvin, J. Chicago Press Corp.

Humphreys, W.J., Spurlock, B.O. and Johnson, J.S. (1975). Transmission electron microscopy of tissue prepared for scanning electron microscopy by ethanol cryofracturing. *Stain Technology* **50**, 119-125.

Hunter, J. (1779). *Philosophical Transactions of the Royal Society* **68**, as cited in Stolinski, C. and Breathnach, A.S. (1975).

James, R. and Branton, D. (1973). Lipid- and temperature-dependent structural changes in *Acholeplasma laidlawii* cell membranes. *Biochimica Biophysica Acta* **323**, 378-90.

Kanno, H., Speedy, R.J. and Angell, C.A. (1975). Supercooling of water to −92°C under pressure. *Science* **189**, 880-81.

Kitajima, Y. and Thompson, G. (1977). *Tetrahymena* strives to maintain the fluidity interrelationships of all its membranes constant. Electron microscope evidence. *Journal of Cell Biology* **72**, 744-55.

Knoll, G., Oebel, G. and Plattner, H. (1982). A simple sandwich - cryogen - jet procedure with high cooling rates for cryofixation of biological materials in the native state. *Protoplasma* **111**, 161-76.

Komoda, T. and Saito, S. (1972). Experimental resolution limit in the secondary mode for a field emission source scanning electron microscope. In *Scanning Electron Microscopy. 1972*, pp. 129-36. Edited by Johari, O. Chicago.

Kretzchner, K.M. and Wilkie, D.R. (1969). A new approach to freezing tissues rapidly. *Journal of Physiology* **202**, 66-67.

de Kruijff, B. and Demel, R.A. (1974). Polyene antibiotic-sterol interactions in membranes of *Acholeplasma laidlawii* cells and lecithin liposomes. III Molecular structure of the polyene antibiotic-cholesterol complexes. *Biochimica Biophysica Acta* **339**, 57-70.

Landis, D.M. (1982). Structure and function at synapses. *Trends in Neurosciences* **5**, 215-16.

Lane, N.J. and Skaer, H.leB. (1980). Intercellular junctions in insect tissues. *Advances in Insect Physiology* **15**, 35-213.

Leeson, E. (1983). Freezing injury in *Chlamydomonas. Cryobiology*, **20**, 726.

Lepault, J., Booy, F.P. and Dubochet, J. (1983). Electron microscopy of frozen biological suspensions. *Journal of Microscopy* **129**, 89-102.

McIntyre, J.A., Gilula, N.B. and Karnovsky, M.J. (1974). Cryoprotectant induced redistribution of intramembranous particles in mouse lymphocytes. *Journal of Cell Biology* **60**, 192-203.

Margaritis, L.H., Elgsaeter, A. and Branton, D. (1977). Rotary replication for freeze-etching. *Journal of Cell Biology* **72**, 47-56.

Martin, C.E., Hiramitsuj, K., Nozawa, Y., Skriver, L. and Thompson, G.A. (1976). Molecular control of membrane properties during temperature acclimation. Fatty acid desaturase regulation of membrane fluidity in acclimating *Tetrahymena* cells. *Biochemistry* **15**, 5218-27.

Mayer, E. and Brüggeller, P. (1982). Vitrification of pure liquid water by high pressure jet freezing. *Nature* **298**, 715-18.

Mazur (1970). Cryobiology: the freezing of biological systems. *Science* **168**, 939–49.

Miller, K.R., Prescott, C.S., Jacobs, T.L. and Lassignal, N.L. (1983). Artifacts associated with quick freezing and freeze-drying. *Journal of Ultrastructure Research* **82**, 123–33.

Moor, H. (1964). Die Gefrier-fixation lebender Zellen und ihre Anwendung in der Elektronenmikroskopie. (Freeze fixation of living cells and its application in electron microscopy). *Zeitschrift für Zellforschung und Mikroskopische Anatomie* **62**, 546–80.

Moor, H. (1971). Recent progress in the freeze-etching technique. *Philosophical Transactions of the Royal Society London series B* **261**, 121–31.

Moor, H. (1973). Cryotechnology for the structural analysis of biological material. In *Freeze-Etching Techniques and Applications*, pp. 11–19. Edited by Benedetti, E.L. and Favard, P. Société Française de Microscopie Electronique, Paris.

Moor, H., Bellin, G., Sandri, C. and Akert, K. (1980). The influence of high pressure freezing on mammalian nerve tissue. *Cell Tissue Research* **209**, 201–216.

Moor, H, and Hoechli, M. (1970). The influence of high-pressure freezing on living cells. In *Proceedings of the International Congress on Electron Microscopy Grenoble* **Vol. 1.** pp. 449–50. Edited by Favard, P. Société Française de Microscopie Electronique, Paris.

Moor, H., Kistler, J. and Müller, M. (1976). Freezing in a propane jet. *Experientia* **32**, 805.

Moor, H. Mühlethaler, K., Walder, H. and Frey-Wyssling, A. (1961). A new freezing-ultramicrotome. *Journal of Biophysical and Biochemical Cytology* **10**, 1–13.

Moor, H. and Riehle, U. (1968). Snap-freezing under high pressure: a new fixation technique for freeze-etching. In *Proceedings of the 4th European Regional Conference of Electron Microscopy, Rome 2*, pp. 33–4. Edited by Bocciarelli, D.S.

Morris, G.J., Coulson, G. and Clarke, A. (1979). The cryopreservation of *Chlamydomonas*. *Cryobiology* **16**, 401–410.

Morris, G.J., Coulson, G.E., Clarke, K.J., Grout, B.W.W. and Clark, A. (1981). Freezing injury in Chlamydomonas: a synoptic approach. In *Effects of Low Temperatures on Biological Membranes*, pp. 285–306. Edited by Morris, A.J. and Clarke, A. Academic Press, London.

Müller, M., Meister, N. and Moor, H. (1980a). Freezing in a propane jet and its application in freeze-fracturing. *Balzers report* BB800 011 DE, 1–8.

Müller, M., Meister, N. and Moor, H. (1980b). Freezing in a propane jet and its application in freeze-fracturing. *Mikroskopie* **36**, 129–40.

Nei, T. (1976). Freezing injury to erythrocytes. II. Morphological alterations of cell membranes. *Cryobiology* **13**, 287–94.

Nermut, M.V. and Williams, L.D. (1980). Freeze-fracture autoradiography of the red blood cell plasma membrane. *Journal of Microscopy* **118**, 453–61.

Ono, T-A and Murata, N. (1982). Chilling-susceptibility of the blue-green alga *Anacystis nidulans*. III. Lipid phase of cytoplasmic membrane.

Plant Physiology **69**, 125–9.

Pal, R., Barenholz, Y. and Wagner, R.R. (1980). Effect of cholesterol concentration on organization of viral and vesicle membranes. *Journal of Biological Chemistry* **255**, 5802–5806.

Pihakaski, K. and Seveus, L. (1980). High polymeric protective additives in cryoultramicrotomy of plants. I. PVP-Infusion of fixed plant tissue. *Cryoletters* **1**, 494–505.

Pinto da Silva, P. and Branton, D. (1970). Membrane splitting in freeze-etching. Covalently bound ferritin as a membrane marker. *Journal of Cell Biology* **45**, 598–605.

Pinto da Silva, P., Douglas, S.D. and Branton, D. (1971). Localization of A antigen sites on human erythrocyte ghosts. *Nature* **232**, 194–6.

Pinto da Silva, P. and Kachar, B. (1980). Quick freezing vs. chemical fixation: capture and identification of membrane fusion intermediates. *Cell Biology International Reports* **4**, 625–40.

Pinto da Silva, P., Moss, P. and Fudenberg, H.H. (1973). Anionic sites on the membrane intercalated particles of human erythrocyte ghost membranes. Freeze-etch localization. *Experimental Cell Research* **81**, 127–38.

Pinto da Silva, P. and Nicolson, G.L. (1974). Freeze-etch localization of concanavalin A receptors to the membrane intercalated particles of human erythrocyte ghost membranes. *Biochimica Biophysica Acta* **363**, 311–19.

Pinto da Silva, P. and Torrisi, M.R. (1982). Freeze-fracture cytochemistry: partition of glycophorin in freeze-fractured human erythrocyte membranes. *Journal of Cell Biology* **93**, 463–9.

Pinto da Silva, P., Parkinson, P. and Dwyer, N. (1981a). Fracture label cytochemistry of freeze-fracture faces in the erythrocyte membrane. *Proceedings of the National Academy of Sciences USA* **78**, 343–7.

Pinto da Silva, P., Torrisi, M.R. and Kachar, B. (1981b). Freeze-fracture cytochemistry: localization of wheat-germ agglutinin and concanavalin A binding sites on freeze-fractured pancreatic cells. *Journal of Cell Biology* **91**, 361–72.

Plattner, H. and Bachmann, L. (1982). Cryofixation: a tool in biological ultrastructural research. *International Review of Cytology* **79**, 237–304.

Plattner, H., Fischer, W.M., Schmitt, W.W. and Bachmann, L. (1972). Freeze-etching of cells without cryoprotectants. *Journal of Cell Biology* **53**, 116–26.

Porter, K.R. and Tucker, J.B. (1981). The ground substance of the living cell. *Scientific American* **244**, 40–51.

Pringle, M.J. and Chapman, D. (1981). Biomembrane structure and effects of temperature. In *Effects of Low Temperatures on Biological Membranes*, 21–37. Edited by Morris, G.J. and Clarke, A. Academic Press, London.

Pscheid, P., Schudt, C. and Plattner, H. (1981). Cryofixation of monolayer cell cultures for freeze-fracturing without chemical pre-treatments. *Journal of Microscopy* **121**, 149–67.

Rash, J.E. (1979). The sectioned-replica technique: direct correlation of freeze-fracture replicas and conventional thin-section images. In *Freeeze-Fracture: Methods, Artifacts and Interpretations*, pp. 153–60. Edited by Rash, J.E. and Hudson, C.S. Raven Press, New York.

Rash, J.E. (1983). The rapid-freeze technique in neurobiology. *Trends in Neurosciences* **6**, 208–212.

Rash, J.E. and Hudson, C.S. (1979). *Freeze-Fracture: Methods, Artifacts and Interpretations*. Raven Press, New York.

Rash, J.E., Hudson, C.S., Alberquerque, E.X., Eldefrawi, M.E., Mayer, R.F., Graham, W.F., Johnson, T.J.A. and Giddings, F.D. (1981). Freeze-fracture, labelled-replica and electrophysiological studies of junctional fold destruction in myasthenia gravis and experimental autoimmune myasthenia gravis. *Annals of the New York Academy of Science* **377**, 38–60.

Rash, J.E., Johnson, T.J.A., Hudson, C.S., Giddings, F.D., Graham, W.F. and Eldefrawi, M.E. (1982). Labelled replica techniques: post-shadow labelling of intramembrane particles in freeze-fracture replicas. *Journal of Microscopy* **128**, 121–38.

Riehle, U., (1968). *Uber die Vitrifizierung verdünter wässeriger Lösungen*. Diss. Nr 4271, ETH Zurich Juris Verlag, Zurich.

Riehle, U. and Hoechli, M. (1973). The theory and technique of high pressure freezing. In *Freeze-Etching. Techniques and Applications*, pp. 31–61. Edited by Benedetti, E.L. and Favard, P. Société Française de Microscopie Electronique, Paris.

Rix, E., Schiller, A. and Taugner, R. (1976). Freeze-fracture autoradiography. *Histochemistry* **50**, 91–101.

Robards, A.W. (1974). Ultrastructural methods for looking at frozen cells. *Science Progress, Oxford* **61**, 1–40.

Robards, A.W., Bullock, G.R., Goodall, M.A. and Sibbons, P.D. (1981). Computer assisted analysis of freeze-fractured membranes following exposure to different temperatures. In *Effects of Low Temperatures on Biological Membranes* pp. 219–38. Edited by Morris, G.J., and Clarke, A. Academic Press, London.

Robards, A.W. and Severs, N.J. (1981). A comparison between cooling rates achieved using a propane jet device and plunging into liquid propane. *Cryoletters* **2**, 135–44.

Roof, D.J. and Heuser, J.E. (1982). Surfaces of rod photoreceptor disk membranes: integral membrane components. *Journal of Cell Biology* **95**, 487-500.

Roof, D.J., Korenbrot, J.I. and Heuser, J.E. (1982). Surfaces of rod photoreceptor disk membranes: light-activated enzymes. *Journal of Cell Biology* **95**, 501–509.

Satir, B., Schooley, C. and Satir, P. (1973). Membrane fusion in a model system. Mucocyst secretion in *Tetrahymena. Journal of Cell Biology* **56**, 153–76.

Saubermann, A.J., Echlin, P., Peters, P.D. and Beeukes, R. (1981). Application of scanning electron microscopy and X-ray analysis of frozen-hydrated sections. I. Specimen handling techniques. *Journal of*

Cell Biology **88**, 257–67.

Schiller, A., Rix, E. and Taugner, R. (1978a). Freeze-fracture autoradiography: the *in vacuo* coating technique. *Histochemistry* **59**, 9–16.

Schiller, A., Rix, E. and Taugner, R. (1978b). Gefrierbruch - Autoradiographie. *Mikroskopie* **34**, 24–8.

Schiller, A., Sonnhof, J. and Taugner, R. (1979). Tissue functions - compatible cryoprotectants in cryo-ultramicrotomy and freeze-fracturing. *Mikroskopie* **35**, 23–30.

Schiller, A. and Taugner, R. (1980). Freeze-fracturing and deep-etching with volatile cryoprotectant ethanol reveals true membrane surfaces of kidney structures. *Cell Tissue Research* **210**, 57–69.

Schiller, A., Taugner, R. and Rix, E. (1978). Die Anwendung Kryoprotektiver Substanzen in der Gefrierbruchtechnik. *Mikroskopie* **34**, 19–23.

Schliwa, M. and van Blerkom, J. (1981). Structural interaction of cytoskeletal components. *Journal of Cell Biology* **90**, 222–35.

Schnapp, B.J. and Reese, T.S. (1982). Cytoplasmic structure in rapid-frozen axons. *Journal of Cell Biology* **94**, 667–79.

Severs, N.J. and Green, C.R. (1983). Rapid freezing of un-pretreated tissues for freeze-fracture electron microscopy. *Biologie Cellulaire* **47**, 193–204.

Severs, N.J. and Robenek, H. (1983). Detection in microdomains in biomembranes. An appraisal of recent developments in freeze-fracture cytochemistry. *Biochimica Biophysica Acta* **737**, 373–408.

Severs, N.J. and Simons, H.L. (1983). Failure of filipin to detect cholesterol-rich domains in smooth muscle plasma membrane. *Nature* **303**, 637–8.

Severs, N.J., Warren, R.C. and Barnes, S.H. (1981). Analysis of membrane structure in the transitional epithelium of rat urinary bladder. 3. Localization of cholesterol using filipin and digitonin. *Journal of Ultrastructure Research* **77**, 160–88.

Sherman, J.K. and Kim, K.S. (1967). Correlation of cellular ultrastructure before freezing, while frozen and after thawing in assessing freeze-thaw-induced injury. *Cryobiology* **4**, 61–74.

Shotton, D.M. (1982). Quantitative freeze-fracture electron microscopy of dystrophic muscle membranes. *Journal of the Neurological Sciences* **57**, 161–90.

Shotton, D.M. (1984). The proteins of the erythrocyte membrane. In *The Electron Microscopy of Proteins*, Vol. 3. Edited by Harris, J.R. Academic Press, London.

Shotton, D., Thompson, K., Wofsy, L. and Branton, D. (1978). Appearance and distribution of surface proteins of the human erythrocyte membrane. An electron microscope and immuno-chemical labelling study. *Journal of Cell Biology* **76**, 512–31.

Singer, S.S. and Nicholson, G.L. (1972). The fluid mosaic model of the structure of the cell membranes. *Science* **175**, 720–31.

Sjöstrand, F.S. (1979). The interpretation of pictures of freeze-fractured biological material. *Journal of Ultrastructure Research* **69**, 378–420.

Skaer, H. le B. (1982). Chemical cryoprotection for structural studies.

Journal of Microscopy **125**, 137–47.

Skaer, H. le B., Franks, F., Asquith, M.H. and Echlin, P.E. (1977). Polymeric cryoprotectants in the preservation of biological ultrastructure. III. Morphological aspects. *Journal of Microscopy* **110**, 257–70.

Skaer, H. le B., Franks, F. and Echlin, P.E. (1978). Non-penetrating polymeric cryofixatives for ultrastructural and analytical studies of biological tissues. *Cryobiology* **15**, 589–602.

Skaer, H. le B., Franks, F. and Echlin, P. (1979). Freeze-fracture studies of ice recrystallization in tissues quenched in aqueous solutions in polyvinyl pyrrolidone. *Cryoletters* **1**, 61–70.

Skaer, H. le B. and Skaer, R.J. (1979). A new approach to freeze-fracture autoradiography. *Cryoletters* **1**, 14–19.

Sleytr, U.B. (1974). Freeze-fracturing at liquid helium temperature for freeze-etching. In *Proceedings of the 8th International Congress on Electron Microscopy*. Canberra, Vol. II, p. 30.

Sleyter, U.B. and Robards, A.W. (1977a). Plastic deformation during freeze-cleaving: a review. *Journal of Microscopy* **110**, 1–26.

Sleyter, U.B. and Robards, A.W. (1977b). Freeze-fracturing: a review of methods and results. *Journal of Microscopy* **111**, 77–100.

Sleyter, U.B. and Robards, A.W. (1982). Understanding the artefact problem in freeze-fracture replication: a review. *Journal of Microscopy* **126**, 101–122.

Small, J.V. and Langanger, G. (1981). Organization of actin in the leading edge of cultured cells: influence of osmium tetroxide and dehydration on the ultrastructure of actin meshworks. *Journal of Cell Biology* **91**, 695–705.

Speth, V. and Wunderlich, F. (1973). Membranes of *Tetrahymena*. II. Direct visualization of reversible transitions in biomembrane structure induced by temperature. *Biochimica Biophysica Acta* **291**, 621–8.

Staehelin, L.A. (1976). Reversible particle movements associated with unstacking and restacking of chloroplast membranes *in vitro*. *Journal of Cell Biology* **71**, 136–158.

Staehelin, L.A. and Arntzen, C.J. (1983). Regulation of chloroplast membrane function: protein phosphorylation changes and spatial organization of membrane components. *Journal of Cell Biology* **97**, 1327–37.

Steere, R.L. (1957). Electron microscopy of structural detail in frozen biological specimens. *Journal of Biophysical and Biochemical Cytology* **3**, 45–60.

Steinbrecht, R.A. (1980). Cryofixation without cryoprotectants. Freeze-substitution and freeze-etching of an insect olfactory receptor. *Tissue and Cell* **12**, 73–100.

Steponkus, P.L., Garber, M.P., Myers, S.P. and Lineberger, P.D. (1977). Effects of cold acclimation and freezing on structure and function of chloroplast thylakoids. *Cryobiology* **14**, 303–321.

Steponkus, P.L., Wolfe, J. and Dowgert, M.F. (1981). Stresses induced by contraction and expansion during a freeze-thaw cycle: a membrane perspective. In *Effects of Low Temperatures on Biological Membranes*, pp. 307-22. Edited by Morris, G.J. and Clarke, A. Academic Press, London.

Stewart, M. and Vigers, G. (1986). Electron microscopy of frozen-hydrated biological material. *Nature* **319**, 631-6.

Stolinski, C. and Breathnach, A.S. (1975). *Freeze-Fracture Replication of Biological Tissues. Techniques, Interpretation and Applications.* Academic Press, London.

Swales, L.S. and Lane, N.J. (1983). Insect intercellular junctions: rapid freezing by jet propane. *Journal of Cell Science* **62**, 223-36.

Tanaka, K. (1980). Scanning electron microscopy of intracellular structures. *International Review of Cytology* **68**, 97-125.

Toivio-Kinnucan, M.A. Chen, H.H., Li, P.H. and Stushnuff, R. (1981). Plasma membrane alterations in callus tissues of tuber-bearing *Solanum* species during cold acclimatization. *Plant Physiology* **67**, 478-83.

Umrath, W. (1983). Berechnung von Gefriertrocknungszeiten für die elektronenmikroskopie Präparation. (Calculation of the freeze-drying time for electron-microscopical preparation). *Mikroskopie* **40**, 9-37.

Van Deurs, B. and Luft, J.H. (1979). Effects of glutaraldehyde fixation on the structure of tight junctions. A quantitative freeze-fracture analysis. *Journal of Ultrastructure Research* **68**, 160-72.

van Harreveld, A. and Crowell, J. (1964). Electron microscopy after rapid freezing on a metal surface and substitution fixation. *Anatomical Record* **149**, 381-6.

van Harreveld, A., Trubatch, J. and Seiner, J. (1974). Rapid freezing and electron microscopy for the arrest of physiological processes. *Journal of Microscopy* **100**, 189-98.

van Venrooji, G.E.P.M., Aertsen, A.M.H.J., Hax, W.M.A., Ververgaert, P.H.J.T., Vermoeven, J.J. and van der Vorst, H.A. (1975). Freeze-etching: freezing velocity and crystal size at different locations in samples. *Cryobiology* **12**, 46-61.

Verkleij, A.J., van Echteld, C.J.A., Gerritsen, W.J., Cullis, P.R. and de Kruijff, B. (1980). The lipidic particle as an intermediate structure in membrane fusion processes and bilayer to hexagonal H_{II} transitions *Biochimica Biophysica Acta* **600**, 620-24.

Verkleij, A.J., Ververgaert, P.H.J., van Deenen, L.L.M. and Elbers, P.F. (1972). Phase transitions of phospholipid bilayers and membranes of *Acholeplasma laidlawii* B visualized by freeze-fracturing electron microscopy. *Biochimica Biophysica Acta* **288**, 326-32.

Willison, J.H. and Brown, R.M. (1979). Pretreatment artifacts in plant cells. In *Freeze-Fracture: Methods, Artifacts and Interpretations*, pp. 51-7. Edited by Rash, J.E. and Hudson, C.S. Raven Press, New York.

Wilson, A.J. and Robards, A.W. (1980). Some limitations of the polymer polyvinyl pyrrolidone for the cryoprotection of barley roots during quench freezing. *Cryoletters* **1**, 416-25.

Wolosewick, J.J. and Porter, K.R. (1976). Stereo high voltage electron microscopy of whole cells of human diploid line W1-38. *American Journal of Anatomy* **147**, 303-24.

Wolosewick, J.J. and Porter, K.R. (1979). Microtrabecular lattice of the cytoplasmic ground subtance. Artifact or reality. *Journal of Cell Biology* **82**, 114–39.

Wunderlich, F., Batz, W., Speth, V. and Wallach, D. (1974). Reversible, thermotropic alteration of nuclear membrane structure and nucleocytoplasmic RNA transport in *Tetrahymena. Journal of Cell Biology* **61**, 633–40.

Wunderlich, F. and Ronai, A. (1975). Adaptive lowering of the lipid clustering temperature within *Tetrahymena* membranes. *FEBS Letters* **55**, 237–41.

Wunderlich, F., Ronai, A., Speth, V., Seelig, J. and Blume, A. (1975). Thermotropic lipid clustering in *Tetrahymena* membranes. *Biochemistry* **14**, 3730–35.

Yu, J. and Branton, D. (1976). Reconstitution of intramembranous particles in recombinants of erythrocyte protein Band 3 and lipid: effects of spectrin-actin association. *Proceedings of the National Academy of Sciences USA* **73**, 3891–5.

6

Temperature-controlled cryogenic light microscopy – an introduction to cryomicroscopy

J.J. McGrath
Bioengineering Transport Processes Laboratory,
Mechanical Engineering Department
Michigan State University

Introduction

Biological systems are very complex and therefore many experimental and analytical techniques must be applied in order to describe and understand the response of living systems to changes in temperature. One method that has proved successful is the direct microscopic observation of biological samples during freezing and thawing.

During the last years of the nineteenth century the first known low temperature microscope system was developed because '. . . the direct observation of the freezing cell is the best means of obtaining information on the causes of death by freezing.' (Molisch, trans. 1982). Having performed ingenious experiments on model solutions (analogous to cell cytoplasm) as well as on a variety of living cells, Molisch concluded that cell death by freezing is in fact due to dehydration caused by the formation of extracellular ice.

These studies clearly demonstrated the occurrence of many phenomena that continue to be of interest. For example, Molisch observed severe dehydration of *Spirogyra* caused by extracellular freezing, noted intracellular ice formation in *Amoeba* and recognized the importance of concentrated solutes in relation to cell death during freezing. Based upon his observations Molisch believed that although cell death may occur in the frozen state, particular combinations of thawing rates and the minimum temperature reached might allow for cell recovery.

Early low temperature microscopy was performed by immersing a microscope in a 'microscope ice box' (Fig. 6.1) or venturing out of doors in the winter with microscope and samples in hand (Molisch, trans. 1982). Studies were primarily qualitative descriptions of the events which occurred and the sequence in which they occurred. However, the earliest low temperature microscopy studies were rather limited with respect to the range of temperatures which could be achieved (Diller, 1982). In addition, the microscopist had virtually no control over the cooling and warming rates.

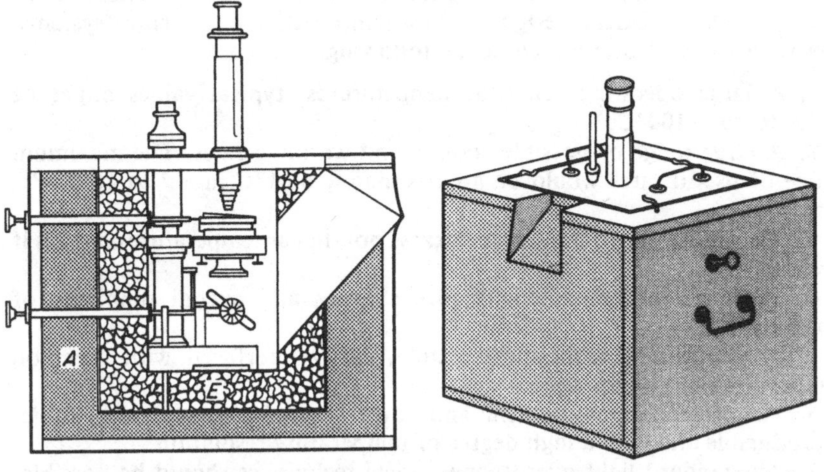

Fig. 6.1 Molisch freezing apparatus for microscope studies (1897). The system consists of an insulated box (*A*) which contains a microscope and a 'freezing mixture' consisting of ice and sodium chloride (*E*). (Reproduced with permission from Molisch, Trans 1982.)

Significant advances have been made in low temperature microscopy since the turn of the century; these include:

(*a*) improved control of temperature;
(*b*) extended range of operating temperatures;
(*c*) extended range of rates of temperature change;
(*d*) application of quantitative optical methods.

As a result of the development of improved technology cryogenic fluids are commonly used as refrigerants for low temperature microscope studies, and such low temperature light microscope systems are called 'cryomicroscopes'. This chapter aims to outline the basic operating principles of a cryomicroscope followed by a discussion of the performance characteristics of such a system. The cryomicroscope used in the Bioengineering Transport Processes (BTP) Laboratory at Michigan State University, USA is given as an example to illustrate various aspects of the discussion. The capabilities and important limitations of current cryomicroscope studies are included. Finally, the chapter includes a selection of photomicrographs illustrating some of the cell responses discussed.

Basic operating and performance characteristics of cryomicroscope systems

Desirable features of cryomicroscope systems

Before considering a detailed description of cryomicroscope systems a list of desirable cryomicroscope characteristics will be outlined. These often constrain the practical design of low temperature microscope systems. Desirable design features include the following.

1. A large operating range of temperatures; typical values might be −196°C to +100°C.
2. A large range of possible cooling and warming rates. The maximum range of useful rates would be approximately 0.01°C min^{-1} to 10 000°C min^{-1}.
3. The ability to programme arbitrary, non-linear temperature changes if desired.
4. Easily programmed temperature changes and overall simplicity of operation.
5. A robust and economical system, easily interfaced with common laboratory light microscopes.
6. Temperature measurement and control should be accurate, stable, reproducible and with a high degree of temperature resolution.
7. All standard light microscope optical techniques should be possible, including bright field, interference contrast, and fluorescent methods.
8. The well-equipped cryomicroscope should be capable of obtaining quantitative optical data using photomultipliers and image analysis devices.
9. Optical and thermal data should be displayed and recorded in convenient formats.

Basic heat transfer concepts

Early low temperature microscopists had limited control of the sample temperature and its rate of change because control was accomplished by placing the microscope and sample in contact with ice or exposing it to winter weather. A sample at temperature T_s is in thermal communication with an

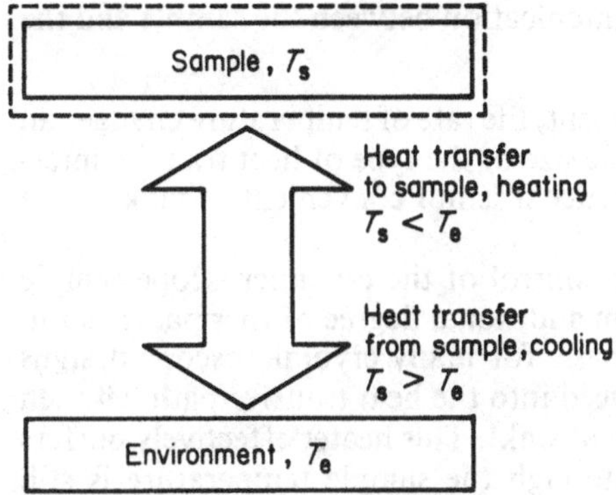

Fig. 6.2 Heat transfer between sample and environment, at temperatures T_s and T_e respectively, schematically represented for an early cryomicroscope.

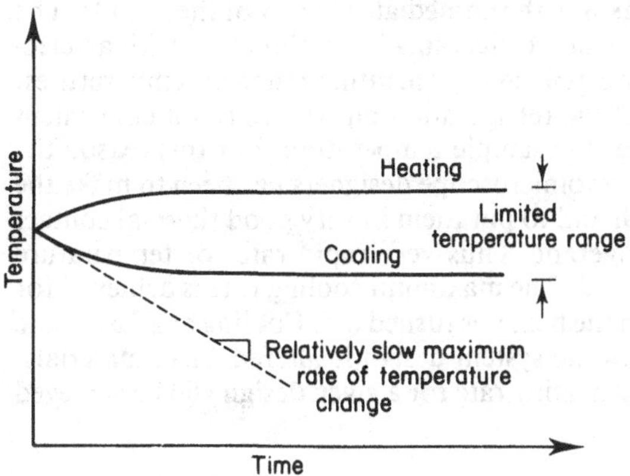

Fig. 6.3 Diagrammatic indication of the relatively slow rates of temperature change possible using an early cryomicroscope. The limited working range of the instrument is illustrated.

environment at temperature T_e. The range of sample temperatures obtainable in this case is dictated solely by the range of environmental temperatures (Fig. 6.2). In the early days of 'cryomicroscopy' the minimum temperature attainable was limited to rather high sub-zero temperatures. To heat or to cool the sample it is necessary to manipulate the environmental temperature above or below the sample temperature.

The rate at which the sample temperature can be changed is also limited in this situation (Fig. 6.3). The factors governing the rate of sample temperature change are:

(*a*) the size of the sample and the thermal characteristics of the sample and environment;

(*b*) the temperature difference between the sample and the environment;

(*c*) the nature of the thermal communication between the sample and the environment.

Although, for this type of arrangement, the rate of temperature change can be manipulated by altering the sample size or the type of heat transfer interaction with the environment, in practice it is not convenient to make these changes.

A significant improvement in the control of the cryomicroscope sample temperature was made by creating an additional degree of thermal freedom (Harmer, 1953). This remains the basis for many cryomicroscope designs today (Fig. 6.4). A heater is introduced into the heat transfer path between the sample and a low temperature 'heat sink'. This heater effectively buffers the sample from the heat sink. Although the sample temperature is still governed by the same factors as in the previous case, it is also affected significantly by the introduction of electrical power, $\dot{W}_{electrical}$, from the heater. The introduction of the heater makes it possible for sample temperatures well above the nominal environmental temperatures to be attained – even when a very low temperature heat sink is in the immediate vicinity of the sample. The development and use of cryogenic refrigerants in combination with an electrical heater therefore greatly extended the operating range of temperatures.

Generally, manipulation of the refrigerant temperature is not convenient and results in slow changes of the sample temperature. For this reason the philosophy adopted by most cryomicroscope designers has been to make the sample and heater quite small and to put them in very good thermal contact with a low temperature refrigerant. Thus very rapid rates of temperature change can be attained if desired. The maximum cooling rate is achieved for any given configuration when the heater is turned off. Cooling at a faster rate than this is impossible, unless the system design is altered (size, materials, etc.). Similarly, the maximum heating rate for a given design will be achieved

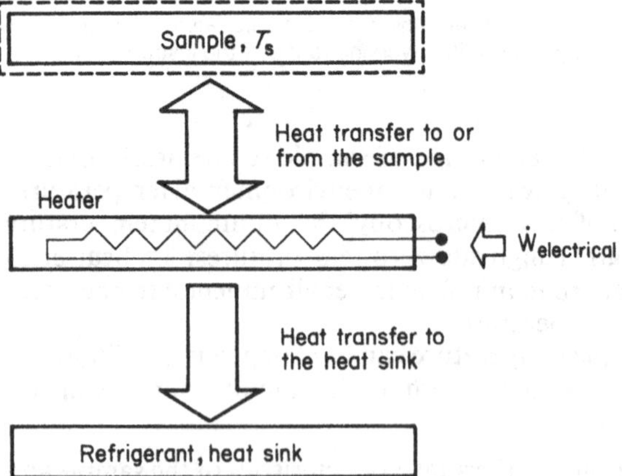

Fig. 6.4 Diagram of a Harmer-type cryomicroscope system, with a sample at temperature T_s and an electrical heating input $\dot{W}$.

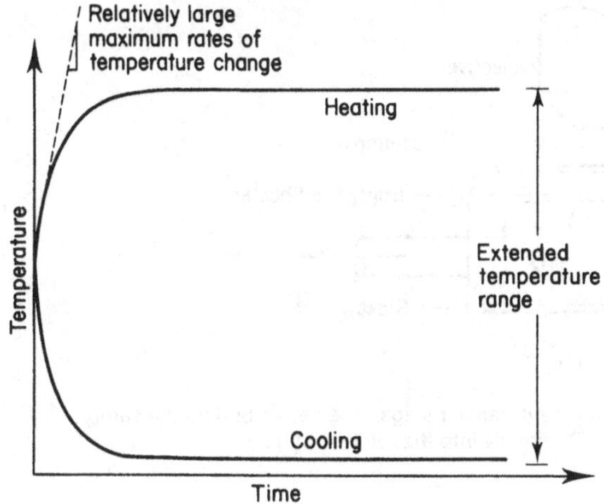

Fig. 6.5 Diagrammatic indication of the relatively fast rates of temperature change, and extended range, that can be achieved using a Harmer-type cryomicroscope.

when the heater is turned on to its maximum output. These maximum rates represent the 'natural response' of the heat transfer system (Fig. 6.5).

The significance of introducing the heater can now be appreciated. Without changing the system design at all, rates of temperature change smaller than the maximum rate described above can be achieved simply by manipulating the heater input, $\dot{W}_{electrical}$, in an appropriate fashion. Required temperature changes can be attained by maintaining the heat sink temperature and varying the heater input, $\dot{W}_{electrical}$, with time. As the input to the system is electrical it is amenable to direct links with electronic controllers and computers.

Convection and conduction of cryomicroscope stages

Two examples of cryomicroscope heat transfer systems will be introduced to illustrate some of the important details involved in such systems.

The subsystems involved in the heat transfer process (Fig. 6.4) are the sample, the heater, and the heat sink (including the refrigerant). These three subsystems are often referred to as the 'cryomicroscope heat transfer stage' or simply the 'stage'. Such a stage is usually attached to the X–Y positioning mechanism of a light microscope.

Two systems, differing in the design of the heat transfer stages, have been developed. The systems are the 'convection heat transfer stage' (Diller and Cravalho, 1971) and the 'conduction heat transfer stage' (McGrath *et al.*, 1975). The distinction between the two is based upon the mechanism of heat transfer in the immediate vicinity of the sample.

If the sample is in intimate thermal contact with the refrigerant stream, convective heat transfer occurs from the sample directly into the refrigerant

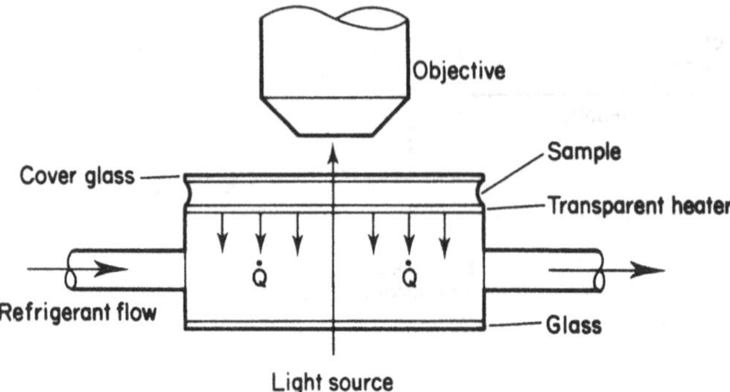

Fig. 6.6 Diagram of a 'convection' heat transfer stage. The heater buffers the sample thermally by supplying a heat flux, $\dot{Q}$, directly into the refrigerant fluid.

(Fig. 6.6). This system is called a 'convection' heat transfer cryomicroscope stage.

Alternatively, if heat transfer from the vicinity of the sample occurs by conduction through a solid medium to a distant peripheral refrigerant fluid, the stage is termed a 'conduction' stage (Fig. 6.7). The term 'distant' here is applied loosely. There is no sharp distinction which can be made between a convection stage and a conduction stage as both mechanisms of heat transfer are present in both systems, rather, it is a matter of degree. For a review of cryomicroscope designs see Diller (1982).

The conduction stage (McGrath *et al.*, 1975) was designed to complement the types of studies that could be performed on the convection stage (Diller and Cravalho, 1971). In particular, the convection stage was too thick to accommodate the use of the very short working distances required for certain types of transmitted light microscopy. The desire to perform transmitted light fluorescence studies with a cryomicroscope necessitated the lateral

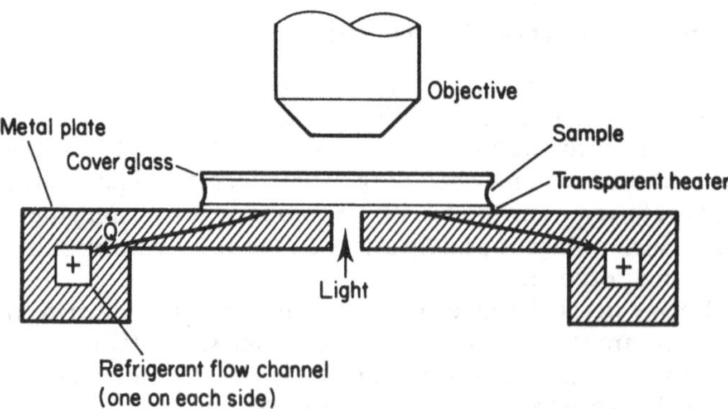

Fig. 6.7 Diagram of a 'conduction' heat transfer stage. Thermal buffering, provided by the heat flux, $\dot{Q}$, occurs with the flow of heat through a metal plate to a peripheral refrigerant system.

displacement of the refrigeration channels. It should be noted that epi-illumination equipment makes it possible to perform fluorescent studies on a convection cryomicroscope system (Morris and McGrath, 1981a). Not all microscopes, above, are easily adapted for epi-illumination.

Although the designation of a cryomicroscope stage as a convection or conduction stage is imprecise, it is worth noting some of the general operating characteristics of these two systems.

The convection system is capable of rapid thermal response, both rapid cooling and warming rates being made possible by using a sample and heater which are extremely thin and by transferring the thermal energy directly into the refrigerant (Fig. 6.7). As the sample is exceptionally thin and the heater buffers the sample from the refrigerant fluid, negligible vertical temperature gradients are produced within the sample even at rapid cooling rates (Diller, 1982). Since the heat transfer is predominantly in the vertical rather than the lateral direction, very little temperature gradient appears in the lateral direction. The lack of significant thermal gradients in the sample is important because normally only a single sample temperature is measured and controlled. This quasi-isothermal characteristic of the convection stage is beneficial because a large fraction of the sample area may be examined and assumed to be at the measured temperature. Disadvantages of this system are:

(*a*) the stage is often too thick to apply optical techniques when short working distance objectives and condensers are required;

(*b*) the thin glass windows and heaters used are somewhat fragile and, when subjected to cyclic extreme temperature variations as well as pressure and shear stress variations from the refrigerant flow, are liable to break.

The major advantages of the conduction stage are:

(*a*) that this system is approximately 5–10 times thinner than the typical convection stage, so that techniques including high resolution, high numerical aperture work can be performed;

(*b*) that the thin sample sandwich and glass heater are not in direct mechanical contact with the refrigerant and therefore are not subjected to the pressure variations or shear stresses imposed by the refrigerant fluid.

An important characteristic of the conduction stage design is that a rather steep lateral temperature gradient is established as a result of the normal operating mode. The convection stage tends to cool and heat in a rather uniform fashion (Fig. 6.8), using a vertical temperature gradient present between the heater and refrigerant stream to transfer the thermal energy. Large lateral temperature gradients occur within the conduction stage because the heat flux in the lateral direction is large (Fig. 6.9).

Obviously, if the lateral temperature gradient is large, substantial errors in temperature measurement may occur if it is assumed that the single temperature measured corresponds to the temperature of a sample displaced a distance away from the temperature-measuring sensor. In all cases, to obtain accurate measurements the temperature sensor should be placed in intimate thermal contact with the sample. Early experimental procedures with the

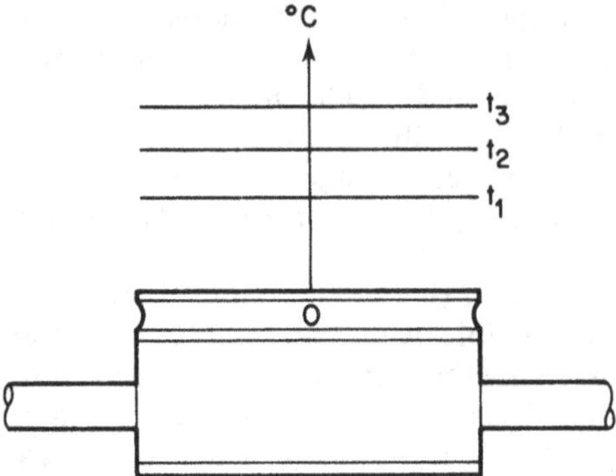

Fig. 6.8 An illustration of the lateral uniformity of temperature which is typical of a 'convection' stage. The profiles of temperature fields that might be measured at times t_1, t_2 and t_3 are indicated.

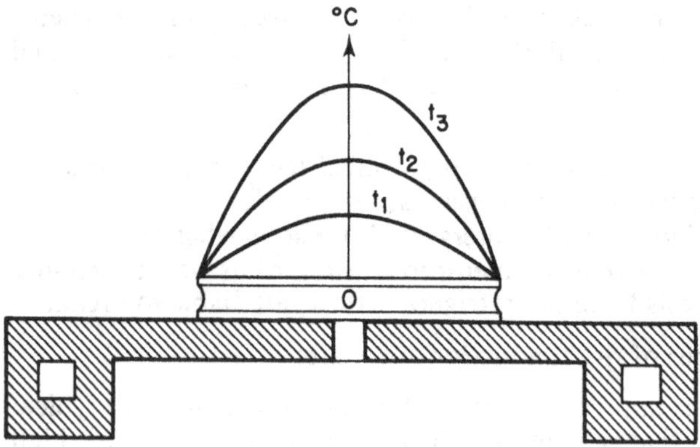

Fig. 6.9 An illustration of the parabolic temperature fields that are typical of a 'conduction' heat transfer stage. The profiles of fields that might be measured at times t_1, t_2 and t_3 are indicated.

convection and conduction stages made use of thermocouples as sensors and placed the thermocouple directly into the thin sample sandwich. The extremely small thermocouples required (1–50 microns diameter wire) are delicate and susceptible to mechanical damage during cleaning of the stage between experiments or during the preparation of a new sample. Most current designs make use of some type of 'thermocouple sandwich' which permanently fixes the thermocouple to the cryomicroscope stage between thin glass sheets and protects it from mechanical trauma (Fig. 6.10). The thermocouple is insulated electrically from the heater and is displaced vertically from the sample by the thickness of a thin glass sheet (Diller, 1982). In many cases, the glass sheet is etched so that the temperature sensor can be placed closer to the sample.

Analysis has shown that in virtually all cases of interest the vertical temperature gradient within the sample is not significant (Diller, 1982; Tu, 1983).

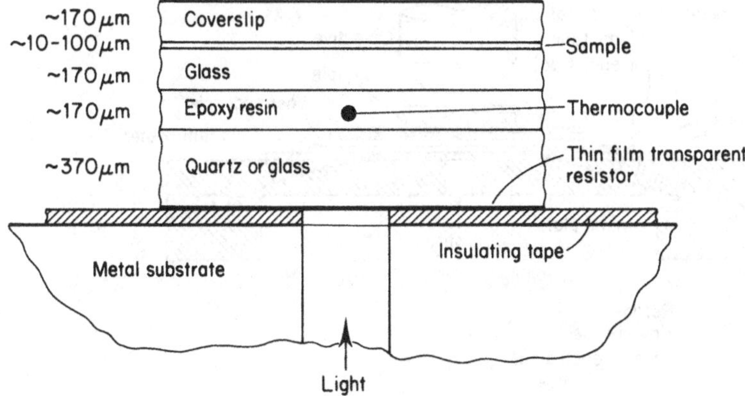

Fig. 6.10 A diagram showing a typical thermocouple sandwich sensor and thin film heater incorporated into the heat transfer stage The dimensions are approximately to scale.

Analyses and experiments suggest that the vertical temperature difference existing in the typical cryomicroscope sample sandwich is less than 1 °C even for the fastest cooling and warming rates attainable (*c.* 1000°C min^{-1}).

The heaters used in most cryomicroscope systems today are transparent resistors formed by coating a glass or quartz substrate with a thin layer of electrically conductive translucent material. Tin oxides (Schwartz and Diller, 1982), vacuum deposition of various metals (Reid, 1978) and commercially available coatings have been used. Typical films may be several thousand angstroms thick and have resistance of 50Ω per square. Typically, such films are deposited uniformly to produce uniform Joule heating in use. They should be permanent films resistant to all forms of degradation, and they should have acceptable optical properties including large, spatially uniform transmissivities.

A typical example of a conduction cryomicroscope heat transfer stage is depicted in Figs. 6.11 and 6.12.

Temperature measurement and control

The next problems to be considered are the measurement and control of the sample temperature.

Although multiple temperature sensors may be placed in the sample for experimental studies, in practice a single sensor is used to monitor the sample temperature. This temperature is used as an input to control the rate of temperature change of the sample. If, however, there is a significant temperature difference between the sample and the measuring sensor, an error in temperature measurement will be introduced.

The basic elements involved in controlling the temperature of a sample on the cryomicroscope (Fig. 6.13) can be divided into six convenient subsystems for the purpose of this discussion. Three of the subsystems have been described previously: the sample, heater, and heat sink (Fig. 6.5).

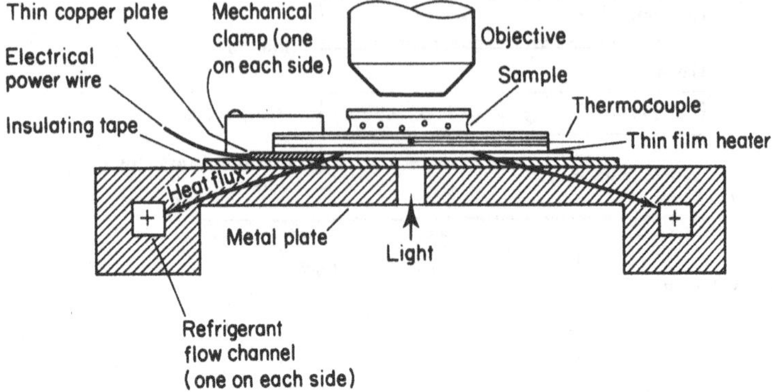

Fig. 6.11 The assembled components of a typical 'conduction' heat transfer stage as used on a modern cryomicroscope.

Fig. 6.12 A 'conduction' heat transfer stage photographed *in situ* on a light microscope.

The desired sample temperature is created by the reference signal generator (Fig. 6.13). This is the part of the system where programming of the desired temperature changes is performed. This signal is compared in the control scheme with the actual sample temperature measured by the temperature sensor. Controllers of various types have been applied to cryomicroscope systems (Diller and Cravalho, 1971; Morris and McGrath, 1981a; Evans and

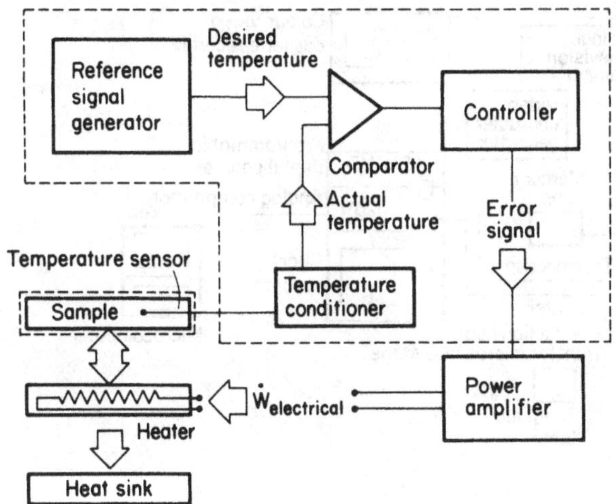

Fig. 6.13 The basic elements of a temperature controller for a modern cryomicroscope.

Diller, 1982). These devices either produce an on-off output or an output which is proportional to the difference between the desired and actual sample temperatures. Alternatively, the controller output may be related to the derivative or integral of the difference between the actual and desired temperatures, or a combination of these three possibilities. Further details concerning the controller are beyond the scope of this description.

Regardless of the control strategy employed the output of the controller is used to drive a power amplifier. The output of this amplifier furnishes $\dot{W}_{electrical}$, the electrical power dissipated in the heater. As the dissipated electrical energy appears as thermal energy which in turn affects the sample temperature, a closed feedback loop has been created (Fig. 6.14).

Methods of generating the desired temperature signal

Temperature control of the cryomicroscope sample is effected by an instantaneous comparison of the actual sample temperature and a signal representing the desired temperature (Fig. 6.14). A number of methods have been employed for generating the desired temperature signal. The voltage drop across a motor-driven potentiometer has been used to programme the desired temperature changes (Reid, 1978). Analog circuits have been used for this purpose (Dillar and Cravalho, 1971) but this method suffers from problems associated with long-term electrical drift and is not extensively used. Digital electronics have the advantage of improved stability and this technology has been incorporated into a programmable digital counter which has been used to produce the desired temperature signal (Morris and McGrath, 1981a).

Microprocessor technology has been incorporated with the cryomicroscope system (Morris and McGrath, 1981a; Diller, 1972; Cosman, 1983). Previous systems were suitable for producing simple, constant cooling and

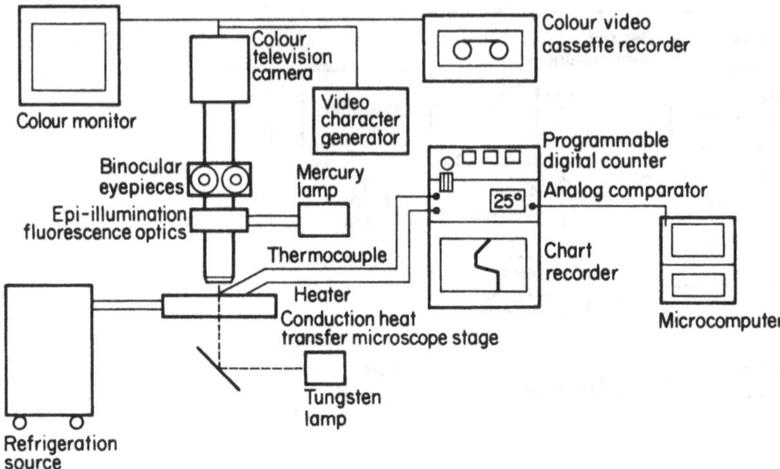

Fig. 6.14 A diagrammatic representation of a complete cryomicroscope system using a conduction heat transfer stage. This system is equipped for both incident and transmitted light microscopy and has video-recording facilities.

warming rates. In contrast, the microprocessor-controlled device is easily programmed to produce quite arbitrary and complex patterns of thermal change which may be of interest. In addition, the operation of such devices is easier than most previous methods. For example, normally temperatures and rates of temperature change can be entered into the microprocessor with a computer keyboard directly, rather than indirect coding of such information (e.g. setting potentiometers and decade resistance boxes).

Similarly, a microcomputer has been interfaced with a conduction cryomicroscope system at Michigan State University, USA (Morris and McGrath, 1981a). Software developed for this system prompts the user to enter a series of target temperatures and rates of temperature change on a computer terminal (Fig. 6.15) such that piece-wise linearization of abitrary non-linear temperature changes may be generated (Fig. 6.16).

The microcomputer used in this instance is equipped with a 10 MHz real-time clock and a 4 channel, 12 bit digital-to-analog converter. In this way the desired temperature change can be generated by the computer, which results in a temperature resolution of 0.05 °C for the desired temperature signal (see Fig. 6.13) over the operating temperature range of − 150°C to + 50°C. This resolution in time and temperature in part accounts for the smooth appearance of the experimental temperature histories produced by the system (Fig. 6.16). This microcomputer is also equipped with a 16 channel, 12 bit analog-to-digital converter operating at 35 KHz for data acquisition. Thus multiple thermocouples including the control thermocouple may be monitored at rapid rates with this system. Other data, for example photometric light levels, may be sampled by the computer during cryomicroscopic studies. One example of this capability would be the determination of the

```
        THIS IS THE DATA ENTRY BLOCK

        The user must specify ramp slopes, target temperatures, and
hold time for each temperature.

        NOTE: Zero hold times ARE acceptable.

        To EXIT from data entry enter a NEGATIVE HOLD TIME after final
temperature.

        NOTE that slope sign (+,-) is NOT important as programme will
determine if ramp is up or down

        Up to 19 target temperatures can be entered.

        ENTER INITIAL TEMPERATURE IN DEGREES CELSIUS: 10

        ******* TEMPERATURE INITIATED AT 10.00 C *******
        ******** CONTROL OFFSET =      0.000 C *******

        ENTER RAMP # 1 SLOPE IN (DEGREES C/MIN): 100
        ENTER TARGET TEMPERATURE # 1 IN (DEGREES C): 0.5
        ENTER TIME OF HOLD AT  0.50 DEGREES C IN (MIN): .25

        ENTER RAMP # 2 SLOPE IN (DEGREES C/MIN): 100
        ENTER TARGET TEMPERATURE # 2 IN (DEGREES C): -9.5
        ENTER TIME OF HOLD AT  -9.50 DEGREES C IN (MIN): .25

        ENTER RAMP # 3 SLOPE IN (DEGREES C/MIN): 25
        ENTER TARGET TEMPERATURE # 3 IN (DEGREES C): -19
        ENTER TIME OF HOLD AT  -19.00 DEGREES C IN (MIN): 1

        ENTER RAMP # 4 SLOPE IN (DEGREES C/MIN): 500
        ENTER TARGET TEMPERATURE # 4 IN (DEGREES C): 10
        ENTER TIME OF HOLD AT  10.00 DEGREES C IN (MIN): -1

        DATA ENTRY TERMINATED; BEGIN DATA VERIFICATION

        INITIAL TEMPERATURE= 10.00 DEGREES C

        N       RAMP SLOPE         TARGET TEMPERATURE        HOLD TIME
                (DEGREES C/MIN)    (DEGREES C)               (MIN)

        1        100.00            0.50                      0.25
        2        100.00            -9.50                     0.25
        3         25.00            -19.00                    1.00
        4        500.00            10.00                    -1.00

        ARE THESE VALUES ACCEPTABLE (Y OR N):

        ******  BEGINNING OF START SEQUENCE  ******


        PUSH "-" (MINUS) TO START SEQUENCE
        PUSH "." (PERIOD) AT ANY POINT FOLLOWING TO STOP SEQUENCE
        PUSH "1" (ONE) TO RETURN TO DATA ENTRY BLOCK
        PUSH "2" (TWO) TO DISPLAY CURRENT VALUES
        PUSH "3" (THREE) TO END PROGRAMME EXECUTION
        PUSH '4" (FOUR) TO RESET CONTROL OFFSET
```

Fig. 6.15 Illustration of computer software used for programming cryomicroscope sample temperature. Total programmer time approximately 60 seconds to set up the programme shown.

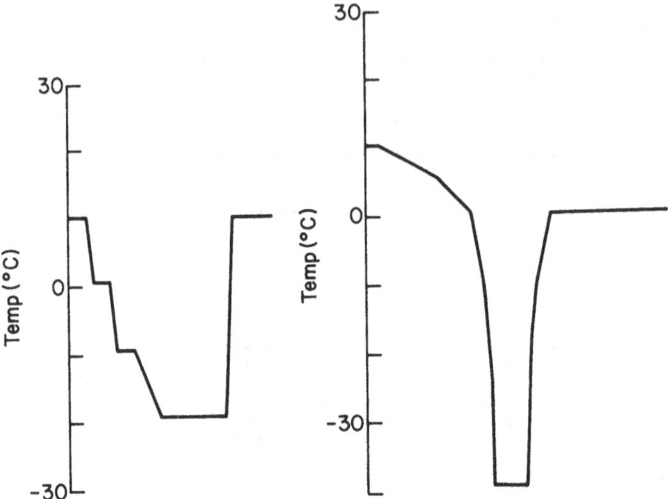

Fig. 6.16 Examples of temperature – time profiles produced by the software programme illustrated in Fig. 6.15.

temperature and/or concentration gradient within or around the sample during cooling and warming at various rates.

Optical techniques available on cryomicroscopes

As mentioned previously the convection cryomicroscope stage may be too thick to be useful when high numerical aperture, short-working distance condensers are required. Epi-illumination techniques have been applied successfully in such cases. Otherwise convection and conduction stages can be expected to include all common light microscopy methods including bright field, phase contrast, differential interference contrast (Nomarski) and fluorescent optical techniques.

Displaying and recording data

Low temperature microscope studies involve defining the relationship between the dynamic response of biological systems to temperature change and rates of temperature change. The thermal information of interest is normally the single sample temperature measured in such studies. A continuous record of the sample temperature is desirable and is usually recorded on a strip chart recorder (Figs. 6.14, 6.16). The instantaneous sample temperature can be digitized and displayed on one or more of the following: a panel meter; the computer terminal; or superimposed upon a videotaped image from the television camera (Fig. 6.17).

Optical data are recorded in a variety of ways. For example, a beam splitter allows simultaneous observation and recording (Fig. 6.14). Images may be recorded by 35 mm still photography, motion picture photography or video-

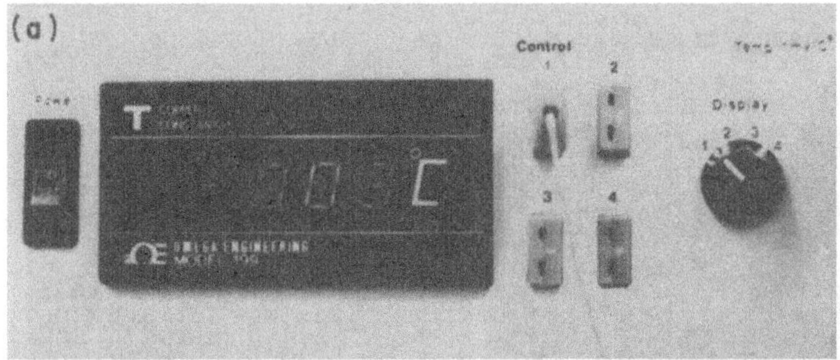

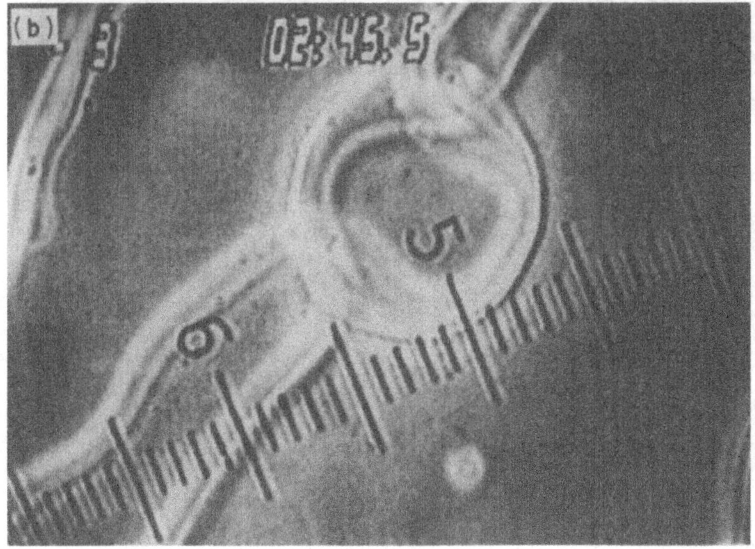

Fig. 6.17 Display and recording of sample temperatures. (a) Digital panel meter displaying instantaneous sample temperature. (b) Photograph taken from a video monitor. At the top left of the screen is the instantaneous sample temperature corresponding to the digital panel meter. This temperature and the elapsed time during the experiment (seen at the top centre of the monitor) are superimposed onto the video image by a video character generator (see Fig. 6.14).

taping. The latter may be augmented with a video character generator (Fig. 6.17) which superimposes the sample temperature and an elapsed time on the video image.

Quantitative microscopic techniques have also been applied - a photomultiplier can be used in place of the camera to perform quantitative fluorescent studies. Image analysis techniques have proved to be very useful for the rapid determination of cell volumes associated with freezing and thawing (Diller, 1982). An example of computer image analysis is illustrated in Fig. 6.18.

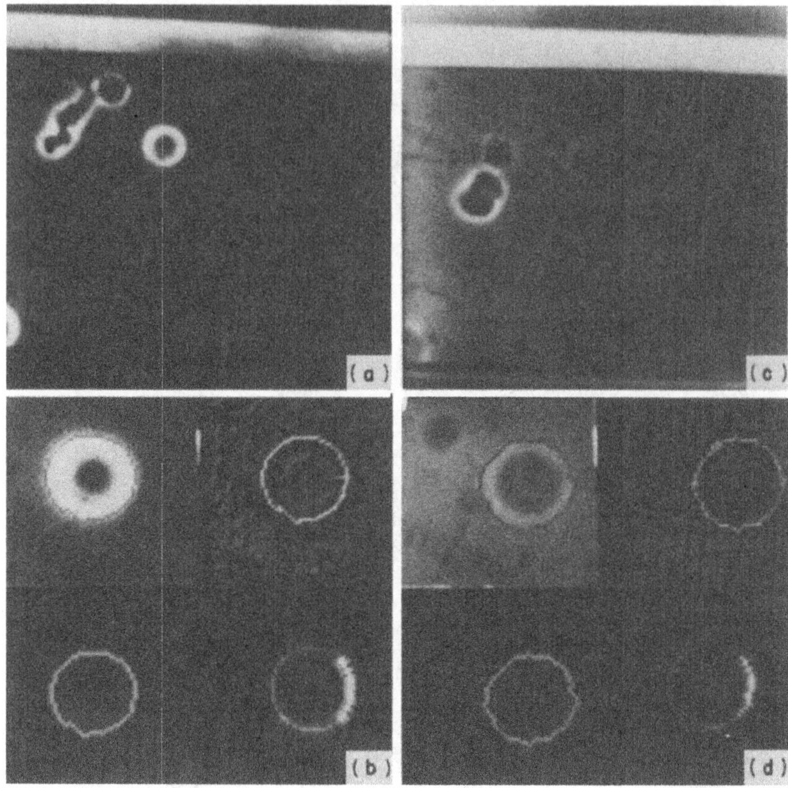

Fig. 6.18 Image analysis in cryomicroscopy. Outlines determined by computer image analysis may be used to calculate volume changes during freezing and thawing. Two digitized images of liposomes were taken before freezing and thawing. (a) was taken shortly after thawing and (b) is of the same liposomes after rehydration. (c) and (d) are the corresponding computer-generated outlines of the liposome in the centre of (a) and (b) on the left. (Courtesy of Dr K. Diller.)

Cryomicroscope capabilities and limitations

The cryomicroscope systems in operation have some impressive capabilities, however, there are limitations that should be recognized.

It should be noted that cryomicroscope systems are not necessarily extremely expensive, but less costly models will contain design compromises which may affect accuracy, reliability, ease of operation and general versatility. For example, inexpensive versions of cryomicroscope systems have been described using Peltier elements (Troyer, 1974) as well as a controlled temperature bath (Mersey *et al.*, 1978).

Temperature range (–196°C–+100°C). The minimum temperature is limited by the refrigerant used to provide the heat sink. As a result of the availability of cryogens such as liquid nitrogen low temperatures are readily achieved. Sample temperatures of –100°C to –125°C are obtained with little insulation of the cryomicroscope stage or the tubing from the dewar. The high temperature limit is defined by the amount of electrical energy dissipated in the heater and the temperature limits of the materials used. As an example of a typical operating condition for the conduction cryomicroscope system, temperatures of +40°C are achieved when gaseous nitrogen at –196°C is used as a refrigerant and approximately 10–15 Watts of electrical power are dissipated in the transparent heater with a resistance of 50–60Ω.

Cooling and warming rates (0.01–1000°C min^{-1}). The physical size and thermal diffusivity of the materials comprising the overall heat transfer stage limit the fastest rates obtainable. The thermal contact between the heater and heat sink, and the heat sink temperature also play an important role in this respect.

Small heater configurations in good contact with a low temperature heat sink have produced cooling rates of 7200°C min^{-1} on the conduction cryomicroscope stage (McGrath *et al.*, 1975). Systems capable of these rapid rates are generally more fragile than the larger, slower-responding designs. The typical conduction cryomicroscope system used in the BTP Laboratory has maximum cooling rates in the range of 500–2000°C min^{-1}.

Heat transfer considerations do not limit slow cooling rates. Limitations at the slower rates are governed by electrical stability, including that of the reference signal generator and the controller. Electronic drift associated with analog circuitry was a problem with earlier designs but is largely eliminated with the use of digital components.

Temperature measurement. Temperature resolution in the range of 0.1–1.0°C is quite easily attained. However the absolute accuracy of most cryomicroscope sample temperatures is not likely to be better than 0.5–2.0°C. Inaccuracies greater than 2.0°C are probable if care is not taken. Common sources of error result from temperature gradients in and around the sample as well as electronic factors involved in temperature measurement. These electronic factors include errors introduced by electronic drift, analog to digital conversion, linearization, cold junction compresentation and frequency response problems.

Programming temperature changes. As cryomicroscope systems have developed to allow temperature changes to be programmed, a number of techniques have been devised to allow a thermal protocol for the sample to be specified. Early designs coded the information into the system by selecting potentiometer settings, decade box resistances, or motor speeds. Generally, this hardware-intensive approach has limited versatility. In contrast, the introduction of the microprocessor and microcomputer results in a very flexible system with computer programs easily developed to specify the desired thermal protocols. Complex patterns of temperature change may be entered into the system by typing straightforward answers to computer-generated prompts on a computer terminal keyboard.

Optical considerations. Optical limitations that have arisen in cryomicroscopic studies are similar to those which occur in standard light microscopy.

Some microscopes have made use of rather thick heat transfer stages. These designs preclude the use of transmitted illumination with short-working distance optics. Most often the condenser, rather than the objective, is the problem in this respect. Epi-illumination techniques can be used to circumvent this problem in many cases.

Low temperature studies may also create problems with condensate on the sample or optics. To reduce this problem cryomicroscopes are often enclosed in a plexiglass housing or plastic hood which may be purged with a non-condensible gas such as dry nitrogen (Diller and Cravalho, 1971). Simpler alternatives include local purging with a dry gas stream and the use of dry cotton ball packing (Schwartz and Diller, 1982).

The use of immersion optics would be expected to limit the range of operating temperatures and rates of temperature change, although they have been used successfully (Diller, 1972; McGrath, 1974). Solidification of the immersion medium may degrade optical quality and damage to the optics could occur as a result of thermal stresses.

Determination of cell volumes during freezing represents an important type of cryomicroscopic study. However, volume determinations are inferred in most cases from two-dimensional projected areas assuming the cell to be spherical (Schiewe and Korber, 1982a). Defining cell boundaries can be a problem particularly with phase contrast optics which produce a 'halo' image around the cell. Attempts to circumvent this problem by sectioning the cell optically may not be possible due to the large depth-of-field of many objectives. These latter problems are, of course, not unique to cryomicroscopy, unlike the problem of the detection of cell boundaries encased in ice crystals.

Summary of representative cryomicroscope studies

Although many of the important cryobiology questions raised by Molisch and other early investigators remain unanswered, there are several significant areas where cryomicroscopy has advanced.

Improved technology and design now make low temperature microscopy a much more powerful research tool. In addition to more convenient operating procedures and precise control over a very wide range of operating conditions, modern cryomicroscopy takes advantage of quantitative optical techniques. Furthermore, modern cryomicroscopy is often conducted in conjunction with thermodynamic modelling and computer simulation of the response of cells to freezing (Knox *et al.*, 1980a). Such modelling attempts to place the understanding of the response of a cell during freezing on a theoretical basis. In some cases quantitative predictions of freezing response are possible.

Molisch did not discuss in detail the effects produced by intracellular compared to extracellular ice formation. Mazur has made such a distinction and formulated a 'two-factor' theory of freezing damage (Mazur *et al.*, 1972) which proposes that intracellular ice formation is responsible for cell death at supra-optimal cooling rates and that so-called 'solute effect' damage caused by prolonged exposure to concentrated solutions occurs at sub-optimal cooling rates. These factors are considered to be independent effects (see Grout and Morris, this volume).

In an effort to quantify these competing effects a thermodynamic model describing the kinetics of water loss during cell freezing has been developed (Mazur, 1963). This predicts cell volume changes during freezing and provides estimates of the cooling rates required for intracellular ice formation. Intracellular as well as extracellular solute concentration may be predicted from the model.

The 'two-factor' hypothesis of freezing injury is valuable because it successfully correlates experimental results for a wide variety of biological cell types. Computer simulations using this theoretical model have defined which factors in the model are most critical in determining the response of a cell to freezing. Cryomicroscope studies can be used to study the validity of the thermodynamic model. Primary areas of interest include:

(*a*) cell volume changes as a function of temperature for various cooling rates;

(*b*) the occurrence of intracellular ice formation as a function of cooling rate.

Before illustrating the response of cells to freezing and thawing as observed with the cryomicroscope it is useful to discuss the interpretation of cryomicroscopic images.

Interpretation of cryomicroscope images

During slow freezing the extracellular solution becomes increasingly concentrated as solidification occurs; and an entrapped cell will lose water due to the consequent transmembrane water chemical potential difference, resulting in cell shrinkage (Fig. 6.19). Rehydration to the original volume occurs upon thawing if no damage to plasma membrane semipermeability has occurred during freezing. Cell volume changes are calculated from the projected areas of cells which remain spherical, or the volume of an irregularly-shaped cell is defined in terms of an equivalent sphere having the same projected area as the irregularly-shaped cell.

At faster rates of cooling cells will often darken at some point after the formation of extracellular ice (Fig. 6.20) (Luyet and Gibbs, 1937; Smith *et al.*, 1951; Leibo *et al.*, 1978; Schiewe and Korber, 1982a; Steponkus and Dowgert, 1981). In some cases investigators have stated that close examination reveals the outlines of small crystals contained within the cells. Electron

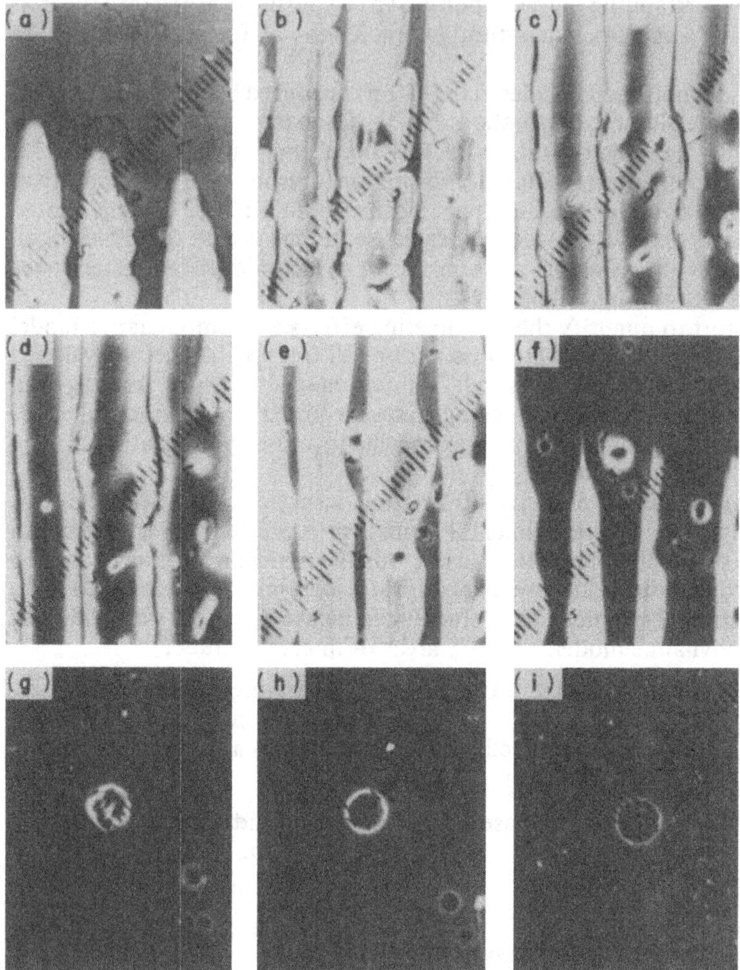

Fig. 6.19 Osmotic shrinkage and swelling of cell-sized, unilamellar liposomes during freezing and thawing. Liposomes were cooled at 2°C min^{-1} from 0°C (a) to −5°C (d) and warmed at 20°C min^{-1} from −5°C to room temperature.

microscopy of cells frozen in such conditions has revealed the presence of intracellular ice crystals of various sizes (Bank, 1973; Bank and Mazur, 1973), which suggests that crystals large enough to be detected with the light microscope could exist inside cells. In addition, theoretical calculations suggest that many small ice crystals should be created at fast cooling rates whereas fewer but larger ice crystals should be created at slower cooling rates (Mazur, 1966). Hence the intracellular darkening, or 'flashing' as it is often called, is commonly taken to be the result of the formation of many small ice

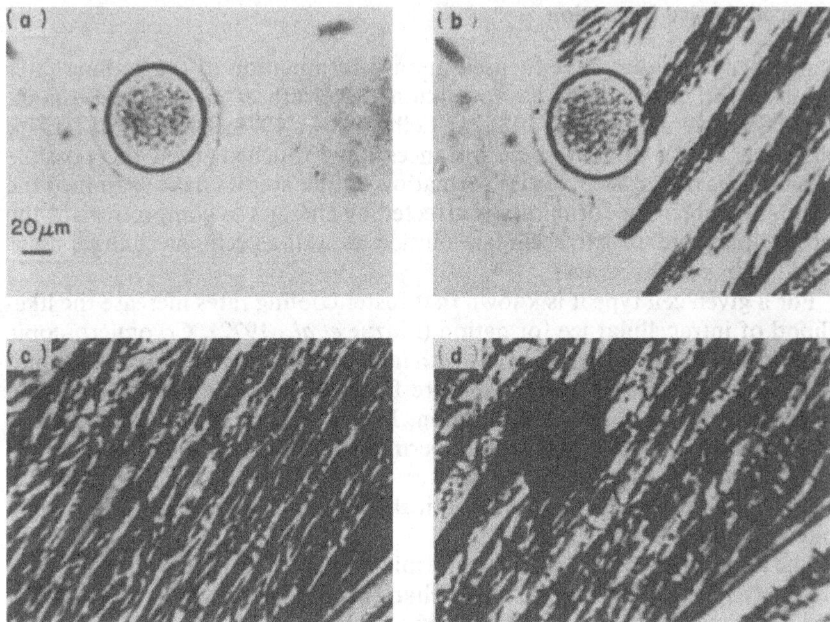

Fig. 6.20 An example of 'flashing' in an unfertilized mouse ovum. This phenomenon is, arguably, a result of intracellular ice formation (for discussion, see text). The ovum was cooled at 32°C min^{-1} to 0°C (a), −3°C (b), where extracellular ice crystals are seen, −20°C (c) and −39°C (d) where a marked increase in contrast, or flashing, is evident within the cell boundary. (Reproduced with permission from Leibo *et al.*, 1978.)

crystals within the cell which scatter much of the incident light which arrives at the cell sample.

However, to a first approximation structures with characteristic sizes matching the wavelength of the incident light will interfere or diffract light most (Feynman *et al.*, 1963). Therefore structures with dimensions in the visible wavelength range (0.4–0.7 μm) might be expected to cause the observed 'black flashing'. It is conceivable that large ice crystals the size of a cell could form intracellularly under some conditions (Franks *et al.*, 1983), such a single crystal being difficult to observe visually. At very rapid rates of freezing extremely small ice crystals would form, as small as 0.03 μm (Shimada and Asahina, 1978), which would not be resolved with the light microscope. Thus at both very slow and very fast rates of cooling 'black flashing' may not be readily discerned with the light microscope even though intracellular ice may be present.

In summary, it is probable that the darkening of cells during freezing is the result of intracellular ice formation. However, the lack of an obvious darkening cannot be taken to mean that intracellular ice is absent.

Intracellular ice formation

A number of studies have focused upon determination of the cooling rates required for intracellular ice formation (McGrath *et al.*, 1975; Cosman, 1983; Schiewe and Körber, 1982a; Leibo *et al.*, 1978; Diller *et al.*, 1976; Shabana, 1983), and in some instances have touched upon the possible mechanisms of intracellular ice formation. Other studies have examined the manner in which ice formation is affected by changes in composition of the intracellular and/or extracellular solution as well as cellular changes associated with cold hardiness.

For a given cell type it is known that faster cooling rates increase the likelihood of intracellular ice formation (Mazur *et al.*, 1972). Cryomicroscopic observation has revealed that cells with large surface area/volume ratios and large water permeabilities will require faster cooling rates to form visible evidence of intracellular ice formation. In contrast, cells with small surface area/volume ratios and small water permeabilities will form intracellular ice at slower cooling rates. It should be stressed that a limitation of such studies is that ice crystals smaller than the optical resolution of the light microscope may form and remain undetected.

The application of low temperature microscopy to cell freezing studies has not made it possible to define the mechanisms of intracellular ice formation, but it is known that for a given cell type changes induced by cold hardiness (Steponkus *et al.*, 1982) or by the stage of cell growth cycle (Morris, unpublished results) affect the cooling rates required for intracellular ice formation.

The cryomicroscope has been used to study the observed, direct correlation between visually-detected intracellular ice and cell survival (McGrath *et al.*, 1975; Schiewe and Körber, 1983). These studies used a fluorescent viability assay in conjunction with cryomicroscopy to define cell survival. It should be noted that the 'flashing' or cell darkening usually taken to represent intracellular ice formation (Luyet and Gibbs, 1937; McGrath *et al.*, 1975) normally observed during freezing has also been reported during thawing (McGrath *et al.*, 1975; Rall *et al.*, 1980; Schiewe and Körber, 1983). The survival of 8-cell mouse embryos in 1.5 M DMSO did not correlate with the 'flashing' temperature during freezing or thawing (Rall *et al.*, 1980); slowly frozen lymphocytes, however, did not exhibit 'flashing' during freezing but did 'flash' upon thawing (Schiewe and Körber, 1983). Decreased survival correlated reasonably well with increased percentages of 'flashing' during freezing but did not correlate with 'flashing' during warming. These results suggest the existence of innocuous intracellular ice. Further comparable data relating cell survival and cooling rate may be determined for lympocytes both fluorometrically on the cryomicroscope and in bulk samples (Schiewe and Körber, 1983). Thus for this biological system at least, survival results determined using the low temperature microscope are analogous (but not identical) to those obtained in a larger scale apparatus typical of the clinical situation. Given that the percentage survival in a cell population will decrease if the 'flashing' type of ice formation occurs during freezing, it is of practical as well as theoretical interest to determine the temperature at which the nucleation event occurs. The nucleation temperatures of a number of systems have been determined with cryomicroscopes (Leibo *et al.*, 1978; Morris and McGrath, 1981b; Rall *et al.*, 1983; Morris, unpublished) ranging between −0.5°C and −15°C for unprotected cell suspensions. For a given sample it is common to observe standard deviations of approximately ±2–3°C about the mean value, presumably as a result of distributions of important but unknown parameters. The addition of cryoprotective chemicals greatly depresses the intracellular nucleation temperature. Although 8-cell mouse embryos freeze intracellularly in the temperature range −10°C– −15°C without additives, these embryos will freeze intracellularly at −38°C– −44°C in the presence of 1.0 M to 2.0 M glycerol or dimethylsulfoxide (Rall *et al.*, 1983). This same trend has been reported for unfertilized hamster ova (Shabana, 1983).

Available data suggests that in most cases the intracellular nucleation temperature during freezing is statistically independent of or very weakly dependent upon cooling rate (Leibo *et al.*, 1978; Schiewe and Körber, 1983; Morris and McGrath, 1981b; Shabana, 1983). This appears to be true whether the cryoprotective compounds are present or absent. The strongest evidence that nucleation temperature may depend on cooling rate is the case of unfertilized hamster ova frozen in the presence of 1.5 M dimethylsulfoxide (Shabana, 1983). In this case mean nucleation temperatures at slower cooling rates (5–10°C min^{-1}) are approximately 20–25°C lower than those at higher cooling rates *c.* 50°C min^{-1}. It should be noted however that the nucleation temperature of this same system is independent of cooling rate when no protective additives are present (Shabana, 1983).

The pronounced effect which extracellular supercooling has upon intracellular ice formation during freezing (and therefore presumably upon cell survival) has been determined (Diller, 1975). These data are relevant to the practice of 'seeding' samples prior to freezing in order to improve recovery. The results show that if a cell in its suspending medium is supercooled prior to the nucleation of extracellular ice, then slower rates of cooling must be applied in order to avoid intracellular ice formation during freezing. This is expected on a theoretical basis since a cell placed in a metastable state prior to extracellular freezing can not withstand a significant additional departure from equilibrium imposed by rapid cooling rates.

Relatively few microscopic data have been obtained describing the effects of cryoprotective additives on various aspects of cell freezing. Increasing concentrations of glycerol and DMSO decrease the nucleation temperature (Rall *et al.*, 1980; Shabana, 1983). Glycerol does not seem to effect the cooling rate required to produce intracellular ice in 100 per cent of a population of erythrocytes (with no extracellular supercooling) (Diller, 1979) but causes intracellular ice formation to begin to appear in the population at much slower cooling rates than would be expected for control cells. The latter result has been explained on the basis of a reduced water permeability (Diller and Lynch, 1983). The inclusion of a cryoprotective additive will depress the freezing point of the solution. Once freezing begins, therefore, the cell will be at a lower temperature than it would have been without the presence of a cryoprotective chemical and the membrane water permeability will be diminished as a result. This will displace the cell further from equilibrium and increase the likelihood of intracellular ice formation.

Increasing concentrations of glycerol (0 M, 2 M, 4 M) dramatically increase the survival of kidney cortex cells (Diller and Schmidt, 1981), the effect being to broaden the range of cooling rates yielding high survival in the population, but the optimum freezing rate appears to be relatively unaffected by the glycerol concentration.

Gas bubble formation

Gas bubble formation which may accompany cell freezing was one of the earliest phenomena observed by cryomicroscopists; Molisch observed such bubble formation in cells during his studies in 1897 (Molisch, trans 1982). Since that time the supersaturation of gas during freezing and the role such a phenomenon may play in cell freezing death has received very little attention. An example of the appearance of gas bubbles as observed with a cryomicroscope system is shown in Fig. 6.21.

Steponkus, using an isolated plant protoplast system, has reported that gas bubbles are a direct result of intracellular ice formation and argues that cell darkening ('flashing') which occurs during freezing is a result of gas bubble formation *per se* and is not a result of the interaction of light and the intracellular ice crystals (Steponkus and Dowgert, 1981).

In experiments with the filamentous alga *Spirogyra* intracellular ice was apparently required for intracellular gas bubble formation (Morris and McGrath, 1981b). Bubbles were not observed at very slow warming rates ($< 0.5°C\ min^{-1}$). In contrast gas bubble formation was not considered responsible for granulocyte and lymphocyte 'flashing' (Schiewe and Körber, 1982c), the investigators reporting that intracellular gas bubbles may occur even when no visible intracellular ice is formed (Schiewe and Körber, 1982b).

This rather confusing state of affairs with respect to gas bubbles, ice crystals and 'flashing' illustrates an important point, namely the limited optical resolution of the cryomicroscope. As both bubbles and ice crystals may exist in cells within the size range expected to interfere most with visible light, the cryomicroscope is probably not capable of resolving this problem of the interpretation of the visual information (Morris and McGrath, 1981b). The significance of the presence of gas bubble formation is also unclear with respect to cell injury. It has been suggested that bubbles formed in one cell may trigger intracellular ice formation in others (Steponkus and Dowgert, 1981), and both mechanical and chemical interactions between the cell and gas bubbles also have been suggested as important considerations (Morris and McGrath, 1981b).

Slow cooling and cellular dehydration

The cryomicroscope has proved to be useful for studying the osmotic response of cells during freezing at slow rates. In this way experimental observations of cell volume changes during cell freezing and thawing may be used to assess the validity of various thermodynamic models. Such studies

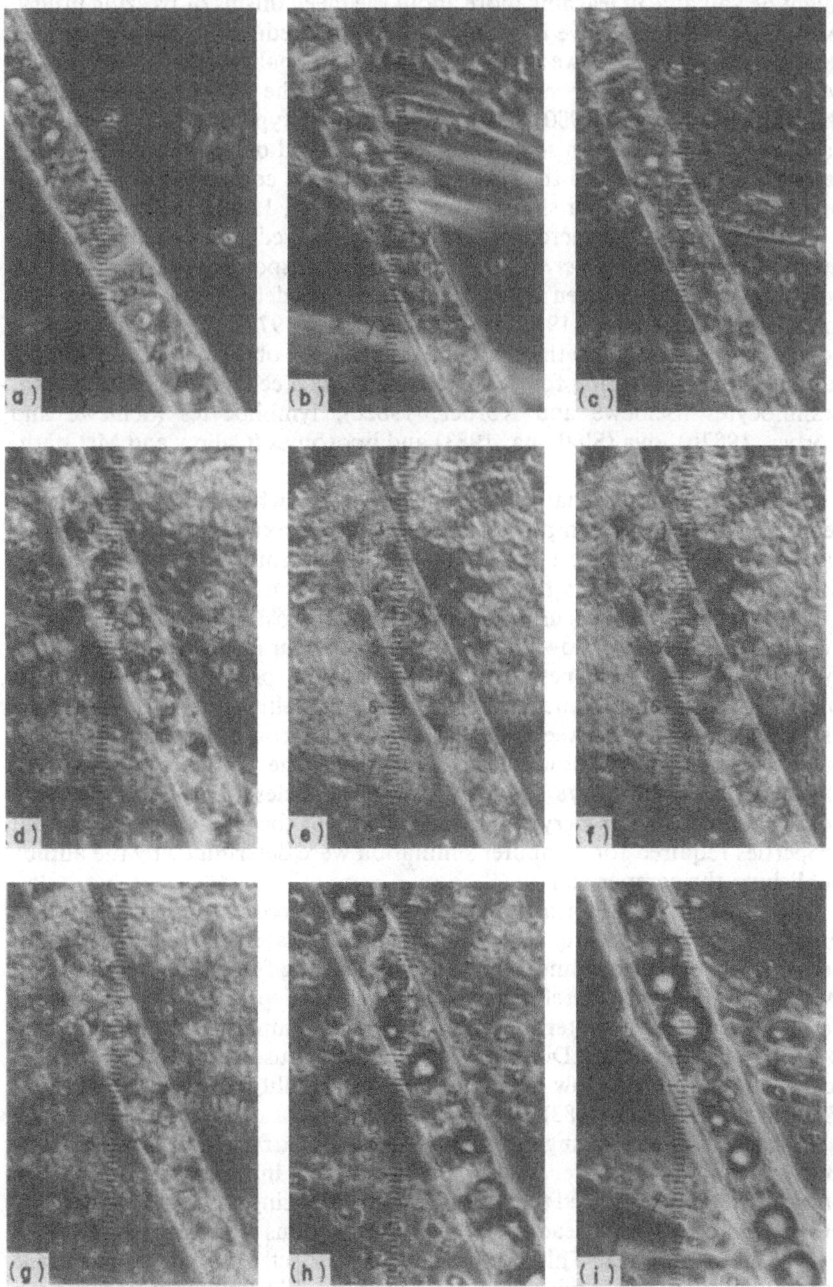

Fig. 6.21 An example of gas bubble formation in the filamentous alga, *Spirogyra*. This phenomenon is, arguably, associated with intracellular freezing (for discussion, see text) with the gas bubbles becoming evident during thawing. Control cells (a), (b) during cooling at 10°C min^{-1} to −2.5, (c) −5, (d), −7.5, (e), −10°C, (f) warming at a rate of 1°C min^{-1} to −5, (g), −3, (h) −2, (i) −1°C.

might be valuable in learning more about the mechanisms of freezing injury. A number of studies have been made of volume reductions during freezing (Knox *et al.*, 1980; Schiewe and Körber, 1982b,c; Shabana, 1983; Callow and McGrath, 1982). Fewer studies have included the cell responses during thawing (Knox *et al.*, 1980; McGrath, 1984) one type of which is virtually complete rehydration. In some cases the published observations of the cell response to freezing and thawing have included a comparison with results predicted from computer simulations (Knox *et al.*, 1980; McGrath, 1974).

In the case of cryomicroscopy of the human red blood cell, shrinkage during freezing (Watson *et al.*, 1973) does not correspond to that predicted by computer simulation even when higher-order modelling modifications are included (Watson *et al.*, 1973; Levin *et al.*, 1976, 1978). On the other hand, good agreement between theory and data has been observed for cell volume response during freezing for yeast cells and HeLa cells (Knox *et al.*, 1980), granulocytes (Schiewe and Körber, 1982c), lymphocytes (Schiewe and Körber, 1982b), ova (Shabana, 1983) and liposomes (Callow and McGrath, 1982).

It should be noted that in virtually all cases when such agreement is realized, one or more computer parameters with unknown values have been varied in order to achieve a fit between the experimental data and the simulation. In short, it appears that in many cases thermodynamic modelling is capable of producing accurate predictions of the osmotic response of cells during freezing using known and/or reasonable parameter values for parameters such as water permeability, and water permeability activation energies. The validity of such thermodynamic modelling might be considered proven if agreement between predictions and data could be obtained using a complete set of well-known cell properties in the computer simulation. However, no results have been published which describe good agreement between simulation and cryomicroscope data for the case where all the cell properties required for computer simulation were determined by the author publishing the comparison.

One of the most important cell properties required to perform computer simulation of cell freezing is the cell membrane water permeability. Water permeability of a membrane is known to be a strong function of temperature (Mazur, 1963), and therefore the temperature dependence of the water permeability must be determined for computer-simulated freezing studies. Recently, Schwartz and Diller have made ingenious use of the cryomicroscope to obtain the low temperature permeability of cell membranes (Schwartz and Diller, 1983).

At rapid rates of freezing one might expect the extracellular solution to be in a non-equilibrium state. As virtually no solute is incorporated into the ice phase it would be expected that at fast rates of freezing the solute rejected by the growing ice would lead to large concentrations of solute close to the ice/solution interface. This non-equilibrium effect could have important consequences for cell survival and, if large, it should be incorporated into the computer simulations used to predict cellular freezing response. When light-absorbing solute is used it is evident that as ice grows from each edge of the cryomicroscope stage (± 1.5 mm) toward the center (0 mm) the solute builds

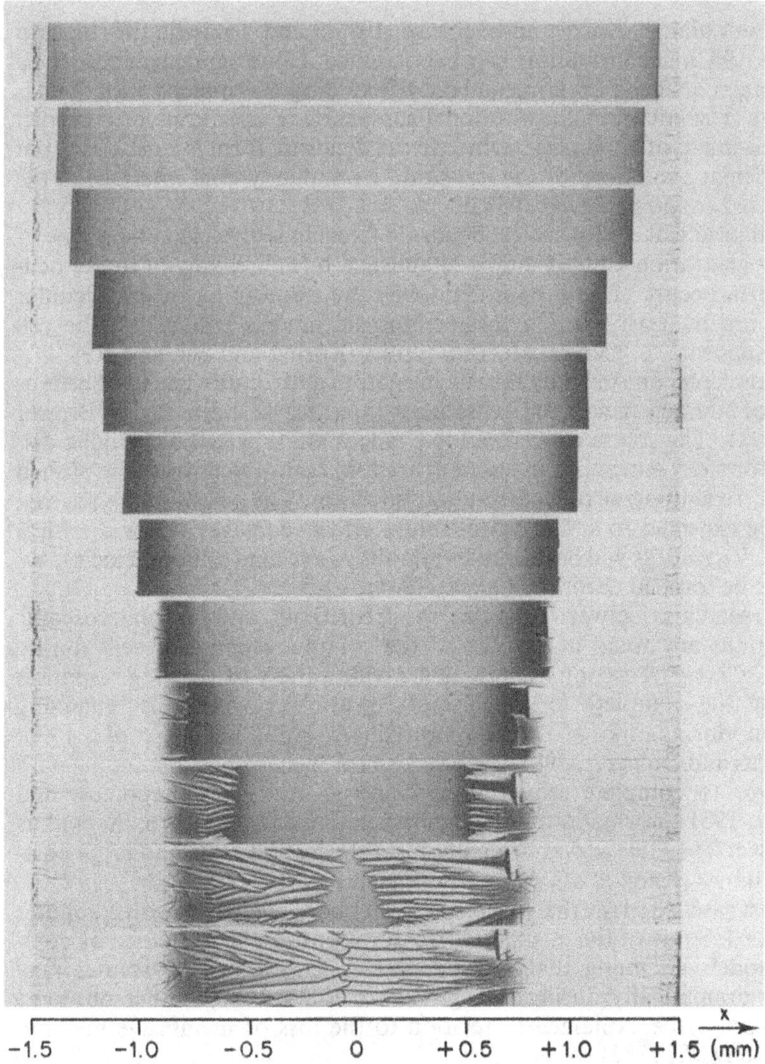

Fig. 6.22 Series of photographs of the freezing process in a $NaMnO_4/H_2O$ solution. Symmetric, inward freezing on a slit-geometry cryomicroscopy stage is illustrated. Note the solute concentration build up at the ice/solution interface evidenced by the darkening. Note also the breakdown from the planar to dendritic freezing. (Reproduced with permission from Körber and Schiewe, 1983b.)

up at the ice/solution interface (Fig. 6.22). This is visualized as the darkening within the remaining liquid at the interface. The breakdown from the initial planar interface to the dendritic interface is also readily noted. A cryomicroscope system capable of quantitative measurement of solute concentration fields in well-defined freezing situations has been developed (Körber *et al.*,

1984). Both planar (Körber and Schiewe, 1983b) and dendritic (Körber and Schiewe, 1983a) ice formation have been studied. Good agreement between a theoretical model and experimental data describing the concentration field is obtained. The common constitutional supercooling criteria used to predict the break-down of a planar interface to the dendritic form is a necessary but not sufficient condition for the transition from one type of interface to the other (Körber and Schiewe, 1983a).

For all practical cooling rates, dendritic freezing is expected and a type of local concentration equilibrium is established once breakdown to the dendritic form occurs. This may explain why the assumption of extracellular equilibrium in most existing thermodynamic models which describe cell freezing appears to yield satisfactory results (Körber and Schiewe, 1983a).

A cryomicroscope to study the encapsulation and/or movement of cells as a result of interaction with the ice/solution interface has been used by Brower *et al.*, 1981. The solute concentration changes are expected to be quite dissimilar for a cell entrapped in ice compared to a cell which would be pushed forward, remaining at the ice/solution interface. The cell in the latter case would be expected to suffer more 'solute effect' damage. Below a critical velocity, V_1, all cells will be rejected while above a second critical velocity, V_2, cells will be trapped (Bronstein *et al.*, 1981).

Disagreements between simulation predictions and cryomicroscopic observations are noted in the case of the volume response of cells during thawing. Cryomicroscopy has revealed a wide variety of thawing responses including the complete loss of osmotic sensitivity with cells remaining shrunken during and after thawing (Diller, 1979; Diller *et al.*, 1976; Steponkus and Dowgert, 1981); normal osmotic response to a point but with lysis prior to complete rehydration (Diller *et al.*, 1976; Steponkus and Dowgert, 1981); normal osmotic rehydration (Diller *et al.*, 1976; Steponkus and Dowgert, 1981); and abnormally fast rehydration rates and large post-thaw volumes (Knox *et al.*, 1980; Schiewe and Körber, 1982b,c).

In most cases cited the thawing response has been compared with computer predictions. Some of the unexpected observations can be simulated by computer models assuming that the membrane permeability to solutes may increase dramatically during storage (Knox *et al.*, 1980). Other observed responses may be explained in relation to the loss of membrane material (McGrath *et al.*, 1983).

Liposome system

Cryomicroscopic studies of liposome model systems appear to provide a promising approach to understanding the mechanisms of cellular freeze-thaw injury. Liposomes are well-established as valuable analogs of cells for a variety of biomedical studies. Specifically as applied to problems of cryobiological interest, liposomes mimic cellular freeze-thaw responses in virtually all respects examined (Siminovitch and Chapman, 1971; Morris and McGrath, 1981a; Morris, 1981; Callow and McGrath, 1984; McGrath, 1984). It is significant that many of the common freeze-thaw responses of biological cells are observed with liposomes made only of lipid. This finding has implications concerning the role played by proteins in freeze-thaw injury.

The advantage of using the liposome model system is closely associated with the experimental flexibility available. Liposomes may be prepared with desired membrane and intraliposomal compositions specifically tailored to study various topics of interest (McGrath, 1983; Morris, 1981).

Other areas of interest

There are several areas of experimental investigation which have received little attention but could benefit from the application of cryomicroscopy. For example, it is known that some cells are sensitive to cold shock (Farrant, 1980), a phenomenon in which cell damage occurs when the temperature of the cell is reduced rapidly but never below the nucleation point of the cell suspension. This type of injury is cooling rate-dependent and occurs in the absence of extracellular ice.

The application of a cryomicroscope to such studies could prove to be valuable (Morris, this volume). Indeed a thermoelectric microscope stage has been used to study the cold shock of human erythrocytes (Williams and Takahashi, 1979; Takahashi and Williams, 1983).

Recent differential scanning calorimetry studies have been conducted using droplet emulsions of cells with the aim of determining the mechanism(s) of the nucleation of undercooled cells (Franks *et al.*, 1983). The use of cryomicroscopy could be of significant value in this type of study, although preliminary work with a cryomicroscope and such emulsion systems suffered from several problems (Franks *et al.*, 1983).

Most cryomicroscopy has been performed on cell suspensions. Several investigators have, however, developed cryomicroscopes to study larger scale systems. Hence, the *in vivo* response of the microcirculation to thermal disturbances, including freezing and thawing, has been examined using cryomicroscopes designed for this purpose. The histological responses to freezing and thawing which have been studied include the stagnation of blood flow; the presence of emboli; and the post-thaw leakage of fluorescently labelled macromolecules from the capillary bed into the surrounding tissue (Hlatky *et al.*, 1973; Hrycaj, 1975; Evans and Diller, 1982).

In summary the cryomicroscope has proved to be an extremely valuable experimental tool. Its major attributes are a result of the ability to control, very precisely, sample temperature and temperature damages. Real-time observation as well as quantitative optical techniques are leading to an improved understanding of the response of living systems to freezing and thawing.

References

Bank, H. and Mazur, P. (1973). Visualization of freezing damage. *Journal of Cell Biology* **57**, 729–42.

Bank, J. (1973). Visualization of freezing damage II. Structural alterations during warming. *Cryobiology* **10**, 157–70.

Bronstein, V.L., Itkin, Y.A. and Ishkov, G.S. (1981). Rejection and capture of cells by ice crystals on freezing aqueous solutions. *Journal of Crystal Growth* **52**, 345–9.

Brower, W.E., Freund, M.J., Baudino, M.D., Ringwald, C. (1981). An hypothesis for survival of spermatozoa via encapsulation during plane front freezing. *Cryobiology* **18**, 277–91.

Callow, R.A. and McGrath, J.J. (1982a). Mass transfer response of cell-sized, semi-permeable vesicles during freezing: a comparison of computer simulations and cryomicroscopic data. In *1982 Advances in Bioengineering*, pp. 104–107. American Society of Mechanical Engineers.

Callow, R.A. and McGrath, J.J. (1982b). Unilamellar liposomes as a model system to study freezing damage to cells. In *Proceedings of the Tenth Annual Northeast Bioengineering Conference*, pp. 269–72. Edited by Hansen, E.W. IEEE 82CH1747-5.

Callow, R.A. and McGrath, J.J. (1985). Thermodynamic modelling and cryomicroscopy of cell-size, unilamellar liposomes. *Cryobiology* 22, 251–67.

Cosman, M.D. (1983). *Effects of Cooling Rate and Supercooling on the Formation of Ice in a Cell Population*. PhD Dissertation, Mechanical Engineering Department, M.I.T.

Diller, K.R. (1972). *A Microscopic Investigation of Intracellular Ice Formation in Frozen Human Erythrocytes*. ScD Dissertation, Mechanical Engineering Department, M.I.T.

Diller, K.R. (1975). Intracellular freezing: effect of extracellular supercooling. *Cryobiology* **12**, 480–85.

Diller, K.R. (1979). Intracellular freezing of glycerolized red cells. *Cryobiology* **16**, 125–31.

Diller, K.R. (1982). Quantitative low temperature optical microscopy of biological systems. *Journal of Microscopy* **126**, 9–28.

Diller, K.R. and Cravalho, E.G, (1971). A cryomicroscope for the study of freezing and thawing processes in biological cells. *Cryobiology* **7**, 191–9.

Diller, K.R. Cravalho, E.G. and Huggins, C.E. (1976). An experimental study of freezing in erythrocytes. *Medical and Biological Engineering* **14**, 321–6.

Diller, K.R. and Lynch. M.E. (1983). An irreversible thermodynamic analysis of cell freezing in the presence of membrane permeable additives. 1. Numerical model and transient cell volume data. *Cryoletters* **4**, 295–308.

Diller, K.R. and Schmitt, K.M. (1981). Freezing responses of rabbit kidney cortex cells. *Cryoletters* **2**, 72–5.

Electrically Conductive Coated Glass. Materials Department, Corning Glass Works, Corning, New York, 14830.

Evans, C.D. and Diller, K.R. (1982). A programmable, microprocessor-controlled temperature stage for burn and freezing in the microcirculation *Microvascular Research* **24**, 314–25.

Farrant, J. (1980). General observations on cell preservation. In *Low Temperature Preservation in Medicine and Biology*, pp. 1–18. Edited by

Ashwood-Smith, M.J. and Farrant, J. University Park Press, Baltimore, MD.

Feynman, R.P., Leighton, R.B. and Danda, M. (1963). In *The Feynman Lectures on Physics*, Volume 1, Addison-Wesley.

Franks, F., Mathias, S.F., Galfre, P., Webster, S. and Brown, D. (1983). Ice nucleation and freezing in undercooled cells. *Cryobiology* **20**, 298–309.

Harmer, J.R. (1953). A specimen holder for low temperature microscopy in biology. *Journal of the Royal Microscopical Society* **73**, 128–33.

Hlatky, M.A., Cravalho, E.G., Diller, K.R., and Huggins, C.E. (1973). *Response of the microcirculation to freezing and thawing*. ASME 73-WA/B10-16.

Hrycaj, T. (1975). *The Effects of Freezing on the Microcirculation of the Hamster Cheek Pouch*. MS Dissertation, Mechanical Engineering Department, M.I.T.

Knox, J.M., Schwartz, G.S. and Diller, K.R. (1980a). Volumetric changes in cells during freezing and thawing. *Journal of Biochemical Engineering* **102**, 91–7.

Körber, C. and Shiewe, M.W. (1983a). Observations on the non-planar freezing of aqueous salt solutions. *Journal of Crystal Growth* **61**, 307–316.

Körber, C. and Schiewe, M. (1983b). In *Forschungsbericht 1981/82*. Helmholtz-Institut für Biomedizinische Technik, Aachen.

Körber, Ch., Schiewe, M.W. and Wollhöver, K. (1984). A Cryomicroscope for the Analysis of Solute Polarization During Freezing, *Cryobiology*, **21**, 68–80.

Körber, Ch., Schiewe, M.W. and Wollhöver, K. (1983). Solute polarization during planar freezing of aqueous salt solutions. *International Journal of Heat and Mass Transfer* **26**, 1241–53.

Lang, W. *Nomarski Differential Interference-Contrast Micorscopy*. Zeiss publication on S41-210.2-5-e, Carl Zeiss, 7082 Oberochen, W. Germany.

Leibo, S.P., McGrath, J.J. and Cravalho, E.G. (1978). Microscopic observation of intracellular ice formation in unfertilized mouse ova as a function of cooling rate. *Cryobiology* **15**, 257–71.

Levin, R.L., Cravalho, E.G. and Huggins, C.E. (1976). Effect of hydration on the water content of human erythrocytes. *Biophysical Journal* **16**, 1411–26.

Levin, R.L., Cravalho, E.G. and Huggins, C.E. (1977). Effect of solution non-ideality on erythrocyte volume regulation. *Biochimica Biophysica Acta* **465**, 179–90.

Levin, R.L., Cravalho, E.G. and Huggins, C.E. (1978). The concentration polarization effect in a multicomponent electrolyte solution - the human erythrocyte. *Journal of Theoretical Biology* **71**, 225–54.

Luyet, B.J. and Gibbs, M.C. (1937). On the mechanism of congelation and of death in the rapid freezing of epidermal plant cells. *Biodynamica* **25**, 1–18.

McGann, L.E. (1979). A versatile microprocessor-based temperature and cooling/warming rate controller. *Cryobiology* **16**, 97–100.

McGrath, J.J. (1974). *The Dynamics of Freezing and Thawing Mammalian Cells: The HeLa Cell.* MS Dissertation, Mechanical Engineering Department, M.I.T.

McGrath, J.J. (1983). Cryomicroscopy of liposome systems as simple models to study cellular freezing response. *Cryobiology* **21**, 81–92.

McGrath, J.J., Callow, R.A. and Melkerson, M. (1983). The response of cell-size unilamellar liposomes to osmotic perturbation and to freezing/thawing. *Cryobiology* **20**, 712.

McGrath, J.J., Cravalho, E.G. and Huggins, C.E. (1975). An experimental comparison of intracellular ice formation and freeze-thaw survival of Hela S-3 cells. *Cryobiology* **12**, 540–50.

McGrath, J.J. and Khompis, V. (1981). A numerical heat transfer analysis of a cryomicroscope conduction stage. *American Society of Mechanical Engineers*, 81-WA/HT 56.

Mazur, P. (1963). Kinetics of water loss from cells at subzero temperatures and the likelihood of intracellular freezing. *Journal of General Physiology* **47**, 347–69.

Mazur, P.J. (1966). Physical and chemical basis of injury in single-celled micro-organisms subjected to freezing and thawing. In *Cryobiology*, pp. 213–315. Edited by Meryman, H.T. Academic Press, New York.

Mazur, P., Leibo, S.P. and Chu, E.H.Y. (1972). A two-factor hypothesis of freezing injury – evidence from chinese hamster tissue culture cells. *Experimental Cell Research* **71**, 345–55.

Mersey, B., McCulley, M.E. and Fatica, E. (1978). An inexpensive controlled temperature stage which allows high resolution optical microscopy. *Journal of Microscopy* **113**, 307–10.

Molisch, H. (1982). Investigations into the freezing death of plants. *Cryoletters* **3**, 331–90.

Morris, G.J. (1981). Liposomes as a model system for investigating freezing injury. In *Effects of Low Temperatures on Biological Membranes*, pp. 241–62. Edited by Morris, G.J. and Clarke, A. Academic Press, London.

Morris, G.J. unpublished research with Chlamydomonas.

Morris, G.J., Coulson, G., Meyer, M.A., McLellan, M.R., Fuller, B.J., Grout, B.W.W., Pritchard, H.W. and Knight, S.C. (1983). Cold shock – a widespread cellular reaction. *Cryoletters* **4**, 179–92.

Morris, G.J. and McGrath, J.J. (1981a). Intracellular ice nucleation and gas bubble formation in Spirogyra. *Cryoletters* **2**, 341–52.

Morris, G.J. and McGrath, J.J. (1981b). The response of multilamellar liposomes to freezing and thawing. *Cryobiology* **18**, 390–8.

Rall, W.F., Mazur, P. and McGrath, J.J. (1983). Depression of the ice nucleation temperature of rapidly cooled mouse embryos by glycerol and dimethylsulfoxide. *Biophysical Journal* **41**, 1–12.

Rall, W.F., Reid, D.S. and Farrant, J. (1980). Innocuous biological freezing during warming. *Nature* **286**, 511–14.

Reid, D.S. (1978). A programmed controller temperature microscope stage. *Journal of Microscopy* **114**, 24–248.

Schiewe, M.W. and Körber, Ch. (1982a). Thermally defined cryomicroscopy and some applications on human leucocytes. *Journal of Microscopy* **126**, 29–44.

Schiewe, M.W. and Körber, C. (1982b). Formation and melting of intracellular ice in lymphocytes. *Cryoletters* **3**, 265–74.

Schiewe, M.W. and Körber, C. (1982c). Formation and melting of intracellular ice in granulocytes. *Cryoletters* **3**, 275–84.

Schiewe, M.W. and Körber, C. (1983). Basic investigations on the freezing of human lymphocytes. *Cryobiology* **20**, 257–73.

Schwartz, G.J. and Diller, K.R. (1982). Design and fabrication of a simple, versatile cryomicroscopy stage. *Cryobiology* **19**, 529–38.

Schwartz, G.J. and Diller, K.R. (1983). Osmotic response of individual cells during freezing II. Membrane permeability analysis. *Cryobiology* **20**, 542–52.

Shabana, M. (1983). *Cryomicroscope Investigation and Thermodynamic Modelling of the Freezing of Unfertilized Hamster Ova*. MS Dissertation, Mechanical Engineering Department, Michigan State University.

Shimada, K. and Asahina, E. (1978). Visualization of intracellular ice crystals formed in very rapidly frozen cells at −27°C. *Cryobiology* **12**, 209–218.

Siminovitch, D. and Chapman, D. (1971). Liposome bilayer model systems of freezing living cells. *FEBS Letters* **16**, 207–212.

Siminovitch, D. and Chapman, D. (1974). Simulation of osmotic stresses of plant cells by lipid liposome membrane systems. *Cryobiology* **11**, 552–3.

Smith, A.U., Polge, C. and Smiles, J. (1951). Microscopic observation of living cells during freezing and thawing. *Journal of the Royal Microscopical Society* **71**, 186–95.

Steponkus, P.L. and Dowgert, M.F. (1981). Gas bubble formation during intracellular ice formation. *Cryoletters* **2**, 42–7.

Steponkus, P.L., Evans, R.Y. and Singh, J. (1982). Cryomicroscopy of isolated rye mesophyll cells. *Cryoletters* **3**, 101–114.

Takahashi, T. and Williams, R.J. (1983). Thermal shock haemolysis in human red cells I. The effects of temperature, time, and osmotic stress. *Cryobiology* **20**, 507–520.

Troyer, D. (1974). A temperature-regulated stage for the light microscope using a peltier element. *Journal of Microscopy* **102**, 215–18.

Tu, S.M. (1983). *Computer Simulation of Two-Dimensional Transient Temperature Field in Cryomicroscope Conduction Stage*. MS Dissertation, Mechanical Engineering Department, Michigan State University.

Watson, W.W., Diller, K.R., Cravalho, E.G. and Huggins, C.E. (1973). Volumetric changes in human erythrocytes during freezing at constant cooling rates. *Cryobiology* **10**, 519.

Williams, R.J. and Takahashi, T. (1979). Microscopic observations of human red cells undergoing thermal shock. *Cryobiology* **16**, 589.

Section III

Environmental low temperature biology

7

Chilling injury in plants

J.M. Wilson
School of Plant Biology
University College of North Wales

Introduction

Low temperature is both a major ecological factor determining the natural distribution of plants and an important economic consideration, as many tropical and sub-tropical crops such as cotton and maize are cultivated near to the limits of their low temperature tolerance.

Chilling injuries are complex phenomena and there is no reason to assume that the same sequence of events leads to injury in widely different species. In some species the most important effect of lowered temperature may be on the fluidity of the membrane lipids but in other species the primary effects may be on the non-membrane proteins, cytoskeleton or cytoplasm. A large number of changes or primary events are therefore possible during chilling, and it is these changes which eventually cause injuries such as imbalance in metabolism, ion leakage or membrane breakdown which may utimately result in cell death.

Definition and symptoms of chilling injury

Chilling sensitive plants are usually of tropical or sub-tropical origin and show signs of injury when exposed to temperatures above the freezing point of the tissue up to approximately 15°C. Determining whether plant tissue is chilling sensitive is difficult as symptoms of injury may take months to occur in some species yet only a few hours in others (e.g. *Episcia reptans*), and there appears to be no sharp distinction between chilling sensitive and chilling resistant plants. Some chilling resistant species, for example cabbages, if grown at high temperature, can show rapid wilting on transfer to low temperatures which is generally regarded as a characteristic symptom of chilling sensitivity. Development of injury depends on the temperature, growth conditions prior to chilling, the age of the plant, light intensity, relative humidity during chilling and the rate of cooling and rewarming. Symptoms of chilling injury usually develop more rapidly if the tissue is returned to the warmth. In many important crop species such as cotton and French bean (*Phaseolus vulgaris*) the first symptoms of chilling injury are rapid leaf wilting within a few hours of the start of chilling and the development of water soaked patches which later form sunken pits due to cell collapse. When the plants are returned to the warmth, the leaf margins and collapsed areas of the leaf usually dry out rapidly so that the brown necrotic patches can be readily seen over the whole leaf, scattered in an apparently random manner. This variability in the sensitivity of different parts of the leaf to low temperature makes quantitative assessment of injury difficult. When some species are chilled at high light intensity (e.g. cucumbers) rapid photo-oxidation of the leaf pigments can occur leading to the bleaching of the leaves (van Hasselt and Strikwerda, 1976). One of the most chilling sensitive species is *Episcia reptans*, from tropical South America, which shows water soaked patches on the leaves within 2 hours of the start of chilling at 5°C, and as little as 4 hours at 5°C can lead to death of the whole plant (Wilson, 1976).

Examples of chilling injury to fruits are the development of sunken pits in cucumber fruit, the browning of the skin and the degeneration of the pulp tissue of bananas and the development of 'blackheart' in pineapples after several days at the chilling temperature. Chilling leads to the development of soft areas on the fruit which aids the penetration and rapid spoiling of the fruit by fungal and bacterial pathogens.

The Lyons-Raison hypothesis of chilling injury

It has long been known that the lipids of cold tolerant plants generally have a higher percentage of unsaturated fatty acids than those of tropical plants. It was considered that this higher degree of unsaturation enabled the membranes to remain more fluid and hence carry out their normal functions at lower temperatures than chilling sensitive species. This idea was developed further by Lyons and Raison (1970) and, in its simplest form, the hypothesis states that at a certain critical temperature within the chilling injury range the membrane lipids of chilling sensitive plants undergo a transition from a liquid crystalline to a solid gel state.

This description of events in a complex mixture of lipids that occurs in biological membranes is an oversimplification. The variety of terms used to describe these lipid changes such as phase change, phase transition and, more commonly, change in molecular ordering reflects a lack of understanding of the precise nature of these events. It is now thought from results obtained by electron spin resonance (ESR) that the membrane does not undergo a phase change but that a change in the molecular ordering of discrete domains of lipid within the membrane takes place (Dalziel and Breidenbach, 1979). Using ESR a correlation has been demonstrated between the transition temperature of the leaf lipids and the habitat temperature of a number of different species (Raison *et al.*, 1979). Temperate plants had transition temperatures in the range 0–2°C whilst tropical plants were nearly always much higher within the range 10–17°C. There is, however, considerable controversy over the correct use and interpretation of data obtained from ESR of complex lipid mixtures. In addition, the transition temperature for a given sample has been shown to vary by several degrees depending on the nature of the spin label used (Schreier *et al.*, 1978). Even more worrying are reports that ESR can detect changes in the molecular ordering of lipids at chilling temperatures in both chilling sensitive and chilling resistant plants (Bishop *et al.*, 1979).

Other techniques such as differential scanning colorimetry (DSC) and fluorescence depolarization using *trans*-parinaric acid as a probe can detect phase transitions, but these have also led to conflicting reports. Several workers (Bishop *et al.*, 1979; Dalziel and Breidenbach, 1979; McMurchie, 1979) have been unable to detect a consistent difference between the membranes of chilling sensitive and chilling resistant plants using DSC. The fluorescent probe *trans*-parinaric acid, which has a strong affinity for the gel phase, has given more consistent results, with isolated phospholipids extracted from a number of different chilling sensitive and chilling resistant plants showing good correlation with their temperature sensitivity (Pike and Berry, 1980). This suggests it is the fatty acid composition of the phospholipids which is important in determining the phase change temperature. Consistent with this is the observation that during chill hardening of a number of sensitive species increases in unsaturation of fatty acids were detected only in the phospholipid fraction (Wilson and Crawford, 1974). Experiments on phospholipids of different fatty acid composition indicate that only those with two saturated fatty acids show phase transitions in the chilling injury range; the substitution of one unsaturated fatty acid for a saturated fatty acid lowers the transition temperature (Morris and Clarke, this volume). It is therefore possible that spin label probes incorporated into membranes were detecting transitions of these di-saturated phospholipids. Such lipids are relatively scarce in plant tissues and occur mainly as dipalmitoyl phosphatidylglycerol and dipalmitoyl sulpholipid. The transition temperatures, as determined by DSC, of phospholipid fractions from chill sensitive mung bean and chilling resistant wheat and pea correlate well with their dipalmitoyl phosphatidylglycerol content (Wright and Raison, 1983). Similar transitions could be induced in the lipids of chilling resistant plants by the addition of small amounts of high melting point, di-saturated phospholipids.

Consequences of membrane lipid phase changes

According to the Lyons-Raison hypothesis phase transitions may lead to injury and death of the tissue as a result of an increase in membrane permeability and an increase in the activation energy of membrane-bound enzymes.

Changes in membrane permeability

The idea that the permeability of the plasmalemma and other membranes to ions and water increases at the temperature of the phase transition as a result of changes in the structure and arrangement of lipids (Trauble and Haynes, 1971) is consistent with the observed leakage of ions, water, sugars, organic acids and other metabolites from a variety of chilling sensitive tissues during chilling (Liebermann *et al.*, 1958; Christiansen *et al.*, 1970; Guinn, 1971; Wright and Simon, 1973; Creencia and Bramlage, 1971; Patterson *et al.*, 1976). In all these studies, however, the rate of leakage only became rapid after many hours or days at the chilling temperature, and this argues against any rapid rise in permeability which can be attributed to lipid phase transitions.

It was speculated that the loss of turgor of the leaves of many chilling sensitive species (e.g. *Phaseolus vulgaris*) in the first few hours of chilling was the result of an increase in the permeability of the plasmalemma of the leaf cells to water and electrolytes. However, this rapid wilting could be prevented simply by enclosing the plant in a polythene bag so that a saturated, 100 per cent relative humidity (r.h.) atmosphere was maintained around the plant (Wilson, 1976). If a phase transition had caused an increase in the permeability of the leaf cells, turgor should be lost when the plants are chilled at both low and high relative humidity as the turgor pressure of the cell would facilitate the movement of water out of the cell leading to loss of turgor. In species such as *P. vulgaris* the rapid wilting of the leaves is caused by locking open of the stomata and the low permeability of the roots to water which results in leaf dehydration and injury. Electrolyte leakage from leaves of *P. vulgaris* during chilling in water at 5°C is slow and can be attributed to degenerative changes in the cells rather than a rapid change in permeability caused by a lipid phase transition (Wilson, 1976).

The leaves of the ultra chilling sensitive plant *Episcia reptans* leak electrolytes at a much faster rate than *P. vulgaris* and lose approximately 25 per cent of their total electrolytes within 5 hours of transfer to water at 5°C. In this species visible signs of cell damage accompany leakage so that it is not possible to determine whether leakage is an immediate effect as a result of a

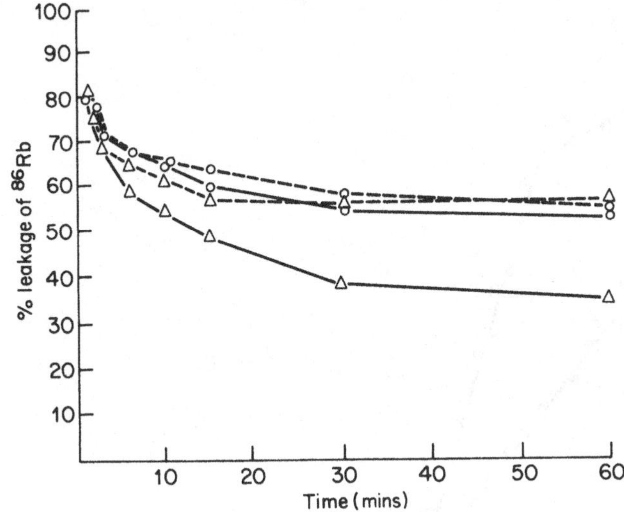

Fig. 7.1 Leakage of ^{86}Rb from leaf slices of *Episcia reptans* (△) and *Vicia faba* (○) at 5°C (———) and 25°C (------). Leaf slices were loaded at 25°C in a solution containing 0.042 mmol l^{-1} ^{86}RbCl and 0.2mmol l^{-1} $CaSO_4$ and placed in unlabelled solution at 5°C or 25°C. Results are expressed as a percentage of total leakage from boiled leaf slices.

phase transition or a secondary effect caused by cell degeneration (Wilson, 1976). To resolve this problem efflux experiments using leaf slices of *E. reptans* loaded with radio labelled rubidium (^{86}Rb) at 25°C were performed. There is a more rapid loss of ^{86}Rb approximately 5 minutes after transfer to 5°C than in the controls of 25°C (Fig. 7.1). By contrast the rate of efflux of ^{86}Rb from leaf slices of a chilling resistant bean (*Vicia faba*) was the same at 5°C and 25°C. Whether this increased rate of leakage of ^{86}Rb from *E. reptans* after only 10 minutes at 5°C can be attributed to a phase transition in the membrane lipids or to a change in membrane protein is unresolved but such experiments do indicate that chilling can cause rapid changes in cell permeability in some species.

Increase in activation energy

Phase transitions can be expected to alter the molecular environment around membrane-associated enzymes, such as succinic dehydrogenase, which may lead to increases in their activation energy at low temperatures in chilling sensitive species (Fig. 7.2). Although an increase in activation energy may not in itself be damaging to the plant, lethal increases in ethanol and acetaldehyde content may occur as a result of metabolic imbalance between cytoplasmic glycolytic reactions and membrane-associated enzymes of the tricarboxylic acid(TCA) cycle. This assumes that no mechanism exists in chilling sensitive plants to adapt to altered activation energies at low temperature and thereby prevent the build up of these metabolites. Few investigations have been able to detect increased ethanol and acetaldehyde levels in chilled

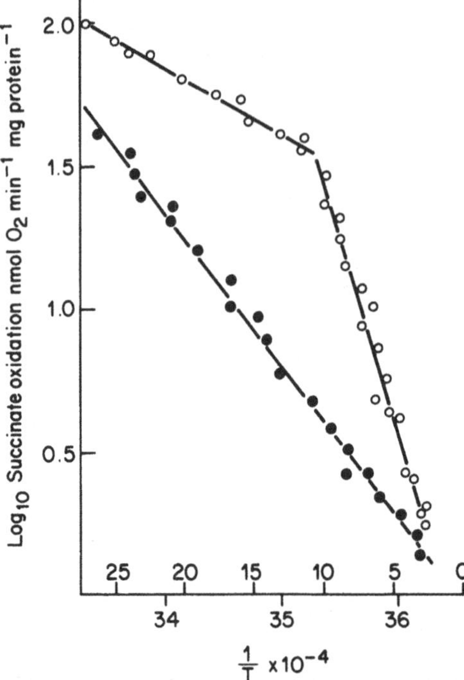

Fig. 7.2 Typical Arrhenius plots for succinic dehydrogenase (SDH) activity in chilling sensitive (O) and chilling resistant (●) plant mitochondria.

tissue, an exception being banana pulp (Murata, 1969). Ethanol and acetaldehyde only accumulate in lemons after rewarming, suggesting that abnormal metabolism results from prior chilling damage (Eaks, 1980).

Arrhenius plots

Perhaps the most contentious area of chilling injury research has been in the use of Arrhenius plots to calculate changes in the activation energies of membrane-associated enzymes at low temperatures. Arrhenius plots of succinic dehydrogenase activity of mitochondrial preparations from chilling sensitive plant tissues were found to be non-linear with a 'break' at the temperature at which it was believed the plant became chill injured. In contrast similar plots of mitochondrial activity from chilling resistant tissue showed no break and hence had a constant activation energy over the whole temperature range (Fig. 7.2). As the temperature at which this break occurred in chilling sensitive species was very similar to the phase change temperature (as determined by breaks in Arrhenius plots of ESR spectra) it was used as further evidence for the idea that the temperature at which species suffer

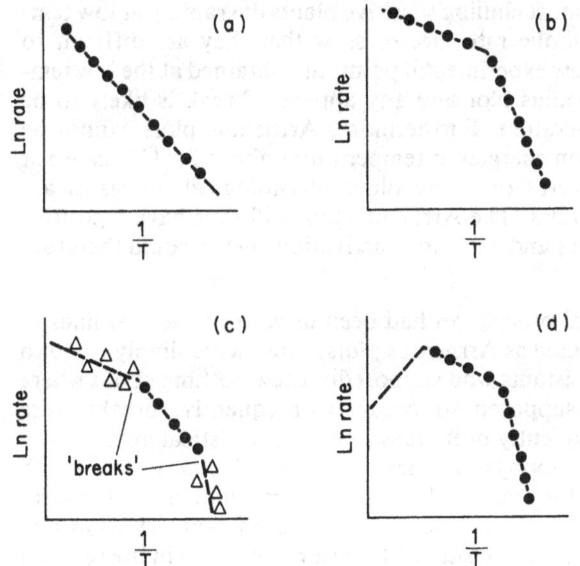

Fig. 7.3 Diagrams illustrating various features of Arrhenius plots. (a) correct Arrhenius plot, all control parameters remain constant throughout the temperature range so that the Arrhenius Law holds true, (b) an Arrhenius plot with a true breakpoint where the rate-limiting step has changed but the Arrhenius Law still holds true either side of the break, (c) an Arrhenius plot showing 2 false break points (arrowed). The points marked with Δ are not linearly related so Arrhenius' Law does not apply, (d) an Arrhenius plot showing a possible negative activation energy (broken line) at temperatures above the optimum

chilling injury is determined by the phase change temperature of the membrane lipids (Lyons, 1973).

The use of Arrhenius plots for presenting changes in mitochondrial activity with temperature and the drawing of breaks in the plots at certain temperatures has been criticized on the following grounds.

1. The Arrhenius law by which the activation energy is calculated is only thought to hold true in reactions when there is only one rate-limiting step (Fig. 7.3a). In a mitochondrial preparation there could be several rate-limiting steps. However, if the rate-limiting step does change at a certain point, the linear relationship will be preserved but the result will be two straight lines joining at a break point coincident with the change in rate-limiting step (Fig. 7.3b).

2. As there may be more than one rate-limiting step, and these may vary independently of each other, an experimentalist does not know if the points are linearly related or not (Fig. 7.3c) and so may draw regression lines through unrelated points.

3. Break point temperatures can be shifted by either increasing or decreasing the length of the Y axis of the Arrhenius plot.

4. Omission of points from the end of an Arrhenius plot will shift any break point as a statistical artefact (Bagnall and Wolfe, 1978). This is of

particular importance as most chilling sensitive plants die rapidly at low temperatures and their metabolic rates are so slow that they are difficult to measure; consequently, few experimental points are obtained at the low temperature end of an Arrhenius plot and any apparent break is likely to be shifted to a higher temperature. Furthermore, Arrhenius plots cannot be used to calculate activation energies at temperatures above 28°C because at these temperatures the rates of many plant physiological processes are decreasing from their optima. The Arrhenius plot will thus have a positive slope at these temperatures and a negative activation energy could therefore be calculated (Fig. 7.3d).

Until 1976 no statistical procedures had been used to fit the best lines to chilling injury data presented as Arrhenius plots. Lines were simply fitted to points by independent assistants who supposedly knew nothing about where the break points were supposed to occur. Consequently, breaks were reported that have subsequently been questioned. A statistical approach to data on growth rates and biophysical measurements of *Vigna radiata* was to work along each point on the graph and calculate the residual sums of squares for the best fitting pair of regression lines from each point (Raison and Chapman, 1976). A break was considered to occur at minima in the residual sums of squares. This procedure has been computerized and is the basis of a maximum likelihood programme (Potter and Ross, 1979). Growth results that have been analyzed using the maximum likelihood programme are illustrated in Fig. 7.4. One of the main problems with this form of analysis is that smooth data with little variability are required otherwise natural variation will cause small minima. In addition it must be assumed *a priori* that straight lines are the best fit and, as an estimation of the break point has to be given,

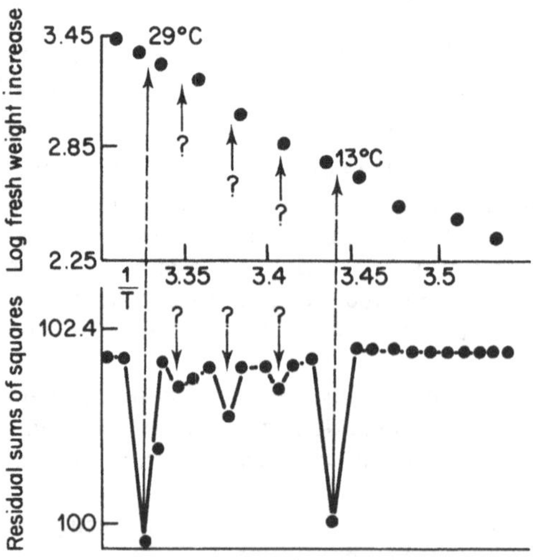

Fig. 7.4 An Arrhenius plot of the increase in fresh weight of cucumber seedlings at different temperatures tested for breaks using the maximum likelihood programme.

the computer is presented with a *fait accompli* in effect saying 'there is a break here, isn't there?' rather than 'is there a break here?'. A better fit obtained by using polynomial curves rather than straight lines has been described (Bagnall and Wolfe, 1978), but the reduction of the residuals of their data is only a matter of 0.00028 when 3rd order polynomial rather than a straight line is used, so this is probably not significant. Other forms of statistical analysis have recently been developed (Willcox and Patterson, 1979; Wolfe and Bagnall, 1979), but these might be thought elaborate when compared with well established techniques, for example the method of least squares.

The shortcoming shared by all these methods is that they attempt to produce a good fit for the data rather than a correct fit. As the existence and position of these breaks is of fundamental importance to the Lyons-Raison hypothesis, it is essential to use a technique of analysis that gives an acceptable (though not necessarily the best) fit to all the available data. A programme has been developed that fits (by maximum likelihood) the best two line, three line and curved model (McMurdo and Wilson, 1980), which then analyses the distribution of the residuals around the line; if these are not random, the fit is incorrect even though it may offer quite low residuals. The programme has been tested using results from growth and respiration of seedlings at different temperatures. The two straight line model offers the slightly better fit in both cases and the distribution of the residuals is more normal and therefore the straight lines are a more correct fit (Fig. 7.5). There is, however, a break in the plot for rye (*Secale cereale*) which is chilling resistant and would not be expected to show a break. Breaks in Arrhenius plots must therefore be used with extreme caution as evidence to support a phase change theory. Furthermore it has been reported that the so-called critical temperature for injury as determined by ESR was not always concomitant with the onset of visible injury as plants can be held at or slightly below this temperature without injury for relatively long periods of time (Bagnall, 1979; Smillie, 1979).

Alternatives to the Lyons-Raison hypothesis

Cold lability of proteins

In some cases breaks in Arrhenius plots of enzyme catalysed reactions could reflect a direct effect of temperature on the protein which is independent of any influence of either the bulk or boundary lipids. For example, pyruvate Pi dikinase (a non-membrane bound enzyme important in C_4 photosynthesis) in maize dissociates reversibly from a tetramer to a dimer at low temperatures, probably as a result of the weakening of the hydrophobic bonds that are important in stabilizing multimeric aggregates (Sugiyama *et al.*, 1979). Phosphoenolpyruvate (PEP) carboxylase is another enzyme which can be obtained lipid free and whose activation energy in chilling sensitive plants increases at low temperature, probably as a result of the weakening of the hydrophobic bonds or the strengthening of the hydrogen bonds and electrostatic interactions at low temperature (Graham *et al.*, 1979). The main evidence for temperature-induced changes in the structure of the enzyme PEP carboxylase has been from changes in the Michaelis constant (K_M) with

Arrhenius plots of the dry weight increase in rye (chilling resistant) and cucumber (chilling sensitive) seedlings

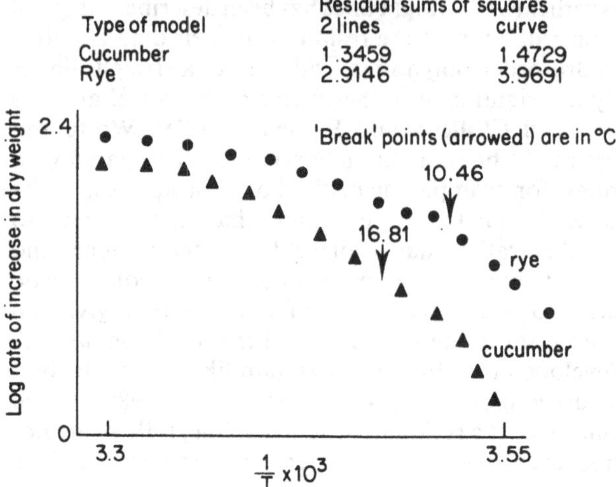

Distribution of the residuals about the model (the more normal the distribution the more correct the fit)

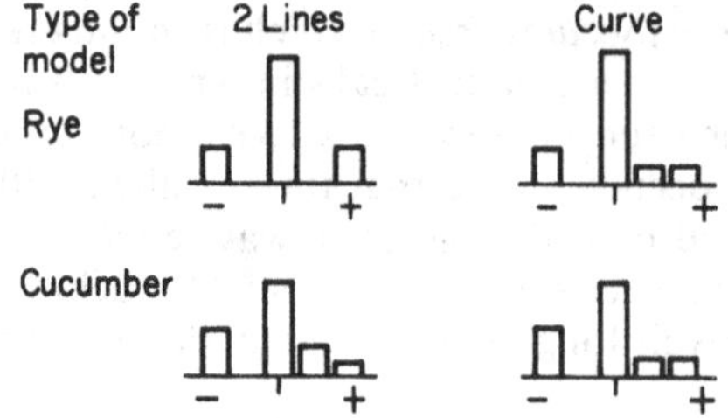

Fig. 7.5 Arrhenius plot of the dry weight increase in Rye (chilling resistant) and Cucumber (chilling sensitive) seedlings.

temperature, as changes in K_M are thought to reflect changes in the ordered structure of the protein around the active site. The K_M values for the chilling resistant species did not change significantly at low temperature, but in the chilling sensitive tomato the K_M increased nine fold on transfer from 20°C to 1.3°C (Graham *et al.*, 1979). These results suggest an alternative or additional mechanism to the Lyons-Raison hypothesis which would lead to the disruption of plant metabolism at low temperatures. For example, a decrease in the activity of PEP carboxylase could result in an imbalance in metabolism as the operation of the TCA cycle in the synthetic mode depends on a continued supply of C_4 acids which are principally produced by this enzyme.

Water relations

The rapid wilting and dehydration that occurs in the first 24 hours of chilling

Phaseolus vulgaris plants can result in 50 per cent necrosis to the leaf on return to the warmth (Wilson, 1976). The rapid wilting of leaves in the first few hours at the chilling temperature can be prevented, in some species, for up to 9 days by enclosing the plant inside a polythene bag thus maintaining a saturated atmosphere. Leaf wilting and dehydration during chilling can also be prevented for up to 9 days by chill hardening the plants for 4 days at 12°C, 80 per cent r.h. Drought hardening plants at 25°C, 40 per cent r.h. by withholding water from the roots so that the leaves wilt is as effective as chill hardening in preventing injury at 5°C, 80 per cent r.h.

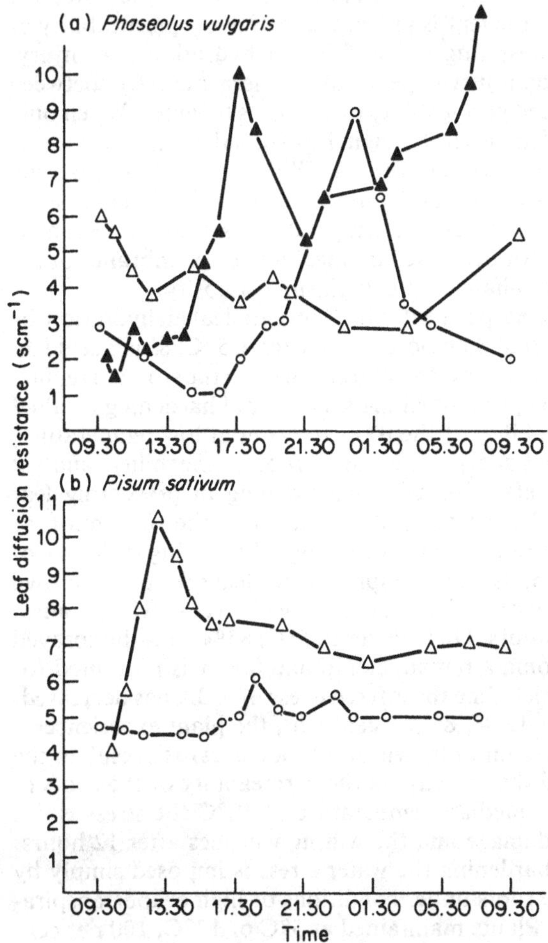

Fig. 7.6 (a) Changes in the leaf diffusion resistance of chill hardened (▲) and non-hardened (△) *Phaseolus vulgaris* during chilling at 5°C, 80% r.h. in comparison to the controls (○) maintained at 25°C, 80% r.h.; (b) Chilling resistant *Pisum sativum* during chilling (△) and in the controls (○) at 25°C.

The primary cause of chilling injury to *P. vulgaris* leaves at 5 °C, 80 per cent r.h. is water loss as a result of the locking open of the stomata when the permeability of the roots to water is low (Wilson, 1976). Fig. 7.6a shows the low leaf diffusion resistance of *P. vulgaris* during chilling in comparison to the normal diurnal sequence of opening and closing in the controls kept at 25 °C and the increase in the leaf diffusion resistance during the chilling of chill hardened plants (Eamus *et al.*, 1982). In contrast, the leaf diffusion resistance of non-hardened chilling resistant *Pisum sativum* plants increased rapidly during chilling (Fig. 7.6b). The open condition of the stomata of *Phaseolus vulgaris* leaves after 2 hours at 5 °C, 80 per cent r.h. is surprising as the leaf is wilted and in nearly all species the stomata close in the early stages of water stress before visible wilting occurs. The replacement of the water lost by evapotranspiration from the leaf is prevented by the low permeability of the roots to water at 5 °C resulting in rapid leaf dehydration and injury. Hence the severity of chilling injury depends on a synergistic effect between stomatal opening and reduced permeability of the roots to water. When only the leaves or roots were chilled the fresh weight loss from the leaves was not as severe and far less damage occurred to the leaf (Wilson, 1976). The permeability of the roots to water decreases far more rapidly in chilling sensitive than in chilling resistant species as the temperature is lowered (Kramer, 1949); this may be the result of decreased permeability of membranes of the endodermal cells to water or changes in cytoplasmic viscosity.

Chill hardening at 12 °C, 85 per cent r.h., prevents leaf dehydration by conditioning the stomata so that they close on transfer to 5 °C, 85 per cent r.h. Similarly, drought hardening causes stomatal closure and the stomata remain closed on transfer to 5 °C, 85 per cent r.h. Although chill hardening resulted in an increase in the permeability of the roots to water at low temperature, drought hardening produced a large decrease in root permeability, and yet drought hardening was as effective as chill hardening in preventing leaf injury. This suggests that the most important factor in the prevention of chilling injury to *P. vulgaris* during chill and drought hardening is closure of the stomata. This can be demonstrated by spraying the leaves of plants grown at 25 °C, 85 per cent r.h., with 100 μm abscisic acid (ABA) which causes stomatal closure within 24 hours. On transfer to 5 °C, 85% r.h., the sprayed leaves do not wilt as the stomata remain closed and injury is prevented for approximately 2 days by which time the effectiveness of ABA has decreased.

During chill hardening at 12 °C, 85 per cent r.h., the plant experiences a water stress (as shown by the temporary wilting of the leaves) as a result of the opening of the stomata and the decrease in the permeability of the roots to water. However, at the intermediate temperature of 12 °C the stress is not severe enough to result in damage and the wilting vanishes after 12 hours. Similarly, during drought hardening the water stress is imposed simply by withholding water from the roots under conditions of high evapotranspiration so that the leaves wilt. Plants maintained at 5 °C or 12 °C, 100 per cent r.h., do not harden because they experience no water stress. Although the stomata are open under these conditions no water can be lost from the leaf so that the stomata remain fully open on transfer to 5 °C, 85 per cent r.h. (Wilson, 1976).

The anomalous behaviour of the stomata of chilling sensitive plants at 5°C has been detected in cotton and cucumber as well as in *P. vulgaris* but not in tomato (Eamus and Wilson, 1983; McWilliam *et al.*, 1982). An explanation for locking open has been sought by investigating the changes in the levels of ABA during chilling and hardening and in the sensitivity of the stomata to ABA before and after hardening. During chill hardening of *P. vulgaris* the ABA level reaches a peak approximately four times that of the control level by day 3 of hardening and then declines (Table 7.1). This decline is thought to reflect the resumption of full turgor by the leaf. On chilling the chill hardened plants at 5°C, 80 per cent r.h., the ABA level decreased further to 112 ng ABA g FW^{-1}. In spite of the low levels of ABA in chill hardened plants, the stomata are able to close on subsequent chilling due to the greater sensitivity of chill hardened stomata to ABA (Eamus and Wilson, 1983).

The synthesis of ABA is not prevented by low temperatures in chilling sensitive plants as chilling non-hardened plants for 24 hours at 5°C, 80 per cent r.h. caused a doubling of the ABA content (Table 7.1). However, the ABA content of plants chilled for the same period of time at 5°C, 100 per cent r.h. decreased. It is therefore concluded that ABA synthesis is induced by water stress at low temperatures rather than the low temperature *per se*. The higher ABA content of the plants chilled directly at 5°C, 80 per cent r.h. does not assist stomatal closure, indeed at low temperatures ABA helps to maintain open stomata in chilling sensitive plants (Eamus and Wilson, 1983). The reduction in the hydraulic conductivity of the roots of chilling sensitive plants at low temperature might be caused by phase transitions in the membrane lipids and similar events may occur in the stomatal guard cells. The rate of efflux of water and potassium from these cells will be slowed and the capacity of the stomata to close in response to decreasing water potential in the chilled leaf will be reduced as a consequence. Such results may indicate that one of the earliest consequences of low temperature is a reduction in the permeability of some membranes to water and ions. This is contrary to the Lyons-Raison hypothesis of injury which would suggest an increase in permeability under such conditions.

Table 7.1 Changes in the ABA content of *Phaseolus vulgaris* leaves during chill hardening at 12°C, 80% r.h. over 4 days followed by chilling at 5°C, 80% r.h. in comparison to plants chilled directly at 5°C, 80% r.h. or 5°C, 100% r.h.

Treatment	ABA content (ng ABA g FW^{-1})
Control	175
Day 1 chill hardened 12°C, 80% r.h.	565
Day 2 chill hardened 12°C, 80% r.h.	536
Day 3 chill hardened 12°C, 80% r.h.	817
Day 4 chill hardened 12°C, 80% r.h.	272
Other treatments	
Chill hardened and chilled 24 hours at 5°C, 80% r.h.	112
Not hardened and chilled 24 hours 5°C, 80% r.h.	440
Not hardened and chilled 24 hours 5°C, 100% r.h.	85

In species such as *Episcia reptans* water loss of the type described above is far less important in the development of chilling injury since the rate of injury cannot be significantly reduced by enclosing the plant inside a polythene bag before chilling (to maintain high humidity). In addition, it is not possible to chill harden or drought harden the leaves of this species to withstand chilling stress (Wilson, 1976). Hence chilling injury to this species is primarily metabolic. Tropical fruits also possess little ability to harden against chilling injury (Wheaton and Morris, 1967).

Cytoplasmic and cytoskeletal changes

Upon cooling trichomes of water melon and tomato to 4°C, the normal thin strands of cytoplasm which span the vacuole were no longer evident and had been replaced by spherical vesicles (Patterson *et al.*, 1979). Provided that chilling was not prolonged, these vesicles disappeared on rewarming and the thin strands of cytoplasm were reformed. These changes were not observed in chilling resistant plants, but a similar sequence can occur in chilling resistant plants when they are exposed to toxic agents, mechanical damage or bacterial infection. These cytoplasmic changes may be caused by the disruption of the microtubules of the cytoskeleton at low temperatures and their re-association on rewarming (Patterson *et al.*, 1979). However, chemical destabilization of microtubules using colchicine has been shown to increase the severity of chilling injury in cotton (Rikin *et al.*, 1980). It is not known whether chilling resistant plants have microtubules that are stable at low temperatures. Further evidence for the importance of cytoplasmic changes in the development of chilling injury is provided by the rapid cessation of protoplasmic streaming in chilling sensitive plants at low temperatures (Patterson and Graham, 1977) and the occurrence of crystalline-like deposits in the cytoplasm of *E. reptans* within a few hours of the start of chilling (Murphy and Wilson, 1981). The reduced rate of propagation of action potentials in chilling sensitive seismonastic plants such as *Biophytum sensitivum* at low temperatures may also be caused by cytoplasmic changes at low temperatures (Jones and Wilson, 1982).

Respiration and ATP supply

The significance of changes in the activation energy of respiration and of breaks in Arrhenius plots of respiration of chilling sensitive plants has already been covered in a previous section. However, several workers have obtained linear Arrhenius plots for the respiration of chilling sensitive tissue (Patterson *et al.*, 1979; Wright *et al.*, 1982). These results indicate that low rates of mitochondrial respiration are not a significant factor in the development of chilling injury in some species, contrasting with early work speculating on a decline in ATP in chilled cotton leaves which may cause cellular injury (Stewart and Guinn, 1969). Such decline in ATP levels is, however, caused mainly by water stress as the ATP and ADP levels in *Phaseolus vulgaris* leaves chilled at 5°C, 100 per cent r.h. increased over the first 24 hours and remained high during the following 7 days. Even after 8 or

9 days of chilling in a saturated atmosphere there was only a slight fall in ATP and ADP levels which coincided with the development of visible signs of chilling injury to the leaf (Wilson, 1978). A decrease in ATP supply below that necessary to maintain the metabolic integrity of the cytoplasm cannot therefore be considered to cause chilling injury to leaves at 5°C, 100 per cent r.h. In agreement with this an increase in the ATP level of *P. vulgaris* leaves at low temperatures has been reported (Jones, 1970). The increase in ATP level of *P. vulgaris* leaves at 5°C may be due to the cold sensitivity of an ATPase which is readily inactivated at low temperatures (Penefsky and Warner, 1965). In contrast, chilling leaves of *P. vulgaris* at 5°C, 85 per cent r.h. resulted in rapid leaf wilting and injury, and the ATP and ADP levels decreased after 12 hours of chilling. Although the leaves chilled at 5°C, 85 per cent r.h., showed *c.* 50 per cent injury after 24 hours, the level of ATP had decreased by less than 33 per cent, suggesting that leaf dehydration, and not a reduction in ATP supply, is the cause of cell death. Impaired phosphorylation of cotton leaves at 5°C (Stewart and Guinn, 1969) can be attributed to the effects of water stress and not to low temperature *per se*. Even in the extremely chilling sensitive species *E. reptans* there was no rapid decrease in ATP levels during the first 5 hours of chilling at 5°C; low ATP supply is therefore unlikely to be the cause of chilling injury to this species (Wilson, 1978).

During the chilling of sensitive tissue the respiration rate often increases initially, for example doubling of the respiration of cucumber fruit after 8 days at 5°C which coincided with the onset and development of injury (Eaks and Morris, 1956). Furthermore in *Episcia reptans* there is a three-fold increase in oxygen uptake within 2 hours of the start of chilling and a lower increase in carbon dioxide evolution (Wilson, 1978).

Lipid and fatty acid changes

In order to maintain membrane structure and function, cells must be able to synthesize, transfer and incorporate fatty acids into the membrane systems throughout the cell. The effect of chilling on these processes in chiling sensitive plants has received little attention. Using ^{14}C acetate as a radio-labelled precursor of fatty acid synthesis the half lives of ^{14}C labelled fatty acids of *Episcia reptans* and *Phaseolus vulgaris* leaves have been shown to be shorter at chilling temperatures in comparison to those of chilling resistant *Hordeum sativum* at 5°C (McMurdo, 1982). The free fatty acid and phosphatidic acid content of *E. reptans* and *P. vulgaris* cells also increases at 5°C reflecting an increased rate of lipid breakdown that did not occur in chilling resistant *H. sativum* at 5°C (Fig. 7.7). The rapid rate of increase in the free fatty acid content of *E. reptans* indicates this could be related to cell injury as free fatty acids are known to have detergent-like properties which can cause further membrane damage and also inhibit phosphorylation and photosynthetic electron transport. By contrast the increase in free fatty acid in *P. vulgaris* at 5°C, 100 per cent r.h. is slow. It is possible that these changes in membrane fatty acids may be initiated by a phase transition and a progressive accumulation on prolonged chilling may lead to further changes in membrane permeability and function, However, it is also possible that enzymes involved in

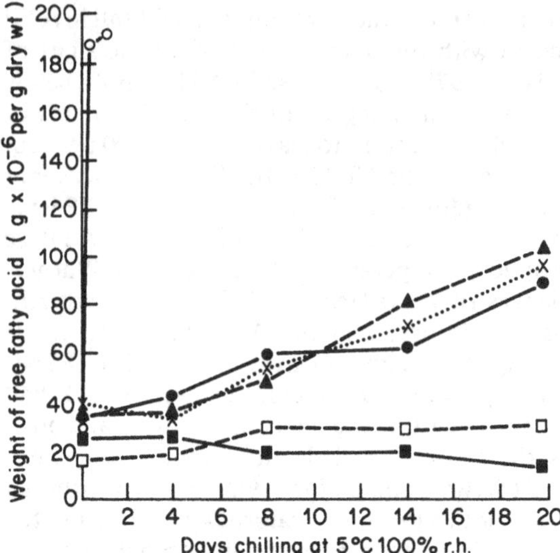

Fig. 7.7 Changes in the free fatty acid content of chilling sensitive *Episcia reptans* (○), *Phaseolus vulgaris* (x), *Cucumis sativus* (▲) and *Vigna radiata* (●) in comparison to chilling resistant *Hordeum sativum* (■) and *Brassica rapa* (□) during chilling at 5°C, 100% r.h.

fatty acid and lipid synthesis and breakdown may show altered activity at low temperatures resulting in release of free fatty acids.

Photo-oxidation of leaf pigments and membrane lipids is an important cause of chilling injury to some species (e.g. cucumber) after 2–3 days chilling at 1°C (van Hasselt and Strikwerda, 1976). Bleaching of *P. vulgaris* leaves after 7 days at 5°C, 100 per cent r.h. can also be severe if the light intensity is high. It is speculated that in chilled plants small amounts of superoxide (O_2^-) ion produced in the chloroplasts and mitochondria peroxidize unsaturated fatty acid chains as superoxide dismutase which normally protects the cell against such damage is inactivated at low temperature (Michalski and Kaniuga, 1981). Light has also been shown to cause the loss of NADP malate dehydrogenase, pyruvate Pi dikinase and catalase activities in maize leaves at 10°C (Taylor, Slack and McPherson, 1974).

Photosynthesis and translocation

Photosynthesis is more sensitive to extremes of temperature than respiration so it is possible that during prolonged chilling starvation of plant tissues may occur. Translocation is inhibited in chilling sensitive species at 5°C (Giaquinta and Geiger, 1973) consequently starch accumulates. The accumulated starch may damage the photosynthetic apparatus (Forde, Whitehead and Rowley, 1975) and further inhibit photosynthesis. The cessation of translocation in chilling sensitive species at 5°C has been attributed to a phase change in the membrane lipids of the plasmalemma of the sieve tube resulting in the collapse of the material lining the cell and the blockage of the sieve plate (Giaquinta and Geiger, 1973).

Metabolic changes during chill hardening

Hardening is a process which increases the resistance of a plant to the stress of chilling. The most important effect of chill hardening is the prevention of the water stress injury described previously. Chill hardening may, however, also involve changes in proteins, lipids and other metabolites which help the plant to withstand chilling temperatures. Chill hardening is known to increase the degree of unsaturation of the phospholipids of several chilling sensitive species (Wilson and Crawford, 1974), but these increases did not appear to impart any greater degree of chilling tolerance as the leaves died at the same rate as those of drought hardened plants where there was no increase in unsaturation during hardening. It has also proved difficult to relate these lipid changes to alterations in chloroplast membrane activity after various hardening treatments (Rosinger, Wilson and Kerr, 1981). However, chill hardening of *Phaseolus vulgaris* can shift the temperature optimum for photosynthesis and respiration and this may be important in prolonging survival at temperatures in the upper end of the chilling range.

Measurement of chilling sensitivity

Measurement of chilling sensitivity is difficult because symptoms of chilling injury vary between different species and are often difficult to quantify. The sensitivity of different parts of the plant may also vary with plant age and previous temperature history which can lead to discrepancies between the results of different investigators. Injury can take a long time to develop and often becomes apparent faster on return of the plant to the warmth. Hence any subjective assessment of damage should pay attention to the environmental conditions during chilling and rewarming as they can profoundly influence the results.

Perhaps the most widely used method of assessing chilling sensitivity has been the leakage of cellular contents, mainly K^+ ions, by conductivity. *Passiflora* species have been ranked in terms of their chilling tolerance on the basis of kinetics of electrolyte loss at 0°C (Patterson *et al.*, 1976). However, electrolyte leakage at low temperature is not solely a property of chilling sensitive tissues as rates of leakage from chilling resistant tissues can be quite rapid at chilling temperatures (Murata and Tatsumi, 1979). A further problem with electrolyte leakage is that visible signs of damage can occur before a significant increase in conductivity resulting from leakage occurs (Murata and Tatsumi, 1979). The sensitivity of such a technique may be increased by monitoring efflux of ^{86}Rb from cells, for example a good correlation between the rate of leakage of ^{86}Rb at 1°C from tomato leaves and the altitude of origin of tomato varieties collected in the Peruvian Andes has been demonstrated (Paull *et al.*, 1979). Other techniques may require the use of sophisticated laboratory procedures such as measuring the change in activity of either isolated mitochondria (Lyons and Raison, 1970) or chloroplasts (Smillie, 1979) extracted from tissue which has been kept at chilling temperatures for various periods of time, the rate of decline in activity being taken as a measure of sensitivity. More simple methods which can be considered are the use of vital stains such as neutral red or dyes that are excluded by intact membranes (e.g. Evans' blue). Assay methods have been developed which use the reduction of triphenyl tetrazolium chloride (Stergios and Howell, 1973) at low temperatures as an indicator of the ability of the tissue to respire. A simple assay based on the ability of etiolated plants to green up when illuminated at temperatures just above the chilling range has also been employed (McWilliam *et al.*, 1979). Other criteria which can be used to assess chilling sensitivity, and which are used in plant breeding programmes, include time to germination at 15°C, minimum germination temperature, seedling growth rate and successful fruit set at low temperature. A new technique involves the measurement of chlorophyll fluorescence in intact leaves (Smillie, 1979). Although this technique has been used to compare the sensitivity of closely related species there can be a large degree of variability between replicates so that 16 replicates are often used for each treatment. The method does not appear to be suitable for all species.

References

Bagnall, D.J. (1979). Low temperature responses of three *Sorghum* species. In *Low Temperature Stress in Crop Plants: The Role of the Membrane*, pp. 67–80. Edited by Lyons, J.M., Graham, D. and Raison, J.K. Academic Press, New York.

Bagnall, D.J. and Wolfe, J.A. (1978). Chilling sensitivity in plants: do the activation energies of growth processes show an abrupt change at a critical temperature? *Journal of Experimental Botany* **29**, 1231–42.

Bishop, D.G., Kenrick, J.R., Bayston, J.H., MacPherson, A.S., Johns, S.R. and Willing, R.I. (1979). The influence of fatty acid unsaturation on fluidity and molecular packing of chloroplast membrane lipids. In *Low Temperature Stress in Crop Plants: The Role of the Membrane*, pp. 375–85. Edited by Lyons, J.M., Graham, D. and Raison, J.K. Academic Press, New York.

Christiansen, M.N., Carns, H.R. and Slyter, D.J. (1970). Stimulation of solute loss from radicles of *Gossypium hirsutum* L. by chilling, anaerobosis and low pH. *Plant Physiology* **46**, 53–6.

Creencia, R.P. and Bramlage, W.J. (1971). Reversibility of chilling injury to corn seedlings. *Plant Physiology* **47**, 389–92.

Dalziel, A.W. and Breidenbach, R.W. (1979). Differential thermal analysis of tomato mitochondrial lipids in low temperature stress. In *Low Temperature Stress in Crop Plants: The Role of the Membrane*, pp. 319–26. Edited by Lyons, J.M., Graham, D. and Raison, J.K. Academic Press, New York .

Eaks, I.L. (1980). Effect of chilling on respiration and volatiles of California Lemon Fruit. *Journal of the American Society for Horticultural Science* **105**, 865–9.

Eaks, I.L. and Morris, L.L. (1956). Respiration of cucumber fruits associated with physiological injury at chilling temperatures. *Plant Physiology* **31**, 308–14.

Eamus, D., Fenton, R. and Wilson, J.M. (1982). Stomatal behaviour and water relations of chilled *Phaseolus vulgaris* L. and *Pisum sativum* L. *Journal of Experimental Botany* **33**, 434–41.

Eamus, D. and Wilson, J.M. (1983). ABA levels and effects in chilled and hardened *Phaseolus vulgaris*. *Journal of Experimental Botany* **34**, 1000–1006.

Forde, B.J., Whitehead, H.C.M. and Rowley, J.A. (1975). Effect of light intensity and temperature on photosynthetic rate, leaf starch content and ultrastructure of *Paspalum dilatatum*. *Australian Journal of Plant Physiology* **2**, 185–95.

Giaquinta, R.J. and Geiger, D.R. (1973). Mechanism of inhibition of translocation by localized chilling. *Plant Physiology* **51**, 372–7.

Graham, D., Hockley, D.G. and Patterson, B.D. (1979). Temperature effects on phosphoenol pyruvate carboxylase from chilling sensitive and chilling resistant plants. In *Low Temperature Stress in Crop Plants: The Role of the Membrane*, pp. 453–62. Edited by Lyons, J.M., Graham, D. and Raison, J.K. Academic Press, New York.

Guinn, G. (1971). Leakage of metabolites from chilled cotton seedlings. *Crop Science* **11**, 101–102.

Jones, C. and Wilson, J.M. (1982). The effects of temperature on action potentials in the chill-sensitive seismonastic plant *Biophytum sensitivum*. *Journal of Experimental Botany* **33**, 313–20.

Jones, P.C.T. (1970). The effect of light, temperature and anaesthetics on ATP levels in the leaves of *Chenopodium rubrum* and *Phaseolus vulgaris*. *Journal of Experimental Botany* **21**, 58–63.

Kramer, P.J. (1949). *Plant and Soil Water Relationship*. McGraw-Hill, New York.

Liebermann, M., Craft, C.C., Audia, W.V. and Wilcox, M.S. (1958). Biochemical studies of chilling injury in sweet potatoes. *Plant Physiology* **33**, 307–311.

Lyons, J.M. (1973). Chilling injury in plants. *Annual Review of Plant Physiology* **24**, 445–66.

Lyons, J.M. and Raison, J.K. (1970). Oxidative activity of mitochondria isolated from plant tissues sensitive and resistant to chilling injury. *Plant Physiology* **45**, 386–9.

McMurchie, E.J. (1979). Temperature sensitivity of ion-stimulated ATPases associated with some plant membranes. In *Low Temperature Stress in Crop Plants: The Role of the Membrane*, pp. 163–76. Edited by Lyons, J.M., Graham, D. and Raison, J.K. Academic Press, New York.

McMurdo, A.C. (1982). *Metabolic aspects of chilling injury in plants*. Ph.D. Thesis, University of Wales.

McMurdo, A.C. and Wilson, J.M. (1980). Chilling injury and Arrhenius plots. *Cryoletters* **1**, 231–8.

McWilliam, J.R., Kramer, P.J. and Musser, R.L. (1982). Temperture induced water stress in chilling sensitive plants. *Australian Journal of Plant Physiology* **9**, 343–52.

McWilliam, J.R., Manokaran, W. and Kipnis, T. (1979). Adaptation to chilling stress in *Sorghum*. In *Low Temperature Stress in Crop Plants:*

The Role of the Membrane, pp. 491-506. Edited by Lyons, J.M., Graham D. and Raison, J.K. Academic Press, New York.

Michalski, W.P. and Kaniuga, Z. (1981). Photosynthetic apparatus of chilling sensitive plants. X. Relationship between superoxide dismutase activity and photoperoxidation of chloroplast lipid. *Biochemica Biophysica Acta* **637**, 159-67.

Murata, T. (1969). Physiological and biochemical studies of chilling injury in bananas. *Physiol. Plantarum* **22**, 401-411.

Murata, T. and Tatsumi, Y. (1979). Ion leakage in chilled plant tissues. In *Low Temperature Stress in Crop Plants: The Role of the Membrane*, pp. 141-52. Edited by Lyons, J.M., Graham, D. and Raison, J.K. Academic Press, New York.

Murphy and Wilson. (1981). Ultrastructural features of chilling injury in *Episcia reptans*. *Plant, cell and Environment* **4**, 261-5.

Patterson, B.D. and Graham, D. (1977). Effect of chilling temperatures on the protoplasmic streaming of plants from different climates. *Journal of Experimental Botany* **28**, 736-43.

Patterson, B.D., Murata, T. and Graham, D. (1976). Electrolyte leakage induced by chilling in *Passiflora* species tolerant to different climates. *Australian Journal of Plant Physiology* **3**, 435-42.

Patterson, B.D., Paull, R. and Graham, D. (1979). Adaptation to chilling: survival, germination, respiration and protoplasmic dynamics. In *Low Temperature Stress in Crop Plants: The Role of the Membrane*, pp. 25-35. Edited by Lyons, J.M., Graham, D. and Raisson, J.K. Academic Press, New York.

Paull, R.E., Patterson, B.D. and Graham, D. (1979). Chilling injury assays for plant breeding. In *Low Temperature Stress in Crop Plants: The Role of the Membrane*, pp. 507-19. Edited by Lyons, J.M., Graham, D. and Raison, J.K. Academic Press, New York.

Penefsky, H.S. and Warner, R.C.J. (1965). Partial resolution of the enzymes catalysing oxidative phosphorylation. *Journal of Biological Chemistry* **240**, 4694-702.

Pike, C.S. and Berry, J.A. (1980). Membrane phospholipid phase separations in plants adapted to or acclimated to different thermal regimes. *Plant Physiology* **66**, 238-41.

Potter, J.F. and Ross, G.J.S. (1979). Maximum likelihood estimation of break points and the comparison of the goodness of fit with that of conventional curves. In *Low Temperature Stress in Crop Plants: The Role of the Membrane*, pp. 535-42. Edited by Lyons, J.M., Graham, D. and Raison, J.K. Academic Press, New York.

Raison, J.K. and Chapman, E.A. (1976). Membrane phase changes in chilling sensitive *Vigna radiata* and their significance to growth. *Australian Journal of Plant Physiology* **3**, 291-9.

Raison, J.K., Chapman, E.A., Wright, L.C. and Jacobs, S.W.L. (1979). Membrane lipid transitions: their correlation with the climatic distribution of plants. In *Low Temperature Stress in Crop Plants: The Role of the Membrane*, pp. 177-86. Edited by Lyons, J.M., Graham, D. and Raison, J.K. Academic Press, New York.

Rikin, A., Atsmon, D. and Gitler, C. (1980). Chilling injury in cotton (*Gossypium hirsutum* L.): effects of antimicrotubular drugs. *Plant Cell Physiology* **21**, 829-37.

Rosinger, C.H., Wilson, J.M. and Kerr, M.W. (1981). Changes in the temperature response of Hill-reaction activity of chilling-sensitive and chilling-resistant plants after hardening. *Journal of Experimental Botany* **33**, 321-31.

Schreier, S., Polnasek, C.F. and Smith, I.C.P. (1978). Spin labels in membranes: problems in practice. *Biochemica Biophysica Acta* **515**, 395–436.

Smillie, R.M. (1979). The useful chloroplast: a new approach for investigating chilling stress in plants. In *Low Temperature Stress in Crop Plants: The Role of the Membrane*, pp. 187–202. Edited by Lyons, J.M., Graham, D. and Raison, J.K. Academic Press, New York.

Stergios, B.G. and Howell, G.S. (1973). Evaluation of viability tests for cold stressed plants. *Journal of the American Society for Horticultural Science*, **98**, 325-30.

Stewart, J.McD. and Guinn, G. (1969). Chilling injury and changes in adenosine triphosphate of cotton seedlings. *Plant Physiology* **44**, 605-608.

Sugiyama, T., Schmitt, M.R., Ku, S.B. and Edwards, G.E. (1979). Differences in cold cability of pyruvate Pi dikinase among C_4 species. *Plant and Cell Physiology* **20**, 965–71.

Taylor, A.O., Slack, C.R. and McPherson, H.G. (1974). Plants under climatic stress. IV. Chilling and light effects on photosynthetic enzymes of *Sorghum* and maize. *Plant Physiology* **54**, 696–701.

Träuble, H. and Haynes, D.H. (1971). The volume change in lipid bilayer lamellae at the crystalline-liquid crystalline phase transition. *Chemistry and Physics of Lipids* **7**, 324–35.

Van Hasselt, P.L.R. and Strikwerda, J.J. (1976). Pigement degradation in discs of the thermophilic *Cucumis sativus* as affected by light, temperature, sugar application and inhibitors. *Physiology of Plants* **37**, 253–7.

Wheaton, T.A. and Morris, L.L. (1967). Modification of chilling sensitivity by temperature conditioning. *Proceedings of the American Society for Horticultural Science* **91**, 529–33.

Willcox, M.E. and Patterson, B.D. (1979). Breaks or curves? A visual aid to the interpretation of data. In *Low Temperature Stress in Crop Plants: The Role of the Membrane*, pp. 523–6. Edited by Lyons, J.M., Graham, D. and Raison, J.K. Academic Press, New York.

Wilson, J.M. (1976). The mechanism of chill- and drought-hardening of *Phaseolus vulgaris* leaves. *New Phytologist* **76**, 257–70.

Wilson, J.M. (1978). Leaf respiration and ATP levels at chilling temperatures. *New Phytologist* **80**, 325–34.

Wilson, J.M. and Crawford, R.M.M. (1974). The acclimatization of plants to chilling temperatures in relation to the fatty-acid composition of leaf polar lipids. *New Phytologist* **73**, 805–820.

Wolfe, J. and Bagnall, D.J. (1979). Statistical tests to decide between straight line segments and curves as suitbale fits to Arrhenius plots or other data.

In *Low Temperature Stress in Crop Plants: The Role of the Membrane*, pp. 527–34. Edited by Lyons, J.M., Graham, D. and Raison, J.K. Academic Press, New York.

Wright, L.C. McMurchie, E.J., Pomeroy, M.K. and Raison, J.K. (1982). Thermal behaviour and lipid composition of cauliflower plasma membranes in relation to ATPase activity and chilling sensitivity. *Plant Physiology* **69**, 1356–60.

Wright, L.C. and Raison, K.J. (1983). Thermal phase transitions in the polar lipids of plant membranes: their induction by disaturated phospholipids and their relation to chilling injury. *Biochemica Biophysica Acta* **731**, 69–78.

Wright, M. and Simon, E.W.J. (1973). Chilling injury in cucumber leaves. *Journal of Experimental Botany* **24**, 400–411.

8

Higher plants at freezing temperatures

B.W.W. Grout
Department of Biological Sciences
Plymouth Polytechnic

Introduction

Low temperature is responsible for two major environmental stresses affecting higher plant performance and distribution. The first, chilling stress, depends upon reduced temperatures *per se* (without freezing) and has been reviewed previously (Wilson, this volume). The second stress, or complex of stresses, is a direct consequence of freezing either of solutions within the plant or adjacent to its external surfaces. Such freezing will bring about physical and chemical changes within the tissues which can cause injury, and may be potentially lethal (Grout and Morris, this volume). To survive, plants must avoid critical injury at the extremes of freezing stress typical of their natural habitat (i.e. they must be resistant either by avoiding or tolerating freezing). The following discussion will examine the freezing stresses experienced by higher plants in the natural environment and consider some of the factors involved in both injury and resistance.

Freezing resistance and plant distribution

The impact of freezing stresses on global plant distribution is necessarily large as a mean annual minimum air temperature below 0°C is typical of 64 per cent of the earth's land masses (Larcher and Bauer, 1981). The severity of environmental freezing stresses can be used to identify four major habitats with respect to plant distribution (Fig. 8.1). In the polar regions there is permanent ice cover and plants are restricted to lower, poikilohydric forms that are extremely resistant to low temperatures. Similar conditions are found at high altitude on mountain slopes. A second region can be outlined with a mean annual minimum air temperature below −40°C that will impose extreme stresses on its native plants. These extremes are likely to fall below the homogeneous nucleation point of internal plant solutions, and so

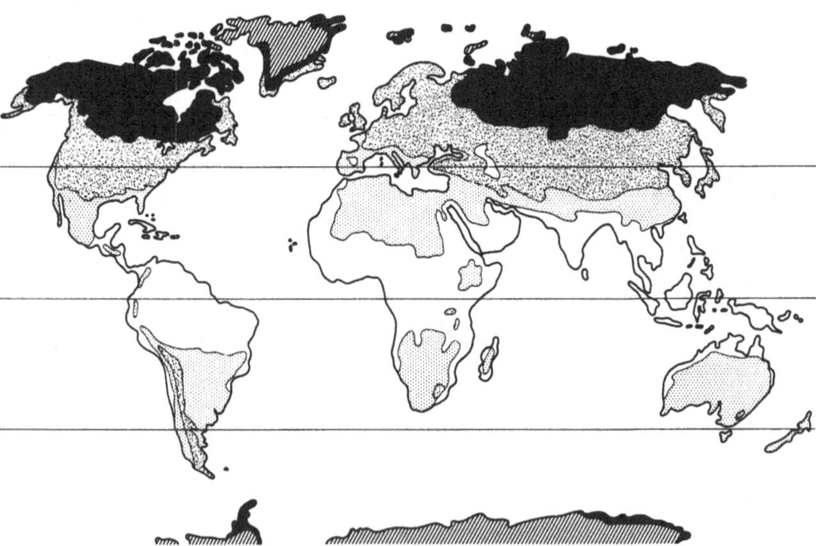

Fig. 8.1 A diagrammatic global map showing areas where environmental freezing will influence the characteristics of the native plant populations and their distribution. (Adapted from Hoffman, 1960.)

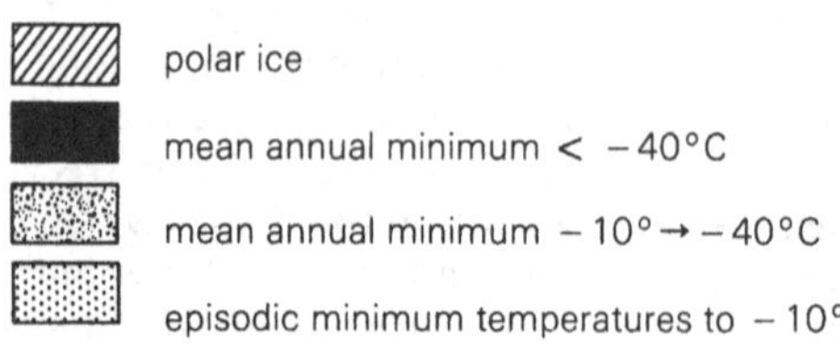

freezing is inevitable, (Rasmussen and MacKenzie, 1972). Plants with over-wintering, vegetative structures that survive this climate must be fully resistant to freezing by being completely tolerant of it.

A third region is characterized by cold winters, with an annual minimum of between −10°C and −40°C, and extends over much of Europe, North America and Asia. Within this region the probability of internal tissue freezing is high, and so native plants must have well-developed freezing resistance mechanisms to ensure survival. In the fourth region sub-zero temperatures to −10°C are episodic, and so plants without some degree of freezing resistance, by either tolerance or avoidance, will be at risk during portions of the cold seasons.

Vegetative plant structures that survive the cold season in each of these regions must have the ability to endure the imposed freezing stress without significant damage to integrated function (i.e. a high survival capacity). Following freezing stress they must be able to grow on successfully and reproduce (i.e. possess high productive and reproductive capacities also).

The seed habit

A reasonably long growing season (conventially, the period for which the mean air temperature is greater than 6°C) offers a survival opportunity for many plants with low survival capacity with respect to freezing. These species would suffer lethal, vegetative injury at the low temperature extremes of the environment, and may have an inadequate adaptive, hardening response.

In terms of survival these plants avoid freezing stress by producing seeds of a low water content (< 15 per cent of their fresh weight) at the end of the warm season (annuals) that are tolerant of freezing to extremely low temperatures under a wide range of conditions (Roberts, 1972; Stanwood and Bass, 1978; Grout and Crisp, 1985). Survival of the population depends upon the persistance of viable seeds through to the next growing season. It should be noted that not all plants produce these *orthodox* seeds and that *recalcitrant* seeds (typically at high water content), that are more commonly associated with tropical and subtropical species, are not naturally freezing resistant (Roberts, 1972).

Natural freezing resistance

The evolutionarily-adapted plants of regions where the risks of freezing are high have developed appropriate levels of resistance (Table 8.1). Many species achieve their maximum resistance following adaptive responses to the environmental changes that signal the onset of the low temperature season. These changes, usually in response to reducing temperature and photoperiod, are known collectively as hardening or acclimatization, and they enhance whole plant survival by postponing the lethal injuries caused by freezing to lower temperatures. The data in Table 8.1 include such adaptations, where they occur, and so represent the most hardy situations for each group.

Table 8.1 The temperature range (°C) below which freezing injury is likely in a number of higher plant groups (based on data reviewed in Larcher and Bauer, 1981)

Herbaceous plants	Arctic	Temperate	Tropical
Graminoids	→ −80*	−13 → −25	−1 → −4
Herb. dicots.	−15 → −30†	−5 → −20	−1 → −2
Woody plants	Conifers	Deciduous	Shrubs
Woody plants of temperate regions	−10 → −20‡ −15 → −30§	−15 → −30†† −20 → −40‡‡	−15 → −30‡,¶ −15 → −35§,¶
Woody plants of regions with severe winters	−40 → −70‡ −50 → −196§	−25 → −35†† −30 → −50‡‡	−30 → −50‡,** −30 → −196§,**

* Arctic sedge; † Rosette ferns; ‡ Leaves; § Stem; ¶ Ericaceous shrubs; ** Arctic dwarf shrubs; †† Buds; ‡‡ Xylem.

Freezing under environmental conditions

The critical freezing events which threaten plant survival are:

(i) freezing of available water in the soil;
(ii) freezing of water within plant tissues.

Freezing of soil water and frost drought

The soil water available to the plant is found in pores (0.2–0.8 μ in diameter) between the soil particles, and contributes to a dilute solution of inorganic salts and organic compounds, with some materials in colloidal suspension. Such solutions have a limited undercooling ability and are likely to freeze at approximately −2°C. Once frozen, the amount of soil water available to the normal transpiration stream of the plant is reduced and, if evaporative loss continues from the aerial plant portions, water deficit will result.

In Alpine locations, for example, freezing soil temperatures can be found at 30 cm depths and may persist for several months (Tranquillini, 1982). Tree roots will penetrate to this depth and so the increased negativity of soil water potential resulting from freezing will impose stress on these plants. The stresses will not be lethal to the majority of the plant population under normal environmental conditions, as the indigenous plants are successfully adapted, but losses can be severe under atypical, more extreme conditions. In a winter with prolonged soil freezing the twigs of *Rhododendron ferrigeneum* (at the Alpine tree line) may lose 60–70% of their water by the spring (Pisek *et al.*, 1973). Extreme, atypical conditions, were also found near Plymouth, UK, close to sea level, where persistent freezing temperatures at a 5 cm soil depth were recorded for 12 consecutive days in January 1985. These conditions of frost drought contributed to significant injury to herbaceous plants that overwinter in this usually mild area. In overwintering, evergreen plants there will be a relatively large foliar surface area for water loss, and despite regulation by stomatal closure, cuticular water loss may be significant if the period of frost drought is prolonged. In deciduous species the available surface area for evaporative loss following leaf-fall is restricted to stem axes.

The water stress imposed by frost drought may be exacerbated by a temperature-related increase in resistance to water transport that may affect the membranes of root cells. In *Picea abies*, for example, 35 per cent less water is taken up at 0°C than at 20°C, without any effects of freezing (Tranquillini, 1982).

Freezing of solutions within the plant

Freezing within the whole plant will have effects on structure and function at both the tissue and cellular level; the effects at a cellular level have been discussed (Grout and Morris, this volume). The following discussion will consider events at the tissue level primarily, and will emphasize the special role of the cell wall in plant cell freezing. This is particularly relevant, as widely accepted models of freezing injury to plant cells are based on isolated protoplasts and take no direct account of cell wall properties.

Where extracellular freezing occurs under environmental conditions, at relatively high sub-zero temperatures, both ice and hypertonic solutions will exist in close proximity to the cell wall. The hypertonic solution may act as a physical barrier to contact between ice and the wall materials, such that the ice crystals act solely as accumulating sites for available water molecules (Olien, 1981). If this physical barrier becomes ineffective and ice contacts the cell wall, adhesions may occur that will impose asymmetric mechanical stresses upon the cell wall, particularly during ice crystal growth, leading to cellular distortion and collapse, particularly if the stress is transmitted to the plasmalemma. Serious cellular lesions may result from these distortions. A similar situation may also occur if the ice interface contacts the plasmalemma directly, particularly as it shrinks away from the cell wall during surface area reduction that follows freeze-induced desiccation. Further, studies of cultured cells and their derived protoplasts have indicated that mechanical contact between the cell wall and plasmalemma, particularly during contractions, can transmit such damaging, asymmetric mechanical stresses (Tao *et al.*, 1983). Experiments using isolated rye mesophyll cells do not, however, support this hypothesis, with no obvious relationship between survival following freezing and the protoplast/cell wall relationship (Steponkus *et al.*, 1982).

The polymeric compounds that are physical components of cell walls (e.g. arabinoxylans) can inhibit ice crystal growth and interfere with the interface during freezing (Shearman *et al.*, 1973; Olien, 1981). Such xylans aggregate during freezing in those species where they are least effective, but where they are most effective the polymers form a cohesive film at the interface that adheres to the ice lattice and inhibits the kinetics of freezing (Olien, 1965). The effects of polymer can be seen microscopically and can be evaluated (Shearman *et al.*, 1973). At high polymer concentrations a network forms that interrupts the ice and traps extracellular solution between a relatively large number of crytals. This reduces the quantity of hypertonic solution available to act as a barrier between ice and cell wall and may therefore increase the chances of direct contact between advancing ice, the cell walls and plasmalemma and thereby increase subsequent mechanical stress.

Plant cells are not rich in effective ice nucleators at relatively high sub-zero temperatures (George and Burke, 1976; Mazur, 1977), and this contributes to the predominance of extracellular ice formation as the primary freezing event under environmental conditions. Tissue nucleators are mainly cell wall structures and not materials suspended in solution (Kaku and Salt, 1968; Levitt, 1980; Rajashekar *et al.*, 1982) and vary in effectiveness according to plant species, stages of development and temperature range. Most higher plants fall into the range −5°C to −15°C (Kaku, 1973; 1975). Undercooling of plant tissues will occur, and persist if the tissue temperature remains above the characteristic temperature of the most effective nucleator in the system (Single and Marcellos, 1981). The duration of exposure to potentially freezing temperatures is also important. For example, bean and tomato tissues may remain undercooled for several hours at −3°C before freezing over the subsequent 24 hours.

Ice formed on the external plant surfaces (hoarfrost) may often be more effective than internal structures in nucleating intercellular freezing, the ice front travelling through ruptures in the epidermis, lenticels and incompletely closed stomata. Formation of hoarfrost on plant surfaces can be promoted by epicuticular nucleators including dust particles (Arny *et al.*, 1976), plant fragments resulting from decomposition and disintegration (Schnell and Varli, 1972) and epiphytic bacteria (Lindow, 1982; 1983). The role of the epiphytic bacteria can be demonstrated in experiments with spring corn where frost injury at −2°C was halved by application of the antibacterial agent, streptomycin as a foliar spray (Single and Marcellos, 1981; Lindow, 1982). It should be noted that data on undercooling ability of aerial plant structures would be more meaningful if the presence or absence of epicuticular ice was also recorded.

Once freezing has been initiated in the plant tissues ice will spread rapidly along the vascular system (rates of 2 cm $second^{-1}$ have been recorded in wheat) and penetrate into extracellular spaces until it meets an effective barrier to ice propagation. Warmer regions, those of a more negative water potential as in plants under water stress, immature stem nodes and the vascular endodermis may act as such a barrier (Kaku, 1971; Single and Marcellos, 1981).

The presence of ice within the tissues will impose mechanical, osmotic and chemical stresses upon the living tissues. These have been discussed previously at the cellular level (Grout and Morris, this volume). Intercellular ice formation may not be uniform and larger aggregates of ice crystals (glaciers) may develop in some regions, particularly between the epidermis and mesophyll of leaves (Idle, 1966). 'Epidermal lifting' as a result of glacier formation may be seen in cabbage and alfalfa (North and Fisher, 1971), and cells in a range of tissues may separate also along the middle lamella (Hudson and Idle, 1962). Much of this type of separation injury is repairable, and may even have a protective effect by containing ice, and mechanical stress, in less vulnerable, rapidly-repairing locations. The flowers of spring wheat may derive protection during late radiation frosts in this way (Olien and Smith, 1981a). Similarly, the skin of immature apple and pear fruits may separate from underlying tissues as a result of ice formation in late spring frosts, 'popping'. If the skin ruptures, desiccation and disease can cause secondary injury, but if it remains intact, the primary ice injury is commonly repairable (Modlibowska, 1946; 1968).

Relatively large trees may also suffer mechanical injury from intercellular ice (Levitt, 1980). Ice formation between the bark and wood can loosen tissue attachment and large cracks through to the outer surface may develop (Modlibowska and Field, 1942; Smith and Olien, 1981a). Such injuries result from tangiential contraction of cooled bark together with pressure exerted by glaciers between the wood and bark. Commonly, the cambial layer is torn and remains affixed, largely to the bark. During repair, callus proliferation from the cambium reseals it to the underlying wood (Modlibowska and Field, 1942). Similar mechanisms may be responsible for shearing injuries in the vascular tissues of wheat crowns (Olien, 1971) and azalea flower buds (Graham and Mullin, 1976). Evidence of frost injury to the cambial tissues can be seen in the 'frost ring' a darkened, relatively small ring of wood. This arises where xylem mother cells and differentiating tracheids have released gum materials as a result of desiccation injury or ice crushing. These gums stain the restricted growth of wood that follows such severe injury (Smith and Olien, 1981).

Freezing effects on photosynthesis and respiration

Photosynthesis

Frost tolerant plants (i.e. those that survive ice formation within their tissues) must minimize injury to the photosynthetic mechanism during freezing and must rapidly resume photosynthetic activity following thawing at a level that sustains integrated metabolism, necessary repair and subsequent growth.

Lower plants such as lichens and mosses may be relatively resistant to frost, in photosynthetic terms, with species reported as assimilating carbon dioxide in the frozen state to −20°C (Larcher *et al.*, 1973). Typically, higher plants show reduced levels of carbon fixation as temperature falls, and drastic reduction as ice forms within the photosynthetic tissues. Although not necessarily damaging the photosynthetic structures *per se*, the ice effectively blocks the pathways for gas exchange needed for photosynthetic activity (Larcher, 1981).

In leaves that survive freezing, the assimilation of carbon dioxide and net photosynthesis are often diminished after thawing and may take some while to recover (Fig. 8.2). The reduction will depend upon the extent and duration of freezing, and there is a deleterious effect of cumulative frosts. This inhibition may be a direct effect of frost dehydration causing injury to the chloroplasts, or the accumulation of soluble carbohydrates following frosting may inhibit photosynthesis (Peoples and Koch, 1978; Rutherford and Whittle, 1981).

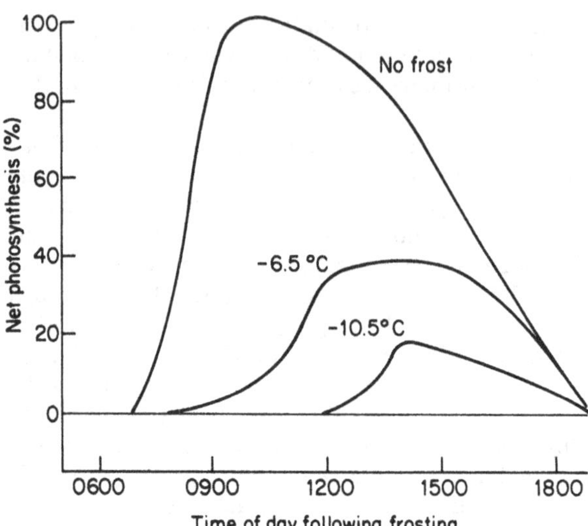

Fig. 8.2 The effects of frosts of differing severity on the net photosynthesis of pine twigs following thawing. (Adapted from Polster and Fuchs, 1963.)

Freezing injury in the chloroplast

Isolated chloroplasts are osmotically active and therefore many of their reponses to freezing stress may be comparable to those of isolated protoplasts, particularly with respect to shrinkage and surface area reduction of limiting membrane (Grout and Morris, this volume). For example, when exposed to hypertonic stress both membrane leakiness and a degree of osmotic non-responsiveness have been observed in isolated spinach chloroplasts (Kaiser and Heber, 1981; Kaiser, 1982). It may be that the chloroplast adjusts to external hypertonicity in the same way as isolated plant protoplasts and that lesions of the limiting membranes result in a similar fashion (Steponkus, 1984; Schmitt *et al.*, 1985). Different species will respond in different ways to hypertonicity and stress on the chloroplast, those conserving the outer membrane during volume reduction will probably suffer less injury than non-conservative chloroplasts (Kaiser, 1982).

The internal, thylakoid system of the chloroplast consists of osmotically active membrane vesicles. This complex is also damaged by the hypertonic stresses associated with freezing, the extent of injury being related to the temperature and duration of freezing and to the composition of solutions surrounding the thylakoid. Compounds such as dextrans and sugars can have a cryoprotective effect on a colligative basis (Schmitt *et al.*, 1985). The injuries may be complex, with separate functions affected differently, for example freezing spinach thylakoids to $-15°C$ (in the presence of sucrose and sodium chloride) may radically inhibit photophosphorylation but will enhance photosynthetic electron transport. The membrane vesicles become leaky under these conditions and are unable to maintain the necessary proton gradient for ATP synthesis. The structural integrity of the electron transport chain, embedded in the membrane matrix, is not disrupted (Heber and Santarius, 1964). Under extremes of hypertonic stress, and in the absence of protective solutes, the thylakoid vesicles may become sufficiently injured to leak plastocyanin, a soluble protein involved in electron transport and usually entrapped within the vesicle interior.

During freezing, proteins may also become detached from thylakoid membranes as a result of structural injury. These include peripheral proteins such as the coupling factor complex and ferrodoxin NADP reductase. Dissociation, or detachment, of these proteins is slowed by reduced temperature and enhanced by increasing ion concentration. Maximum injury during freezing, therefore, results from the interaction of these factors (Mollenhauller *et al.*, 1983). These proteins may rebind to the thylakoid membranes if they are dislodged by altering ionic concentration alone, but they do not respond in this way to freeze-damaged membranes. This may be due to structural alterations in the membrane induced by freezing rather than hypertonicity *per se*, and possibly may be the result of conformational change in the proteins themselves. When unprotected higher plant leaves are killed by freezing, and the thylakoids are isolated rapidly, peripheral proteins are detached (Schmitt *et al.*, 1985).

Respiration

Higher plant respiration will decline in a predictable way with temperature above the tissue freezing point, and will show an abrupt cessation as gas exchange is restricted by ice formation within the tissues (Larcher *et al.*, 1973; Larcher, 1981). At this point the mitochondria and related structures will experience the stresses imposed by extracellular ice and may be injured to a greater or lesser extent according to the severity of the stress (Palta and Li, 1978; Heber *et al.*, 1981; Morris *et al.*, 1981). A further problem for some plants may be maintenance of a balance between respiration (carbon utilization) and photosynthesis (carbon assimilation). Photosynthesis is characteristically more temperature-sensitive than respiration (e.g. in *Pinus sylvestris* net photosynthesis stops at $-4°C$, whereas respiration continues to $-18°C$) (Ungerson and Scherdin, 1965). If such an imbalance was prolonged as in periodic frosts or subarctic regions, the plant would begin to deplete stored nutrient which might affect post-thaw growth and recovery from stress.

Following thawing, a marked increase in respiratory activity, known as 'respiratory overshoot' is commonly observed in higher plant tissues (Fig. 8.3). As much as a five-fold increase in carbon dioxide output compared to control tissues can be measured, depending upon the temperature and duration of freezing (Larcher, 1981). The overshoot is usually apparent within a few hours of thawing, and it may take several days for respiratory gas exchange to fall back to within normal limits. This response is a consequence of accumulated respiratory substrate at low temperatures and may contribute to repair mechanisms.

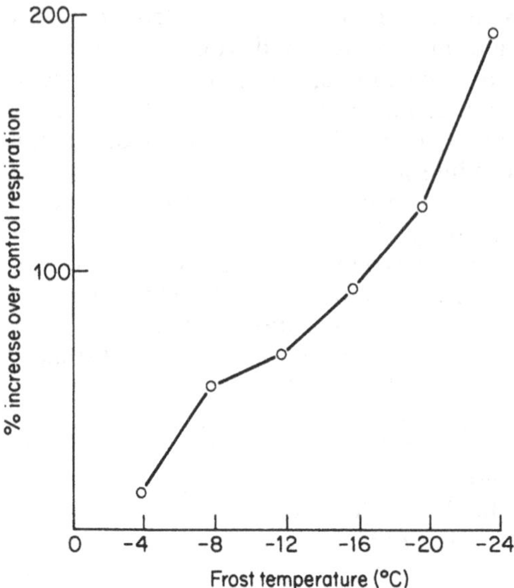

Fig. 8.3 Respiratory overshoot demonstrated by pine needles thawed to 20°C following freezing stress. This data is for winter needles. (Adapted from Bauer *et al.*, 1969.)

Assessment of freezing injury

Assay techniques that assess plant injury following freezing are of two types:

(i) those that assess injury and viability of the entire plant;
(ii) tests that are more appropriate to assessing function of an isolated organ or tissue piece.

The most meaningful assay of freezing injury in the intact plant is one that gives an estimate of survival capacity (i.e. the ability to grow following freezing stress). This means that assessment cannot be made until secondary injuries and any post-thaw infections have had their maximum effect, often several weeks after the initial freezing event. Although useful as a measure of freezing impact upon population survival (commonly as LD 50), a field regrowth assay provides little or no measure of specific tissue injury with respect to an individual plant. To provide information on the sites and extent of localized freezing injuries, more precise assay techniques are used. At the simplest level these rely upon visual assessment of waterlogging or necrosis in foliar organs following a freezing treatment (Chen *et al.*, 1976). Other simple tissue assays are based upon observation of osmometric function (plasmolysis) and the use of vital stains such as neutral red (Levitt, 1980).

The most widely used assays depend upon measurement of uncharacteristic leakage of ionic, intracellular components across the plasmalemma following freeze-related injury (Dexter, 1932; Steponkus, 1978). These assays commonly record the conductivity of a defined solution (often de-ionized water), in which the leaking tissues have been incubated, to quantify tissue injury. This type of information can be used to compile an index of freezing injury for comparative purposes (Sukumarar and Weiser, 1972). Leakage of other types of compounds such as amino acids can also be measured (Siminovitch *et al.*, 1964). Other parameters such as changes in xylem potential, chlorophyll fluorescence and tissue impedance to alternating current have also been used as the basis of a tissue assay (Brown *et al.*, 1977; Levitt, 1980). This group of assays for freezing injury rely upon altered plasmalemma properties being a primary event, but they may not be a particularly reliable indicator of whole plant viability, for injury can be localized (Olien and Smith, 1981a).

Biochemically-based assays will look at a single, specific function within the tissue (e.g. assessment of dehydrogenase activity by recording the reduction of triphenyl tetrazolium chloride to a coloured formazan – the TTC assay) in either a qualitative or quantitative way (Steponkus, 1978). The major disadvantage of this type of technique is that it measures the activity of a single enzyme in a single tissue type and cannot be taken as a reliable indicator of integrated metabolism. Importantly, dehydrogenase activity is often preserved below the lethal freezing temperature for the tissue, and thus the TTC assay is likely to provide an overestimate of viability. The breakdown of intracellular compartmentalization as a facet of freezing injury can disturb the normal enzyme/substrate interactions and may even result in an enhancement of activity of enzymes such as the dehydrogenases. An assay based on this enzyme activity is therefore a poor indicator of the state of integrated, cellular metabolism (Fennema, 1978).

Acclimatization and frost resistance

Resistance of higher plants to freezing stress has two components:

(i) avoidance of freezing;
(ii) tolerance of freezing within the tissue.

Each component may contribute to survival at times of environmental freezing stress, and they will be discussed separately.

Avoidance of freezing

When undercooled, aqueous plant solutions are in metastable equilibrium and may exist substantially below their melting points without ice nucleation. A metastable system is one which will undergo a spontaneous transition on addition of the stable phase (e.g. undercooled water nucleates if an ice crystal is added) (Taylor, this volume). For water, undercooling can extend to a homogeneous nucleation temperature of approximately −38°C (Rasmussen and MacKenzie, 1972), and for typical plant solutions undercooling can extend to −47°C (George *et al.*, 1982). Extracellular solutions will remain undercooled depending upon the effectiveness of available nucleators at the capillary space that contains the solutions. This means that ice nucleation in intact plants is most likely to occur in the larger cavities of the vascular system and then spread rapidly throughout the plant (Asahina, 1956; Burke *et al.*, 1976). The probability of ice nucleation and the rate of crystal growth increases with the extent of prior undercooling (Diller, 1975).

Undercooling of an entire plant as a way of avoiding freezing injury provides little resistance, for the majority of species show no significant undercooling (e.g. most agricultural plants do not undercool by more than 4–6°C) (Burke *et al.*, 1976; Levitt, 1980; Rajashekar, 1982). Under conditions of moderate, episodic frosts extremes of undercooling may, however, have some advantage in a limited range of species, for example wintergreen leaves do not freeze until −32°C and olive leaves do not freeze until −10°C (Levitt, 1980). Persistent undercooling of a restricted range of tissues within plants that also contain extracellular ice may provide a survival strategy against extremes of environmental low temperature. The plants relying upon such a strategy are cold-hardy species, native to temperate regions of North America and Asia, and they survive in regions where the minimum winter temperature is not likely to fall below −40°C, but may almost reach this level. The low temperatures achieved by the undercooled tissues have suggested a description of 'deep undercooling' for the phenomenon (George *et al.*, 1982).

In many woody species deep undercooling is restricted to living cells within the xylem tissues, especially parenchyma of the xylem rays (George *et al.*, 1982); extracellular freezing occurs in the remainder of the tissues. When environmental conditions are sufficiently severe to freeze these cells, ice formation is intracellular and potentially lethal. Although it comprises only a small proportion of the total vegetative tissue, the 'black heart' injury thus caused can be lethal to the whole plant. Deep undercooling may also occur in vegetative buds of a range of species (particularly conifers) and also in floral buds (Table 8.2). In each of these three systems (floral buds, vegetative buds and xylem), the undercooled tissues are effectively isolated from the ice in surrounding structures, so that they do not lose water by freeze-dehydration but retain their physiological hydration levels.

In those species where xylem rays undercool, the barrier to water transport is believed to be part of the cell wall structure, including the microcapillaries between the wall components that are small enough to inhibit the transgression of ice. The nature of specific wall polymers may also have a role (Olien, 1965; Burke *et al.*, 1976; Olien, 1981). In vegetative buds it is a pith cavity that exists between the bud's axis and the promordial shoot that prevents ice propagation. In species without this cavity there is no deep

Table 8.2 Examples of species which exhibit deep undercooling as an aspect of freezing resistance

Floral buds	Vegetative buds	Xylem
Rhododendron japonicum	*Larix leptolepis*	*Pyrus caucasia*
Vaccinium smallii	*Pseudotsuga menziesii*	*Malus pumila*
Cornus officinalis	*Abies homolepis*	*Quercus velutina*
Prunus nigra	*Abies vetchii*	*Carya ovata*

undercooling (Sakai, 1979; George *et al.*, 1982). In floral buds, the primordia also have barriers to ice to enable them to remain hydrated and undercooled (Quanne, 1978; George *et al.*, 1982). The cuticle of the primordia prevents nucleation by ice that they form on the surface of the bud scales, and the tissues at the base of the primordia inhibit ice movement as they are at a low water content. The micropore structure of the cell walls may also be important in this case.

Artificial techniques for freezing avoidance

There are circumstances where plants, particularly valuable agricultural/horticultural plants, without a mechanism of freezing resistance may be exposed to short, potentially lethal frosts (e.g. spring frosts and their effects on temperate fruit trees in flower). Artificial techniques for insulating these plants have been developed, which at the simplest level have involved air heating, air mixing and laying straw over ground level plants. More recently, a liquid foam that solidifies at reduced temperatures has been used to insulate both strawberry and citrus flowers (Bartholic, 1972). An alternative technique is to spray tissues with surface water, in the hope that when it freezes the exothermic heat and the ice layer will protect against further frost injury. This method is fraught with risk because surface ice may, in fact, nucleate the tissues if undercooling becomes significant. A typical example of this technique is the sprinkling with water of peach trees in flower, if a night frost is predicted, which can hold sprayed bud temperatures at -4°C (undercooled, but coated in a film of ice) whereas the air temperature and unsprinkled buds reach -8°C. The flower buds are likely to be frozen if they reach -8°C, and are lethally injured when thawed (Brazee *et al.*, 1984).

Tolerance of freezing

The freezing tolerance of a higher plant depends upon its genetic constitution, stage of development and response to environmental conditions. A tolerant plant will demonstrate resistance to the mechanical and osmotic stresses imposed by extracellular freezing (see above; Grout and Morris, this volume), and a low probability of intracellular freezing at typical environmental temperatures.

If a plant, or organ, is killed when ice forms in the tissues, it has no freezing tolerance, regardless of undercooling ability. Such an absence of tolerance may reflect the inherent genetic makeup of the plant or indicate inappropriate development or environmental conditions for the expression of tolerance.

The changes in environment that signal the onset of winter will induce, or enhance, tolerance in capable plants at the correct stage of development. The complex of processes involved is described as acclimatization or hardening, and has the effect of postponing lethal freezing events to lower temperatures. The details of acclimatization in specific plant systems have been described at the cellular and tissue levels earlier in this volume and in a number of recent reviews and edited volumes (Li and Sakai, 1978; Steponkus, 1978; Levitt,

1980; Olien and Smith, 1981c; Franks and Mathias, 1982; Li and Sakai, 1982; Steponkus *et al.*, 1983; Steponkus 1984). The aim of this discussion is not to précis this catalogue of events but to highlight the major adaptive mechanisms involved in terms of structure and function of the whole plant. Extensive detail will be found by reference to the literature cited above.

The changes that are made during acclimatization are dependent upon active photosynthetic and respiratory metabolism; the major adaptive changes must therefore take place before plant metabolism is drastically affected by reduced temperature. The environmental factors that regulate acclimatization include:

(i) light;
(ii) temperature;
(iii) available water.

Light will act as the energy source in accumulation of photosynthate and in the synthesis of ATP and NADPH via the photosynthetic electron transport chain. The provision of energy currency for synthesis and modification in adaptive metabolism is therefore assured. The central role of photosynthesis is indicated by studies in which germinating seeds acclimatize without light, as they have tissue-stored nutrient, and various plant tissues acquire hardiness by incubation in sugar solutions (Andrews, 1958). The relative partitioning of assimilate at reduced temperatures, typical of acclimatizing conditions, is described in Fig. 8.4. The reduction in total assimilate is a function of temperature, but the low temperature coefficient of the 'light' reactions will ensure that ATP and NADPH synthesis remain significantly closer to levels monitored at higher temperature (Levitt, 1980). In acclimatizing plants a high proportion of the stored carbohydrates are soluble compounds contributing to the increase in negativity of cytoplasmic osmotic potential (Siminovitch, 1981).

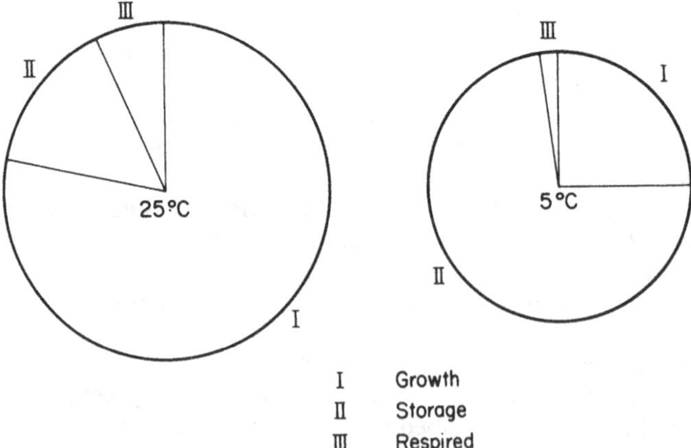

Fig. 8.4 The relative proportions of carbon assimilated at high and low temperatures for a typical higher plant (area of circles). The metabolic distribution of the assimilate is also indicated (Adapted from Levitt, 1980.)

There is an extensive, and often conflicting, literature on the role of photoperiod in acclimatization. A much reduced day length need not be critical *per se*, for example in Alaska native plants acclimatize under a regimen of 15 hours daylight reducing to 11 hours (Steponkus, 1978). Changing photoperiod is, however, closely associated with plant hormonal control and direction of metabolism and differentiation, and so may be important in bringing a plant, developmentally, to a stage at which acclimatization events can occur.

Reduction in temperature may be the most important single factor in acclimatization. Temperatures below 10°C are necessary and there may be a sequence of hardening events relating to temperatures below this. The final stages will require freezing temperatures to promote maximum acclimatization of the plants. Tolerance of freezing will respond to temperature throughout the winter, increasing as mean daily temperature falls, and decreasing as it rises. A typical pattern for temperature-related acclimatization is shown in Fig. 8.5.

Reduction in water content is also commonly associated with an increase in hardiness, and a degree of water stress tolerated by a plant during acclimatization will probably enhance maximum freezing tolerance. Freeze-desiccation, once ice has formed within the tissues, may also improve the final extent of freezing-tolerance.

In summary, as acclimatization begins a reduction of temperature below a critical level, changes in light intensity and photoperiod and a degree of water stress will conspire to reduce growth rate, alter gene expression (via growth-regulated activity) and shift the emphasis of metabolism. This shift will result in the changes at the cellular level that are crucial to effective acclimatization.

If tolerance of ice within tissues is to be enhanced, the mechanical injuries associated with freeze-induced protoplast shrinkage must be postponed to

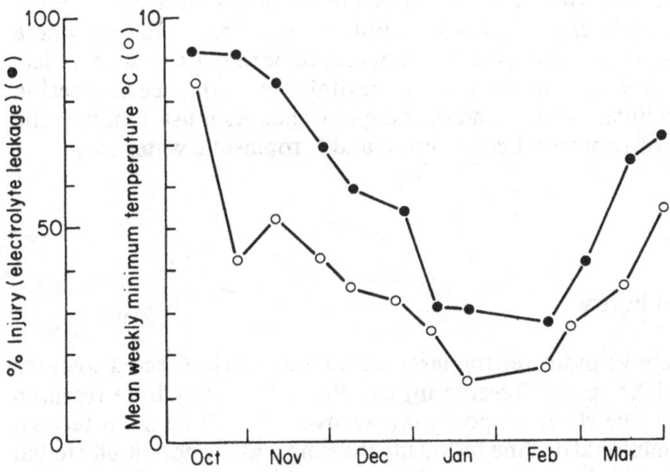

Fig. 8.5 The hardening pattern of overwintering cauliflower leaves grown in the extreme south-west of England. Weekly minimum temperatures are also indicated.

lower temperatures, below those typical of the particular plant environment. Similarly, any injurious effects related to a more concentrated cytoplasm must be postponed and intracellular nucleation effectively prevented.

Osmotic adjustment is a typical consequence of acclimatization; a doubling of cytoplasmic solute concentration is not uncommon. This increase, which is at least in part the result of photosynthate accumulation, would decrease the amount of water lost by freeze-dehydration at a given temperature by 50 per cent. As less water is lost from the protoplast in reaching thermodynamic equilibrium under such conditions, the contraction of the protoplast will also be reduced. This, in turn, will mean that contraction-induced deletion of plamalemma materials (and possibly from those membranes limiting organelles) will be reduced. The proportion of irreversible deletions will also be reduced by adaptive changes in the structure of the plasmalemma of acclimatized cells (Grout and Morris, this volume; Steponkus, 1984).

Toxic effects may result from the concentration of specific cytoplasmic components during freeze-dehydration, resulting in injury. In such instances there may be some compensating, protective effect of other compounds acting in a colligative fashion to dilute the toxins. Specific compounds may also have protective effects such as sugars (particularly trehalose), stabilizing membranes and proline and betaine, stabilizing proteins.

The probability of intracellular ice formation following extracellular freezing will also be reduced in acclimated plants, as an increase in soluble, cytoplasmic solutes (e.g. sugars, organic acids and amino acids) will depress the intracellular freezing point. This probability will also depend upon there being no disadvantageous alteration in intracellular nucleator properties and the effectiveness of the plasmalemma as a barrier to ice propagation being maintained.

In the simplest terms, therefore, the mechanisms underlying improved freezing tolerance following acclimatization are cellular, adaptive changes to minimize effects of freeze-dehydration upon the protoplast. Except where there are barriers to water movement, as appears to be the case in deep undercooling, some degree of desiccation is an inevitable consequence of freezing within the extracellular matrix, and the adaptive changes must minimize the injurious effects of protoplast contraction and cytoplasmic water loss.

Recovery from injury

The discussion above indicates the large volume of work directed towards understanding higher plant freezing injury. Proportionately little research has been done on the study of post-thaw recovery (in whole plant terms), structural and functional decline following thawing and the potentially lethal effects of secondary injury. This imbalance may give cause for concern, for in many situations the primary freezing injuries of the plant may be relatively slight and ostensibly repairable until secondary events, such as microbial infection, impose lethal stresses. Recovery from freezing injury is thus at two levels.

(i) Repair of intracellular structures (e.g. membranes) and the restoration of the normal metabolic balance. The injured cell that undergoes such repair may be adversely affected by extracellular events (e.g. toxin secretion from degenerating cells or pervasive micro-organisms). Survival may therefore depend more on the post-thaw circumstances than on either the extent of freezing injury or the cellular capacity for repair.

(ii) Restoration of full function to tissues which will involve some replacement of cells that have been killed by either primary freezing events or other, secondary stresses and will rely upon the activity of cambial and terminal meristems. Post-thaw events such as spread of infection from freeze-damaged to healthy tissues such as meristems will put recovery at risk and may threaten plant survival. The extent of the original freezing injury *per se* may not be lethal, and it is the nature of secondary stresses that are of importance.

It is clear that the spread of pathogens, originating in freeze-injured tissues, can be a major problem. These organisms may enter the plant through external pores (stomata, lenticels, lesions caused by freezing) or may be systemic infections. The populations may divide rapidly, promoted initially by lytic products from dead and injured cells in many instances, and may spread to uninjured tissues with serious effect. This occurs particularly where transport is via the vascular tissue, and distant meristems, essential to whole plant survival, can be put at risk (Smith and Olien, 1978; Olien and Smith, 1981b).

Non-aggressive pathogens, that are not seriously damaging to a particular species (Schoeneweiss, 1973), may behave differently when incubating in proximity to freeze-injured tissues, and may inflict serious injury, for example some *Fusarium* infections of barley (Smith and Olien, 1981b) and the spread of *Pseudomonas* in peaches (Chandler and Daniel, 1974). Microbial activity may also generate toxic compounds from endogenous plant material (e.g. amygdalin and emulsin released from peach bark tissue can be converted to hydrogen cyanide and benzaldehyde respectively).

Autotoxicity may also have serious post-thaw effects when cell and tissue compartmentalization may have broken down as a result of freezing. Unusual combinations and concentrations of enzyme and substrate, and other reactive materials, may give rise to novel, toxic products (e.g. oxidized phenols). Similarly, toxic levels of a normally innocuous cell constituent may be generated. The release of lytic compounds from lysosomes is obviously a major contributor to this effect.

Once freezing injury has occurred, there are few management techniques available to limit secondary injury and enhance repair. In one case, however, the application of plant growth regulators, particularly auxins, as a spray has been employed to limit crop losses on frosted fruit trees. It is believed that the exogenous regulators compensate for imbalance due to death and injury of developing seeds (Howell and Dennis, 1981).

References

Andrews, J.E. (1958). Controlled low temperature tests of sprouted seeds as a measure of cold hardiness of winter wheat varieties. *Canadian Journal of Plant Science* **38**, 1–7.

Arny, D.C., Lindow, S.E. and Upper, C.D. (1976). Frost sensitivity of *Zea Mays* increased by application of *Pseudomonas syringae. Nature* **262**, 282–4.

Asahina, E. (1956). The freezing process of plant cells. *Contrib. Inst. Low Temperature Science (Hokkaido Univ.)* **10**, 83–126.

Bartholic, J.F. (1972). Thin layer foam for plant freeze protection. *Proceedings of the Florida State Horticultural Society* **85**, 299–302.

Brazee, R.D., Fox, R.D., Ferree, D.C. and Cahoon, C.A. (1984). Advective and radiative frost control with irrigation in peaches. *Ohio State University Agric. Res. Circular* **284**, 54–62.

Brown, G.N., Bixby, J.A., Melcarek, P.K., Hinckley, T.M. and Rogers, R. (1977). Xylem pressure potential and chlorophyll fluorescence as indicators of freezing survival in black locust and western hemlock seedlings. *Cryobiology* **14**, 94–9.

Burke, M.J., Gusta, L.V., Quamme, H.A., Weiser, C.J. and Li, P.H. (1976). Freezing and injury in plants. *Ann. Rev. Plant. Physiol* **27**, 507–28.

Chandler, W.A. and Daniel, J.W. (1974). Effect of leachates from peach soil and roots on bacterial canker and growth of peach seedlings. *Phytopathology* **64**, 1281–84.

Chen, P.M., Burke, M.J. and Li, P.H. (1976). The frost hardness of several Solanum species in relation to the freezing of water, melting point depression and tissue water content. *Botanical Gazette* **137**, 313–17.

Dexter, S.T. (1932). Studies of the hardiness of plants: a modification of the Newton pressure method for small samples. *Plant Physiology* **7**, 721–6.

Diller, K.R. (1975). Intracellular freezing: effect of extracellular supercooling. *Cryobiology* **12**, 480–85.

Fennema, O. (1978). Enzyme kinetics at low temperature and reduced water activity. In *Dry Biological Systems*, pp. 297–322. Edited by Crowe, J.H. and Clegg, J.S. Academic Press, London.

Franks, M. and Mathias, S. (1982). *Biophysics of water* (eds.) pp. 400, John Wiley, New York.

George, M.F., Becwal, M.R. and Burke, M.J. (1982). Freezing avoidance by deep undercooling of tissue water in winter-hardy plants. *Cryobiology* **19**, 628–39.

George, M.F. and Burke, M.J. (1976). The occurrence of deep supercooling in cold hardy plants. *Current Advances in Plant Science* **22**, 349–60.

Graham, P.R. and Mullin, R. (1976). A study of flower bud hardiness in azalea. *Journal of the American Society of Horticultural Science* **101**, 4–7.

Grout, B.W.W. and Crisp, P.C. (1985). Germination as an unreliable indicator of the effectiveness of cryopreservative procedures for imbibed seeds. *Ann. Bot.* **55**, 289–292.

Heber, U.W. and Santarius, K.A. (1964). Loss of adenosine triphosphate synthesis caused by freezing and its relationship to frost hardiness problems. *Plant Physiol* **39**, 712–19.

Heber, U.W., Schmitt, J.M. Krause, G.H., Klosson, R.J. and Santarius, K.A. (1981). Freezing damage to the thylakoid membranes *in vitro* and *in vivo*. In *Effects of Low Temperature on Biological Membranes*, pp. 263–84. Edited by Morris, G.J. and Clarke, A.C. Academic Press, London.

Howell, G.S. and Dennis, F.G. (1981). Cultural management of perennial plants to maximise resistance to cold stress. In *Analysis and Improvement of Plant Cold Hardiness*, pp. 175–204. Edited by Olien, C.R. and Smith, M.N. CRC Press, Boca Raton, Florida.

Hudson, M.H. and Idle, D.B. (1962). The formation of ice in plant tissues. *Planta* **57**, 718–30.

Idle, D.B. (1966). The photography of ice formation in plant tissues. *Ann. Bot.* **30**, 199–206.

Kaiser, W.M. (1982). Correlation between changes in photosynthetic activity and changes in total protoplast volume in leaf tissue from hygro-, meso- and xerophytes under osmotic stress. *Planta* **154**, 538–45.

Kaiser, W.M. and Heber, U. (1981). Photosynthesis under osmotic stress. Effect of high solute concentrations on the permeability properties of the chloroplast envelope and on activity of stroma enzymes. *Planta* **153**, 423–29.

Kaku, S. (1971). A possible role of the endodermis as a barrier for ice propagation in the freezing of pine needles. *Plant and Cell Physiology* **12**, 941–8.

Kaku, S. (1973). High ice nucleating ability in plant leaves. *Plant and Cell Physiology* **14**, 1035–38.

Kaku, S. (1975). Analysis of freezing temperature distribution and plants. *Cryobiology* **12**, 154–9.

Kaku, S. and Salt, R.W. (1968). Relation between freezing temperature and length of conifer needles. *Canadian Journal of Botany* **46**, 1211–13.

Larcher, W. (1981). Effects of low temperature stress and frost injury on productivity. In *Physiological Processes Limiting Plant Productivity*, pp. 253–69. Edited by Johnson, C.B. Butterworths, London.

Larcher, W. and Bauer, H. (1981). Ecological significance of resistance to low temperature. In *Physiological Plant Ecology I*, pp. 403–38. Edited by Lauge, O.L., Nobel, P.S., Osmons, C.B. and Ziegler, H. Springer Verlag, Berlin.

Larcher, W., Heber, U. and Santarius, K. (1973). Limiting temperatures for life functions. In *Temperature and Life*, pp. 195–263. Edited by Precht, H., Christopherson, J., Hensel, H. and Larcher, W. Springer Verlag, Berlin.

Levitt, J. (1980). *Responses of Plants to Environmental Stresses. Vol. I. Chilling, freezing and high temperature stresses*, pp. 497. Academic Press, London.

Li, P.H. and Sakai, A. (1978). *Plant Cold Hardiness and Freezing Stress:*

Mechanisms and crop implications Vol. I. (eds.), pp. 416, Academic Press, London.

Li, P.H. and Sakai, A. (1982). *Plant Cold Hardiness and Freezing Stress: Mechanisms and crop implications Vol. II.* (eds.), pp. 694, Academic Press, London.

Lindow, S.G. (1982). Population dynamics of epiphytic ice nucleation active bacteria on frost sensitive plants and frost control by means of antagonistic bacteria. In *Plant Cold Hardiness and Freezing Stress; Mechanisms and crop implications Vol. 2*, pp. 395–416. Edited by Li, P.H. and Sakai, A. Academic Press, London.

Lindow, S.G. (1983). The role of bacterial ice nucleation in frost injury to plants. *Annual Review of Phytopathology* **21**, 363–84.

Mazur, P. (1977). The role of intracellular freezing in the death of cells cooled at supraoptimal rates. *Cryobiology* **14**, 251–72.

Modlibowska, I. (1946). Frost injury to apples. *Journal of Pomology* **22**, 46–51.

Modlibowska, I. (1968). Effects of some growth regulators on frost damage. *Cryobiology* **5**, 175–87.

Modlibowska, I. and Field, C.J. (1942). Winter injury to fruit trees by frost in England, 1939–1940. *Journal of Pomology* **19**, 197–204.

Mollenhauler, A., Schmitt, J.M., Coughlan, S. and Heber, U. (1983). Loss of membrane proteins from thylakoids during freezing. *Biochim. Biophys. Acta* **728**, 331–38.

Morris, G.J., Coulson, G.E., Clarke, K.J., Grout, B.W.W. and Clarke, A.C. (1981). Freezing injury in *Chlamydomonas*: a synoptic approach. In *Effects of Low Temperature on Biological Membranes*, pp. 285–306. Edited by Morris, G.J. and Clarke, A.C. Academic Press, London.

North, J. and Fisher, T.C. (1971). Anatomical study of freezing injury in hardy and non-hardy alfalfa varieties treated with cytosine and quanine. *Cryobiology* **8**, 420–30.

Olien, C.R. (1965). Interference of cereal polymers and related compounds with freezing. *Cryobiology* **2**, 47–54.

Olien, C.R. (1971). Freezing stress in barley. In *Barley Genetics II*, pp. 356–63. Edited by Nilan, R.A. Washington State University Press, Washington.

Olien, C.R. (1981). Analysis of midwinter freezing stress. In *Analysis and Improvement of Plant Cold Hardiness*, pp. 35–39. Edited by Olien, C.R. and Smith, M.N. CRC Press, Boca Raton (USA).

Olien, C.R. and Smith, M.N. (1981a). Protective systems that have evolved in plants. In *Analysis and Improvement of Plant Cold Hardiness*, pp. 61–87. Edited by Olien, C.R. and Smith, M.N. CRC Press, Boca Raton, (USA).

Olien, C.R. and Smith, M.N. (1981b). Extension of localized freezing injury in barley by acute post-thaw bacterial disease. *Cryobiology* **18**, 404–9.

Olien, C.R. and Smith, M.N. (1981c). Analysis and improvement of plant cold hardiness (eds.), pp. 248, CRC Press, Boca Raton (USA).

Palta, J.P. and Li, P.H. (1978). Cell membrane properties in relation to freezing injury. In *Plant Cold Hardiness and Freezing Stress*, pp. 93–116. Edited by Li, P.H. and Sakai, A. Academic Press, London.

Peoples, T.R. and Koch, D.W. (1978). Physiological response of the alfalfa cultivars to one chilling night. *Crop Science* **18**, 255–58.

Pisek, A., Larcher, W., Vegis, A. and Napp-Zinn, K. (1973). The normal temperature range. In *Temperature and Life*, pp. 102–194. Edited by Precht, H., Christopherson, J., Hensell, H. and Larcher, W. Springer Verlag, Berlin.

Quanne, H.A. (1978). Mechanism of supercoding in overwintering peach flower buds. *Journal of the American Society of Horticultural Science* **103**, 57–61.

Rajashekar, C., Li, P.H. and Carter, J.V. (1982). Frost injury in supercooled leaves of tuber-bearing *solanum* species. In *Plant Cold Hardiness and Freezing Stress*: Mechanisms and crop implications Vol.2, pp. 379–94. Edited by Li, P.H. and Sakai, A. Academic Press, London.

Rasmussen, D.H. and MacKenzie, A.P. (1972). Effect of solute on ice-solution interfacial free energy: calculations from measured homogenous nucleation temperatures. In *Water Structure at the Water-Polymer Interface*, pp. 126–45. Edited by Jellinek, H.H.G. Plenum Press, New York.

Roberts, E.H. (1972). Storage environment and the control of viability. In *Viability of Seeds*, pp. 14–58. Edited by Roberts, E.H. Chapman and Hall, London.

Rutheford, P.P. and Whittle, R. (1981). Changes in sucrose levels as an indicator of the marketable quality of cauliflowers exposed to freezing temperatures. *Scientia Hort.* **14**, 117–23.

Sakai, A. (1979). Freezing avoidance mechanism of primordial shoots of conifer buds. *Plant Cell Physiology* **20**, 1381–90.

Schmitt, J.G., Schramm, M.J., Hardy, P., Coughlan, S. and Heber, U. (1985). Damage to chloroplast membranes during dehydration and freezing. *Cryobiology* **22**, 93–104.

Schnell, R.C. and Varlin, G. (1972). Atmospheric ice nuclei from decomposing vegetation. *Nature* **236**, 163–5.

Schoeneweiss, D.F. (1973). Predisposition, stress and plant disease. *Annual Review of Phytopathology* **13**, 193–211.

Shearman, L.L., Olien, C.R., Marchetti, B.L. and Everon, E.H. (1973). Characterization of freezing inhibitors from winter wheat cultivars. *Crop Science* **13**, 514–19.

Siminovitch, D. and Cloutier, Y. (1981). Drought and freezing tolerance and adaptation in plants: some evidence of near equivalences. *Cryobiology* **20**, 487–503.

Siminovitch, D., Therrier, H., Gfeller, F. and Rheaume, B. (1964). *Canadian Journal of Botany* **42**, 637–49.

Single, W.V. and Marcellos, H. (1981). Ice formation and freezing injury in actively growing cereals. In *Analysis and Improvement of Plant Cold Hardiness*, pp. 17–33. Edited by Olien, C.R. and Smith, M.N. CRC Press, Boca Raton, (USA).

Smith, M.N. and Olien, C.R. (1978). Pathological factors affecting survival of winter barley following controlled freeze tests. *Phytopathology* **68**, 773–7.

Smith, M.N. and Olien, C.R. (1981). Recovery from winter injury. In *Analysis and Improvement of Plant Cold Hardiness*, pp. 117–38. Edited by Olien, C.R. and Smith, M.N. CRC Press, Boca Raton, (USA).

Stanwood, P.C. and Bass, L.N. (1978). Ultracold preservation of seed germplasm. In *Plant Cold Hardiness and Freezing Stress*, pp. 361–72. Edited by Li, P.H. and Sakai, A. Academic Press, London.

Steponkus, P.L. (1978). Cold hardiness and freezing injury of agronomic crops. *Adv. Agron.* **30**, 51–98.

Steponkus, P.L. (1984). Role of the plasma membrane in freezing injury and cold acclimation. *Annual Review of Plant Physiology* **35**, 543–84.

Steponkus, P.L., Evans, R.Y. and Singh, J. (1982). Cryomicroscopy of isolated rye mesophyll cells. *Cryoletters* **3**, 101–14.

Steponkus, P.L., Dowgert, M.F. and Gordon-Kamm, W.J. (1983). Destabilization of the plasma membrane of isolated plant protoplasts during a freeze: thaw cycle - the influence of cold acclimation. *Cryobiology* **20**, 448–65.

Sykumaran, N.P. and Weiser, C.J. (1972). An excised leaflet test for evaluating potato frost tolerance. *Horticultural Science* **7**, 464–68.

Tao, D., Li, P.H. and Carter, J.V. (1983). Role of cell walls in freezing tolerance of cultured potato cells and their protoplasts. *Physiol. Plant.* **58**, 527–32.

Tranquillini, W. (1982). Frost-drought and its ecological significance. In *Physiological Plant Ecology II*, pp. 379–400. Edited by Large, O.L., Nobel, P.S., Osmond, C.B. and Zeigler, H. Springer Verlag, Berlin.

Ungerson, J. and Scherding, G. (1965). Unter suchungenuber photosynthese and atmung unter naturlichen bedingungen wahrend des Winter halbjahres bei *Pinus sylvestris.* L., *Picea excelsis* Link and *Juniperus communis* L. *Planta* **67**, 136–7.

9

The adaptation of aquatic animals to low temperatures

A. Clarke
British Antarctic Survey NERC
Cambridge

Introduction

The temperature of seawater varies from almost −2°C in polar waters to some 30°C in some tropical surface waters. Very hot water can be found close to hydrothermal vents, and bacterial growth has been reported from waters at 250°C (Baross and Deming, 1983, but see Trent *et al.*, 1984); but in general marine organisms are faced with a maximum range of just over 30°C. Even in the tropics, however, warm temperatures are limited to well mixed surface waters. Seawater temperature falls rapidly through the thermocline (which

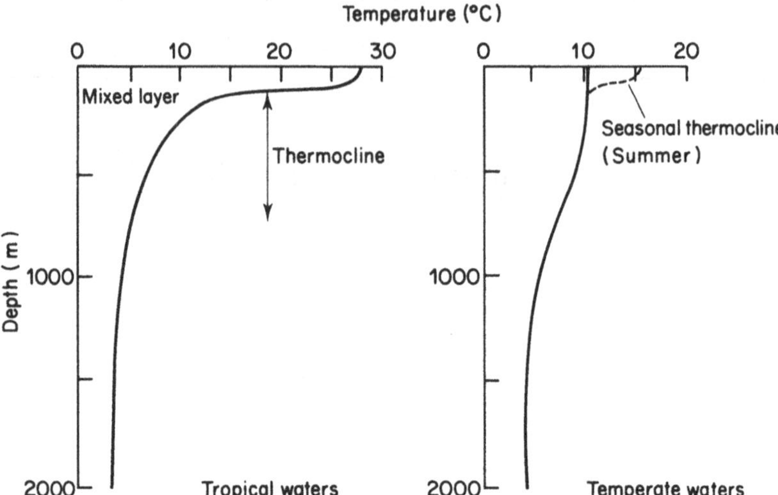

Fig. 9.1 Typical mean temperature/depth profiles for tropical and mid-latitude waters. There is a pronounced seasonal thermocline in temperate latitudes. At very great depths (> 4000 m) there is a slight increase in *in situ* temperature caused by the increasing pressure.

typically starts at approximately 100–200 metres) below which seawater is generally 5°C or colder (Fig. 9.1), and it has been calculated that over 90 per cent of the global seawater is below 5°C (Morita, 1968). For oceanic animals cold water is therefore the norm, but these cold waters are rich in life. All known marine phyla are represented, and in some Antarctic shelf communities, biomass and diversity are as great as anywhere else in the world (Richardson and Hedgpeth, 1977). Physiological adaptation to low temperature would not, therefore, seem to have been an insurmountable evolutionary problem.

In shallow seas, particularly those in temperate latitudes, there is no stable thermocline and surface temperatures extend right to the sea-bed. These areas frequently exhibit a strong seasonal variation in temperature (Fig. 9.2). The physiological problems which must be overcome under these conditions are very different from those in stable, cold environments. Nevertheless these seas too are rich in life and traditionally have supported diverse fisheries.

Any consideration of adaptation to temperature by marine organisms must take into account these differing patterns of temperature variation. Since many other ecological factors vary, as well as temperature, care must also be taken to distinguish adaptation to temperature *per se* from adaptation to other parameters (e.g. light, food availability or predation).

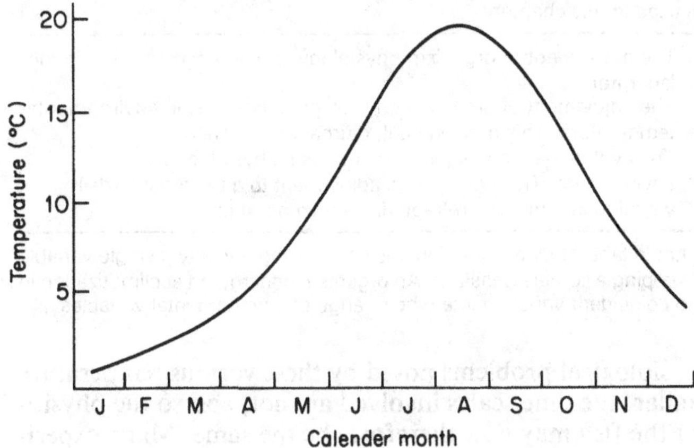

Fig. 9.2 The annual cycle of seawater temperature for a typical temperate water site.

Adaptation, acclimation and acclimatization

An excellent example of physiological problems posed by differing water temperatures is the maintenance of locomotor ability in fish.

Fish living in seas close to the Antarctic continent, which are at −1.9°C throughout the year, must adjust the structure and function of their muscle to swim effectively at low temperatures. As they do not have to contend with seasonal or diurnal changes in temperature, the fish have evolved a locomotor system which is finely tuned to low temperature and, if this adaptation means that survival at slightly higher temperatures is not possible, then this is of little consequence (at least in the short term) with respect to survival of the species.

Fish living in shallow temperate seas face a different problem. They must have a locomotor system capable of operating effectively at low temperatures in winter and at high temperatures in summer. In theory at least, two solutions are feasible. One is for organisms to have eurythermal, highly flexible systems capable of activity both at high and low temperatures, the other would involve some form of seasonal switching from a low temperature system to a high temperature and back again.

Many open-ocean fish undertake vertical migrations, often of several hundred metres, in search of food. These migrations may be daily, and can carry the fish through the thermocline; these fish may thus experience changes in temperature of up to 10°C every day. An even more rapid temperature change is experienced by the estuarine killifish *Fundulus heteroclitus* from the coastal salt marshes of the eastern USA. These fish defend territories in the salt marsh creeks and, when the tide ebbs, water temperatures may exceed 30°C. As the tide floods, temperatures may fall by almost 15°C in less than 1 hour and yet killifish maintain their ability to swim against the inrushing tide (Sidell, *et al.*, 1983).

Table 9.1 Definitions used in this chapter

Acclimation	The adjustment of organism physiology to a new temperature in the laboratory.
Acclimatization*	The adjustment of organism physiology to changes in environmental temperature. This may be tidal, diurnal or seasonal.
Adaptation	The evolutionary adjustment of organism physiology to environment. This can include adjustment to a seasonal or daily variation in temperature requiring acclimatization.

* It should be noted that in laboratory acclimation it is usual to modify only a single variable (e.g. temperature), keeping all others constant. An organism undergoing acclimatization in the field is subject to coincident variation in a whole range of environmental variables.

Although the physiological problems posed by these various temperature changes may be similar, the time scales involved are not, and so the physiological responses of the fish may not, therefore, be the same. Much experimental work in the laboratory has been concerned with eurythermal fish, particularly goldfish *Carassius auratus* and sunfish *Lepomis cyanellus*. The ways in which these eurythermal freshwater fish cope with experimental changes in temperature are not necessarily the same as those which species have evolved to live at sub-zero temperatures, in hot lakes, to migrate through the thermocline, or to cope with the flooding tide. To distinguish the major types of response to temperature change the terms adaptation, acclimation and acclimatization will be used; these terms are defined in Table 9.1.

Compensation

The idea of compensation is central to any consideration of the way organisms adjust to temperature. The need for compensation in locomotor ability is obvious; polar, temperate and tropical organisms must all be able to catch prey, find mates and evade predators. Even casual observation indicates that some degree of compensation has evolved. At first sight, however, processes such as growth or respiration do not appear to have compensated for the low temperature, for polar organisms generally grow very slowly and have low rates of oxygen consumption compared with related species from warmer water. Many explanations have been suggested for this apparent difference, mostly based on the supposed rate-limiting effect of low temperatures on growth and metabolism. Why organisms should be able to evolve compensation for temperature in one aspect of their physiology (locomotion) but not in others (growth, metabolic rate) was not usually made clear. However, this dichotomy was apparent to Fox (1936) and was critical to the development of the concept of metabolic cold adaptation (see below).

Many studies of temperature compensation have involved the measurement of oxygen consumption and have revealed a bewildering array of responses. Two major schemes have been developed to describe these various types of response, those of Prosser (1973) and Precht *et al.*, (1955). Prosser's classification involves the comparison of the rate/temperature (R/T) curves

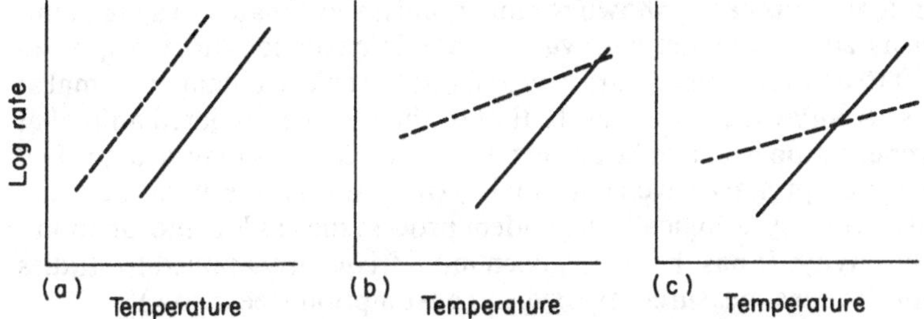

Fig. 9.3 The major types of temperature acclimation as described by Prosser (1973). (a) Translation of the rate/temperature (R/T) curve. (b) Rotation of the R/T curve. (c) Both translation and rotation. Note that the rate axis is logarithmic and so the R/T curve is essentially straight for much of its length.
--- = cold acclimated animal. —— = warm acclimated animal.

for organisms acclimated to high and low temperature (Fig. 9.3). If the R/T curves of cold and warm acclimated animals are similar, no compensation has occurred. If the curves are parallel, but at any given environmental temperature the cold acclimated animal has a higher oxygen consumption, there has been a translation of the R/T curve (Fig.9.3a). If slopes differ, the R/T curve has undergone rotation (Fig. 9.3b). It is also possible for translation and rotation to occur together (Fig. 9.3c).

Prosser's scheme involves comparison of complete R/T curves; the scheme of Precht *et al.*, (1955) may involve only two observations (Fig. 9.4). The rate

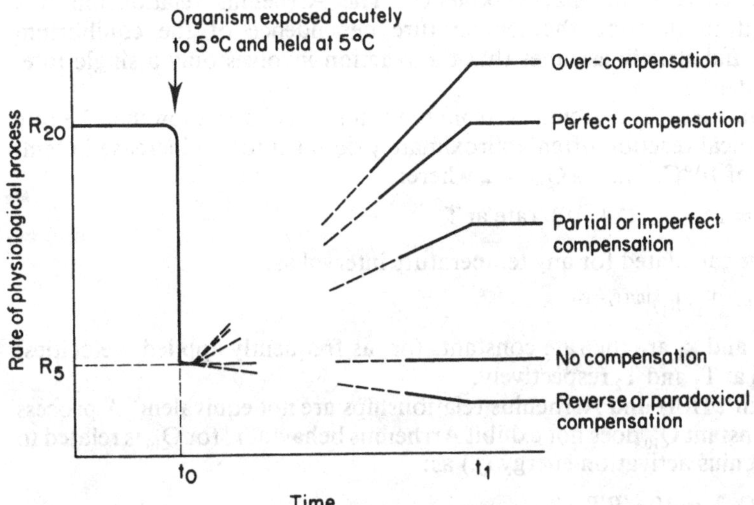

Fig. 9.4 An idealized time course of the change in the rate of a physiological process of a hypothetical eurythermal organism acclimated to 20 °C and then exposed to 5 °C. R_{20} is the rate at 20 °C and R_5 is the rate immediately on exposure to 5 °C at t_0. By time t_1 the rate has stabilized and the response may be classified by the scheme devised by Precht *et al.* (1955).

of a physiological process is shown for an organism acclimated to a high temperature (say 20°C) together with various possible compensatory responses after a shift to a lower temperature (say 5°C). If the cold acclimated animal at 5°C shows activity which is similar to that of the warm acclimated animal at 20°C, compensation is said to be perfect. Other stable levels may be described as partial, superoptimal or even paradoxical (inverse). This terminology may be applied to any physiologically dependent process such as locomotor ability or enzyme activity. It has, however, been most frequently applied to studies of metabolic rate as measured by oxygen consumption (see below).

Temperature and biological processes

The two mathematical approaches which have been used most frequently to describe the relationship between the rate of a biological process and temperature are the Arrhenius relationship and the van't Hoff rule. Both approaches were developed for simple chemical systems but have been widely applied to biology.

The Arrhenius relationship describes the relationship between reaction rate (ν) and absolute temperature (T):

$$\nu = A \exp(-\mu/RT)$$

where R is the gas constant, A is a constant and μ (or E_a) is the Arrhenius activation energy. A plot of $\ln \nu$ against T^{-1} is therefore linear, with a slope of $-\mu/R$ (or μ/kB, where B is Boltzmann's constant and k a further constant). This is an Arrhenius plot and it is a standard way of estimating activation free energies in enzyme kinetics. The Arrhenius relationship was developed to describe the temperature dependence of the equilibrium constant and strictly assumes that the reaction involves only a single rate-limiting step.

The van't Hoff rule stemmed from the empirical observation that the rate of a chemical reaction often approximately doubled for an increase in temperature of 10°C. That is $Q_{10} = 2$ where:

$$Q_{10} = \text{rate at } (T + 10)/\text{rate at } T$$

Q_{10} can be calculated for any temperature interval as:

$$Q_{10} = (k_2/k_1)^{10/(T_2-T_1)}$$

where k_2 and k_1 are the rate constants (or, as frequently applied, velocities) observed at T_2 and T_2 respectively.

The van't Hoff and Arrhenius relationships are not equivalent. A process with a constant Q_{10} does not exhibit Arrhenius behaviour, for Q_{10} is related to the Arrhenius activation energy (μ) as:

$$\ln(Q_{10}) = 10\mu/RT_1T_2$$

In strict terms application of either of these expressions to biological systems depends on the assumption that nothing in the system under investigation changes, apart from temperature. This assumption is often invalid both for *in vitro* systems and when dealing with whole organism physiology. In the latter case alteration of temperature (e.g. during an investigation of respiratory physiology) triggers a whole battery of physiological, biochemical and even behavioural responses. To treat the resultant data on the assumption that the organism is the same at different temperatures may be very misleading. Nevertheless, the Arrhenius plot and the Q_{10} calculation remain firmly entrenched in the literature of temperature physiology.

Temperature and enzyme activity

Temperature can affect enzyme activity in two distinct but interrelated ways. Firstly, there is a direct effect of temperature on protein structure (Morris and Clarke, this volume). Secondly, there is the influence of temperature on the rate of enzyme catalysed reactions.

Consider a simple isolated system catalysed by a single soluble enzyme. The fraction of total substrate molecules having a certain energy at any given temperature can be calculated from the Maxwell–Boltzmann equation (Pauling, 1970). Energy distribution curves are shown for two temperatures (Fig. 9.5.) If E_a is the Arrhenius activation energy (strictly related only to the enthalpy of activation, $\Delta H^{\ddagger}$) required for the formation of an activated enzyme–substrate complex, then reduction of the temperature from T_2 to T_1 will result in a large decrease in the proportion of reactive substrate molecules (represented by the shaded areas in Fig. 9.5). In other words the reaction rate will slow down greatly. This conceptual approach explains why a temperature drop of 10°C which decreases mean substrate kinetic energy by only about 3 per cent can halve the rate of reaction.

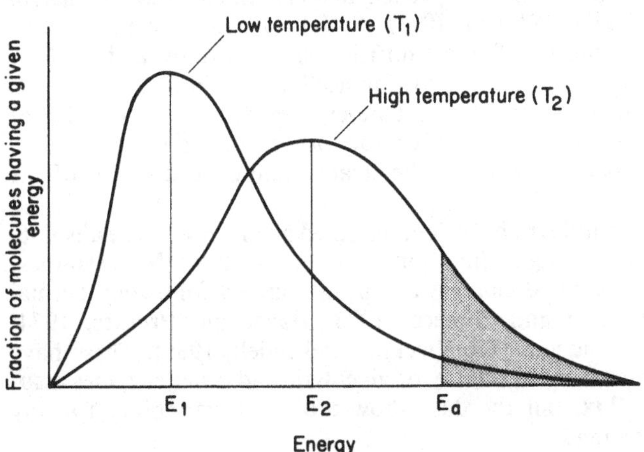

Fig. 9.5 Energy distribution curves for populations of substrate molecules at two temperatures, T_1 and T_2. The form of these curves is given by the Maxwell–Boltzmann distribution. E_a is the activation energy; only molecules which have an energy equal to or greater than E_a are reactive. These molecules are indicated by shading. It can be seen that a greater proportion of substrate molecules are reactive at the higher temperature.

Such sensitivity of enzyme catalysed reactions to changes in temperature presents organisms with a potentially severe physiological problem. The major adaptive strategies which might counteract the effect of temperature are:

1. an increase in the amount of active enzyme at low temperatures;
2. production of new enzyme variants which are better able to function at low temperatures;
3. alteration of the immediate environment of the enzyme to minimize the direct thermodynamic effect of temperature.

Evidence from aquatic organisms suggests that all three strategies may be employed, depending on circumstances; these strategies are considered below.

Increase in enzyme concentration

Increasing the amount of enzyme in the cell will increase the rate of reaction for a given substrate concentration (Law of Mass Action) whilst preserving the critical relationship between K_m and substrate concentration (Sidell, 1983). Disadvantages of this strategy are that there is a limit to the solvent capacity of the cell and that a time delay of hours, days or even weeks may be involved.

In only two cases has it proved feasible to demonstrate an increase in the absolute amount of enzyme following acclimation to a low temperature; these are cytochrome oxidase in goldfish muscle and cytochrome c in green sunfish skeletal muscle.

In goldfish, *Carassius auratus*, the cytochrome oxidase content of skeletal muscle was found to be greater by 66 per cent in fish acclimated to 5°C than in those acclimated to 25°C (Wilson, 1973). An increase in the cytochrome c content of the skeletal muscle of green sunfish, *Lepomis cyanellus*, from 0.98 to 1.51 nmol g^{-1} (wet wt) occurs following acclimation from 25°C to 5°C (Sidell, 1977). Transfer to the lower temperature decreased the rate of synthesis by 40 per cent but the rate of degradation decreased by 66 per cent. As a result there was an increase in the steady state concentration of the enzyme.

These are the only studies to have demonstrated an increase in the absolute amount of enzyme following acclimation to low temperature. Many studies, however, have demonstrated changes in enzyme activity following thermal acclimation (Hochackha and Somero, 1973; Hazel and Prosser, 1974; Shaklee *et al.*, 1977; Sidell, 1983; Kleckner and Sidell, 1985). These have often revealed increases in the activity of glycolytic and other enzymes associated with energy flux, but the data show too much variability for any generalizations to be made.

A general increase in mitochondrial enzymes occurs following the acclimation of eurythermal fish to low temperatures (Sidell, 1983), which suggests that the number of mitochondria increases, during cold acclimation. Acclimation of crucian carp, *Carassius carassius*, from 28°C to 2°C results in an increase in mitochondrial density in all muscle types (Johnston and

Maitland, 1980) and a similar increase occurs in goldfish acclimated from 25°C to 5°C (Sidell, 1983; Tyler and Sidell, 1984).

All of the studies described above have involved acclimation of eurythermal fish. Very few studies have investigated whether similar responses occur in fish adapted evolutionarily to low temperature, although the volume fractions of mitochondria in slow and cardiac muscle cells in the Antarctic haemoglobinless fish *Chaenocephalus aceratus* are toward the upper range reported for teleosts (Johnston, 1981; Johnston *et al.*, 1983).

New enzyme variants

Not all enzyme molecules will be equally affected by low temperatures; some may remain viable, others lose activity and a few may even be denatured. Clearly, in the case of some enzymes new variants will probably be needed in order for cells to continue to function at low temperatures. Such variants would generally be detectable by high resolution electrophoresis and latitudinal clines have been demonstrated in the relative frequency of different alleles of three enzymes in the killifish *Fundulus heteroclitus* (Powers and Place, 1978; Palumbi *et al.*, 1980), and two enzymes in the sea anemone *Metridium senile* (Hoffman, 1981b). Both studies were performed on the eastern seaboard of North America where the variation of mean seawater temperature with latitude is particularly severe. In both studies temperature was implicated as the selective force maintaining the cline in gene frequency.

In order to demonstrate that such genetic changes are relevant to physiological compensation, kinetic studies are also necessary (Johnston, 1983). Crude preparations of brain acetylcholinesterase (AChE) from rainbow trout, *Salvelinus fontinalis*, acclimated to 2°C and 17°C differed in electrophoretic mobility (Fig. 9.6), and the temperature of minimum K_m of AChE

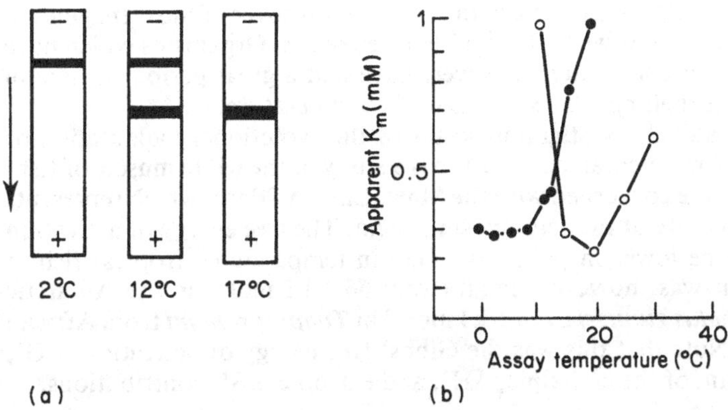

Fig. 9.6 Variation in acetylcholinesterase (AChE) of rainbow trout, *Salmo gairdneri*, with acclimation temperature. (a) Electrophoretic patterns of brain AChE isolated from fish acclimated to 2°C, 12°C and 17°C. (b) The variation of the apparent K_m of AChE for acetylcholine with temperature for enzyme isolated from fish acclimated to 2°C (●) and 18°C (○). (Redrawn from data in Baldwin and Hochachka, 1970 and Baldwin, 1971.)

for acetylcholine was very close to the acclimation temperature (Baldwin, 1971). These data suggest that rainbow trout can switch from a low temperature to a high temperature variant of AChE, and that these two forms are functionally different. Interestingly, at 12°C both isoenzymes were visible on the electrophoresis gel. AChE, however, requires associated membrane for activity, and alterations in membrane composition may explain, at least in part, variations in enzyme activity. Rainbow trout in their natural environment experience a seasonal temperature variation from just above 0°C to almost 20°C. The laboratory data indicate that rather than a whole suite of enzymes adapted to different temperatures, rainbow trout used only two; at intermediate temperatures a mixture of the two were present. Unfortunately, no seasonal studies of AChE isoenzymes in wild rainbow trout were performed, and we do not know the time course of appearance of the new variant following transfer of the trout to a new temperature.

In *Metridium senile* the two allozymes of phosphoglucose isomerase showed similar pH optima, K_m sensitivity to pH, and sensitivity of K_m and V_{max} to temperature (Hoffman, 1981a, 1983). Differences were apparent, however, in specific activity and the ratio V_{max}/K_m. In both cases these were greater in the fast (Pgi^f) allozyme, and the difference between the two forms was greatest at 25°C. The kinetic data would suggest that the Pgi^f variant would have to be selectively favoured at higher temperatures, and this matched the observed cline in gene frequency where the Pgi^f allozyme increased towards the more southerly (and hence warmer) latitudes.

Despite much work, examples of seasonal or acclimation isoenzyme induction are few. In the green sunfish, *Lepomis cyanellus*, acclimation to 5°C or 25°C resulted in changes in the activity of several enzymes but changes in isoenzyme pattern were observed only in the esterases of liver and the eye (Shaklee *et al.*, 1977).

Organisms faced with a seasonal variation in temperature are exposed to any given temperature for a relatively short period of time. Nevertheless, the winter period is sufficiently long with respect to turnover of most enzymes to allow the expression of whole new suites of enzymes. Organisms which have evolved to live in cold water, however, have had a great period of time in which to adjust their enzyme systems to the temperature.

The most detailed investigation so far of the evolutionary adaptation of enzymes to environmental temperature is a study of the white muscle of fish. Early studies were concerned with the Mg^{2+}-Ca^{2+}-ATPase, which represents the catalytic activity of the actomyosin system. The free energy of activation was found to be lower in polar fish than in temperate or tropical fish, a difference that was, however, small (from 66.1 kJ mol^{-1} in the Antarctic *Champsocephalus gunnari* to 74.5 kJ mol^{-1} in *Tilapia grahami* from African hot springs). Note that this was the Gibbs' free energy of activation, $\Delta G^{\ddagger}$, which is the sum of the enthalpic, $\Delta H^{\ddagger}$, and entropic, $\Delta S^{\ddagger}$, contributions:

$$\Delta G^{\ddagger} = \Delta H^{\ddagger} - T\Delta S^{\ddagger}$$

where T is absolute temperature. It is the enthalpy of activation ($\Delta H^{\ddagger}$) which is related to the Arrhenius activation energy. In essence $\Delta H^{\ddagger}$ is a measure of the contribution of the kinetic energy of substrate and enzyme, that is the cell temperature, to the total free energy required for activation. $\Delta S^{\ddagger}$ is the entropic contribution, essentially a measure of the structural input (molecular flexibility, number of weak bonds broken and reformed during the reaction and so on) to total free energy. Many investigations are limited to $\Delta H^{\ddagger}$, thereby missing a vital component.

Despite the small variation in $\Delta G^{\ddagger}$ with environmental temperature, there is an enormous variation with respect to *in vitro* activity of the Mg^{2+}-Ca^{2+}-ATPase isolated from different fish. This is accompanied by a shift in the relative proportions of the enthalpic, $\Delta H^{\ddagger}$, and entropic, $\Delta S^{\ddagger}$, contributions to the total free energy of activation. In fish from low environmental temperatures, where less kinetic energy is available within the cell, the contribution of $\Delta H^{\ddagger}$ is less, and that of $\Delta S^{\ddagger}$ is greater (Fig. 9.7). The increased contribution of $\Delta S^{\ddagger}$ at low temperatures is associated with a difference in molecular structure. The actomyosin contains fewer disulphide bridges, suggesting a more open and flexible molecular structure. Such a difference would be compatible with an increase in $\Delta S^{\ddagger}$, and is also associated with a decreased thermal stability. The ATPases from cold water fish are far more sensitive (by a factor of about 500) to thermal denaturation at 50°C than those from warmer water fish (Johnston and Walesby, 1977; Cossins *et al.*, 1981). The differences in thermodynamic activation parameters and thermal stability are assumed to be caused by changes in the tertiary structure

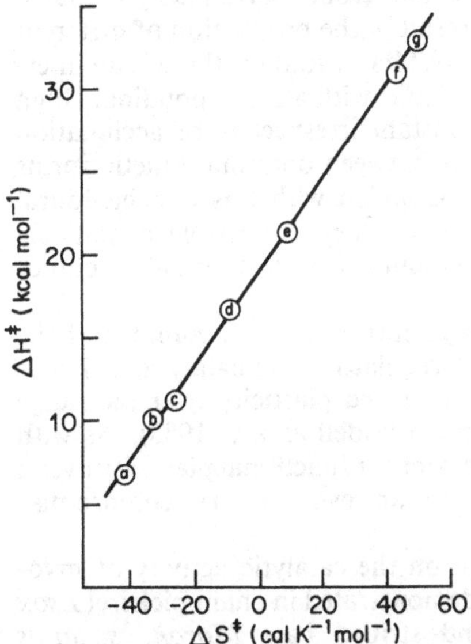

Fig. 9.7 Relationship between enthalpy of activation ($\Delta H^{\ddagger}$) and entropy of activation ($\Delta S^{\ddagger}$) for fish living at different environmental temperatures. Species are, in order of increasing enthalpy of activation (with approximate environmental temperatures in parentheses): (a) *Notothenia rossii* (0–2°C), (b) *Gadus virens* (5–14°C), (c) *Gadus morhua* (5–14°C), (d) *Amphiprion sebea* (23–25°C), (e) *Carassius carassius* (acclimated to 26°C), (f) *Tilapia nigra* (23–31°C), (g) *Tilapia grahami* (35–38°C). (Reproduced with permission from Johnston and Goldspink, 1975).

of the molecule, as a result of alterations in primary structure (that is, amino acid sequence) fixed by natural selection.

Detailed interpretations of the changes in activation parameters are prevented by a lack of precise information about the kinetics of the ATPase reaction. Recent studies of the contractile characteristics of individual muscle fibres, however, reveal that the power output of white and red muscle from Antarctic fish at 0°C is comparable to that of temperate species measured at 15°C or 25°C, indicating a significant degree of compensation for temperature (Johnston and Harrison, 1985).

Natural actomyosin is a multi-protein complex which consists of both catalytic (actin, myosin) and regulatory (troponin, tropomyosin) subunits; the regulatory subunits confer Ca^{2+} sensitivity on the hydrolysis of ATP. Evolutionary adaptation of the actomyosin complex to low temperature involves modification of the myosin heavy and light chains; the calcium regulatory proteins do not appear to be involved (Johnston and Walesby, 1977).

Acclimation of goldfish actomyosin to different temperatures in the laboratory, however, involves not the myosin but Ca^{2+}-regulatory proteins (Johnston, 1979), accompanied by a decrease in $\Delta G^{\ddagger}$, $\Delta H^{\ddagger}$ and $\Delta s^{\ddagger}$ at the low temperatures (Johnston and Lucking, 1978). This difference may be related to the more rapid turnover of troponins (Funabiki and Cassens, 1972) which would allow a more rapid response to a change in temperature.

A different strategy is employed by brook trout, *Salvelinus fontinalis*. Acclimation to 4°C and 22°C does not result in the production of different kinetic forms of myofibrillar Mg^{2+}-Ca^{2+}-ATPase. Rather, there is an intermediate value of $\Delta H^{\ddagger}$, and a low value of $\Delta G^{\ddagger}$ with a correspondingly high turnover number (K_{cat}) which remain constant irrespective of acclimation temperature. This suggests a compromise between optimum kinetic forms for the higher and lower temperatures. Coupled with this is behavioural thermoregulation by the fish which will select, wherever possible, a relatively narrow thermal range, although still retaining a wide thermal tolerance (Walesby and Johnston, 1981).

A similar response is shown by the myofibrillar protein complex of the killifish *Fundulus heteroclitus*. The Ca^{2+}-regulatory and catalytic activities of the myofibrillar complex show a pronounced plasticity over the range 12–35°C and are unaffected by acclimation (Sidell *et al.*, 1983). As with *Salvelinus*, it would appear that a requirement for functional plasticity over a wide temperature range has resulted in an evolutionary compromise (Somero, 1978).

The lack of any effect of acclimation on the catalytic activity of myofibrillar ATPase has also recently been demonstrated in chain pickerel *Esox niger* (Sidell and Johnston, 1985) and striped bass *Morone saxatilis* (Moerland and Sidell, in Sidell and Johnston, 1985). It is possible that a significant compensatory response to temperature acclimation by myofibrillar ATPase, as for example in goldfish *Carassius auratus* (Johnston *et al.*, 1975; Sidell, 1980), may be limited to polyploid species. The major difference between all of these species and those shown in Fig. 9.7 is that the latter (with the exception of *Carassius carassius*) have adapted over evolutionary time to their respective, relatively fixed, environmental temperatures. The former group are all temperate or cold temperate fish subject to severe changes in temperature.

Finally, care must be taken in extrapolating from the thermal and catalytic behaviour in *in vitro* preparations of isolated enzymes in dilute media to the situation *in vivo*. In at least two species, *Esox niger* (Sidell and Johnston, 1985) and bull trout, *Myxocephalus scorpius* (Johnston and Sidell, 1984) the temperature dependence of myofibrillar ATPase activity *in vitro* and that of muscle contractile properties have been found to differ. Enzyme activity *in vitro* is very sensitive to conditions such as pH, ionic strength, substrate concentration and the presence of inhibitors or competitors. The enzyme environment *in vivo* is a complex unknown, therefore extrapolation from the behaviour of isolated enzymes in dilute media to the tissue must be undertaken with care.

Alteration of the immediate environment of the enzyme

When the temperature of a cell is reduced, the decrease in kinetic energy is not the only significant change in physical properties (Taylor, this volume). Factors such as the pH of the solvent environment or the fluidity of membranes will also alter, and these may have dramatic effects on enzyme activity. There is evidence that organisms may regulate both intracellular pH and membrane fluidity, thereby minimizing the effects of temperature change.

Water is a very weak electrolyte and ionization is influenced by both temperature and pressure. Over the temperature range of interest to physiologists, the rate of change of pH of pure water (frequently termed the pH of neutrality, pN) with temperature, $\Delta pH/\Delta T$, is -0.0171 pH units per degree centigrade.

The pH of the body fluids of marine organisms varies with temperature. Blood pH is actively regulated, primarily through control of respiratory gas exchange, and the $\Delta pH/\Delta T$ is similar to that of water, thus preserving a roughly constant $[OH^-]/[H^+]$ ratio (relative alkalinity; Heisler *et al.*, 1976; Reeves, 1972; Malan *et al.*, 1976; Hazel *et al.*, 1978). Most studies have examined acclimation of eurythermal organisms, but two polar organisms, the fish *Dissostichus mawsoni* (Qvist *et al.*, 1977) and the isopod *Glytonotus antarcticus* (Jokumsen *et al.*, 1981), both show relatively high body fluid pH values, suggesting that a similar control of body fluid pH may be involved in evolutionary adaptation to low temperature.

Within the physiological pH range the net charge of many proteins is determined by the dissociation of the imidazole moiety of histidine residues (Hazel *et al.*, 1978). The dissociation constant of the imidazole moiety, $K_{imidazole}$ is influenced by temperature in the opposite direction to that of water (K_w) and, as a result, over the range of 0°C–30°C the fraction of histidine imidazole moieties present in the dissociated state remains approximately constant as temperature drops. Thus for a protein whose net charge is determined largely by histidine imidazole residues, this charge will vary only slightly as temperature falls.

Reeves (1972, 1977) has postulated that poikilotherms regulate their body fluids to maintain a constant net protein charge rather than a constant net alkalinity; this he termed alphastat control. Regulating body fluid pH to maintain constant alkalinity, and regulating pH to maintain constant net protein charge (alphastat control) are different processes. In the case of proteins whose net çhange is determined mostly by histidine imidazole residues, however, it is difficult to distinguish the two effects. The shift in optimal pH with temperature has been examined for five enzymes from tissues of rainbow trout, *Salmo gairdneri* (Hazel *et al.*, 1978). In all cases the pH optimum shifted towards higher pH values as temperature decreased, and the values of $\Delta H_{opt}/\Delta T$ were not distinguishable from that of -0.0218 previously reported for trout arterial blood $\Delta pH/\Delta T$ (Randall and Cameron, 1973). This value is itself statistically indistinguishable from the value for water; thus, it is not possible to distinguish between the hypotheses of alphastat control and constant relative alkalinity with these data. However, the pH of intracellular fluid from the skeletal muscle of the turtle *Pseudemys scripta* diverged increasingly from the pH of water at higher temperatures, suggesting that α_{im} rather than relative alkalinity was being regulated (Malan *et al.*, 1976). In both plasma and skeletal muscle, α_{im} remained constant over the temperature range 10–30°C (Fig. 9.8). But, it should be noted that *Pseudemys* is an air-breather with a very different mechanism for adjusting acid–base balance than an aquatic-breather such as a fish.

Clearly, *in vitro* investigations of the properties of enzymes must take these factors into account. For example, the epidermal Na^+-K^+-ATPase of the toad *Bufo marinus*, assayed at pH 7.4, has an apparent K_m that increased dramatically with decreasing temperature, whereas at a constant $[OH^-]/[H^+]$ ratio of 16, apparent K_m was independent of temperature in the range 8–37°C, and there was no appreciable variation in activity over the

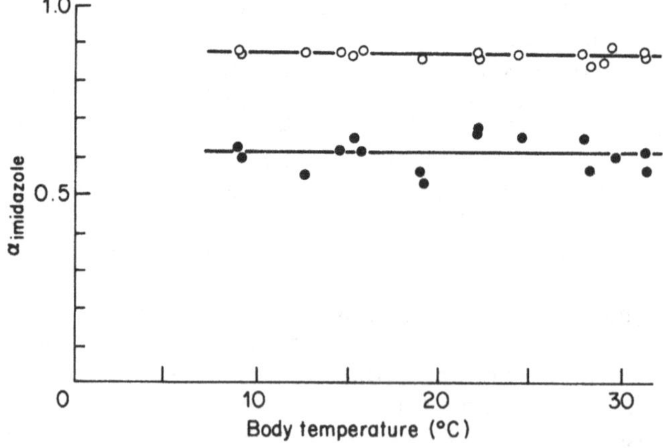

Fig. 9.8 Dissociation ratio ($\alpha_{imidazole}$) of protein imidazole groups as a function of temperature in the plasma (●) and skeletal muscle (○) of the turtle *Pseudemys scripta*. Individual turtles were acclimated to various ambient temperatures and muscle $\alpha_{imidazole}$ calculated from intracellular pH. (Redrawn from Malan, 1980.)

range 20–37°C (Park and Hong, 1976). These results suggest that ideas on positive thermal modulation (i.e. an increase in apparent K_m with temperature) may be based on artefacts of inadequate pH/temperature regulation during the assay (Yancey and Somero, 1978). This is another example of the dangers inherent in extrapolating to the whole organism from unrepresentative assay conditions *in vitro*.

For many membrane-bound enzymes the physical properties of the supporting membrane are of paramount importance in regulating catalytic activity. A clear example of this is the Na^+-K^+-ATPase of chinese hamster ovary plasma membrane (Sinensky *et al.*, 1979), where an alteration of the membrane lipid acyl chain order was induced by cholesterol substitution in a mutant cell-line deficient in the cholesterol synthesis pathway (Fig. 9.9).

An increase in membrane phospholipid fatty acid unsaturation at low temperatures is a widely described phenomenon (Morris and Clarke, this volume). It is usually assumed that this change in unsaturation offsets the increase in membrane viscosity and the possibility of a phase separation at low temperatures. Very few studies, however, have measured both enzyme activity and membrane composition in acclimation experiments. Usually only one or other is measured and a functional connection assumed. Hazel (1972a, b), however, demonstrated a functional connection for succinate

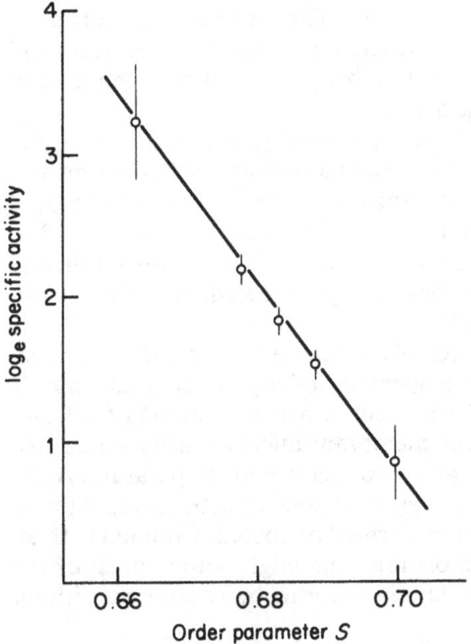

Fig. 9.9 The relationship between the specific activity of the Na^+-K^+-ATPase of Chinese hamster ovary cell plasma membranes and bulk membrane viscosity (expressed as the order parameter *S* determined from the motion of an electron-spin resonance probe inserted into the membrane). Different values of *S* were produced by cholesterol supplementation of a cell line defective in cholesterol synthesis. (Redrawn from data in Sinensky *et al.*, 1979.)

dehydrogenase (SDH) isolated from cold (5°C) and warm (25°C) acclimated goldfish. SDH isolated from the 5°C fish had 1.3 and 2.2 times the activity of SDH from 25°C fish although no difference in the protein could be detected by electrophoresis. Enzyme activity was reduced three- or four-fold by lipid extraction of crude SDH preparations and partially restored by addition of purified mitochondrial lipids. The more unsaturated lipids isolated from 5°C fish produced a higher V_{max} regardless of whether the SDH was isolated from a 5°C or 25°C goldfish.

Enzymes and temperature: conclusions

It is apparent from the previous sections that a number of potential mechanisms exist to counteract the rate-depressing effect of low temperature on enzyme reactions. Changes in enzyme concentration, the expression of new enzyme variants and a delicate tuning of the enzyme microenvironment have all been demonstrated, during both acclimation and evolutionary adaptation to temperature. What are the advantages and disadvantages of each process?

Both the solvent capacity of the cell and the amount of membrane available to support membrane-bound enzymes are finite. This would limit the effectiveness of increasing enzyme (or substrate) concentration at low temperature. Nevertheless, an increase in the number of mitochondria does occur in both acclimation and evolutionary adaptation to low temperature in fish. In the case of the contractile proteins of muscle increasing concentration will not be effective for an increase in fibril number will affect only the tension developed and not the speed of contraction.

The cell contains many different kinds of protein (perhaps 10 000 at any one time in a typical vertebrate cell) involved in a complex web of pathways. To switch between different suites of enzyme variants seasonally would involve a massive alteration in genome expression, even if this is limited to those enzymes believed to regulate metabolic fluxes. This seems intuitively unlikely, and it is perhaps no surprise that it has proved so difficult to demonstrate such changes by electrophoresis.

Enzyme activity may also be extremely sensitive to variations in local microenvironment; for example, the specific activity of goldfish muscle phosphofructokinase may vary by 60 per cent or a change in pH of 0.1 unit (Freed, 1971). The intracellular pH and membrane microviscosity would also have to be controlled to minimise the affects of decreased temperature. Acid-base regulation in particular allows a relatively rapid adjustment, and this is frequently seen when poikilotherms are warmed or cooled. Coupled with an adjustment of membrane lipid composition, this might allow much of the acute rate-depressing effects of low temperature to be countered without recourse to new types of enzyme.

If the organism is adjusting to a long-term cooling of the environment (that is, an evolutionary adjustment to temperature) then a fine tuning of enzyme structure, and hence function, may follow. Only one set of enzymes is therefore necessary, although by the time adjustment to a local temperature of approximately zero is complete, these enzymes may be very different from those present in Eocene ancestors living at 15°C. There may also be widespread restructuring of the tissues.

Organisms that experience seasonal change in temperature will also be able to adjust the intracellular environment as winter approaches. Further compensation may be achieved by increasing enzyme concentration (sunfish cytochrome c) or perhaps by switching to different enzyme variants in a few instances (trout AChE). Seasonal changes in temperature are sufficiently long to allow changes involving alterations in genome expression.

Organisms subjected to daily or tidal changes in temperature face much more severe problems. Acid-base regulation is feasible, but alteration of membrane composition may take 20–30 days in fish (Cossins, 1982). Expression of new enzyme variants over such a short time scale is unlikely. More likely strategies would be a compromise enzyme with low temperature sensitivity, or expression of a suite of variants with different kinetic characteristics so that activity can be maintained over a wide range of temperatures. The latter idea suggests the hypothesis, which can be tested, that organisms experiencing rapid changes of temperature might be genetically more polymorphic than stenothermal organisms, at least for those enzymes believed to be rate-determining for metabolic fluxes. Some evidence for this comes from studies of killifish *Fundulus heteroclitus*, where the maximum degree of heterozygosity (assessed over four loci) corresponds with the zone of maximum seasonal temperature fluctuation (Powers and Place, 1978). There is a similar suggestion of a correlation between heterozygosity and temperature stability in the anemone *Metridium senile* (Shick *et al.*, 1979).

These various strategies are outlined in Fig. 9.10. This diagram is speculative but it does serve to illustrate that the processes employed by organisms to offset the effects of a change in temperature are liable to vary with the time scale of the change. In is therefore necessary to be very careful when extrapolating the results of acclimation experiments on eurythermal organisms to evolutionary adaptation (or *vice versa*).

Temperature and metabolic pathways

The section above highlighted two major strategies to cope with the physiological effects of low temperature, alterations in the amount or type of individual enzymes and an overall adjustment of enzyme microenvironment at the cellular level (pH and membrane viscosity). Consideration of the overall physiology of the cell underlines this separation. Athough a few cell types can be given over wholly to a few metabolic pathways (e.g. adipocytes or certain liver cells producing vitellogenin in maturing fish), the majority of cells exhibit a balanced array of pathways.

Not all enzymes, and hence not all metabolic pathways, will be affected equally by a lowering of cell temperature. In order to avoid a catastrophic metabolic imbalance during a change of temperature, the relative importance of the various pathways must be regulated. This need not mean that the relative balance must remain the same, for in both *Fundulus heteroclitus* (Moerland and Sidell, 1981) and *Morone saxatilis* (Stone and Sidell, 1981;

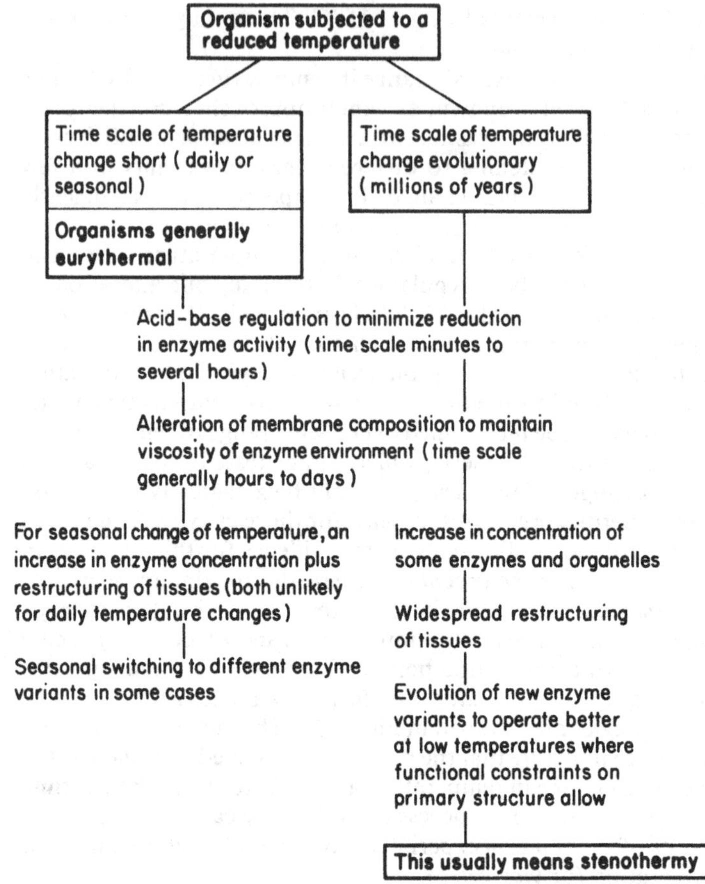

Fig. 9.10 A hypothetical scheme showing processes involved in both acclimation and evolutionary adaptation to low temperatures.

Jones and Sidell, 1982) there is an increase in the activity of the pentose shunt relative to glycolysis at low temperatures. Although the adaptive significance of this shift is not clear, it does demonstrate a regulation of metabolic balance during temperature acclimation.

It seems intuitively unlikely that adjustment to a new temperature will involve a wholesale switching to new enzyme variants, and no such widespread occurrence of isoenzymes has been detected by electrophoresis. Taken together the evidence suggests some form of overall homeostatic mechanism, such as acid-base balance, within the cell to cope with an alteration in temperature. Athough regulation of intracellular pH in the face of temperature change has been reported from a wide variety of organisms, there is not yet sufficient evidence to claim that this is a universal response.

The classical view of the regulation of metabolic flux has centred on the behaviour of key flux-regulating enzymes. Equally, consideration of the way cells maintain metabolic balance has also concentrated upon the way certain 'key' enzymes compensate for temperature (e.g. Newsholme and Paul, 1983). An alternative view is to consider metabolic flux as a character of a metabolic pathway as a whole, rather than being dictated by the properties of selected regulatory enzymes (Kacser and Burns, 1973, 1979; Kacser, 1983). Such a view of metabolic regulation also suggests that cells would probably regulate metabolic balance through whole-cell homeostatic means. This would simplify temperature compensation and allow rapid homeostasis. To a certain extent the more traditional view of metabolic flux regulation is simply an approximation of the Kacser and Burns approach, merely ignoring the (often small) degree of control exerted by all the non-equilibrium steps in the pathway (Johnston, personal communication). Nevertheless, considering metabolic flux as a property of a metabolic pathway as a whole suggests once more the possibility of whole-cell homeostasis. Unfortunately, most consideration of temperature compensation at the cellular level has been concerned with either molecular adjustment or membrane function. Very few studies have examined whole-cell homeostasis.

Protein synthesis

The rate of protein synthesis measured at any given time will be the sum of requirements for protein for:

1. basal (maintenance) turnover;
2. growth;
3. gametogenesis.

In maturing female organisms the protein demands of gamete production may exceed all other requirements. Under more normal circumstances, however, a measurement of protein synthesis will include contributions from basal turnover and growth.

Measurements of rates of protein synthesis in Antarctic fish during summer are much lower than in tropical fish (Fig. 9.11). These fish were all starved before measurements were made, and so it is probable that the bulk of the synthesis measured represented basal turnover, although there may have been some contribution from growth.

Why are the rates of protein synthesis so low in polar fish? Two possible explanations are that:

1. protein synthesis is unavoidably limited by temperature; in other words compensation for the effects of temperature on protein synthesis is not possible.
2. the requirement for basal turnover is much reduced in polar fish.

The first explanation is unlikely. Tropical fish show a wide variability in protein synthesis rates, though all live at the same temperature. These differences are associated with ecology and are mediated through variations in the activity of EF-1 (an enzyme involved in the binding of aminoacyl-tRNA to the codon on mRNA - ribosome complex). Evidence of compensation for temperature in the embryonic development of copepod eggs also makes this explanation unlikely (McLaren *et al.*, 1969; Clarke, 1983).

If the requirement for basal protein turnover is greatly reduced in polar fish (the second option), this also requires an explanation. The turnover rate of a given protein will be related to the inherent stability of that molecule at the cell temperature to which it is adapted. Evidence from studies of muscle and brain ATPases in dilute solution *in vitro* suggests that enzymes from polar organisms are more unstable than those from tropical or temperate organisms, when measured at the same temperature. Although extrapolations from data obtained *in vitro* to the situation *in vivo* must be made with care, the difference in stability may not be the same when proteins are compared at the temperatures at which they are adapted to operate. If a slow turnover rate is the underlying explanation for the reduced protein synthesis observed in polar fish, measurements of enzyme turnover should demonstrate this. To date no such measurements have been made.

Temperature and metabolic rate

Whole organism metabolic rate is a concept that is quickly defined and easily measured, but proves elusive when examined in depth. Metabolic rate is the sum of all the energetic processes active at the time of measurement. Historically, it has been treated as a process distinct in itself, and it is still frequently regarded as an aspect of organism physiology which must compensate for changes in temperature. Temperature compensation is therefore often examined by measuring the response of whole organism oxygen demand to a change in temperature, and many discussions of temperature compensation start with a treatment of respiratory rate. It is not always clear, however, exactly what is being measured or described. In particular it is not necessarily the case that measures of oxygen consumption under two different conditions (e.g. high and low temperature) are strictly comparable.

What does oxygen consumption measure?

Most cellular activities are fuelled by ATP generated from the tricarboxylic acid cycle; this requires oxygen and therefore metabolic rate may be estimated by oxygen consumption. (Exceptions are anaerobiosis in intertidal organisms and burst escape activity; in these situations, however, the oxygen debt incurred is usually cleared later when oxygen is once more available). Very few comparisons of total metabolic rate (by calorimetry) and oxygen consumption have been made. Where this has been done a small, though significant, contribution to total metabolism by anaerobic processes has been revealed (Hammen, 1979; Shick, 1981).

Processes which may contribute to the measured oxygen consumption of a marine organism during a respiration experiment can be summarized (Table 9.2). If it is intended to measure basal or standard metabolic rate, it is usual to minimize feeding and other activity. Recent evidence has suggested that the pulse of oxygen uptake frequently observed after feeding (the specific

Table 9.2 Metabolic processes which may be operational during a measurement of respiration rate in a marine organism

Basal metabolism (as defined in Clarke, 1980, 1983)
- Protein turnover (including both synthesis of new protein and the activity of deamination and urea synthesis pathways)
- Ion pump activity
- Membrane turnover
- Nucleic acid turnover
- Basal neurocircuitry function
- Maintenance of muscle tonus
- Movement of respiratory apparatus

Processes also included when standard (resting) respiration is measured
- Growth this includes all processes associated with the production of new tissue, including skeletal tissue. A major component may be those metabolic processes classified as the specific dynamic action (SDA) – see Jobling, 1983
- Synthesis of reproductive tissue
- Processes involved with moulting (ecdysis) in crustaceans

Processes usually classified as activity (and hence included in measurements of both routine and active metabolism)
- Feeding activity (ciliary movement, action of feeding apparatuses, pumping of water, intestinal work)
- Locomotor activity (including behavioural activity)

dynamic action, SDA) is not the cost of intestinal work or urea synthesis but rather the respiratory cost of growth (Jobling, 1983). In other words growth does not proceed steadily, rather it appears to proceed in pulses following meals. If the time between successive meals is sufficiently long, the post-prandial rise in oxygen consumption can be seen to return to prefeeding levels (Vahl, 1983). In an organism that feeds rapidly or continuously there will therefore be a general elevation of standard metabolic rate related solely to the synthetic activity associated with growth. Similar arguments apply to vitellogenesis, although in both cases there will also be a higher level control (usually hormonal) resulting possibly in a general elevation of turnover and synthesis at certain times of the year. These considerations alone mean that standard respiratory rates measured in a growing and non-growing season will not be comparable (Vahl, 1978; Parry, 1983 and below).

Other activities may also increase the measured oxygen consumption, but no theoretical or practical problems are present if respiration rate is regarded simply as the net demand for oxygen by all those metabolic processes active at the time of measurement. If, however, respiration is regarded as a process in itself, and one which must compensate for any alteration in temperature, then interpretation of experimental data is difficult. In essence the effect, and not the cause is being measured.

Temperature compensation and metabolic rate

There are two different aspects to temperature compensation, short term (acclimation) or long term (seasonal acclimatization and evolutionary

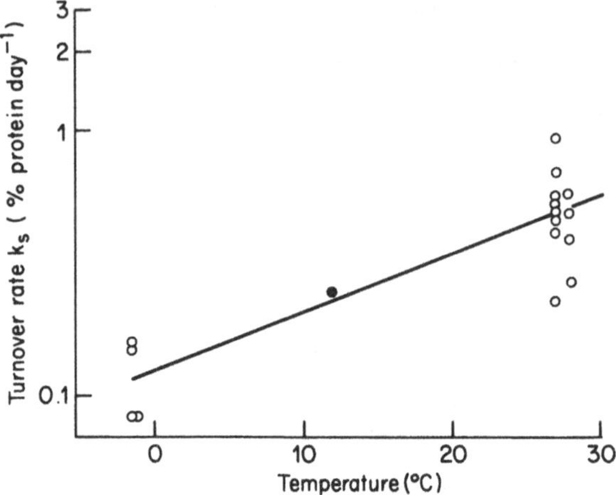

Fig. 9.11 Relationship between white muscle protein synthesis (as K_s fraction of total protein synthesized per day) and estimated environmental temperature for polar and tropical fish (○). (●), data for rainbow trout, *Salmo gairdneri*, acclimated to 12°C. (Data replotted with permission, from Smith and Haschemeyer, 1980 and Smith *et al.*, 1980.)

adaptation). Metabolic rate has been used frequently to assess both forms of compensation.

If a eurythermal fish such as a goldfish is transferred from 25°C to 5°C, there is an immediate decrease in oxygen uptake. This is usually assumed to be caused by purely thermodynamic effects, a general slowing of all cellular reactions at the lower temperature. After a short time oxygen consumption increases to a more or less stable value. This increase in metabolic rate is frequently termed compensation and qualified as perfect, incomplete, partial and so on, depending on the new level of oxygen consumption attained (Fig. 9.4). Why a level of oxygen demand which is less than that at 25°C should be incomplete is not clear; if a process such as locomotion or nervous transmission was being described, the terminoloy would be acceptable. Oxygen consumption is not, however, a single process which must compensate for changes in temperature. It is merely the net cost of all those processes induced by the lowering of temperature; these may include respiratory adjustment of acid–base balance, increased synthesis of enzymes or the restructuring of tissue. Clearly, the higher the oxygen demand, the more expensive the compensation; from an energetic standpoint it is the lower respiratory cost that is adaptive, not the 'perfect' compensation of Precht *et al.*, (1955) (Fig. 9.4). Assessment of the degree of compensation for temperature by the level of oxygen demand misses the point.

Seasonal acclimatization to temperature is also frequently expressed in terms of oxygen consumption. Thus an oxygen consumption of a summer animal measured at a high temperature may be combined with a measurement of a winter animal at low temperature to provide an index of acclimatization. This procedure, however, is problematical in that the metabolic processes being measured in summer and winter animals may differ. Summer animals are likely to be growing and/or undergoing gametogenesis; both processes are liable to increase oxygen consumption.

A demonstration of the difficulties associated with interpreting seasonal changes in respiratory rate is found in the intertidal limpet *Cellana tramoserica* (Parry, 1978, 1983). This organism has a higher oxygen consumption in winter than in summer (so-called adaptive acclimation), which cannot be explained by a direct temperature effect on basal metabolism because higher respiratory rates are observed at the lower environmental temperatures. An alternative explanation is that basal metabolism is relatively insensitive to the seasonal variation in temperature, and that the increased oxygen consumption observed in winter is simply the metabolic cost of growth.

Organisms with a predictable seasonal variation in temperature may be able to modify internal organization and enzyme forms to compensate for temperature. The observed seasonal variation in oxygen consumption (even if allowance could be made for the effects of growth or gametogenesis) will therefore give little information about the evolutionary relationship between basal respiratory rate and temperature.

Metabolic cold adaptation

The relationship between respiratory rate and temperature has interested physiologists for over 60 years. In a much quoted work, Krogh (1916) suggested that animals living at low temperatures would be expected to have an elevated metabolic rate. Krogh was referring, though not explicitly, to basal metabolic rate and his expectation arose from comparing the active behaviour of polar organisms at low temperatures with the torpor shown by temperate creatures cooled to the same temperatures. In other words polar organisms had obviously evolved some form of compensation for low temperature in their locomotor apparatus, and so a similar compensation should be expected in metabolic rate. If metabolic rate is viewed as a process distinct in itself which must compensate for temperature, this seems a reasonable conclusion (and it must be remembered that Krogh proposed his view long before the underlying physiology of respiration had been elucidated). Clearly, cold water organisms have indeed evolved compensation for temperature in their locomotor apparatus (see above), although as far as is known there has been no significant alteration in the metabolic cost of locomotor activity. Viewing respiratory rate as merely the sum metabolic cost of all those physiological processes active at the time of measurement suggests that there would be no selective advantage in 'compensation' (i.e. a high respiratory rate at low temperature); indeed such a strategy would be energetically wasteful.

Nevertheless, the idea that polar organisms should somehow compensate for the low temperature by an elevated oxygen consumption became enshrined in the literature as the concept of *metabolic cold adaptation*. (For a history of this key concept *see* Dunbar [1968] and Clarke [1983]). However, recent data have indicated that this hypothesis must be abandoned. Measurements of the respiratory rates of isopods (White, 1975; Luxmoore, 1984), bivalves (Ralph and Maxwell, 1977a, b) and fish (Holeton, 1973, 1974) have shown that there is a general trend for the standard or resting respiratory rate to decrease at lower temperatures. This would suggest that there may be a general underlying trend between basal metabolic rate and environmental temperatures.

An approximately exponential decrease in respiratory rate of marine crustaceans with temperature has been demonstrated (Fig. 9.12). These data have been collected from many studies, made under a variety of experimental conditions. Despite the extensive procedural and biological variation in the data set, an underlying trend is apparent; there is a clear tendency for the respiratory rate of marine crustaceans to decrease with temperature. It is apparent that living in cold water brings a distinct energetic advantage, for the basal metabolic rate of a marine crustacean living at 25°C is over six times that of a marine crustacean living at 0°C. As basal metabolism is energy wasted in the sense that it cannot be diverted into ecologically useful areas such as growth or reproduction, clearly for any given energy intake relatively less is wasted at low temperatures.

Temperature and metabolic rate: conclusions

Temperature affects all physiological processes and it therefore seems

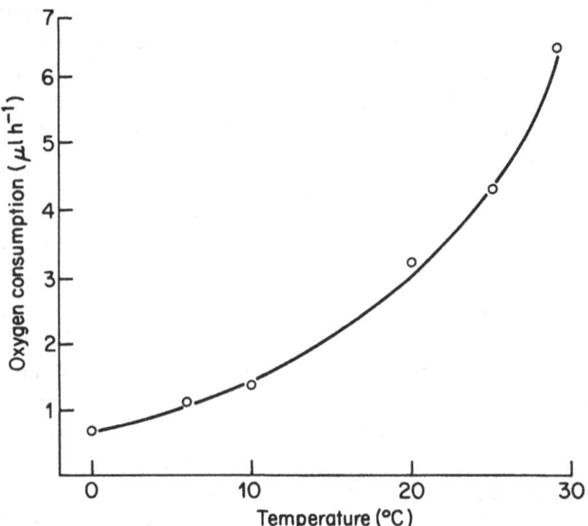

Fig. 9.12 Relationship between the oxygen consumption of a representative marine crustacean of 1 mg dry wt (in μl h^{-1}) and estimated mean environmental temperature. These data were obtained from the parameters a and k in the relationship $QO_2 = aW^k$ (where QO_2 is oxygen uptake in μl h^{-1} and W is the dry wt in mg). Values of a and k were calculated by Ivleva (1980) from all respiration data for marine crustaceans available from the literature, pooled into six temperature blocks between 0°C and 30°C. These six values of a and k were then used to calculate the estimated oxygen consumptions plotted here, taking $W = 1$ mg dry wt. The total number of determinations pooled was > 5000. For further details see Ivleva (1980) and Clarke (1983).

reasonable to suppose that evolutionary adaptation to low temperature will also involve all these processes. The relative balance of these various processes will not, however, necessarily remain the same.

Basal metabolic rate decreases markedly with temperature (Fig. 9.12). A full explanation for this observation is lacking but, at least in fish, it is associated with a decrease in protein turnover at low temperatures. Together these observations suggest that proteins efficient at low temperatures require relatively slow turnover; in other words the patterns in Figs. 9.11 and 9.12 are the result of a three-way interaction between structure, function and temperature. This does not necessarily represent a general slowing of protein metabolism in all tissues; adaptation is likely to be tissue-specific and may involve subtle shifts in the relative balance of metabolic pathways.

Physiological processes other than basal metabolism (i.e. locomotion, growth and vitellogenesis) must also adapt to temperature but direct evidence of such adaptation is still largely lacking. The best evidence for compensation of locomotion comes from the study of fish white muscle (Johnston, 1985); evidence for compensation of vitellogenesis is almost non-existent (but see Garwood and Olive, 1978). Adaptation of these processes to temperature is distinct from adaptation of basal metabolism in that their metabolic costs will be broadly similar to those in warmer waters. In other words a given amount of locomotor activity in a tropical, temperate or polar fish will cost roughly the same amount of ATP; differences between species in the amount of energy diverted to locomotion will have more to do with ecology than with habitat temperature (although the increased viscosity of seawater at low temperatures may well have an energetic consequence).

The only process where this is not the case is basal metabolism, and it is the relatively low energetic cost that gives polar organisms that energetic advantage (note that this argument applies only to evolutionary adaptation, it cannot apply to seasonal acclimation of temperate organisms unless their protein turnover also slows at low temperatures). The energetic advantages of living in cold water, and their ecological consequences, have been discussed elsewhere (Clarke, 1980; 1983).

Freezing resistance

The body fluids of most marine invertebrates are isosmotic or hyperosmotic to seawater and so the avoidance of tissue freezing poses no particular physiological problem; if the seawater remains unfrozen, so do they. Occasionally, sessile organisms become trapped within anchor ice and freeze, but this is an infrequent hazard for most marine invertebrates.

Freezing is, however, a potential hazard for sublittoral or intertidal organisms exposed to very low air temperatures at low tides. Several intertidal molluscs have been shown to withstand temporary freezing, including *Littorina littorea* (Kanwisher, 1955; Sømme, 1966; Murphy, 1979), *Nassarius obsoletus* (Murphy, 1979), *Acmaea digitalis* (Roland and Ring, 1977), *Mytilus edulis* (Williams, 1970) and *Modiolus demiosus* (Murphy, 1977a, b; Murphy and Pierce, 1975). Freezing is extracellular, with between

64 per cent and 75 per cent of body water being removed from the cells osmotically during the process of freezing (*see* Grout and Morris, this volume). The rates of cooling experienced in the natural environment are unlikely to result in intracellular freezing. Very little is known of the physiological mechanisms underlying freezing tolerance in intertidal organisms, although a peptide which reduced the freezing point of a solution by 0.4 °C at a concentration of 4 per cent has been isolated from *Mytilus edulis* (Theede, 1972; Theede *et al.*, 1976). For a recent review of freezing resistance in marine invertebrata see Aarset (1982).

The body fluids of teleost fish are hyposmotic to seawater and for many polar species the temperature of the seas in which they live can be as much as 1 °C below the equilibrium freezing point of their blood serum. A few polar fish avoid freezing by living in deep water where they avoid any contact with ice, and live permanently supercooled. These include the Antarctic liparid *Paraliparus devriesii* and zoarcid *Rhigophila dearborni* (although the latter can tolerate surface waters in summer), and *Liparus koefoedi*, *Gymnocanthus tricuspis*, and *Icelus spatula* from the fjords of northern Labrador (Scholander *et al.*, 1957; DeVries, 1974). Some cold-temperate fish such as *Myoxocephalus octodecemspinosus* move into deeper water in winter to avoid contact with ice (Leim and Scott, 1966), and in McMurdo Sound *Notothenia kempi* lives permanently in a tongue of warm water (DeVries and Lin, 1977a). How this species survives elsewhere is unknown.

A more common way of avoiding freezing in polar fish is a lowering of the freezing point of the body fluid. A substantial fraction of the lowered freezing point is a result of an increase in NaCl (40–70 per cent according to the species). The remainder is the result of the presence of macromolecular antifreezes. These antifreezes may be either peptides or glycopeptides; both types are found in Arctic and Antarctic species, though never together in the same fish (Clarke, 1983).

The majority of Antarctic fish have glycopeptide antifreezes. These usually consist of a repeating sequence of the tripeptide alanine-alanine-threonine, with the disaccharide galactose-N-acetylgalactosamine attached to each threonine residue. Depending upon the number of repeating units, these antifreeze molecules exist in a series of seven discrete molecular weights (DeVries and Lin, 1977a). The blood serum concentration is about 4 per cent w/v and so their contribution to osmotic strength is negligible; their antifreeze action must therefore be noncolligative. The glycopeptide antifreezes of the Arctic *Boreogadus saida* and *Gadus ogac* are similar, but those in *Eleginus gracilis* contain arginine in place of threonine (O'Grady *et al.*, 1982). Peptide antifreezes have been demonstrated in several species from Arctic and northern cold-temperatures seas. The winter flounder *Pseudopleuronectes americanus* contains three discrete peptides. These consist of only eight amino acids, of which alanine accounts for two-thirds of the total residues (Duman and DeVries, 1976). The evolution of the different forms of antifreeze has been discussed by Scott *et al.* (1986).

How these antifreeze molecules work is still far from clear. There is no evidence at all, however, that they poison small ice crystals which must then be melted elsewhere in the fish's body. Rather they must act by preventing the growth of very small ice nuclei before these have grown sufficiently large to be stable; thermal agitation then breaks up the nucleus, freeing the antifreeze molecule. Antifreeze molecules thus cause a hysteresis effect in that they decrease the temperature of ice nucleation (both homogeneous and heterogeneous) but do not affect the melting point of frozen serum. Possible modes of antifreeze action have been discussed (Vandenheede *et al.*, 1972; DeVries and Lin 1977b; DeVries, 1980, Ahmed *et al.*, 1981; Burcham *et al.*, 1982; Knight *et al.*, 1984).

In the Antarctic temperatures remain at approximately − 1.9°C throughout the year and antarctic fish need antifreeze all the time. Some high Arctic fish (e.g. *Myoxocephalus scorpius*) also need antifreeze throughout the year. In northern cold-temperate waters, however, there are large seasonal fluctuations in seawater temperature and antifreezes are needed only in winter; in January winter flounder blood contains 3 per cent w/v peptide antifreeze, whereas in September antifreeze cannot be detected (Lin, 1979).

Acknowledgements

I am grateful to Drs B.D. Sidell and I.A. Johnston for particularly helpful comments on an early draft of this chapter.

References

Aarset, A.V. (1982). Freezing tolerance in intertidal invertebrates (a review). *Comparative Biochemistry and Physiology* **73A**, 571–80.

Ahmed, A.I. Osuga, D.T., Yeh, Y., Bush, C.A., Matson, E.M., Yamasaki, R.B. and Feeney, R.E. (1981). Tools for studying the function of antifreeze glycoproteins. *Cryoletters* **2**, 263–8.

Baldwin, J. (1971). Adaptation of enzymes to temperature: acetylcholinesterases in the central nervous system in fishes. *Comparative Biochemistry and Physiology* **40**, 181–7.

Baldwin, J. and Hochachka, P.W. (1970). Functional significance of isoenzymes in thermal acclimation. Acetylcholinesterase from trout brain. *Biochemical Journal* **116**, 883.

Baross, J.A. and Deming, J.W. (1983). Growth of 'black smoker' bacteria at temperatures of at least 250°C. *Nature (London)* **303**, 423–6.

Burcham, T.S., Osuga, D.T., Yeh, Y. and Feeney, R.E. (1982). Antifreeze glycoprotein activity as a function of ice crystal habit. *Cryoletters* **3**, 173–6.

Clarke, A. (1980). A reappraisal of the concept of metabolic cold adaptation in polar marine invertebrates. *Biological Journal of the Linnean Society* **14**, 77–92.

Clarke, A. (1983). Life in cold water: the physiological ecology of polar marine ectotherms. *Oceanography and Marine Biology: an Annual Review* **21**, 341–453.

Cossins, A.R. (1982). The adaptation of membrane structure and function to changes in temperature. In *Cellular Acclimatisation to Environmental Change*, pp. 3–32. Edited by Cossins, A.R. and Sheterline, P. Cambridge University Press, Cambridge.

Cossins, A.R., Bowler, K. and Prosser, C.L. (1981). Homeoviscous adaptation and its effects upon membrane-bound enzymes. *Journal of Thermal Biology* **6**, 183–7.

DeVries, A.L. (1974). Survival at freezing temperatures. In *Biochemical and Biophysical Perspectives in Marine Biology* **Vol. 1**, pp. 289–330. Edited by Sargent, J.R. and Mallins, D.C. Academic Press, London.

DeVries, A.L. (1980). Biological antifreezes and survival in freezing environments. In *Animals and Environmental Fitness*, pp. 583–607. Edited by Gilles, R. Pergamon, Oxford.

DeVries, A.L. (1984). Role of glycopeptides and peptides in inhibition of crystallization of water in polar fishes. *Philosophical Transactions of the Royal Society London B* **304**, 575–88.

DeVries, A.L. and Lin, Y. (1977a). The role of glycoprotein antifreezes in the survival of antarctic fishes. In *Adaptation within Antarctic Ecosystems*, pp. 439–58. Edited by Llano, G.A. Gulf Publishing Co., Houston.

DeVries, A.L. and Lin, Y. (1977b). Structure of a peptide antifreeze and mechanisms of adsorption to ice. *Biochimica et Biophysica Acta* **495**, 388–92.

Duman, J.G. and DeVries, A.L. (1976). Isolation, characterisation and physical properties of protein antifreezes from the winter flounder, *Pseudopleuronectes americanus. Comparative Biochemistry and Physiology* **53B**, 375–80.

Dunbar, M.J. (1968). *Ecological Development in Polar Regions.* Prentice-Hall, Englewood cliffs, N.J.

Fox, H.M. (1936). The activity and metabolism, of poikilothermal animals in different latitudes. *Proceedings of the Zoological Society of London* **98**, 945–55.

Freed, J.M. (1971). Properties of muscle phosphofructokinase of cold- and warm-acclimated *Carassius auratus. Comparative Biochemistry and Physiology* **39B**, 747–64.

Funabiki, R. and Cassens, R.G. (1972). Heterogeneous turnover of myofibrillar protein. *Nature New Biology* **236**, 249–50.

Garwood, P.R. and Olive, P.J.W. (1978). Environmental control of reproduction in the polychaetes *Eulalia viridis* and *Harmothoe imbricata.* In *Physiology and Behaviour of Marine Organisms*, pp. 331–99. Edited by McClusky, D.S. and Berry, A.J. (*Proc. 12th Europ. Mar. Biol. Symp.*), Pergamon, Oxford.

Hammen, C.C. (1979). Metabolic rates of marine bivalve molluscs determined by calorimetry. *Comparative Biochemistry and Physiology* **62A**, 955–9.

Hazel, J.R. (1972a). The effect of temperature acclimation upon succinic dehydrogenase activity from the epaxial muscle of the common goldfish (*Carassius auratus* L.) – I. Properties of the enzyme and the effect of lipid extraction. *Comparative Biochemistry and Physiology* **43B**, 837–61.

Hazel, J.R. (1972b). The effect of temperature acclimation upon succinic dehydrogenase activity from the epaxial muscle of the common goldfish (*Carassius auratus* L.) – II. Lipid reactivation of the soluble enzyme. *Comparative Biochemistry and Physiology* **43B**, 863–82.

Hazel, J.R. and Prosser, C.L. (1974). Molecular mechanisms of temperature compensation in poikilotherms. *Physiological Reviews* **54**, 620–77.

Hazel, J.R., Garlich, W.S. and Sellner, P.A. (1978). The effect of assay temperature upon the pH optima of enzymes from poikilotherms: a test of the imadazole alphastat hypothesis. *Journal of Comparative Physiology* 97–104.

Heisler, N., Weitz, H. and Weitz, A.M. (1976). Extracellular and intracellular pH with changes of temperature in the dogfish *Scyliorhinus stellaris. Respiration Physiology* **26**, 249–63.

Hochackha, P.W. and Somero, G.N. (1973). *Strategies of Biochemical Adaptation* Saunders, W.B. Philadelphia.

Hoffman, R.J. (1981a). Evolutionary genetics of *Metridium senile*. I. Kinetic differences in phosphoglucose isomerase allozymes. *Biochemical Genetics* **19**, 129–44.

Hoffman, R.J. (1981b). Evolutionary genetics of *Metridium senile* II. Geographic patterns of allozyme variation. *Biochemical Genetics* **19**, 145–54.

Hoffman, R.J. (1983). Temperature modulation of the kinetics of phosphoglucose isomerase genetic variants from the sea anemone *Metridium senile. Journal of Experimental Zoology* **227**, 361–70.

Holeton, G.F. (1973). Respiration of arctic char (*Salvelinus alpinus*) from a high arctic lake. *Journal of the Fisheries Research Board of Canada* **30**, 717–23.

Holeton, G.F. (1974). Metabolic cold adaptation of polar fish: fact or artefact? *Physiological Zoology* **47**, 137–52.

Ivleva, I.V. (1980). The dependence of crustacean respiration on body mass and habitat temperature. *Internationale Revue der Gesamten Hydrobiologie* **65**, 1–47.

Jobling, M. (1983). Towards an explanation of specific dynamic action (SDA). *Journal of Fish Biology* **23**, 549–55.

Johnston, I.A. (1979). Calcium regulatory proteins and temperature acclimation of actomyosin from a eurythermal teleost (*Carassius auratus* L.). *Journal of Comparative Physiology* **129**, 163–7.

Johnston, I.A. (1981). Structure and function of fish muscles. *Symposia of the Zoological Society of London* **48**, 71–113.

Johnston, I.A. (1983). Cellular responses to an altered body temperature: the role of alterations in the expression of protein isoforms. In *Cellular Acclimatisation to Environmental Change*, pp. 121–43. Edited by Cossins, A.R. and Sheterline, P. Society for Experimental Biology, Seminar Series, 17. Cambridge University Press, Cambridge.

Johnston, I.A. (1985). Temperature adaptation of enzyme function in fish muscle. In *Physiological Adaptations of Marine Animals*, pp. 95–122. Edited by Laverack, M.S. Society for Experimental Biology, Seminar Series, 39. The Company of Biologists Ltd, Cambridge.

Johnston, I.A. and Goldspink, G. (1975). Thermodynamic activation parameters of fish myofibrillar ATPase enzyme and evolutionary adaptations to temperature. *Nature, London* **257**, 620–22.

Johnston, I.A. and Harrison, P. (1985). Contractile and metabolic characteristics of muscle fibres from Antarctic fish. *Journal of Experimental Biology* **116**, 223–36.

Johnston, I.A. and Lucking, M. (1978). Temperature induced variation in the distribution of different types of muscle fibres in the goldfish (*Carassius auratus*). *Journal of Comparative Physiology* **124**, 111–16.

Johnston, I.A. and Maitland, B. (1980). Temperature acclimation in crucian carp: a morphometric analysis of fish muscle fibre ultrastructure. *Journal of Fish Biology* **17**, 113–25.

Johnston, I.A. and Sidell, B.D. (1984). Differences in the temperature dependence of muscle contractile properties and myofibrillar ATPase activity in a cold-temperate teleost. *Journal of Experimental Biology* **111**, 179–89.

Johnston, I.A. and Walesby, N.J. (1977). Molecular mechanisms of temperature adaptation in fish myofibrillar adenosine triphosphatases. *Journal of Comparative Physiology* **119**, 195–206.

Johnston, I.A., Davison, W. and Goldspink, G. (1975). Adaptations in Mg^{++} – activated myofibrillar ATPase activity induced by temperature acclimation. *FEBS Letters* **50**, 293–5.

Johnston, I.A., Fitch, N., Zummo, G., Wood, R.E., Harrison, P. and Tota, B. (1983). Morphometric and ultrastructural features of the ventiscular myocardium of the haemoglobin-less icefish *Chaenocephalus aceratus*. *Comparative Biochemistry and Physiology* **76A**, 475–480.

Jokumsen, A., Wells, R.M.G., Ellerton, H.P. and Weber, R.E. (1981). Haemocyanin of the giant Antarctic isopod, *Glytonotus antarcticus*: structure and effects of temperature and pH on its ogygen affinity. *Comparative Biochemistry and Physiology* **70A**, 91–5.

Jones, P.L. and Sidell, B.D. (1982). Metabolic responses to striped bass (*Morone saxatilis*) to temperature acclimation. II. Alterations in metabolic carbon sources and distributions of fiber types in locomotory muscle. *Journal of Experimental Zoology* **219**, 163–71.

Kacser, H. (1983). The control of enzyme systems *in vivo*: elasticity analysis of the steady state. *Biochemical Society Transactions* **11**, 35–40.

Kacser, H. and Burns, J.A. (1973). The control of flux. In *Rate of Control of Biological Processes*, pp. 65–104. Edited by Davies, D.D. Cambridge University Press (Symposium of the Society for Experimental Biology, No. 27), Cambridge.

Kacser, H. and Burns, J.A. (1979). Molecular democracy: who shares the controls? *Biochemical Society Transactions* **7**, 1149–60.

Kanwisher, J.W. (1955). Freezing in intertidal animals. *Biological Bulletin Marine Biological Laboratory Woods Hole, Mass.* **109**, 56–63.

Kleckner, N.W. and Sidell, B.D. (1985). Comparison of maximal activities of enzymes from tissues of thermally acclimated and naturally acclimatized chain pickerel (*Esox niger*). *Physiological Zoology* **58**, 18–28.

Knight, C.A., DeVries, A.L. and Oolman, L.D. (1984). Fish antifreeze protein and the freezing and recrystallisation of ice. *Nature*, (*London*) **308**, 295–6.

Krogh, A. (1916). *The Respiratory Exchange of Animals and Man*. Longmans, London.

Leim, A.H. and Scott, W.B. (1966). Fishes of the Atlantic Coast of Canada. *Bulletin of Fish Research Bd Can*. No. **155**.

Lin, Y. (1979). Environmental regulation of gene expression: *in vitro* translation of winter flounder antifreeze mRNA. *Journal of Biological Chemistry* **254**, 1422–6.

Luxmoore, R.A. (1984). A comparison of the respiration rates of some Antarctic isopods with species from lower altitudes. *British Antarctic Survey Bulletin* **62**, 53–65.

Malan, A. (1980). Enzyme regulation, metabolic rate and acid–base state in hibernation. In *Animals and Environmental Fitness*, pp. 487–501. Edited by Gilles, R. Pergamon, Oxford.

Malan, A., Wilson, T.L. and Reeves, R.B. (1976). Intracellular pH in cold-blooded vertebrates as a function of body temperature. *Respiration Physiology* **28**, 29–47.

McLaren, I.A., Corkett, C.J. and Zillioux, E.J. (1969). Temperature adaptation of copepod eggs from the arctic to the tropics. *Biological Bulletin Marine Biological Laboratory, Woods Hole, Mass.* **137**, 486–93.

Moerland, T.S. and Sidell, B.D. (1981). Characterisation of metabolic carbon flow in hepatocytes isolated from thermally acclimated killifish *Fundulus heteroclitus*. *Physiological Zoology* **54**, 379–89.

Morita, R.Y. (1968). The basic nature of marine psychrophilic bacteria. *Bulletin of the Misaki Marine Biological Institute. Kyoto University* **12**, 163–77.

Murphy, D.J. (1977a). Metabolic and tissue solute changes associated with changes in the freezing tolerance of the bivalve mollusc *Modiolus demissus*. *Journal of Experimental Biology* **69**, 1–12.

Murphy, D.J. (1977b). A calcium-dependent mechanism responsible for increasing the freezing tolerance of the bivalve mollusc *Modiolus demissus*. *Journal of Experimental Biology* **69**, 13–21.

Murphy, D.J. (1979). A comparative study of the freezing tolerances of the marine snails *Littorina littorea* (L.) and *Nassarius obsoletus* (Say). *Physiological Zoology* **52**, 219–30.

Murphy, D.J. and Pierce, S.K. (1975). The physiological basis for changes in the freezing tolerance of intertidal molluscs. I. Response to subfreezing temperatures and the influence of salinity and temperature acclimation. *Journal of Experimental Zoology* **193**, 313–22.

Newsholme, E.A. and Paul, J.M. (1983). The use of *in vitro* enzyme activities to indicate the changes in metabolic pathways during acclimation. In *Cellular Acclimatisation to Environmental Change*, pp 81–107. Edited by Cossins, A.R. and Sheterline, P. Society for Experimental Biology, Seminar Series, No. 17. Cambridge University Press, Cambridge.

O'Grady, S.M., Clarke, A. and DeVries, A.L. (1982). Characteristics of glycoprotein antifreeze biosynthesis in isolated hepatocytes from

Pagothenia borchgrevinki. *Journal of Experimental Zoology* **220**, 179–89.

Palumbi, S.R., Sidell, B.D., Van Beneden, R., Smith, G.D. and Powers, D.A. (1980). Glucosephosphate isomerase (GPI) of the teleost *Fundulus heteroclitus* (Linnaeus): isozymes, allozymes and their physiological roles. *Journal of Compartive Physiology* **138**, 49–57.

Park, Y.S. and Hong, S.K. (1976). Properties of toad skin Na, K-ATPase with special reference to the effect of temperature. *American Journal of Physiology* **231**, 1356–63.

Parry, G. (1978). Effects of growth and temperature acclimation on metabolic rate in the limpet *Cellana tramoserica* (Gastropoda: Patellidea). *Journal of Animal Ecology* **47**, 351–68.

Parry, G. (1983). The influence of the cost of growth on ectotherm metabolism. *Journal of Theoretical Biology* **107**, 453–77.

Pauling, L. (1970). *General Chemistry*, 3rd edition. W.H. Freeman, San Francisco.

Powers, D.A. and Place, A.R. (1978). Biochemical genetics of *Fundulus heteroclitus* (L.) I. Temporal and spatial variation in gene frequencies of Ldh-B, Mdh-A, Gpi-B, and Pgm-A. *Biochemical Genetics* **16**, 593–607.

Precht, H., Christophersen, J. and Hensel, H. (1955). *Temperature und Leben*. Springer-Verlag, Berlin.

Prosser, C.L. (editor). (1973). *Comparative Animal Physiology*. W.B. Saunders, Philadelphia.

Qvist, J., Weber, R.E., DeVries, A.L. and Zapol, W.M. (1977). pH and haemoglobin oxygen affinity in blood from the Antarctic cod *Dissostichus mawsoni*. *Journal of Experimental Biology* **67**, 77–88.

Ralph, R. and Maxwell, J.G.H. (1977a). The oxygen consumption of the Antarctic limpet *Nacella* (*Patinigera*) *concinna*. *British Antarctic Survey Bulletin* **45**, 19–23.

Ralph, R. and Maxwell, J.G.H. (1977b). The oxygen consumption of the Antarctic lamellibranch *Gaimardia trapesina trapesina* in relation to cold adaptation in polar invertebrates. *British Antarctic Survey Bulletin* **45**, 41–6.

Randall, D.J. and Cameron, J.N. (1973). Respiratory control of arterial pH as temperature changes in rainbow trout, *Salmo gairdneri*. *American Journal of Physiology* **225**, 997–1002.

Reeves, R.B. (1972). An imidazole alphastat hypothesis for vertebrate acid–base regulation: tissue carbon dioxide content and body temperature in bullfrogs. *Respiration Physiology* **28**, 49–63.

Reeves, R.B. (1977). The interaction of acid–base balance and body temperature in ectotherms. *Annual Review of Physiology* **39**, 559–86.

Richardson, M.D. and Hedgpeth, J.W. (1977). Antarctic soft-bottom macrobenthic community adaptations in a cold, stable, highly productive, glacially affected environment. In *Adaptations within Antarctic Ecosystems*, pp. 181–90. Edited by Llano, G.A. The Smithsonian Institution, Washington, D.C.

Roland, W. and Ring, R.A. (1977). Cold, freezing and desiccation tolerance of the limpet *Acmaea digitalis* (Eschscholtz). *Cryobiology* **14**, 228–35.

Scholander, P.F., Vandan, L., Kanwisher, J.W., Hammel, H.T. and

Gordon, M.S. (1957). Supercooling and osmoregulation in Arctic fish. *Journal of Cellular Comparative Physiology* **49**, 5–24.

Scott, G.K., Fletcher, G.L. and Davies, P.L. (1986). Fish antifreeze proteins: recent gene evolution. *Canadian Journal of Fisheries and Aquatic Sciences* **43**, 1028–34.

Shaklee, J.B., Christiansen, J.A., Sidell, B.D., Prosser, C.L. and Whitt, G.S. (1977). Molecular aspects of temperature acclimation in fish: contributions of changes in enzymic activities and isoenzyme patterns to metabolic reorganisation in the green sunfish. *Journal of Experimental Zoology* **201**, 1–20.

Shick, J.M. (1981). Heat production and oxygen uptake in intertidal sea anemones from different shore heights during exposure to air. *Marine Biology Letters* **2**, 225–36.

Shick, J.M., Hoffman, R.J. and Lamb, A.N. (1979). Asexual reproduction, population structure, and genotype-environment interactions in sea anemones. *American Zoologist* **19**, 699–713.

Sidell, B.D. (1977). Turnover of cytochrome c in skeletal muscle of green sunfish (*Lepomis cyanellus*, R.) during thermal acclimation. *Journal of Experimental Zoology* **199**, 233–50.

Sidell, B.D. (1980). Responses of goldfish (*Carassius auratus*. L.) muscle to acclimation temperature: alterations in biochemistry and proportions of different fiber types. *Physiological Zoology* **53**, 98–107.

Sidell, B.D. (1983). Cellular acclimation to environmental change by quantitative alterations in enzymes and organelles. In *Cellular Acclimatisation to Environmental Change*, pp. 103–20. Edited by Cossius, A.R. and Sheterline, P. Society for Experimental Biology, Seminar Series 17, Cambridge University Press, Cambridge.

Sidell, B.D. and Johnston, I.A. (1985). Thermal sensitivity of contractile function in chain pickerel (*Esox niger*). *Canadian Journal of Zoology* **63**, 811–16.

Sidell, B.D., Johnston, I.A., Moerland, T.S. and Goldspink, G. (1983). The eurythermal myofibrillar protein complex of the mummichog (*Fundulus heteroclitus*) adaptation to a fluctuating thermal environment. *Journal of Comparative Physiology* **153**, 167–73.

Sinensky, M., Pinkerton, F., Sutherland, E. and Simon, F.R. (1979). Rate limitation of (Na^+ + K^+)-stimulated ATPase by membrane acyl chain ordering. *Proceedings of the National Academy of Sciences USA* **76**, 4893–7.

Smith, M.A.K. and Haschemeyer, A.E.V. (1980). Protein metabolism and cold adaptation in Antarctic fishes. *Physiological Zoology* **53**, 373–82.

Smith, M.A.K., Mathews, R.W., Hudson, A.P. and Haschemeyer, A.E.V. (1980). Protein metabolism of tropical reef and pelagic fish. *Comparative Biochemistry and Physiology* **65B**, 415–18.

Somero, G.N. (1978). Temperature adaptation of enzymes: biological optimisation through structure–function compromises. *Annual Review of Ecology and Systematics* **9**, 1–29.

Sømme, L. (1966). Seasonal changes in the freezing tolerance of some intertidal animals. *Nytt. magasin for zoologi.* **13**, 52–5.

Stone, B.B. and Sidell, B.D. (1981). Metabolic responses of striped bass (*Morone saxatilis*) to temperature acclimation. I. Alterations in carbon sources for hepatic energy metabolism. *Journal of Experimental Zoology* **218**, 371–9.

Theede, H. (1972). Vergleichende okologische-physiologische Untersuchungen zur zellularen Katteresistenz marinen Evertebraten. *Marine Biology (Berlin)* **15**, 160–91.

Theede, H., Schneppenheim, R. and Beress, L. (1976). frostschutz-glycoproteine bei *Mytilus edulis*? *Marine Biology (Berlin)* **36**, 183–9.

Trent, J.D., Chastain, R.A. and Yayanos, A.A. (1984). Possible artefactual basis for apparent bacterial growth at 250°C. *Nature (London)* **307**, 737–40.

Tyler, S. and Sidell, B.D. (1984). Changes in mitochondrial distribution and diffusion distances in muscle of goldfish upon acclimation to warm and cold temperatures. *Journal of Experimental Zoology* **232**, 1–9.

Vahl, O. (1978). Seasonal changes in oxygen consumption of the iceland scallop (*Chlamys islandica*) [O.F. Muller] from 70°N. *Ophelia* **17**, 143–54.

Vahl, O. (1983). The relationship between specific dynamic action (SDA) and growth in the common starfish. *Asterias rubens* L. *Oecologia*, **61**, 122–5.

Vandenheede, J.R., Ahmed, A.I. and Feeney, R.E. (1972). Structure and role of carbohydrate in freezing point-depressing glycoproteins from an Antarctic fish. *Journal of Biological Chemistry* **247**, 7885–9.

Walesby, N.J. and Johnston, I.A. (1981). Temperature acclimation of Mg^{2+} Ca^{2+} – myofibrillar ATPase from a cold-selective teleost, *Salvelinus fontinalis*: a compromise solution. *Experientia* **37**, 716–8.

White, M.G. (1975). Oxygen consumption and nitrogen excretion by the giant Antarctic isopod *Glyptonotus antarcticus* Eights in relation to cold-adapted metabolism in marine polar poikilotherms. In *Proceedings of the 9th European Marine Biology Symposium*, pp. 707–24. Edited by Barnes, H.

White, M.G. (1977). Ecological adaptations by Antarctic poikilotherms to the polar marine environment. In *Adaptations within Antarctic ecosystems*, pp. 197–208. Edited by Llano, G.A. Gulf Publishing Co., Houston.

Williams, R.J. (1970). Freezing tolerance in *Mytilus edulis. Comparative Biochemistry and Physiology* **35**, 145–61.

Wilson, F.R. (1973). *Enzyme changes in the goldfish (Carassius auratus L.) in response to temperature acclimation. I. An immunochemical approach. II. Isozymes.* PhD thesis, University of Illinois.

Yancey, P.H. and Somero, E.N. (1978). Temperature dependence of intracellular pH: its role in the conservation of pyruvate apparent K_m values of vertebrate lactate dehydrogenases. *Journal of Comparative Physiology* **125**, 129–34.

10

Mammalian hibernation

L.C.H. Wang
Department of Zoology
University of Alberta

Introduction

Natural torpidity in endotherms (e.g. mammals and birds) in response to cold and/or shortage of food or water is characterized by a profound reduction in body temperature and other physiological functions which may last from a few hours to a few weeks. Unlike ectotherms (e.g. frogs and snakes), torpid endotherms are capable of leaving the depressed metabolic state at any time, using exclusively endogenously produced heat to restore normal body temperature. Extensive studies of the biochemical, physiological,

neurophysiological and neuro-endrocrine aspects of this phenomenon have established that torpor in mammals represents an advanced form of thermoregulation rather than a reversion of primitive poikilothermy (Hudson, 1973; Lyman *et al.*, 1982).

Members of at least six mammalian orders are capable of exhibiting torpor and these have been described in detail (Lyman, 1982). Some examples are the echidna of *Monotremata*, many dasyurids of *Marsupialia*, tenrecs, shrews of the subfamily *Crocidurinae* (Vogel, 1980) and at least one of *Soricinae* (Lindstedt, 1980), hedgehogs of Insectivora, many bats of both sub-orders *Megachiroptera* and *Microchiroptera*, dwarf and mouse lemurs of *Primates* and the various sciurids, cricetids, heteromyids, murids and zapodids of *Rodentia*.

Depending upon timing, duration and depth, torpor in mammals may be characterized as seasonal or nonseasonal. Seasonal torpor is represented by estivation and hibernation, although in many animals these may be combined into one torpor season lasting more than 8 months (e.g. Richardson's ground squirrel; Wang, 1979). Some common groups which exhibit seasonal torpor include the hedgehogs (*Erinaceus*), marmots and woodchucks (*Marmota*), ground squirrels (*Spermophilus* [*Citellus*]), bats (*Eptesicus* and *Myotis*), and jumping mice (*Zapus*). In these animals, two distinct physiological states exist annually:

(i) *the non-hibernating phase*, during which body weight is relatively constant and exposure to cold results in both increased heat production and maintenance of euthermia, typical of other non-hibernating species;

(ii) *the hibernating phase*, during which a drastic, rapid weight gain occurs until a plateau is reached, followed by a gradual weight loss. Exposure to cold results in hibernation with body temperature capable of decreasing to near 0°C.

The physiological mechanisms underlying the transition from the non-hibernating to the hibernating phase are not understood, but alternation of the two phases is apparently regulated by an endogenous circannual rhythm (Fig. 10.1). This has a free-running periodicity of, typically, less than 365 days (Pengelley and Asmundson, 1974) and may be entrained to synchronize with the natural environment by zeitgebers (Germ: *zeit*-time; *geber*-giver) such as photoperiod and temperature (Mrosovksy, 1978; Joy and Mrosovsky, 1983).

Non-seasonal torpor, by definition, indicates torpor that can be induced at any time of the year when the proper stimuli are presented to the animals (i.e. cold and/or shortage of food). Typical is daily torpor, the duration of which is less than 24 hours with the lowest level of body temperature tolerable generally between 10°C and 22°C (considerably higher than that found during hibernation). Diverse groups such as marsupials, insectivores, chiropterans, primates and rodents all contain members which exhibit daily torpor (Hudson, 1978). Another type of non-seasonal torpor is found in the Syrian (golden) hamster (*Mesocricetus auratus*), which exhibits deep torpor (body temperature decreasing to 5°C and lasting a few days) but requires a few weeks of cold exposure before torpor.

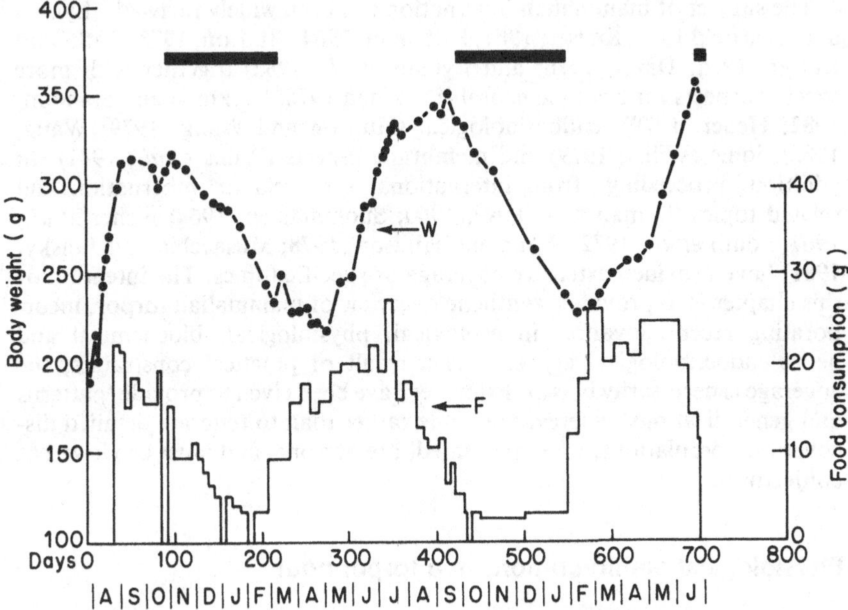

Fig. 10.1 Circannual rhythms of body weight (W), food consumption (F) and hibernation season (▬ ; periodic arousals not indicated) in the gold-mantled ground squirrel under constant temperature (22°C) and photoperiod (12L:12D) in the laboratory over 2 years. The non-hibernating phase is characterized by the declining and low body weight whereas a rapid weight gain typically signals the transition from the non-hibernating to the hibernating phase. (Reproduced with permission from Pengelley and Fisher [1963] *Can. J. Zool.* **41**., 1103–20.)

Despite differences in pattern of torpor, studies of metabolism, heart rate, respiratory rate and body temperature during entry into, maintenance of, and arousal (rewarming) from the torpid state have shown that the patterns are basically similar, differing only in quantitative aspects (Hudson, 1973). Thus, in spite of the polyphyletic origin in evolution of torpor in mammals (e.g. marsupials vs. rodents; Hudson, 1973), the specific ecological constraint of the environments (e.g. desert vs. alpine), and the great diversity of patterns in torpor, there is physiological convergence in achieving a depressed metabolic state for energy/water conservation. This does not necessarily mean, however, that the regulatory mechanisms which govern torpor are also similar. On the contrary, very different aspects of neuroendocrine regulation (Hudson and Wang, 1979; Wang, 1982) and intermediary metabolism (Willis, 1982a) have been observed amongst different species. Thus, one must be cautious when extrapolating mechanistic findings from one species to another, even though they may share similarities in metabolic and body temperature depressions during torpor. This may be problematic when attempting to generalize about aspects of torpor, but such differences do underscore the heterogeneity in selection pressure that different groups have been subjected to in their ecological niches.

The subject of mammalian hibernation has been widely reviwed, (Lyman and Chatfield 1955; Kayser, 1961; Hoffman, 1964; Hudson, 1973; Raths and Kulzer, 1976; Davis, 1976; and Lyman *et al.*, 1982) together with more specific aspects for example neural (Beckman 1978; Beckman and Standon, 1982; Heller, 1979), endocrinological (Hudson and Wang, 1979; Wang, 1982), ionic (Willis, 1979) and membrane aspects (Willis *et al.*, 1981). In addition, proceedings from international symposia on hibernation and related topics (Lyman and Dawe, 1960; Suomalainen, 1964; Fisher *et al.*, 1967; South *et al.*, 1972; Wang and Hudson, 1978; Musacchia and Jansky, 1981) have provided extensive coverage of specific topics. The intention of this chapter is to provide a synthetic overview of mammalian torpor, incorporating recent advances in ecological, physiological, biochemical and neuro-endocrinological aspects. As a result of practical constraints, the coverage is necessarily biased. Priorities have been given to provide 'patterns and generalizations' wherever possible rather than to reiterate detailed discoveries. Speculations, when presented, are personal and must be viewed as conjectural.

Physiological manifestations in a torpor bout

A torpor bout consists of entry into, maintenance of, and arousal from torpor. Following complete arousal a period of euthermia lasting from a few hours to a few days ensues before the animal enters another bout of torpor. To date, all mammalian torpor follow this pattern regardless of whether it is hibernation, estivation, daily torpor, or others (e.g. torpor in Syrian hamster). Thus, in a hibernation season, torpor is not continuous but is composed of many individual torpor bouts and inter-torpor euthermic periods (Fig. 10.2).

During entry into torpor, there is marked inhibition of heart rate, occurrence of arrhythmia as a result of enhanced parasympathetic activity, a progressive increase in vasoconstriction to maintain blood pressure while heart rate decreases, a decrease in respiratory rate with irregular periods of apnoea, and a significant decrease in oxygen consumption. Body temperature falls following the decrease of heat production with periodic shivering serving to temporarily raise or stabilize the body temperature against too fast a decrease. These deliberated controls indicate that entry into torpor is an actively regulated process rather than a passive abandonment of the euthermic state (Lyman, 1965; Hudson, 1973). It therefore takes approximately three to five times longer to cover the same range of body temperature change during entry into torpor than during arousal.

During torpor, which may last from a few hours to a few weeks, all physiological functions are at minimum. Heart rate may decrease to 1/30 or less and oxygen consumption to 1/100 or less of their respective euthermic levels. Prolonged apnoea lasting up to 40 minutes has been observed in the ground squirrel (Steffen and Riedesel, 1982) and up to 150 minutes in the hedgehog (Kristoffersson and Soivio, 1964). Cheyne–Stoke breathing (apnoea followed by bursts of breathing) is typical for the hedgehog, dormouse,

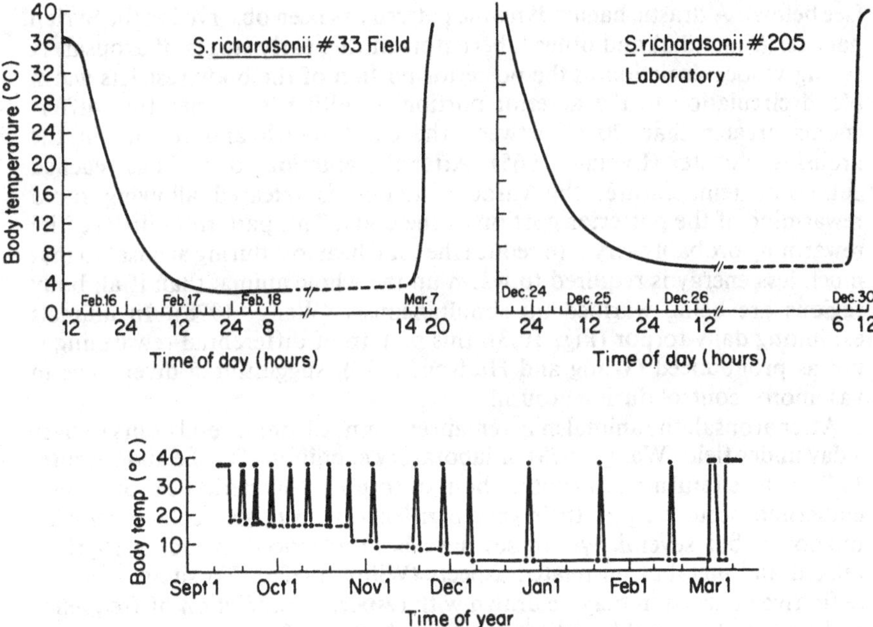

Fig. 10.2 (a) Typical dynamic change of body temperature (via radiotelemetry) during a torpor bout in the Richardson's ground squirrel under field (#33) and laboratory (#205) conditions. Note the relatively long duration of entry into torpor (24 hours or longer) and the relatively short duration of arousal from torpor (less than 4 hours). Note also the similarities between the two recordings both qualitatively and quantitatively. (b) A complete torpor season in a juvenile male Richardson's ground squirrel under field conditions. Note the seasonal variations in duration of torpor and the decrease of minimum body temperature during torpor. The inter-torpor euthermic period is too short to be represented accurately except in early March when the euthermic period exceeded 24 hours. (Reproduced from Wang [1978] with permission.)

hamster, and the ground squirrel but has not been observed in the marmot. Body temperature is typically within 2°C of the ambient down to a 'critical' level, below which, a further decrease in ambient temperature results in one of three possibilities:

(i) temperature stays at the 'critical' level at the expense of increased metabolism,
(ii) initiation of arousal,
(iii) death if body temperature falls below the 'critical' level rendering the animal incapable of spontaneous arousal.

Arousal from torpor is an explosive event requiring from 20 minutes in small rodents and bats to a few hours in squirrels and marmots. The sympathetic drive is at maximum during arousal, stimulating substrate mobilization for energy production, the cardiovascular system for tissue perfusion and non-shivering thermogenesis in the brown adipose tissue for heat production (see below). A drastic haemodynamic pattern has been observed in the Syrian hamster (Fig. 10.3) and other hibernators during early stages of arousal. A strong vasoconstriction in the posterior portion of the body restricts warm blood circulation to the anterior portion, resulting in temperature differentials greater than 20°C between the check-pouch and rectum in the arousing hamster (Lyman, 1965). After the anterior portion has reached euthermic temperature, the vasoconstriction is released allowing rapid rewarming of the posterior portion of the body. This pattern of differential rewarming probably serves to reduce the total heat loss during arousal so that much less energy is required to warm up the whole animal than if all body regions are being warmed up simultaneously (Wang, 1978). In animals exhibiting daily torpor (Fig. 10.3), this pattern of differential rewarming is not as pronounced (Wang and Hudson, 1970), suggesting a divergence in vasomotor control during arousal.

After arousal, the animal may remain euthermic from several hours to over a day under field (Wang, 1979) or laboratory conditions (Torke and Twente, 1977) before returning to another bout of torpor. Why periodic arousal and euthermia occur despite their very high energy costs (see below) remains unknown, but several hypotheses have been advanced including rhythm-related and metabolism-related aspects (Willis, 1982a). To summarize, no definitive conclusion may be drawn with respect to prediction of frequency and timing of arousal based on circadian rhythms of body temperature and sleep-wakefulness as observed in euthermia. With regard to the metabolism-related aspects, the need to eliminate nitrogenous wastes accumulated during torpor has been dispelled (Pengelley *et al.*, 1971). The need for replenishing carbohydrate supplies, diminished as a result of their continuous use during torpor (Galster and Morrison, 1975), remains a possibility in the Arctic ground squirrel, although it is apparenty not the case in the golden-mantled ground squirrel (Zimmerman, 1982). Finally, the need for restoring ionic balance altered during torpor (Willis *et al.*, 1971, and below) remains a possibility, although experimental testing may be difficult (Willis, 1982a).

Carnivorean lethargy

A current controversy concerns whether 'winter dormancy' in bears should also be included as true hibernation (Nelson, 1980; Lyman, 1982). During winter denning, which lasts more than 100 days, bears do not feed, urinate or defecate and they use fat exclusively as their energy and water source. The body temperature of the black bear during denning ranges from 31°C to

Fig. 10.3 (a) Changes of oxygen consumption (○), cheek-pouch (●), and rectal temperature (◑) during arousal from torpor in the Syrian hamster. Note the large temperature differential between the cheek-pouch and rectum. (Reproduced with permission from Lyman [1948] *Journal of Experimental Zoology* **109**, 55–78.) (b) Changes in oxygen consumption (○), neck (●) and abdomen temperature (◑) during arousal from daily torpor in the pocket mouse. Note the relatively small temperature differential between the neck and abdomen. (Redrawn from Wang and Hudson, 1970).

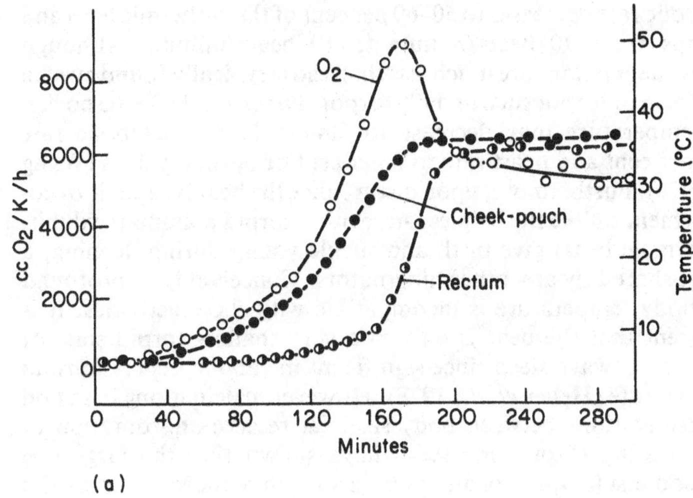
O2
Cheek-pouch
Rectum
8000
6000
4000
2000
0
cc O2/K/h
50
40
30
20
10
Temperature (°C)
0
40
80
120
160
200
240
280
Minutes
(a)

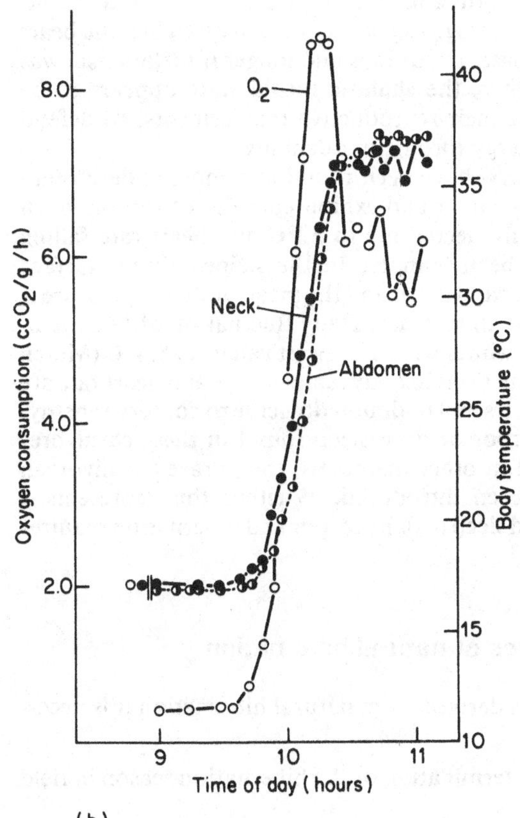
O2
Neck
Abdomen
8.0
6.0
4.0
2.0
Oxygen consumption (ccO2/g/h)
40
35
30
25
20
15
10
Body temperature (°C)
9
10
11
Time of day (hours)
(b)

33 °C, its metabolic rate decreases to 50–60 per cent of the euthermic level and heart rate drops from 40 beats/minute to 10 beats/minute. Although impressive, these depressions are much less than those typically found even in the shallowest form of torpor such as daily torpor. For example, in the pocket mouse body temperature may decrease to almost 12°C, metabolic rate decreases to 3 per cent and heart rate to 6 per cent of normal values (Wang and Hudson, 1970). Furthermore, upon disturbance the bear is capable of co-ordinated movement unlike the helpless state other torpid mammals exhibit. Additionally, female bears give birth and suckle young during denning, a unique trait not shared by any other hibernators. Conceivably, a profound depression in body temperature is incompatible with these activities. It is therefore apparent that the bear is only in a very shallow torpid state or perhaps in deep slow wave sleep since significant metabolic depressions do occur during this state (Heller *et al.*, 1978). However, calculations based on theoretical considerations between body size, fat reserve and duration of survival under fasting (Morrison, 1960) have shown that the large size (300 kg) and good insulation of bears result in a much reduced lower critical temperature for its thermal neutral zone (approximately − 50°C) and a low thermal conductance. Consequently, reduction of basal metabolic rate and cooling of body temperature towards ambient temperature could be difficult. With only a moderate fat reserve (e.g. 13 per cent of body weight), the bears could survive 100 days under basal conditions and longer if metabolism was further reduced. Taken together, the shallow torpid state appears to be optimal for bears to carry out their reproductive requirements, to defend themselves and to conserve energy throughout denning.

Other carnivores (e.g. badgers) have been found to remain underground for 70 consecutive days during winter and exhibit episodes of torpor which last 29 hours, body temperature decreasing to 29°C and heart rate falling from 55 beats/minute to 25 beats/minute. In the striped skunk surface activity is nil between December and April and the mean body temperature is 3.3°C lower during winter than in summer. Daily fluctuation of body temperature is also greatest during winter with the lowest value at 28.4°C (Mutch and Aleksiuk, 1977). These reductions in body temperature and heart rate are similar to those found in the bears and undoubtedly serve to conserve energy. To distinguish the shallow torpor or deep sleep found in these carnivores from those typically observed in other mammals, the phrase 'carnivorean lethargy' (Hock, 1960) has been introduced. Whether this represents a continuum from euthermia and sleep to daily torpor and hibernation requires further elucidation.

Time pattern and energetics of natural hibernation

To quantify the energy savings derived from natural hibernation it is necessary to document:

(i) the timing of onset and termination of the hibernation season in field animals;

(ii) the duration of each hibernation bout and its seasonal variation;

(iii) the energy consumption during entry into, maintenance of, and arousal from hibernation at different body and ambient (burrow) temperatures;

(iv) the duration of euthermia between successive hibernation bouts and the energy consumption associated with this period at different ambient temperatures.

A combined field and laboratory study, employing temperature-sensitive radiotransmitters (Wang, 1979) has provided information about these aspects in the Richardson's ground squirrel (*Spermophilus richardsonii*). In this species, adults commence estivation/hibernation from early to mid-July and the young start estivation/hibernation in mid-September. Emergence from hibernation is similar for all age classes and occurs around mid-March. The average duration of torpor ranges from 3 to 4 days in July and August and gradually lengthens to 15–19 days in December–January, shortening again to 14 days in February and 6 days in March. The body temperature during torpor is within 1-4°C of burrow temperature and ranges from 16.4°C in July to 2.1°C in January-February. The duration of euthermia between successive hibernation bouts ranges from 5 to 25 hours depending on the season, being longer in February–March than in November–December. A typical individual record for the torpor season is shown in Fig. 10.2.

By using the time pattern, body and ambient temperatures observed in hibernating squirrels under field conditions and incorporating metabolic measurements made throughout hibernating bouts in the laboratory, an energy budget for the torpor season can be derived. The proportional energy costs for different stages of a hibernation bout are: (a) entry into torpor 12.8 per cent, (b) maintenance of torpor 16.6 per cent, (c) arousal from torpor 19.0 per cent and (d) euthermia between torpor 51.6 per cent. The monthly energy savings by utilization of torpor vs. resting but maintaining euthermia range from 38 per cent in July to 51.2 per cent in March but are 81.6–96.0 per cent when considered between August and February. For the complete torpor season which lasts approximately 8 months in adult Richardson's ground squirrels, the energy saving comes to 87.8 per cent (Wang, 1979).

Physiological adaptations

Central nervous system regulation of body temperature

A variety of neuronal and neurochemical models have been proposed for the regulation of body temperature in euthermic as well as in hibernating mammals (South *et al.*, 1978 for review). The neurochemical adjustments which allow the transition from euthermia to hibernation are unknown. It has been suggested, however, that on theoretical considerations two additional interactive neuronal pools which are active only during the hibernating phase could exert active inhibition or facilitation on thermogenesis in the cold allowing the occurrence and termination, respectively, of hibernation (South *et al.*, 1978). Evidence favouring this proposition is presented below.

It has been shown conclusively that ground squirrels and marmots enter hibernation through extension of slow wave sleep (Heller *et al.*, 1978). During slow wave sleep, the body temperature is regulated at a lower level (lowered set-point temperature, T_{set}) than during wakefulness and the thermosensitivity (magnitude of thermoregulatory response to thermal stimulation) of the hypothalamus is decreased. Consequently, metabolic rate during slow wave sleep is reduced. During entry into hibernation, there is a progressive decline in T_{set} and thermosensitivity, resulting in further decreases of metabolic rate and body temperature. Electrophysiological recordings from the diencephalon in hedgehogs during entry into hibernation indicate suppression of neuronal activities prior to any significant decrease in body temperature (Wünnenberg *et al.*, 1978). Other studies have also shown that during entry into hibernation suppression of electrical activities first occur in the neocortex, later in the midbrain reticular formation and lastly in areas of the limbic system (Wünnenberg *et al.*, 1978; Beckman and Stanton, 1982). During hibernation, at a brain temperature of 6.1°C, spontaneous electrical activity remains in most, if not all, brain areas but especially in the motor and sensory cortex, medial pre-optic area, septum, and the ventromedial nucleus of the hypothalamus, although the magnitude of electroencephalograph has decreased to about 10 per cent of that observed in euthermia (Strumwasser, 1960). Of particular interest is the typical electrical silence in the midbrain reticular formation during hibernation. However, when animal is touched with a glass rod, bursts of activity appear immediately, indicating that this area may be inhibited during hibernation but is capable of responding when stimulated. Summarizing the electrophysiological evidence, it has been proposed that increased inhibitory influence on midbrain reticular formation from the hippocampus facilitates sleep and this may be extended to initiate entry into and the maintenance of deep hibernation (Heller, 1979; Beckman and Stanton, 1982). The release of hippocampal inhibition or an activation of the midbrain reticular formation re-instates wakefulness or arousal from hibernation. A schematic representation of this hypothesis is depicted in Fig. 10.4.

Neurochemically, the regulation of body temperature in mammals is mediated by brain biogenic amines, in particular norepinephrine (noradrenaline) and serotonin (5-HT) acting on the pre-optic/anterior hypothalamic area (Myers, 1980). The possibility that amines, plus others, may be involved in the regulation of hibernation has been demonstrated by studies in which micro-injections of norepinephrine, 5-HT and acetylcholine into the pre-optic/anterior hypothalamic area of the golden-mantled ground squirrel trigger arousal, whereas only acetylcholine injected into the midbrain reticular formation elicits the same effect (Beckman, 1978; Beckman and Stanton, 1982). These results suggest that aminergic and cholinergic neurons in these areas may be involved in the mechanisms of arousal but how they act is, as yet, unknown.

Norepinephrine elicits increased heat production in the Richardson's ground squirrel when injected into the lateral ventricle, regardless of whether the squirrel is in its non-hibernating or hibernating phase (Glass and Wang, 1979a). In the 13-lined ground squirrel (Draskoczy and Lyman, 1967) and the

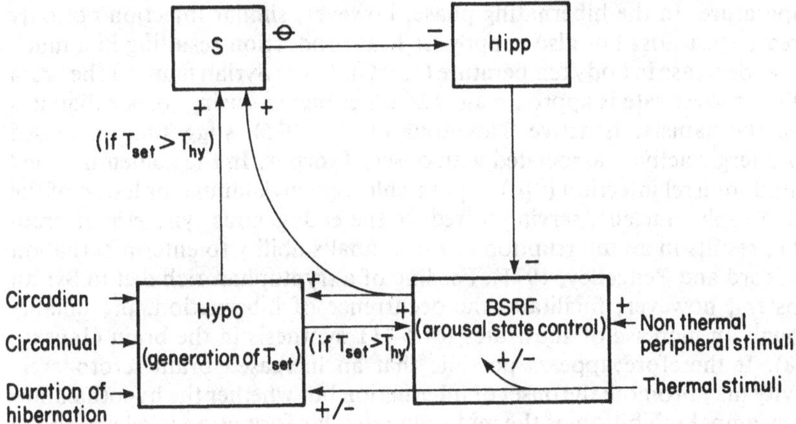

Fig. 10.4 Possible interactions between the brain-stem reticular formation (BSRF), the hypothalamus (Hypo), the limbic system as represented by the septum (S) and the hippocampus (Hipp) in the control of sleep/wakefulness cycle and the hibernation/euthermia cycle. Excitatory influences are indicated by '+' and inhibitory influences by '−'. T_{set} refers to the hypothalamic temperature threshold (set-point) for the metabolic heat production response. T_{hy} is the actual hypothalamic temperature. $T_{set} > T_{hy}$ indicates a 'thermogenic drive' for the thermoregulatory response. At the onset of sleep, excitatory influence from BSRF to Hypo is reduced, resulting in decreased T_{set} and decreased ($T_{set} - T_{hypo}$) which further decreases the excitatory influence on BSRF from Hypo (an internal positive feedback loop). Declining BSRF activity also decreases excitation to the septum, curtailing the activity of the septal theta-generating cells (θ), resulting in disinhibition of the hippocampus. Increasing hippocampal inhibition on the BSRF maintains the sleeping state. For entry into hibernation, the T_{set} is lowered further than that found in sleep. The circannual changes including the increased sleep time, lowered T_{set} for level of euthermic body temperature, and the low extrahypothalmic thermosensitivity during the hibernating phase of the animal may contribute to the lowering of T_{set} for entry into hibernation. Awakening or arousal from hibernation may be initiated by peripheral stimuli or internal influences (e.g. circadian drive in body temperature cycle) which are sufficient to overcome the tonic hippocampal inhibition during sleep or hibernation. (Reproduced from Heller, [1979] with permission.)

European hedgehog (Sauerbier and Lemmer, 1977), brain norepinephrine turnover rate is essentially nil during hibernation. In the Richardson's ground squirrel arousing from hibernation intraventricular injections of norepinephrine at body temperature of 10°C, 20°C and 30°C, increase the magnitude of thermogenesis and rate of rewarming, indicating that the thermogenic effect of norepinephrine persists even at depressed body temperature (Glass and Wang, 1979a). It would therefore appear that brain norepinephrine exerts mainly thermogenic effects and promotes exit from hibernation.

In contrast, intraventricular injection of 5-HT elicits a decrease in body temperature in the Syrian hamster (Jansky and Novotona, 1976). In the Richardson's ground squirrel (Glass and Wang, 1979b), intraventricular 5-HT elicits differential thermoregulatory effects depending on the physiological state of the animal. In the non-hibernating phase, 5-HT increases heat loss in the cold resulting in a slight decrease of body temperature (0.4°C) before compensatory increase of heat production restores normal body temperature. In the hibernating phase, however, similar injection not only increases heat loss but also suppresses heat production resulting in a much greater decrease in body temperature (1.6°C). In the Syrian hamster the brain 5-HT turnover rate is approximately 24 times higher during torpor than it is when the hamster is active (Novotona *et al.*, 1975), suggesting increased serotonergic activity associated with onset of torpor. In the golden-mantled ground squirrel injection (i.p.) of para-chlorophenylalanine, or lesion of the medial raphe nucleus, serving to reduce the endogenous synthesis of brain 5-HT, results in an interruption of the animal's ability to enter hibernation (Spafford and Pengelley, 1971). Feeding of a tryptophan-rich diet to Syrian hamsters, however, facilitates the occurrence of hibernation, presumably through an increase of substrates for 5-HT synthesis in the brain (Jansky, 1978). It therefore appears possible that an increased brain serotonergic activity may promote the onset of hibernation but whether the hypothalamo-hippocampal inhibition of the midbrain reticular formation is related to the heightened serotonergic activity remains to be investigated.

Cardiovascular adaptations

Cardiovascular adaptations related to hibernation have been extensively reviewed (Lyman, 1965; South *et al.*, 1972; Hudson, 1973), and only selected highlights will be presented here. Studies on the isolated, perfused organ indicate that hearts from species which exhibit deep torpor (body temperature typically 2–7°C, e.g. woodchuck, ground squirrel, chipmunk and hamster) continue to beat at −0.5°C to 7°C whereas hearts from animals which do not exhibit torpor (e.g. tree squirrel, white rat, cotton rat and mountain beaver) stop beating at 10 to 16°C (Lyman and Blinks, 1959). In animals which exhibit shallow, daily torpor (body temperature 8–22°C; e.g. pocket mice, white-footed mice), isolated hearts can beat at 1.5–7.2°C (Hudson, 1967). However, a significant increase in loss of intracellular K^+ and a decrease in ventricular tension have been observed in the white-footed mouse below 15°C, corresponding to the lowest tolerable body temperature during torpor in this species. In the ground squirrel atrial preparation both the membrane and action potentials are maintained between 37°C and 6°C (Marshall and Willis, 1962) whereas in the non-hibernating species both potentials decline rapidly below 20°C (Marshall, 1957). In the torpid hamster, the resting membrane potential is maintained at 12°C but is significantly decreased below 20°C in the non-torpid state (Jacobs and South, 1976). A torpid hamster also shows greater excitability, peak tension and intensity of active state between 38°C and 5°C in the cardiac trabeculae than a non-torpid animal (South and Jacobs, 1973). These two latter studies also indicate that transitional changes in cardiac function occur at the cellular level during the cold-acclimating period which hamsters require prior to exhibiting torpor.

Respiratory adaptations

As well as the typical reductions in ventilation during hibernation, a physiological trait unique to hibernators is the maintenance of a relatively constant extracellular pH (approximately 7.40) between euthermia and hibernation in spite of the 30°C or more change in body temperature (Malan, 1982). This is in strong contrast to the maintenance of constant alkalinity rather than pH in ectotherms under fluctuating temperatures (Reeves, 1977). Since neutral water alters from a pH of approximately 6.90 at 37°C to 7.40 at 5°C, the maintenance of pH at 7.40 during hibernation represents relative acidosis when compared to the euthermic state. The acidosis is respiratory in nature resulting from the intermittent breathing pattern during hibernation. The intracellular pH during hibernation is also low, ranging from 6.8 in the brain to 7.3 in the liver, with intermediate values in the diaphragm and skeletal muscle (Malan, 1982). The physiological significance of respiratory acidosis during hibernation is uncertain but may be related to inhibition of certain metabolic functions. For example, in a euthermic hamster respiratory acidosis induced by breathing 10 per cent CO_2 results in a depression of hypothalamic neuronal activity involved in thermoregulation (Wünnenberg and Baltruschat, 1982). This could facilitate the lowering of T_{set} and entry into hibernation as demonstrated recently for hamsters using 5 per cent CO_2, 21 per cent O_2 and 74 per cent N_2 (Kuhnen *et al.*, 1983). Glycolysis is also severely depressed by acidosis due to the inhibition of phosphofructokinase at low pH (Hand and Somero, 1983). The norepinephrine-stimulated non-shivering thermogenesis in the brown adipose tissue is also significantly depressed by acidosis (Friedli *et al.*, 1978). Consequently, a metabolic suppression beyond the level dictated by low temperature alone may be realized during hibernation (Malan, 1982). Since respiratory acidosis is easily correctable simply by increasing ventilation, the inhibition of neuronal and metabolic functions can be quickly reversed. In this regard, it is significant that during the early stages of arousal when thermogenesis is required, hyperventilation has been observed in the marmot resulting in a decrease in blood P_{CO_2} and an increase in pH (Malan, 1982).

Shivering and non-shivering thermogenesis

Two major types of heat production are involved during arousal from torpor, shivering and non-shivering thermogenesis (NST). Shivering is involuntary and is manifested as rhythmic, simultaneous contractions of flexor and extensor muscles, following an ordered or co-ordinated pattern in activation of different parts of the body. The primary center for regulating shivering is in the dorsomedial area of hypothalamus (Hemingway, 1963), and activation of shivering appears to be determined by both the skin and the spinal cord temperature (Brück and Wünnenberg, 1970). Non-shivering thermogenesis is a heat producing mechanism without muscle contraction. The major site of NST in mammals is the brown adipose tissue (BAT) which is well developed in neonates, many small mammals after cold-acclimation or winter acclimatization and in particular, adults of species which exhibit torpor (Janksy, 1973). The BAT is characterized by its multilocular lipid droplets, numerous mitochondria, high cytochrome oxidase activity, extensive vascularization and rich sympathetic innervation. In the cold-acclimated rat BAT can account for 65–80 per cent of the total thermogenesis during cold exposure, although the total amount is less than 2 per cent of body weight (Foster and Frydman, 1979). This extraordinary thermogenic capability of BAT is due to its biochemical properties which allow extensive uncoupling of oxidative phosphorylation, resulting in the production of heat instead of adenosine triphosphate (ATP). A BAT-specific inner mitochondrial membrane protein, thermogenin, which is 32 000 Daltons in weight and has specific binding sites for purine nucleotides (e.g. GDP, GTP, ADP, ATP) has been isolated (Cannon and Nedergaard, 1983; Nicholls and Locke, 1984). Thermogenin forms a part of the natural protonophore of the inner mitochondrial membrane. When bound by GDP, the channel is impermeable to proton flow and the proton gradient generated by the electron transport chain provides the energy needed for ATP synthesis (the coupled state). When the BAT is stimulated by sympathetic discharge, norepinephrine stimulates intracellular cyclic adenosine monophosphate (cAMP) formation via the beta-receptor + adenylate cyclase system located on the brown adipocyte membrane. The activation of lipase by cAMP leads to lipolysis and the formation of fatty acids and acyl-CoA. The fatty acids and acyl-CoA competitively bind to thermogenin and displace GDP. Consequently, the proton channel becomes permeable to proton flow and a proton short-circuit is formed which dissipates the energy gradient into heat instead of ATP synthesis (the uncoupled state). The metabolism of fatty acids sustains this uncoupled state.

The proportional contribution of shivering and NST during arousal from torpor differs with species. In the little brown bat (*Myotis lucifugus*), 80 per cent of heat production comes from NST (Hayward, 1968) whereas in the hamster and dormouse approximately 40 per cent and 20 per cent respectively, are derived in this way (Hayward and Lyman, 1967). In Richardson's ground squirrel the maximum thermogenic capacity does not vary seasonally but NST is significantly greater during the hibernating season, presumably to facilitate heat production during arousal (Wang and Abbotts, 1981).

Biochemical adaptations

Intermediary metabolism

Seasonal changes in activity of enzymes involved in fat metabolism have been observed. In the dormouse the activity of the adipose tissue glucose-6-phosphate dehydrogenase, a key enzyme in the pentose phosphate shunt which generates NADPH for lipogenesis, is six times greater in the autumn than it is in the spring and summer (Castex and Sutter, 1981). In the 13-lined ground squirrel the activities of liver enzymes involved in lipogenesis are twice as great in June as they are in September (Whitten and Klain, 1969). Furthermore, the incorporation of ^{14}C-glucose into adipose tissue lipids increases approximately 88–108-fold between June and August in the juvenile Richardson's ground squirrels in preparation for hibernation under field conditions (Bintz and Strand, 1983). Lipogenesis during the hibernation season is generally reduced as judged by the greatly decreased glucose-6-phosphate dehydrogenase activity in the liver of the 13-lined ground squirrel (Whitten and Klain, 1969) and in the adipose tissue of the hedgehog (Olsson, 1972), and the reduced incorporation of ^{14}C-glucose into total lipids in the adipose tissue of the hibernating golden-mantled ground squirrel (Tashima *et al.*, 1970). These observations suggest an increased lipogenic capacity during the summer–Autumn fattening and a decrease during the hibernating season. With respect to energy utilization during hibernation, fat is the primary fuel (Willis, 1982b). Since fatty acids are bound to serum albumin in their transport to tissues, the four-fold increase in serum albumin concentration in winter compared to its summer level observed in the arctic ground squirrel also reflects the increased reliance on fat metabolism during the hibernation season. In the hibernating big brown bat, *Eptesicus fuscus* fat metabolism supports nearly all the energy needs and glucose oxidation is severely inhibited; muscle mitochondria preferentially oxidize fatty acids over pyruvate and physiological concentrations of palmitoyl-carnitine inhibit pyruvate oxidation (Yacoe, 1983a). The inhibition of glucose oxidation during hibernation could be the result of the inactivation of phosphofructokinase, a key regulatory enzyme controlling glycolytic flux. Phosphofructokinase may be reversibly converted from a catalytically active tetrameric form to an inactive dimeric form by the combination of low intracellular pH and low temperature which prevail during the hibernating state (Hand and Somero, 1983). This serves to conserve muscle glycogen during hibernation but allows it to be used for shivering thermogenesis during arousal when acidosis is reversed (see above).

Although the primary fuel during hibernation is fat, the central nervous system continues to utilize glucose especially in areas receiving thermal afferents (e.g. the paratrigeminal nucleus) as illustrated by the tracer study utilizing [^{14}C]2-deoxyglucose (Kilduff *et al.*, 1983). During periodic arousal intense shivering also requires glucose oxidation. Without feeding, gluconeogenesis from amino acid precursors and glycerol become the only sources for carbohydrate replenishment in seasonal torpor. Significantly greater gluconeogenic capacities have been observed in liver (Whitten and Klain, 1968) and kidney cortex slices (Burlington and Klain, 1967) of hibernating ground squirrels as compared to their summer active counterparts. In the arctic ground squirrel (Galster and Morrison, 1975), glycerol released from triglycerides accounts for two-thirds of the replenishment of carbohydrate reserves depleted during hibernation and arousal, whilst amino acid precursors from protein catabolism account for less than 20 per cent of total gluconeogenesis. However, a decrease in tissue protein with the progression of hibernation is typical for many other species: for example, the reduction in protein content of 39 per cent for skeletal muscle, 12 per cent for heart and 22 per cent for liver in the hibernating golden-mantled ground squirrel (Tashima *et al.*, 1970) and the 47 per cent decrease in total muscle protein in the hibernating big brown bat (Yacoe, 1983b). In the latter, liver and pectoris muscle protein synthesis and degradation essentially cease during torpor but rates of protein degradation increase significantly during periods of arousal (Yacoe, 1983c), indicating that tissue protein catabolism accounts for the major supply of gluconeogenic precursors. In the arctic ground squirrel, an elevation in phosphoenolpyruvate carboxylase, a key enzyme involved in hepatic gluconeogenesis, is observed during hibernation presumably favoring gluconeogenesis during this state (Behrisch *et al.*, 1981). In terms of glucose utilization, seasonal variants of liver pyruvate kinase have also been identified in the arctic ground squirrel (Behrisch, 1978), the hibernating form having a higher isoelectric point, higher temperature sensitivity with respect to V_{max} and a much lower temperature dependency of binding affinity to allosteric substrates (phosphoenolpyruvate, ADP), activator (fructose diphosphate) and inhibitor (ATP) than those found in the non-hibernating form. Similar findings have also been reported for the flight muscle and liver pyruvate kinase of the hibernating little brown bat (Borgmann and Moon, 1976). The significance of the relatively temperature-insensitive binding affinities with respect to ligands and the temperature sensitivity of V_{max} of the hibernating enzymes is that as body temperature decreases during entry into hibernation, the glycolytic flux can be greatly suppressed whereas as body temperature rises during arousal, glycolytic flux ca be greatly increased.

In non-seasonal torpor (e.g. Syrian hamster), animals feed between successive torpor bouts. The blood glucose level during hibernation is maintained at a level which is similar to that found in euthermia mainly through glycogenolysis from the liver (Musacchia and Deavers, 1981). Pre-torpor cold exposure in Syrian hamsters results in inhibition of lipogenesis in the liver (Denyes and Carter, 1961) but a four- to sixfold increase in the *in vitro* capacity of lipogenesis from acetate in the white adipose tissue (Baumber and Denyes, 1963). Furthermore, this increased lipogenic capacity is retained during torpor unlike the inhibition of lipogenic capacity with the onset of hibernation in ground squirrels and hedgehogs described above. During arousal from torpor, hamsters use predominantly carbohydrate as reflected by the profound depletion of liver and muscle glycogen in late stages of arousal (Lyman and Leduc, 1953). This is in contrast to the use of mainly fat, but some carbohydrate, in ground squirrels arousing from hibernation (Willis, 1982b). These data again illustrate the various metabolic strategies different species use to cope with the same demand for heat production during arousal.

Ionic regulation

Cells surviving at low temperatures (approximately 5°C) will maintain a differential distribution of key ionic species across the cell membrane, namely, high intracellular K^+ and low intracellular Na^+. Tissues (kidney cortex, liver, skeletal muscle, aortic smooth muscle) and cells (kidney cortex, red blood cells) from hibernators (e.g. 13-lined ground squirrel, hamster) retain intracellular K^+ much more effectively than those from non-hibernators (e.g. guinea pig, rat) after several days exposure to 5°C indicating a superior cold tolerance of the hibernators with respect to ion compartmentalization (Willis, 1979).

Using red blood cells as a model (Kimzey and Willis, 1971), it has been demonstrated that the greater ability of the hibernators to retain intracellular K^+ is due to both a greater capacity for (Na^+ + K^+) pump-mediated K^+ influx and a much reduced passive K^+ leak at 5°C when compared to non-hibernators. In a systematic survey, the (Na^+ + K^+) pump activity ratio at 5/37°C measured as ouabain-sensitive K^+ influx, lies between 1.9 per cent and 3.5 per cent in six of seven hibernators and between 0.18 per cent and 0.78 per cent in eight of nine non-hibernators (Willis *et al.*, 1980). These results suggest that a greater Na^+ pump activity is retained at low temperature in the hibernators, although this is not universally true. Further studies (Ellory and Willis, 1982) into the kinetics of the Na^+ pump in the cold-sensitive guinea pig and cold-tolerant 13-lined and Columbian ground squirrels indicate that a temperate decrease from 37°C to 5°C increases the Na^+ pump affinity for external K^+ and internal Na^+ by three- and fivefold, respectively. There are no significant differences between the guinea pig and Columbian ground squirrel in the number of ouabain binding sites per cell or the amount of ouabain bound per cell at both 37°C and 5°C. The turnover numbers for Na^+ pumps at 37°C are about equal between the two groups, but at 5°C, the turnover numbers are three to fivefold higher in the hibernators. The reason for this difference is unknown; it is not related to a blockage in conversion of E2 (K^+ binding) to the E1 (Na^+ binding) form at low temperature but it may be related to alteration in partial fluxes between K^+:K^+ and Na^+:Na^+ exchange (Ellory and Willis, 1982).

In addition to the Na^+ pump, two other transport systems also govern the electrodiffusive (leak) pathways for K^+ (Ellory and Willis, 1983). These are the Cl^--dependent Na^+ + K^+ co-transport, which is inhibited by 'loop' diuretics (e.g. furosemide or bumetanide), and the intracellular Ca^{2+}-dependent K^+ efflux (Gardos channel), which is inhibited by quinine. The Cl^--dependent Na^+ + K^+ co-transport is absent at 5°C in the guinea-pig red blood cell but present in the 13-lined ground squirrel (Hall and Willis, 1984). The existence of this system is, however, highly variable amongst different species and is unlikely to be a universal mechanism for cellular cold tolerance. On the other hand, the Ca^{2+}-dependent K^+ efflux increases significantly in the presence of external Ca^{2+} at 5°C in the guinea-pig but not in the 13-lined ground squirrel (Hall and Willis, 1984). This indicates an activation of the Gardos channel at low temperature in the guinea-pig red blood cells by the influx of external Ca^{2+} which increases intracellular Ca^{2+} concentration and causes loss of intracellular K^+. The lack of such an activation at 5°C in the 13-lined ground squirrel indicates that either external Ca^{2+} does not enter the red blood cells or that excess Ca^{2+} can be pumped outward by the Ca^{2+} pump. Initial studies have indicated that, at 5°C, passive permeability to Ca^{2+} entry into the red blood cells is similar between guinea-pig and hedgehog, but the active Ca^{2+} pump for Ca^{2+} efflux in the guinea-pig is inhibited (Ellory and Hall, 1983; Fig. 10.5). Thus, the ability to retain intracellular K^+, and the enhanced cell survival at low temperature in hibernators may reside in their superior ability to regulate intracellular Ca^{2+} concentration when compared to their non-hibernating counterparts.

Membrane aspects

The temperature sensitivity of a membrane enzyme is typically assessed with the aid of an Arrhenius plot (log maximum reaction velocity vs. the reciprocal

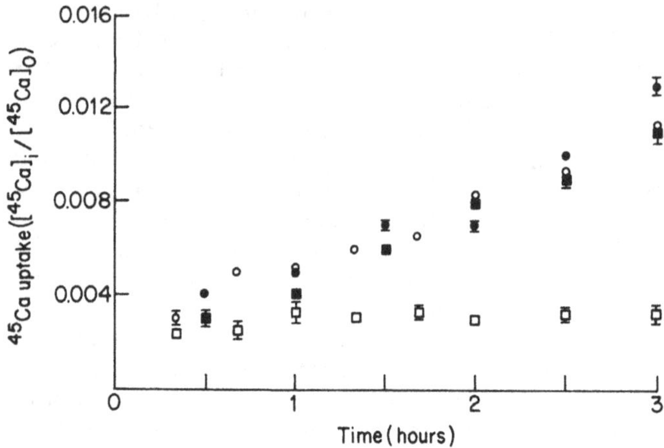

Fig. 10.5 The time course of Ca^{2+} uptake in fresh (○) and ATP-depleted (●) guinea-pig (cold-sensitive) and fresh (□) and ATP-depleted (■) hedgehog (cold-resistant) red blood cells at 5 °C. $[^{45}Ca]_i$ and $[^{45}Ca]_o$ represent calcium (isotope) concentrations inside and outside of red cells, respectively. Results are mean ± SE ($n = 3$). Note ^{45}Ca accumulates inside cells of guinea-pig regardless of whether ATP is present but in hedgehog cells only when ATP is depleted. The difference in red cell calcium uptake between cold-sensitive and cold-resistant species suggests the maintenance of calcium pump activity (active transport of calcium outward) in the hedgehog but not in the guinea-pig at 5 °C. (From Ellory and Hall [1983] with permission.)

of absolute temperature). A plot with constant slope across the physiological temperature range for hibernation (2–37 °C) indicates a constant activation energy (Ea) for the particular enzymic reaction, whereas a discontinuous plot signifies changes of Ea at a critical transition temperature (Tc). A discontinuous Arrhenius plot may also suggest that there has been a conformational change in the enzyme protein, or a transition of the order-disorder in mobility of membrane lipids surrounding the enzyme, or both, at the critical temperature (Charnock, 1978). The possibility, and potential significance, of lipid phase transitions in biological membranes has attracted analyses of thermotropic properties (by differential scanning calorimetry), fluidity (by fluorescence polarization, electron spin resonance) and compositions (class of phospholipids and their fatty acyl groups) of membranes in relation to hibernation. Functionally, a discontinuous Arrhenius plot is often interpreted as being symptomatic of cold sensitivity whereas a continuous plot would indicate cold tolerance. Earlier studies on membrane adaptations at low temperature have been extensively reviewed by Charnock (1978) and Willis *et al.*, (1981). Only a brief summary of this area and some recent studies will be presented here.

The Arrhenius plots of $(Na^+ + K^+)$-ATPase activity from kidney, brain and heart are discontinuous in hibernators, non-hibernators and hibernators between the hibernating and non-hibernating phase (Charnock, 1978), suggesting no special change in this enzyme or its lipid environment in conjunction with hibernation. The sarcoplasmic reticulum Ca^{2+}-ATPase from the leg muscles of rats and awake 13-lined ground squirrels also show discontinuous Arrhenius plots (Charnock, 1978). In the Syrian hamster (Houslay and Palmer, 1978), the Arrhenius characteristics of eight liver plasma membrane-bound enzymes show differences between the torpid (hibernating) and non-torpid state which appear to be related to the location of the enzyme in the lipid bilayer. Of the three enzymes located on the cytoplasmic side of the membrane (fluoride-stimulated adenylate cyclase, basal adenylate cyclase and cAMP phosphodiesterase) the Tc remained constant at near 25°C regardless of the torpid state. For the single enzyme (phosphodiesterase I) on the outer surface of the membrane, the Tc decreased from 13°C to 4°C in torpid hamsters. Of the remaining four enzymes (glucagon-stimulated adenylate cyclase, 5′-nucleotidase, $[Na^+ + K^+]$-ATPase and Mg^{2+}-ATPase) which span both lipo-protein bilayers, values for Tc (25°C and 13°C) are observed in the non-torpid state, the lower value decreasing to 4°C during torpor whereas the higher value remains unchanged. Addition of a membrane lipid fluidizer (benzyl alcohol) to this system decreases Tc by 7–8°C in the glucagon- and fluoride-stimulated adenylate cyclase activities, suggesting a dependence of enzymic rates on membrane fluidity. Fluorescence polarization (DPH) measurements also indicate changes in membrane fluidity associated with torpor, for the Tc values decrease from 25°C and 13°C in the active, to 25°C and 4°C, respectively, in the torpid hamsters. Taken together, these changes in Tc for different enzymes at different, known, cellular locations indicate that a lipid phase separation may occur at 25°C in the inner membrane and at 13°C in the outer membrane in active hamsters. When torpid, only the outer membrane shows a decrease in lipid phase separation temperature from 13°C to 4°C. Thus, depending on the physical location of the particular enzyme in the lipid bilayer, the critical temperature, or temperatures, for alteration in activation energy is either unchanged or lowered with the onset of torpor (Houslay and Palmer, 1978).

In the cardiac membrane of Richardson's ground squirrel analysis by differential scanning calorimetry and electron spin labelling indicate increased membrane fluidity during hibernation as compared to the active state (Charnock *et al.*, 1980; Raison *et al.*, 1981). A small increase in fluidity as detected by fluorescence probe (DHP) has also been observed in the brain microsomal membrane of hibernating Syrian hamsters (Goldman and Albers, 1979) and European hamsters (Montaudon *et al.*, 1984). However, using the same technique, such an increase in membrane fluidity has not been consistently observed in the brain synaptosomes and kidney microsomes of the hibernating Syrian hamster (Cossins and Wilkinson, 1982). It is possible that as the fluorescence probe measures only the average fluidity of the membrane, any small change in the microenvironment surrounding the proteins may not be detectable. By the same argument, it may also be concluded that bulk phase lipid transitions do not occur in these membranes in conjunction with hibernation.

In the mitochondrial membrane, Arrhenius plots of succinate-stimulated oxygen consumption in summer-active ground squirrels are typically discontinuous whereas in the hibernating animals they are continuous (Raison and

Lyons, 1971; Pehowich and Wang, 1981). In a recent study succinate-stimulated mitochondrial proton ejection, calcium uptake and oxygen consumption in the Richardson's ground squirrel were compared under spring, summer, winter (both 20°C and 4°C acclimated), hibernating and aroused conditions (Pehowich and Wang, 1984). Discontinuous Arrhenius plots for these processes were typical of spring and summer animals becoming continuouus in the hibernating and aroused animals (Fig. 10.6). More interestingly, winter animals (both 20°C and 4°C acclimated) which do not show significant weight gain (presumably in their non-hibernating phase) exhibit discontinuous Arrhenius plots whereas those with significant gain (presumably in their hibernating phase) exhibit continuous plots (Fig. 10.6),

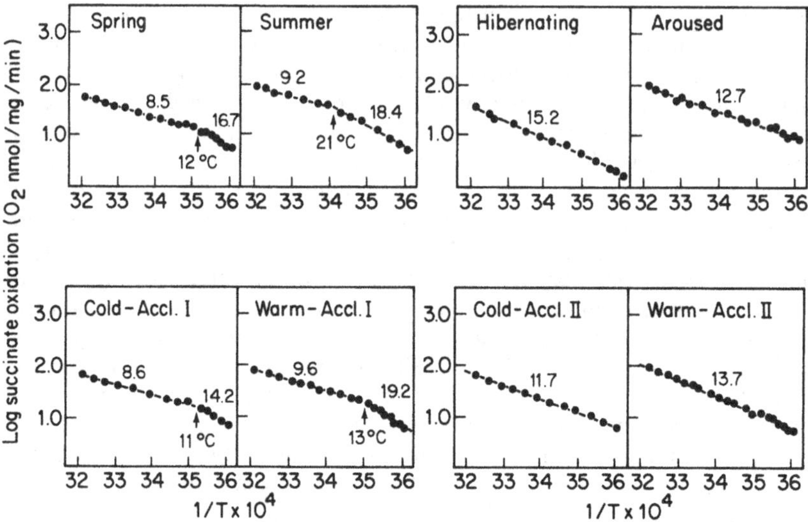

Fig. 10.6 Arrhenius plots of oxygen consumption as a measure of succinate dehydrogenase activity in the Richardson's ground squirrel under different physiological states. Each point is the average of *n* animals in each group. Spring animals ($n = 5$) were freshly captured in April, and Summer animals ($n = 6$) were freshly captured in July. Cold-Accl. I ($n = 4$) and Warm-Accl. I ($n = 7$) were animals acclimated to 20°C and 4°C, respectively, for a minimum of 4 weeks prior to sacrifice in October–November; their body weights were relatively constant and low (presumably in their non-hibernating phase) and had not exhibited hibernation prior to sacrifice. Cold-Accl. II ($n = 4$) and Warm-Accl. II ($n = 4$) were animals acclimated to the same conditions as above except these animals had shown rapid weight gain (presumably in their hibernating phase) but like the above group, had not exhibited hibernation prior to sacrifice. Hibernating animals ($n = 8$) were sacrificed during torpor at body temperature near 5°C and Aroused animals ($n = 9$) were sacrificed 12–14 hours after the initiation of arousal with body temperature near 37°C at time of sacrifice. Numbers below the arrows are the critical transition temperatures for enzymic function. The number beside each line is the Arrhenius activation energy (kcal/mol). Lines are fitted statistically with a computer program for best fit of two lines or a single line. Note the decrease in critical transition temperature from summer to winter (represented by Cold-Accl. I and Warm-Accl. I) and the lack of thermally-induced transition in rate function in the Cold-Accl. II, Warm-Accl. II, Hibernating, and Aroused animals. (Reproduced from Pehowich and Wang [1984] with permission.)

although these latter animals have not yet experienced any decrease in body temperature. Such observations suggest that the seasonal differences in the thermal behaviour of mitochondrial membrane functions are independent of the ambient and body temperatures as well as the dietary lipids, which are constant throughout the year, but are related to the endogenous circannual rhythm for torpor. Furthermore, the membrane lipid changes occur prior to hibernation and persist throughout the hibernating phase irrespective of the changes in body temperature during periodic arousal.

Based on compositional analysis of lipids from the brain and kidney of hamsters, liver and heart mitochondria of ground squirrels (Aloia and Pengelley, 1979; Willis *et al.*, 1981), and liver mitochondria of European hamsters (Cremel *et al.*, 1979), there is no consistent relationship between the unsaturation index of membrane fatty acids and the onset of hibernation. The unsaturation index is often taken as a membrane fluidity indicator but may not be entirely reliable (Willis *et al.*, 1981). It has been shown recently that the change in Arrhenius characteristics of liver and heart mitochondrial membrane enzyme systems induced by dietary lipids is not related to the unsaturation index but to the $\omega 6$ to $\omega 3$ unsaturated fatty acid ratio (McMurchie *et al.*, 1983). Thus, future studies relating changes in Tc and fluidity to hibernation should perhaps include this particular measurement.

In summary, based on the temperature responses of membrane-bound enzyme activity, physical measurements of membrane fluidity, and compositional analysis of membrane lipids, a firm conclusion concerning the role of changes in membrane lipids in survival at low body temperature during hibernation cannot be reached. It is apparent, however, that present evidence does not favour a bulk lipid phase transition (e.g. similar to that observed in ectotherms following temperature acclimation; Cossins, 1981) as a special adaptation for hibernation. There remains, however, the possibility that local changes in lipids surrounding the proteins may account for the observed changes in Arrhenius kinetics of enzymes. This is supported by the enzyme functions of the liver plasma membrane (Houslay and Palmer, 1978) and the bioenergetics of mitochondrial membrane (Pehowich and Wang, 1981, 1984). Future studies incorporating functional, physical and compositional analysis on the same membrane preparation are needed to advance understanding of the fundamental aspects of membrane lipids and biochemical homeostasis at low body temperature.

Neuroendocrine aspects

Although numerous studies are available (Hudson and Wang, 1979; Wang, 1982), a causal relationship between neuroendocrine functions and torpor, in particular hibernation, has yet been established. Aside from the difficulties associated in with endocrine research, for example, techniques for obtaining 'physiological' blood samples for hormone assays, proper dosages for hormonal stimulation, and interpretation of functional aspects based on histology of endocrine glands and tissue or plasma concentrations of hormones, three other factors may have also contributed to the slow progress in this area.

First, many endocrine glands show endogenous cyclicity, despite the constant environmental temperature and photoperiod imposed on the animal. For example, in the male (Licht *et al.*, 1982) and ovariectomized (Zucker and Licht, 1983) euthermic golden-mantled ground squirrels kept at 23°C and 14L:10D photoperiod, the plasma luteinizing hormone concentration shows a seasonal cycle with increasing values in March and peaks in April; the times corresponded to late hibernation season and spring emergence when reproductive recrudescence and readiness also occur in field animals. Furthermore, circannual testicular cycles have also been observed under 12L:12D and 23°C in the golden-mantled ground squirrel (Kenagy, 1980) and the western chipmunk (Kenagy, 1981a); both species are discouraged to hibernate by the high ambient temperature. A similar circannual testicular cycle has also been observed under similar conditions in the desert ground squirrel, *Ammospermophilus leucurus* (Kenagy, 1981b), which is not a hibernator. These studies indicate that the circannual rhythms of hibernation and reproduction are independent. In addition, a circannual rhythm (i.e. under constant laboratory conditions) in thyroid activity independent of that for torpor has been observed in the round-tailed ground squirrel (Hudson and Wang, 1969). Thus, a seasonal cycle in endocrine activity does not necessarily correlate with the cycle for hibernation.

The second difficulty is to separate the changes in endocrine activities due to depressed body temperature during hibernation from those which result from depressed body temperature alone. Many available studies have reported differences in plasma hormonal levels between euthermic and hibernating states (Wang, 1982). However, under depressed body temperature, the depressed, elevated, or unaltered hormone levels may simply be the result of the inability to secrete (e.g. insulin in hedgehogs; Hoo-Paris and Sutter, 1980), to utilize (e.g. adrenal glucocorticoids in hedgehogs; Saboureau *et al.*, 1980), or to do both (e.g. adrenal glucocorticoids in garden dormice; Boulouard, 1972). In addition, changes in hormone-plasma carrier binding (Young *et al.*, 1979) and hormone-receptor binding (Hoo-Paris and Sutter, 1980) under depressed body temperature could alter the free/total ratio of plasma hormones as well as the 'apparent tissue responsiveness' to hormones. Future experiments must therefore take into consideration the debilitating effects of low body temperature alone on endocrine functions.

Third, as a result of the polyphyletic origin of torpor in mammals (Hudson, 1973), different adaptive strategies are likely to have been evolved for species occupying different ecological niches. For example, endocrine changes in species exhibiting seasonal torpor (e.g. marmot) may be quite different from those in species exhibiting non-seasonal torpor (e.g. Syrian hamster) as a result of differences in energy storage (fat vs. food cache) and utilization. Generalizations on neuroendocrine functions for torpor may not, therefore, always be possible.

With these qualifications, it may be stated that in general polyglandular involution occurs before hibernation. During the first half of the hibernation season, neuroendocrine functions are generally depressed; this is followed by a reactivation of all functions sometime after the mid-hibernation season, and peak activities are reached at the time of, or shortly after the spring emergence. For example, based on the weight and histology, and the hormone levels of the target glands, the pituitary in ground squirrels and bats is least active in December–January, more active in January–March and most active in April–May following spring emergence (Wang, 1982). Histological evidence also suggests that the hypothalmo-hypophyseal interaction is low during hibernation (Wang, 1982) and accumulation of neurosecretory materials is typical. However, immunocytochemical analysis has shown that the somatostatin- and Met-enkephalin-containing neurons are more active during hibernation than in euthermia in the hedgehog (Nürnberger, 1983).

As to the thyroid gland, cooling of the hypothalamus, which typically activates the pituitary–thyroid axis, does not invoke increased release of ^{125}I from thyroid during the hibernating season in the 13-lined ground squirrel (Hulbert and Hudson, 1976). In many other species of ground squirrels and the woodchuck, decreased cell height, accumulation of colloid in follicles and low organic iodine release (Hudson, 1981), decreased synthetic, reabsorptive and secretory activities (Winston and Henderson, 1981), and high plasma thyroxine and triiodothyronine concentrations but low free/bound hormone ratios (Demeneix and Henderson, 1978; Young *et al.*, 1979) have been observed. These observations indicate that an inactive thyroid is typical prior to and during hibernation. In contrast to the thyroids of these seasonal hibernators, the thyroids of Syrian, Turkish, and European hamsters remain active during torpor as do the thyroids of chipmunks (Hudson, 1981). This dichotomy in thyroid activity with regard to hibernation may be related to the mode of energy storage for torpor. As hamsters and chipmunks feed between successive torpor bouts, the assimilation and utilization of fresh nutrient intake may require the thyroid to remain active throughout (Hudson, 1981).

The endocrine pancreas is inactive during hibernation in the hedgehog; plasma levels of immunoreactive insulin (Hoo-Paris and Sutter, 1980) and glucagon (Hoo-Paris *et al.*, 1982) are undetectable, and glucose challenge results in hyperglycaemia whereas glucagon administration elicits no hyperglycaemia. This suppression may be due to low body temperature alone on secretory and hormone-receptor binding aspects, as above a body temperature of 15–25°C both the hypoglycaemic action of insulin and the hyperglycaemic action of glucagon become evident (Hoo-Paris and Sutter, 1980; Hoo-Paris *et al.*, 1982). With regard to calcium metabolism, hypercalcaemia has been observed in the torpid hedgehog and hamster (Wang, 1982). The winter activity of the thyroid parafollicular cells, which secrete calcitonin for calcium deposition in bones, is decreased whereas the activity of the parathyroid gland, which secretes parathormone for calcium mobilization via bone resorption, is increased. These coupled activities may explain the observed osteoporosis (loss of bone) in the torpid bats, arctic ground squirrels, 13-lined squirrels and the Syrian hamster (Wang, 1982).

The role of the pineal gland in regulating torpor is uncertain. Pinealectomy decreases the incidence of torpor in the white-footed mouse (Lynch *et al.*, 1980) and Syrian hamster (Jansky *et al.*, 1981) but has no effect in the adult golden-mantled squirrel and the Richardson's ground squirrel (Harlow *et al.*,

1980). The duration of torpor is also unaffected by pinealectomy in the 13-lined squirrel (Sinnamon and Pivorun, 1982) and the Richardson's ground squirrel (Harlow *et al.*, 1980). However, pinealectomy appears to temporarily compress the hibernation season in the golden-mantled ground squirrel in the second year following pinealectomy; the terminal arousal and testicular recrudescence occur approximately 6 weeks earlier than in controls (Phillips and Harlow, 1982), suggesting a possible effect of pineal hormones on the expression of endogenous circannual cycles for hibernation and reproduction. Administration of melatonin, one of the pineal hormones, increases the incidence of torpor in the white-footed mouse (Lynch *et al.*, 1980), the Syrian hamsters (if kept under constant light but not under constant darkness; Jansky *et al.*, 1981), and the golden-mantled ground squirrel (Palmer and Riedesel, 1976), but has no effect in the 13-lined squirrel (Sinnamon and Pivorun, 1981) or on the Richardson's ground squirrel (Ralph *et al.*, 1982). The majority of evidence on ground squirrels therefore seems to suggest only minor effects of melatonin on induction of hibernation.

The annual cycles of gonadal activity have been demonstrated to be independent of the hibernation cycle in many hibernators (Kenagy, 1980, 1981a, b). In the ground squirrels, castration in both sexes does not affect the circannual rhythmicity for hibernation (Pengelley and Asmundson, 1974). In the Turkish hamster, however, castration prolongs the torpor season from 5–6 months to more than 18 months (Hall and Goldman, 1980) and chronical implantation of testosterone inhibits torpor in the male Syrian hamster (Jansky *et al.*, 1981) and in both sexes of the Turkish hamster (Hall and Goldman, 1980). Thus, in the hamsters, gonadal activity appears to exert a regulatory role on torpor, although the mechanisms are unknown.

The role of the adrenal gland in torpor is unclear. In the ground squirrel (Popovic, 1960) and Syrian hamster (Jansky *et al.*, 1981) adrenalectomy prevents hibernation but grafting of cortico tissue into the anterior eye chamber or supplementation of corticosteroids by subcutaneous implants or injection restores hibernation. In view of the wide-ranging metabolic and thermogenic effects of glucocorticoids (Musacchia and Deavers, 1981), it is difficult to identify the mechanisms for their pro-torpor action. Adrenal activity during torpor varies with species; it is inactive in the garden dormouse, hedgehog, woodchuck (Wang, 1982) and a Russian ground squirrel, the red-cheeked suslik (Popova and Koryakina, 1981), but more active than in euthermia in the little brown bat (Gustafson and Belt, 1981). The significance of this latter change awaits further investigation. There appears to be increased secretion and utilization of aldosterone during torpor in the Syrian hamster and marmot, and normal or even increased plasma renin and aldosterone concentrations are found during hibernation in the 13-lined ground squirrel and marmot (Wang, 1982). In view of the lack of feeding and salt intake in these latter species during the hibernation season and the disturbance of Na^+/K^+ distributions during hibernation (Willis, 1982a), the increased secretion of aldosterone is consistent with the physiological need for electrolyte conservation. The adrenal medulla is apparently not essential for hibernation as demedullated animals can enter and arouse from hibernation (Popovic, 1960). However, in the woodchuck (Florant *et al.*, 1982), plasma norepinephrine and epinephrine (adrenaline) levels increase more than 30-fold during arousal from hibernation, indicating a substantial contribution by the adrenal medulla to the cardiovascular and calorigenic needs for the rewarming process.

Antimetabolic peptides and the 'hibernation induction trigger'

It has been shown that peptides extracted from the crude brain homogenates of hibernating 13-lined ground squirrels can induce a 35 per cent reduction in metabolism and a 3°C decrease in body temperature when given intravenously to white rats. However, recent studies employing the same preparations given intracerebroventricularly were ineffective (Swan, 1981). Thus, the status of the antimetabolic peptides is uncertain. The existence of a blood-borne 'hibernation induction trigger' in hibernating rodents has been demonstrated by Dawe and Spurrier (1969). Transfusion of serum from a hibernator (which may be from other species of ground squirrels or woodchuck) into a 13-lined ground squirrel induces hibernation in the recipient during the summer (Dawe, 1978). Attempts at a similar approach with other species of ground squirrels (Galster, 1978; Abbots *et al.*, 1979) or hamsters (Minor *et al.*, 1978) as the recipient have not been so successful. These results seem to suggest that species specificity might exist in the recipient but not in the donor; the reason for this remains to be elucidated. Biochemical efforts have not yet produced purified preparations but have shown that the 'hibernation induction trigger' is a thermolabile peptide tightly bound to the serum albumin. Intravenous injection of this lyophilized albumin fraction into the 13-lined ground squirrel induces summer torpor 2 days to 5 weeks later (Oeltgen and Spurrier, 1981). Recently, this lyophilized fraction has also been shown to cause bradycardia, hypothermia, behavioural depression and aphagia in macaque monkeys when given into the cerebral ventricle (Myers *et al.*, 1981). These effects may be retarded or reversed by the opiate antagonist, naloxone or naltrexone, suggesting the possibility that the 'hibernation induction trigger' may be an endogenous opioid (Oeltgen *et al.*, 1982).

The possible involvement of endogenous opioids in hibernation has been shown by the increased brain levels of Met- and Leu-enkaphalins (Kramarova *et al.*, 1983) and increased Met-enkaphalin immunoreactivity in specific hypothalamic areas during hibernation (Nürnberger, 1982), the reduction of incidence (Kromer, 1980) and duration of hibernation (Llados-Eckman and Beckman, 1983) by naloxone and the initiation of premature arousal by naloxone (Margulis *et al.*, 1977). Furthermore, in the golden-mantled ground squirrel, physical dependence on morphine fails to develop during hibernation but does develop in euthermia (Beckman *et al.*, 1981), suggesting greater occupation of opiate receptors by endogenous opioids during hibernation. It is therefore possible that the maintenance of hibernation may be associated with an increased activity of the brain opioids. Whether this increase may be manifested in the peripheral circulation and appears as the 'hibernation induction trigger' is unknown.

Summary

Natural torpidity in mammals in response to food and water shortage and/or cold is characterized by a profound reduction in body temperature and other physiological functions lasting from a few hours to a few weeks. This ends in spontaneous arousal during which body temperature is restored to normal using heat generated exclusively from within the animal by shivering and non-shivering thermogenesis. The wide occurrence of torpidity in at least six unrelated mammalian orders suggests that it is polyphyletic in origin, and although patterns of torpor vary, they are physiologically similar. The energy savings by exhibiting hibernation can be substantial; up to 88 per cent is saved in the Richardson's ground squirrel whose torpor season lasts approximately 8 months in the Canadian North. Entry into torpor is an extension of slow wave sleep but the set-point for body temperature regulation continues to decline to much lower levels. This results in regulated reductions of heart rate, respiration, and oxygen consumption followed by body temperature. During torpor, thermoregulation persists; a decrease in ambient temperature below the 'critical' level typically triggers arousal. During arousal, shivering and non-shivering thermogenesis using brown adipose tissues provide the heat required to restore euthermia. Why torpid animals arouse periodically remains unknown.

The physiological/biochemical adaptations associated with hibernation have been examined from the central nervous system to the membrane level. The maintenance of membrane transport for ionic regulation appears to be fundamental to cell survival and function at depressed body temperature. The difference in cold tolerance for Na^+/K^+ transport between the hibernating and non-hibernating species could be the Na^+ pump itself, or more probably, it could be related to the interaction betwen Ca^{2+} and K^+ transport (Gardos effect). This in turn could be related to both the bioenergetics of the cell and to the alteration of membrane lipids. This latter effect may alter protein–lipid interaction in both enzymic reactions and ion transport. Athough not all membranes so far examined show specific changes with respect to hibernation, some, such as the liver plasma membrane and the inner mitochondrial membrane, show a change towards greater membrane fluidity at low temperatures. However, the *in vivo* significance of these membrane changes remains to be elucidated.

The neurochemical mechanisms regulating the onset and termination of hibernation remain largely unknown. Brain monoamines, because of their roles as neurotransmitters in thermoregulation, have been implicated in torpor. In particular, an enhanced 5-HT activity may exert inhibition on heat production thereby facilitating the maintenance of torpor. The precise role of neuroendocrine regulation of hibernation remains undefined; it is complicated by the diverse tactics different species use to meet their metabolic and thermal needs during hibernation, the inherent circannual rhythmicity of many endocrine functions independent of that for hibernation, and the difficulty in separating endocrine changes brought about by hibernation *per se* versus those by low temperature alone. However, recent evidence has suggested that brain opioids may be involved in regulating hibernation, either independently or in concert with other neurotransmitters or neuromodulators.

Acknowledgements

Literature search for this review has been aided by an Operating Grant No. A6455 from the Natural Sciences and Engineering Research Council of Canada to L. Wang.

References

Abbotts, B., Wang, L.C.H. and Glass, J.D. (1979). Absence of evidence for a hibernation 'trigger' in blood dialyzate of Richardson's ground squirrel. *Cryobiology* **16**, 179–83.

Aloia, R.C. and Pengelley, E.T. (1979). Lipid composition of cellular membranes of hibernating mammals. In *Chemical Zoology*, Vol. 11, pp. 1–47. Edited by Florkin, M. and Scheer, B.T. Academic Press, New York.

Baumber, J. and Denyes, A. (1963). Acetate-1-C14 metabolism of white fat from hamsters in cold exposure and hibernation. *American Journal of Physiology* **205**, 905–8.

Beckman, A.L. (1978). Hypothalamic and midbrain function during hibernation. In *Current Studies of Hypothalamic Function* Vol. 2, pp. 29–43. Edited by Veale, W.L. and Lederis, K. Karger, Basel.

Beckman, A.L. and Stanton, T.L. (1982). Properties of the CNS during the state of hibernation. In *The Neural Basis of Behaviour*, pp. 19–45. Edited by Beckman, A.L. Spectrum, New York.

Beckman, A.L., Llados-Eckman, C., Stanton, T.L. and Adler, M.W. (1981). Physical dependence on morphine fails to develop during the hibernating state. *Science* **212**, 1527–9.

Behrisch, H.W. (1978). Metabolic economy at the biochemical level: the hibernator. In *Strategies in Cold: Natural Torpidity and Thermogenesis*, pp. 461–79. Edited by Wang, L.C.H. and Hudson, J.W. Academic Press, New York.

Behrisch, H.W., Smullin, D.H. and Morse, G.A. (1981). Life at low and changing temperatures: molecular aspects. In *Survival in the Cold*, pp. 191–205. Edited by Musacchia, X.J. and Jansky, L. Elsevier/North Holland, Amsterdam.

Bintz, G.L. and Strand, C.E. (1983). Radioglucose metabolism by Richardson's ground squirrels in the natural environment. *Physiological Zoology* **56**, 639–47.

Borgman, A.I. and Moon, T.W. (1976). Enzymes of the normothermic and hibernating bat, *Myotis lucifugus*: temperature as a modulator of pyruvate kinase. *Journal of Comparative Physiology* **107**, 185–99.

Boulouard, R. (1972). Adrenocortical function in two hibernators: the garden dormouse and the hedgehog. *Proc. Int. Symp. Envoron. Physiol.* (Bioenergetics) FASEB, pp. 108–12.

Brück, K. and Wünnenberg, W. (1970). Meshed control of two effector systems: nonshivering and shivering thermogenesis. In *Physiological and Behavioural Temperature Regulation*, pp. 562–80. Edited by Hardy, J.D., Gagge, A.P. and Stolwijk, J.A.J. Thomas, Springfield, IL.

Burlington, R.F. and Klain, G.J. (1967). Gluconeogenesis during hibernation and arousal from hibernation. *Comparative Biochemistry and Physiology* **22**, 701–8.

Cannon, B. and Nedergaard, J. (1983). Biochemical aspects of acclimation to cold. *Journal of Thermal Biology* **8**, 85–90.

Castex, Ch. and Sutter, B. Ch. J. (1981). Insulin binding and glucose oxidation in edible dormouse (*Glis glis*) adipose tissue: seasonal variations. *General and Comparative Endocrinology* **45**, 273–8.

Charnock, J.S. (1978). Membrane lipid phase-transitions: a possible biological response to hibernation? In *Strategies in Cold: Natural Torpidity and Thermogenesis*, pp. 417–60. Edited by Wang, L.C.H. and Hudson, J.W. Academic Press, New York.

Charnock, J.S., Gibson, R.A., McMurchie, E.J. and Raison, J.K. (1980). Changes in the fluidity of myocardial membranes during hibernation: relationship to myocardial adenosinetriphosphatase activity. *Molecular Pharmacology* **18**, 476–82.

Cossins, A.R. (1981). The adaptation of membrane dynamic structure to temperature. In *Effects of Low Temperature on Biological Membranes*, pp. 83–106. Edited by Morris, G.J. and Clarke, A. Academic Press, New York.

Cossins, A.R. and Wilkinson, H.L. (1982). The role of homeoviscous adaptation in mammalian hibernation. *Journal of Thermal Biology* **7**, 107–10.

Cremel. G., Robel, G., Canguilheim, B., Rendon, A. and Waksman, A. (1979). Seasonal variation of the composition of membrane lipids in liver mitochondria of the hibernator *Cricetus cricetus*. Relation to intramitochondrial intermembranal protein movement. *Comparative Biochemistry and Physiology* **63A**, 159–67.

Davis, D.E. (1976). Hibernation and circannual rhythms of food consumption in marmots and ground squirrels. *Quarterly Review of Biology* **51**, 477–514.

Dawe, A.R. (1978). Hibernation trigger research updated. In *Strategies in Cold: Natural Torpidity and Thermogenesis*, pp. 541–63. Edited by Wang, L.C.H. and Hudson, J.W. Academic Press, New York.

Dawe, A.R. and Spurrier, W.A. (1969). Hibernation induced in ground squirrels by blood transfusion. *Science* **163**, 298–9.

Demeneix, B.A. and Henderson, N.E. (1978). Serum T_4 and T_3 in active and torpid ground squirrels, *Spermophilus richardsoni*. *General Comparative Endocrinology* **35**, 77–85.

Denyes, A. and Carter, J.D. (1961). Utilization of acetate-1-C14 by heptic tissue from cold exposed and hibernating hamsters. *American Journal of Physiology* **200**, 1043–6.

Draskoczy, P.R. and Lyman, C.P. (1967). Turnover of catecholamines in active and hibernating ground squirrels. *Journal of Pharmocology and Experimental Therapeutics* **155**, 101–11.

Ellory, J.C. and Hall, A.C. (1983). Ca^{2+} transport in hibernator and non-hibernator species red cells at low temperature. *Journal of Physiology (London)* **344**, 148p.

Ellory, J.C. and Willis, J.S. (1982). Kinetics of the sodium pump in red cells of different temperature sensitivity. *Journal of General Physiology* **79**, 1115–30.

Ellory, J.C. and Willis, J.S. (1983). Adaptive changes in membrane-transport systems of hibernators. *Biochemical Society Transactions* **11**, 330–2.

Fisher, K.C., Dawe, A.R., Lyman, C.P., Schönbaum, E. and South, F.E. (eds.) (1967). *Mammalian Hibernation III*. Oliver and Boyd, Edinburgh.

Florant, G.L., Weitzman, E.D., Jayant, A. and Cote, L.J. (1982). Plasma catecholamine levels during cold adaptation and hibernation in woodchucks (*Marmota monax*). *Journal of Thermal Biology* **7**, 143–6.

Foster, D.O. and Frydman, M.L. (1979). Tissue distribution of cold-induced thermogenesis in conscious warm- or cold-acclimated rats reevaluated from changes in tissue blood flow: The dominant role of brown adipose tissue in the replacement of shivering by nonshivering thermogenesis. *Canadian Journal of Physiology and Pharmacology* **57**, 257–70.

Friedli, C., Chinet, A. and Girardier, L. (1978). Comparative measurements of *in vitro* thermogenesis of brown adipose tissue from control and cold-adapted rats. In *Effectors of Thermogenesis*, pp. 259–66. Edited by Girardier, L. and Seydoux, J. Birkhaeuser, Basel.

Galster, W.A. (1978). Failure to initiate hibernation with blood from the hibernating arctic ground squirrel, *Citellus undulatus*, and eastern woodchuck, *Marmota monax*. *Journal of Thermal Biology* **3**, 93.

Galster, W. and Morrison, P.R. (1975). Gluconeogenesis in arctic ground squirrels between periods of hibernation. *American Journal of Physiology* **228**, 325–30.

Glass, J.D. and Wang, L.C.H. (1979a). Effects of central injection of biogenic amines during arousal from hibernation. *American Journal of Physiology* **236**, R162–7.

Glass, J.D. and Wang, L.C.H. (1979b). Thermoregulatory effects of intracerebroventricular injection of serotonin and a monoamine oxidase inhibitor in a hibernator, *Spermophilus richardsonii*. *Journal of Thermal Biology* **4**, 149–56.

Goldman, S.S. and Albers, R.W. (1979). Cold resistance of the brain during hibernation: changes in the microviscosity of the membrane and associated lipids. *Journal of Neurochemistry* **32**, 1139–42.

Gustafson, A.E. and Belt, W.D. (1981). The adrenal cortex during activity and hibernation in the male little brown bat, *Myotis lucifugus*: annual rhythm of plasma cortisol level. *General Comparative Endocrinology* **44**, 269–78.

Hall, A.C. and Willis, J.S. (1984). Differential effects of temperature on three components of passive permeability to potassium in rodent red cells. *Journal of Physiology* **348**, 629–43.

Hall, V. and Goldman, B. (1980). Effects of gonadal steroid hormones on hibernation in the Turkish hamster (*Mesocricetus brandti*). *Journal of Comparative Physiology* **135B** 107–14.

Hand, S.C. and Somero, G.N. (1983). Phosphofructokinase of the hibernator *Citellus beecheyi*: temperature and pH regulation of activity via influences on the tetramer-dimer equilibrium. *Physiological Zoology* **56**, 380–88.

Harlow, H.J., Phillips, J.A. and Ralph, C.L. (1980). The effect of pinealectomy on hibernation in two species of seasonal hibernators, *Citellus lateralis* and *C. richardsonii*. *Journal of Experimental Zoology* **213**, 301–3.

Hayward, J.S. (1968). The magnitude of noradrenaline-induced thermogenesis in the bat (*Myotis lucifugus*) and its relation to arousal from hibernation. *Canadian Journal of Physiology and Pharmacology* **46**, 713–18.

Hayward, J.S. and Lyman, C.P. (1967). Non-shivering heat production during arousal from hibernation and evidence for the contribution of brown fat. In *Mammalian Hibernation III*, pp. 346–355. Edited by Fisher, K.C., Dawe, A.R., Lyman, C.P., Schonbaum, E. and South, F.E. Oliver and Boyd, Edinburgh.

Heller, H.C. (1979). Hibernation: neural aspects. *Annual Review of Physiology* **41**, 305–21.

Heller, H.C., Walker, J.M., Florant, G.L., Glotzbach, S.F. and Berger, R.J. (1978). Sleep and hibernation: electrophysiological and thermoregulatory homologies. In *Strategies in Cold: Natural Torpidity and Thermogenesis*, pp. 225–65. Edited by Wang, L.C.H. and Hudson, J.W. Academic Press, New York.

Hemingway, A. (1963). Shivering. *Physiological Reviews* **43**, 397–422.

Hock, R.J. (1960). Seasonal variations in physiologic functions of arctic ground squirrels and black bears. *Bull. Mus. Comp. Zool.* Harvard University **124**, 155–71.

Hoffman, R.A. (1964). Terrestrial animals in cold: hibernators. In *Handbook of Physiology, Sec. 4, Adaptation to the Environment*, pp. 379–403. American Physiological Society, Washington, D.C.

Hoo-Paris, R. and Sutter, B. Ch. J. (1980). Blood glucose control by insulin in the lethargic and arousing hedgehog (*Erinaceus europaeus*). *Comparative Biochemistry and Physiology* **66A**, 141–3.

Hoo-Paris, R., Hamsany, M., Sutter, B. Ch. J., Assan, R. and Biollol, J. (1982). Plasma glucose and glucagon concentrations in the hibernating hedgehog. *General Comparative Endocrinology* **46**, 246–54.

Houslay, M.D. and Palmer, R.W. (1978). Changes in the form of Arrhenius plots of the activity of glucagon-stimulated adenylate cyclase and other hamster liver plasma-membrane enzymes occurring on hibernation. *Biochemical Journal* **174**, 909–19.

Hudson, J.W. (1967). Variations in the patterns of torpidity of small homeotherms. In *Mammalian Hibernation III*, pp. 30–46. Edited by Fisher,

K.C., Dawe, A.R., Lyman, C.P., Schonbaum, E. and South F.E. Oliver and Boyd, Edinburgh.

Hudson, J.W. (1973). Torpidity in Mammals. In *Comparative Physiology of Thermoregulation* Vol. III, pp. 97-165. Edited by Whittow, G.C. Academic Press, New York.

Hudson, J.W. (1978). Shallow, daily torpor: a thermoregulatory adaptation. In *Strategies in Cold: Natural Torpidity and Thermogenesis*, pp. 67–108. Edited by Wang, L.C.H. and Hudson, J.W. Academic Press, New York.

Hudson, J.W. (1981). Role of the endocrine glands in hibernation with special reference to the thyroid gland. In *Survival in the Cold*, pp. 33–54. Edited by Musacchia, X.J. and Jansky, L. Elsevier/North-Holland, Amsterdam.

Hudson, J.W. and Wang, L.C.H. (1969). Thyroid function in desert ground squirrels. In *Physiological Systems in Semiarid Environments*, pp. 17–33. Edited by Hoff, C.C. and Riedesel, M.L. University of New Mexico Press, Albuquerque, New Mexico.

Hudson, J.W. and Wang, L.C.H. (1979). Hibernation: endocrinologic aspects. *Annual Review of Physiology* **41**, 287–303.

Hulbert, A.J. and Hudson, J.W. (1976). Thyroid function in a hibernator, *Spermophilus tridecemlineatus. American Journal of Physiology* **230**, 1138–43.

Jacobs, H.K. and South, F.E. (1976). The effect of temperature on the electrical potentials of hibernating hamster ventricular strips. *American Journal of Physiology* **230**, 403–9.

Jansky, L. (1973). Non-shivering thermogenesis and its thermoregulatory significance. *Biological Review* **48**, 85–132.

Jansky, L. (1978). Time sequence of physiological changes during hibernation: the significance of serotonergic pathways. In *Strategies in Cold: Natural Torpidity and Thermogenesis*, pp. 299–326. Edited by Wang, L.C.H. and Hudson, J.W. Academic Press, New York.

Jansky, L. and Novotona, R. (1976). The role of central aminergic transmission in thermoregulation and hibernation. In *Regulation of Depressed Metabolism and Thermogenesis*, pp. 64–80. Edited by Jansky, L. and Musacchia, X.J. Thomas, Springfield, IL.

Jansky, L., Kahlerova, Z., Nedoma, J. and Andrews, J.F. (1981). Humoral control of hibernation in golden hamsters. In *Survival in the Cold*, pp. 13–32. Edited by Musacchia, X.J. and Jansky, L. Elsevier/North-Holland, Amsterdam.

Joy, J.E. and Mrosovsky, N. (1983). Circannual cycles in golden-mantled ground squirrels: lengthenning of period by low temperatures in the spring phase. *Journal of Comparative Physiology* **150A** 233–8.

Kayser, C. (1961). *The Physiology of Natural Hibernation*. Pergamon Press, New York.

Kenagy, G.J. (1980). Interrelation of endogenous annual rhythms of reproduction and hibernation in the golden-mantled ground squirrel. *Journal of Comparative Physiology* **135A**, 333–9.

Kenagy, G.J. (1981a). Effects of day length, temperature, and endogenous control on annual rhythms of reproduction and hibernation in chipmunks (*Eutamias* spp.). *Journal of Comparative Physiology* **141**, 369–78.

Kenagy, G.J. (1981b). Endogenous annual rhythm of reproductive function in the non-hibernating desert ground squirrel, *Ammospermophilus leucurus*. *Journal of Comparative Physiology* **142A** 251–8.

Kilduff, T.S., Sharp, F.R. and Heller, H.C. (1983). Relative 2-deoxyglucose uptake of the paratrigeminal nucleus increases during hibernation. *Brain Research* **262**, 117–23.

Kimzey, S.L. and Willis, J.S. (1971). Temperature adaptation of active sodium-potassium transport and of passive permeability in erythrocytes of ground squirrels. *Journal of General Physiology* **58**, 634–49.

Kramarova, L.I., Kolaeva, S.H., Yukhananov, R.Yu. and Rozhanets, V.V. (1983). Content of DSIP, enkephalins and ACTH in some tissues of active and hibernating ground squirrels (*Citellus suslicus*). *Comparative Biochemistry and Physiology* **74C**, 31–33.

Kristoffersson, R. and Soivio, A. (1964). Hibernation in the hedgehog (*Erinaceus europaeus* L.). Changes of respiratory pattern, heart rate and body temperature in response to gradually decreasing or increasing ambient temperature. *Ann. Acad. Sci. Fenn. Ser.* **A4 82** 3–17.

Kromer, W. (1980). Naltrexone influence on hibernation. *Experientia* **36**, 581–2.

Kuhnen, G., Petersen, P. and Wünnenberg, W. (1983). Hibernation in golden hamsters (*Mesocricetus auratus*, W.) exposed to 5% CO_2. *Experientia* **39**, 1346–7.

Licht, P., Zucker, I., Hubbard, G. and Boshes, M. (1982). Circannual rhythms of plasma testosterone and luteinizing hormone levels in golden-mantled ground squirrels (*Spermophilus lateralis*). *Biology of Reproduction* **27**, 411–18.

Lindstedt, S.L. (1980). The smallest insectivores: coping with scarcities of energy and water. In *Comparative Physiology: Primitive Mammals*, pp. 163–9. Edited by Schmidt-Nielsen, K., Bolis, L. and Taylor, R.C. Cambridge University Press, Cambridge.

Llados-Eckman, C. and Beckman, A.L. (1983). Reduction of hibernation bout duration by icv infusion of naloxone in *Citellus lateralis*. *Proceedings of the Society of Neurosci* **9**, 796.

Lyman, C.P. (1948). The oxygen consumption and temperature regulation of hibernating hamsters. *Journal of Experimental Zoology* **109**, 55–78.

Lyman, C.P. (1965). Circulation in mammalian hibernation. In *Handbook of Physiology, Sec. 2, Circulation 3*, pp. 1967–89. American Physiological Society, Washington, D.C.

Lyman, C.P. (1982). Who is who among the hibernators. In *Hibernation and Torpor in Mammals and Birds*, pp. 12–36. Edited by Lyman, C.P., Willis, J.S., Malan, A. and Wang, L.C.H. Academic Press, New York.

Lyman, C.P. and Blinks, D.C. (1959). The effect of temperature on the isolated hearts of closely related hibernators and non-hibernators. *Journal of Cellular and Comparative Physiology* **54**, 53–63.

Lyman, C.P. and Chatfield, P.O. (1955). Physiology of hibernation in mammals. *Physiological Reviews* **35**, 403–25.

Lyman, C.P. and Dawe, A.R. (eds.) (1960). Mammalian Hibernation. *Bull. Mus. Comp. Zool.* Harvard University **124**.

Lyman, C.P. and Leduc, E.H. (1953). Changes in blood sugar and tissue glycogen in the hamster arousal from hibernation. *Journal of Cellular and Comparative Physiology* **41**, 471–92.

Lyman, C.P., Willis, J.S., Malan, A. and Wang, L.C.H. (eds.) (1982). *Hibernation and Torpor in Mammals and Birds*. Academic Press, New York.

Lynch, G.R., Sullivan, J.K. and Gendler, S.L. (1980). Temperature regulation in the mouse, *Peromyscus leucopus*: effects of various photoperiods, pinealectomy and melatonin administration. *International Journal of Biometeorology* **24**, 49–55.

McMurchie, E.J., Gibson, R.A., Abeywardena, M.Y. and Charnock, J.S. (1983). Dietary lipid modulation of rat liver mitochondrial succinate: cytochrome c reductase. *Biochimica et Biophysica Acta* **727**, 163–9.

Malan, A. (1982). Respiration and acid-base state in hibernation. In *Hibernation and Torpor in Mammals and Birds*, pp. 237–82. Edited by Lyman, C.P., Willis, J.S., Malan, A. and Wang, L.C.H. Academic Press, New York.

Margulis, D.L., Goldman, B. and Finck, A. (1979). Hibernation: an opioid-dependent state? *Brain Research Bulletin* **4**, 721–4.

Marshall, J.M. (1957). Effects of low temperatures on transmembrane potentials of single fibers of the rabbit atrium. *Circulation Research* **5**, 664–9.

Marshall, J.M. and Willis, J.S. (1962). The effects of temperature on the transmembrane potentials in isolated atria of the ground squirrel, *Citellus tridecemlineatus*. *Journal of Physiology* (London) **164**, 64–76.

Minor, J.D., Bishop, D.A. and Badger, C.R. (1978). The golden hamster and the blood-borne hibernation trigger. *Cryobiology* **15**, 557–62.

Morrison, P. (1960). Some interrelations between weight and hibernation function. *Bull. Mus. Comp. Zool.* Harvard University **124**, 75–91.

Montaudon, D., Robert, J. and Canguilhem, B. (1984). Fluorescence polarization study of lipids and membranes prepared from brain hemisphere of a hibernating mammal. *Biochemistry and Biophysics Research Communications* **119**, 396–400.

Mrosovsky, N. (1978). Circannual cycles in hibernators. In *Strategies in Cold: Natural Torpidity and Thermogenesis*, pp. 21–65. Edited by Wang, L.C.H. and Hudson, J.W. Academic Press, New York.

Musacchia, X.J. and Deavers, D.R. (1981). The regulation of carbohydrate metabolism in hibernators. In *Survival in the Cold*, pp. 55–75. Edited by Musacchia, X.J. and Jansky, L. Elsevier/North Holland, Amsterdam.

Musacchia, X.J. and Jansky, L. (eds.) (1981). *Survival in the Cold*. Elsevier/North Holland, Amsterdam.

Mutch, G.R.P. and Aleksiuk, M. (1977). Ecological aspects of winter dormancy in the striped skunk (*Mephitis mephitis*). *Canadian Journal of Zoology* **55**, 607–15.

Myers, R.D. (1980). Hypothalmic control of thermoregulation: neurochemical mechanisms. In *Handbook of Hypothalamus*, pp. 83–210. Edited by Morgane, P. and Panksepp, J. Marcel Dekker, New York.

Myers, R.D., Oeltgen, P.R. and Spurrier, W.A. (1981). Hibernation 'trigger' injected in brain induces hypothermia and hypophagia in the monkey. *Brain Research Bulletin* **7**, 691–5.

Nelson, R.A. (1980). Protein and fat metabolism in hibernating bears. *Federation Proceedings* **39**, 2955–8.

Nicholls, D.G. and Locke, R.M. (1984). Thermogenic mechanisms in brown fat. *Physiological Reviews* **64**, 1–64.

Novotona, R., Jansky, L. and Drahota, Z. (1975). Effect of hibernation on turnover of serotonin in the brain stem of golden hamster (*Mesocricetus auratus*). *General Pharmacology* **6**, 23–6.

Nürnberger, F. (1983). *Der hypothalamus des igels (Erinaceus europaeus L.) unter besonderer berucksichtigung des winterschlafes. Cytoarchitektonische und immuncytochemische studien*. Ph. D. Dissertation, University of Marburg, Federal Republic of Germany.

Oeltgen, P.R. and Spurrier, W.A. (1981). Characterization of a hibernation trigger. In *Survival in the Cold*, pp. 139–57. Edited by Musacchia, X.J. and Jansky, L. Elsevier/North Holland, Amsterdam.

Oeltgen, P.R., Walsh, J.W., Hamann, S.R., Randall, D.C., Spurrier, W.A. and Myers, R.D. (1982). Hibernation 'trigger': opioid-like inhibitory action on brain function of the monkey. *Pharmacology Biochemistry and Behaviour* **17**, 1271–4.

Olsson, S.O.R. (1972). Dehydrogenases (LDH, MDH, G-6-PDH, and a-GPDH) in the heart, liver, white and brown fat. *Acta Physiologica Scandinavica* **Suppl. No. 380**, 62–95.

Palmer, D.L. and Riedesel, M.L. (1976). Response of whole-animal and isolated hearts of ground squirrels, *Citellus lateralis*, to melatonin. Comparative Biochemistry and Physiology **53C**, 69–72.

Pehowich, D.J. and Wang, L.C.H. (1981). Temperature dependence of mitochondrial Ca^{2+} transport in a hibernating and nonhibernating ground squirrel. *Acta Universitatis Carolinae Biol.* **1979**, 291–3.

Pehowich, D.J. and Wang, L.C.H. (1984). Seasonal changes in mitochondrial succinate dehydrogenase activity in a hibernator. *Spermophilus richardsonii. Journal of Comparative Physiology* **154** B, 495–501.

Pengelley, E.T. and Asmundson, S.J. (1974). Circannual rhythmicity in hibernating mammals. In *Circannual Clocks: Annual Biological Rhythms*, pp. 95–160. Edited by Pengelley, E.T. Academic Press, New York.

Pengelley, E.T., Asmundson, S.J. and Ulhman, C. (1971). Homeostasis during hibernation in the golden-mantled ground squirrel, *Citellus lateralis. Comparative Biochemistry and Physiology* **38A**, 645–53.

Pengelley and Fisher (1963). The effect of temperature and photoperiod on the yearly hibernating behavior of captive golden-mantled ground squirrels (*Citellus lateralis tescorum*). *Canadian Journal of Zoology* **41**, 1103–20.

Phillips, J.A. and Harlow, H.J. (1982). Long-term effects of pinealectomy on the annual cycle of golden-mantled ground squirrel, *Spermophilus lateralis*. *Journal of Comparative Physiology* **146B**, 501–5.

Popova, N.K. and Koryakina, L.A. (1981). Seasonal changes in pituitary-adrenal reactivity in hibernating Spermophiles. *Endocrinologia Experimentalis* **15**, 269–76.

Popovic, V. (1960). Endocrines in hibernation. *Bull. Mus. Comp. Zool.* **124**, 104–30.

Ralph, C.L., Harlow, H.J. and Phillips, J.A. (1982). Delayed effect of pinealectomy and hibernation in the golden-mantled ground squirrel. *International Journal of Biometeorology* **26**, 311–28.

Raison, J.K. and Lyons, J.M. (1971). Hibernation: alteration of mitochondrial membranes as a requisite for metabolism at low temperature. *Proceedings of the National Academy of Sciences U.S.A.* **68**, 2092–4.

Raison, J.K., McMurchie, E.J., Charnock, J.S. and Gibson, R.A. (1981). Differences in the thermal behaviour of myocardial membranes relative to hibernation. *Comparative Biochemistry and Physiology* **69B**, 169–74.

Raths, P. and Kulzer, E. (1976). Physiology of hibernation and related lethargic states in mammals and birds. *Bonn. Zool. Monogr.* **9**, 1–93.

Reeves, R.B. (1977). The interaction of body temperature and acid-base balance in ectothermic vertebrates. *Annual Review of Physiology* **39**, 559–86.

Saboureau, M., Bobet, J.P. and Boissin, J. (1980). Activite cyclique de la fonction corticosurrennalienne et variations saisonnieres du metabolisme peripherique du cortisol chez un mammifere hibernant, le herisson (*Erinaceus europaeus* L.). *Journal of Physiology* (Paris) **76**, 617–29.

Sauerbier, I. and Lemmer, B. (1977). Seasonal variations in the turnover of noradrenaline of active and hibernating hedgehogs (*Erinaceus europaeus*). *Comparative Biochemistry and Physiology* **57C**, 61–3.

Sinnamon, W.B. and Pivorun, E.B. (1981). Effect of chronic melatonin administration on the duration of hibernation in *Spermaphilus tridecemlineatus*. *Comparative Biochemistry and Physiology* **70A**, 435–7.

Sinnamon, W.B. and Pivorun, E.B. (1982). Effects of pinealectomy, melatonin injections and melatonin antibody production on the mean duration of individual hibernation bouts in *Spermophilus tridecemlineatus*. *Journal of Thermal Biology* **7**, 243–9.

South, F.E. and Jacobs, H.K. (1973). Contraction kinetics of ventricular muscle from hibernating and non-hibernating mammals. *American Journal of Physiology* **225**, 444–9.

South, F.E., Miller, V.M. and Hartner, W.C. (1978). Neuronal models of temperature regulation in euthermic and hibernating mammals: an alternate model for hibernation. In *Strategies in Cold: Natural Torpidity and Thermogenesis*, pp. 187–224. Edited by Wang, L.C.H. and Hudson, J.W. Academic Press, New York.

South, F.E., Hannon, J.P., Willis, J.S., Pengelley, E.T. and Alpert. (eds.) (1972). *Hibernation and Hypothermia: Perspectives and Challenges*. Elsevier, Amsterdam.

Spafford, D.C. and Pengelley, E.T. (1971). The influence of the neurohumor serotonin on hibernation in the golden-mantled ground squirrel, *Citellus lateralis*. *Comparative Biochemistry and Physiology* **38A**, 239–50.

Steffen, J.M. and Riedesel, M.L. (1982). Pulmonary ventilation and cardiac activity in hibernating and arousing golden-mantled ground squirrels (*Spermophilus lateralis*). *Cryobiology* **19**, 83–91.

Strumwasser, F. (1960). Some physiological principles governing hibernation in *Citellus beecheyi*. In *Mammalian Hibernation*, pp. 285–320. Edited by Lyman, C.P. and Dawe, A.R. Bull. Mus. Comp. Zool. Harvard Univeristy. 124.

Suomalainen, P. (ed.) (1964). *Mammalian Hibernation II*. Ann. Acad. Scient. Fenn. Ser. A. IV. Biologica 71.

Swan, H. (1981). Neuroendocrine aspects of hibernation. In *Survival in the Cold*, pp. 121–38. Edited by Musacchia, X.J. and Jansky, L. Elsevier/North Holland, Amsterdam.

Tashima, L.S., Adelstein, S.J. and Lyman, C.P. (1970). Radioglucose utilization by active, hibernating, and arousing ground squirrels. *American Journal of Physiology* **218**, 303–9.

Torke, K.G. and Twente, J.W. (1977). Behavior of *Spermophilus lateralis* between periods of hibernation. *Journal of Mammalogy* **58**, 385–90.

Vogel, P. (1980). Metabolic levels and biological strategies in shrews. In *Comparative Physiology: Primitive Mammals*, pp. 170–80. Edited by Schmidt-Nielsen, K., Bolis, L. and Taylor, C.R. Cambridge University Press, Cambridge.

Wang, L.C.H. (1978). Energetics and field aspects of mammalian torpor: the Richardson's ground squirrel. In *Strategies in Cold: Natural Torpidity and Thermogenesis*, pp. 109–145. Edited by Wang, L.C.H. and Hudson, J.W. Academic Press, New York.

Wang, L.C.H. (1979). Time patterns and metabolic rates of natural torpor in the Richardson's ground squirrel. *Canadian Journal of Zoology* **57**, 149–55.

Wang, L.C.H. (1982). Hibernation and the endocrine. In *Hibernation and Torpor in Mammals and Birds*, pp. 206–36. Edited by Lyman, C.P., Willis, J.S., Malan, A. and Wang, L.C.H. Academic Press, New York.

Wang, L.C.H. and Abbotts, B. (1981). Maximum thermogenesis in hibernators: Magnitudes and Seasonal Variations. In *Survival in the Cold* pp. 77–97. Edited by Musacchia, X.J. and Jansky L. Elsevier/North Holland, Amsterdam.

Wang, L.C.H. and Hudson, J.W. (1970). Some physiological aspects of temperature regulation in the normothermic and torpid hispid pocket mouse, *Perognathus hispidus*. *Comparative Biochemistry and Physiology* **32**, 275–93.

Wang, L.C.H. and Hudson, J.W. (eds.) (1978). *Strategies in Cold: Natural Torpidity and Thermogenesis*. Academic Press, New York.

Whitten, B.K. and Klain, G.J. (1968). Protein metabolism in hepatic tissues of hibernating and arousing ground squirrels. *American Journal of Physiology* **214**, 1360–2.

Whitten, B.K. and Klain, G.J. (1969). NADP-specific dehydrogenases and hepatic lipogenesis in the hibernator. *Comparative Biochemistry and Physiology* **29**, 1099–1104.

Willis, J.S. (1979). Hibernation: Cellular aspects. *Annual Review of Physiology* **41**, 275–86.

Willis, J.S. (1982a). The mystery of the periodic arousal. In *Hibernation and Torpor in Mammals and Birds*, pp. 92–103. Edited by Lyman, C.P., Willis, J.S., Malan, A. and Wang, L.C.H. Academic Press, New York.

Willis, J.S. (1982b). Intermediary metabolism in hibernation. In *Hibernation and Torpor in Mammals and Birds*, pp. 124–39. Edited by Lyman, C.P., Willis, J.S., Malan, A. and Wang, L.C.H. Academic Press, New York.

Willis, J.S., Ellory, J.C. and Cossins, A.R. (1981). Membranes of mammalian hibernators at low temperatures. In *Effects of Low Temperatures on Biological Membranes*, pp. 121–42. Edited by Morris, G.J. and Clarke, A. Academic Press, New York.

Willis, J.S., Ellory, J.C. and Wolowyk, M.W. (1980). Temperature sensitivity of the sodium pump in red cells from various hibernators and nonhibernator species. *Journal of Comparative Physiology* **138**, 43–47.

Willis, J.S., Goldman, S.S. and Foster, R.F. (1971). Tissue K concentration in relation to the role of the kidney in hibernation and the cause of periodic arousal. *Comparative Biochemisry and Physiology* **39A**, 437–45.

Winston, B.W. and Henderson, N.E. (1981). Seasonal changes in morphology of the thyroid gland of a hibernator, *Spermophilus richardsonii. Canadian Journal of Zoology* **59**, 1022–31.

Wünnenberg, W. and Baltruschat, D. (1982). Temperature regulation of golden hamsters during acute hypercapnia. *Journal of Thermal Biology* **7**, 83–6.

Wünnenberg, W., Merker, G. and Speulda, E. (1978). Thermosensitivity of preoptic neurons and hypothalamic integrative function in hibernators and nonhibernators. In *Strategies in Cold: Natural Torpidity and Thermogenesis*, pp. 267–97. Edited by Wang, L.C.H. and Hudson, J.W. Academic Press, New York.

Yacoe, M.E. (1983a). Adjustments of metabolic pathways in the pectoralis muscles of the bat, *Eptesicus fuscus*, related to carbohydrate sparing during hibernation. *Physiological Zoology* **56**, 648–58.

Yacoe, M.E. (1983b). Maintenance of the pectoris muscle during hibernation in the big brown bat, *Eptesicus fuscus. Journal of Comparative Physiology* **152B**, 97–104.

Yacoe, M.E. (1983c). Protein metabolism in the pectoralis muscle and liver of hibernating bats, *Eptesicus fuscus. Journal of Comparative Physiology* **152B**, 137–44.

Young, R.A., Danforth, E., Jr., Vagenakis, A.G., Krupp, P.P., Frink, R. and Sims, E.A.H. (1979). Seasonal variation and the influence of body temperature on plasma concentrations and binding of thyroxine and triiodothyronine in the woodchuck. *Endocrinology* **104**, 996–9.

Zimmerman, M.L. (1982). Carbohydrate and torpor duration in hibernating golden-mantled ground squirrel (*Citellus lateralis*). *Journal of Comparative Physiology* **147B**, 129–35.

Zucker, I. and Licht, P. (1983). Circannual and seasonal variations in plasma luteinizing hormone levels of ovariectomized ground squirrels (*Spermophilus lateralis*). *Biology of Reproduction* **28**, 178–85.

Section IV

Applications

11

The low temperature preservation of plant cell, tissue and organ cultures and seed for genetic conservation and improved agricultural practice

L.A. Withers
Department of Agriculture and Horticulture
University of Nottingham

Plant germplasm – a threatened natural resource

There is a common misconception that conservation is a pedestrian, if ideological pursuit, divorced from practical reality. However, plant genetic conservation in particular can be technically, scientifically and logistically demanding; it is in every sense a critical discipline. Furthermore, it has a vital role in underpinning all work directed towards crop improvement.

If agricultural production is to meet the needs of the world's increasing population and avoid a frightening projected number of undernourished individuals in the developing world, breeding must be directed towards increasing yields. However, in so doing, it is important that a relatively narrow range of genotypes should not be allowed entirely to replace the vast spectrum embodied in the gene-pool of crop plants and their relatives.

To avoid conflicts between conservation and the demands of agricultural productivity, the strategies and methods adopted must take into account the realities of germplasm utilization. Material committed to storage should not only be that which is easy to collect and store but should be representative of crop gene-pools and should be evaluated sufficiently thoroughly to facilitate the selection of material for breeding.

Plant genetic conservation – the current position

Identifying types of germplasm

Plant genetic conservation is more than just a matter of storing seeds. The types of plant germplasm requiring conservation are best revealed by considering the historical phases of agricultural plant breeding (Wilkes, 1983). Firstly, in the Old World, most major crops emerged into cultivation, little different from their truly wild relatives. With the development of the New World, agricultural practice expanded and germplasm was transported between countries, permitting the introduction and utilization of new genetic diversity.

A third phase was marked by the development of strategies for crop breeding, based upon Mendelian principles. The latter takes us virtually to the present where we are on the threshold of the fourth phase involving genetic engineering and the use of tissue culture/*in vitro* technology. By such means, it should be possible to transfer genes between unrelated, otherwise incompatible species, thereby vastly increasing the scope for crop improvement (Cocking, 1986; Cocking *et al.*, 1981; Ingle, 1982).

The practices of traditional plant breeding and crop production covered by the first three phases utilize seeds, pollen, and vegetative propagules such as tubers, corms, cuttings and grafting material. In contrast, *in vitro* work involves various types of aseptic laboratory culture. Single celled units (protoplasts, pollen and fine cell suspensions) are the subjects of genetic manipulation, whereas organized structures (shoots, meristems, embryos and plantlets) are the means by which newly developed genotypes are multiplied and released into *in vivo* growth for evaluation and eventual crop production. Specific applications of the various systems are noted on pp. 392–400.

This new ability to manipulate and even create genotypes may appear to detract from the value of pre-existing germplasm. However, the germplasm storage needs of *in vitro* technology itself should support, rather than divert efforts and resources away from, conservation work.

Conventional methods of germplasm storage

Ideally, much plant germplasm should be conserved *in situ*, in its natural environment co-existing with the other living elements of that environment.

However, a number of reservations, both practical and scientific require that germplasm be conserved in other, additional ways (i.e. *ex situ*). At present, *ex situ* reserves take the form of seed-banks and plantations.

Seed-banks are divided into base and active collections. The base collections involve storage at temperature as low as −20°C and it is intended that the material should spend decades or centuries in storage. Medium and short-term active collections involve storage at below or above *c.* 5°C respectively for periods ranging from a few years to *c.* 20 years. All the seed is stored in a partially desiccated state and is not truly frozen at any of the temperatures used. Over 30 major institutes worldwide now use refrigerated storage for seeds (Plucknett *et al.*, 1983).

Seed stores require relatively simple but consistent maintenance giving stability of temperature and humidity. Periodic testing of viability and replenishment of stocks have to be carried out. Unfortunately, however, the seeds of a number of tropical and temperate crops, especially fruits, timber species and aquatic species (over 50 in all; King and Roberts, 1979; Roberts and King, 1982) cannot be stored in the conventional way as they lose viability when dried below a certain moisture content and/or when exposed to low temperatures. The seeds are termed 'recalcitrant' or 'short-lived' (seeds which can be stored are termed 'orthodox').

At best, recalcitrant seeds can be stored for a few months in the moist state and protected by an osmoticum (King and Roberts, 1979; Roberts and King, 1982). For recalcitrant seeds, and vegetative crops which are either sterile or produce seeds which are not representative of valuable clones, recourse is taken to conservation in plantations in the medium to long term, and storage of vegetative propagules (a very unsatisfactory method) in the short term.

Plantations have disadvantages in that they are geographically restricted, expensive to maintain and are open to ecological, pathological and organizational threats which may lead to sudden losses of major portions of a gene-pool ('genetic wipeout'; Wilkes, 1983).

Pollen has received relatively little attention but it appears that again there are 'recalcitrant' and 'orthodox' species. For orthodox species, storage under conditions of controlled humidity at ambient or subzero temperatures can be carried out (Roberts, 1975).

In vitro cultures are always subject to conservation in that they are maintained in carefully controlled, artificial environments. However, practical problems in their storage combine some of those of seed-banks and plantations, plus the added risks of microbial contamination and cryptic genetic instability.

Faced with the situation where a significant amount of 'conventional germplasm' plus the materials of the most novel approach to plant breeding cannot be conserved satisfactorily, it is clear that developmental work is urgently needed. For some time, it has been proposed that *in vitro* technology might itself provide the answer to persistent problems in genetic conservation.

In the following sections an assessment will be made of the progress to date in developing *in vitro* storage techniques for vegetative material, pollen and embryos, and in refining existing seed storage technology. Only limited detail can be given here. The interested reader is referred to a number of recent publications for further information, including crop lists (King and Roberts, 1979; Stanwood, 1985; Stanwood and Bass, 1978; Stanwood and Roos, 1979; Styles *et al.*, 1982; Withers, 1980a; 1982a; 1982b; 1985a; 1985b; 1986).

In vitro storage techniques

General approaches

The challenge in developing storage methods is to modify the growth pattern of the culture in a way which minimizes necessary inputs of energy and nutrients without jeopardizing viability or genetic stability. Attempts to store unorganized cultures under conditions which slow down their rate of growth have been relatively unsuccessful. The only noteworthy examples are the maintenance of callus of *Nicotiana tabacum* under low oxygen tension (Bridgen and Staby, 1981) and callus and pro-embryos of *Daucus carota* in the partially desiccated state (Jones, 1974; Nitzsche, 1978).

The picture is very different for organized shoot cultures, in which form a large number of species can be stored either at a reduced temperature or under the influence of growth retardants. Examples include the storage of:

(i) *Solanum tuberosum* and other potato genotypes for periods of 1 year or longer at temperatures in the range of 6–12°C or in the presence of Alar 85, abscisic acid or mannitol as growth inhibitors (Henshaw *et al.*, 1980a; 1980b; Mix, 1984; Westcott, 1981);

(ii) *Manihot esculenta* for similar periods at a rather higher temperature (*c.* 20°C; CIAT, 1978);

(iii) a range of forage grasses and legumes at a temperature of 2–6°C, for periods of *c.* 1 year (Dale *et al.*, 1980).

Much relevant work remains unpublished but is recorded in a recent survey (Withers, 1982b; Wheelans and Withers, 1984).

The total storage period can be very long, being limited only by the necessary labour and materials required at each transfer interval, some progressive deterioration or accidental loss, and the risk (albeit low) of genetic variation in storage. This approach to storage is best considered as providing active *in vitro* collections of germplasm. Such collections cannot equate to the base collections necessary for adequate long-term genetic conservation.

Thus, for the long-term storage of shoot cultures and the adequate preservation on all time scales of other types of culture, an alternative approach is required. The obvious candidate for this role is cryopreservation.

Cryopreservation techniques – common points

The principles underlying cryopreservation of biological materials have been described in detail elsewhere in this volume and will not be reiterated here. It is useful, however, to outline generalizations which can be made regarding the handling of cultured plant material. Specific detail for the very diverse range of specimens involved will then be given, system-by-system.

The intrinsic freeze-tolerance of *in vitro* cultured plant material is very low and measures must be taken at all stages before, during and after freezing to support viability. A number of 'pregrowth' treatments may be used to precondition cultures, including cold-hardening (in the classical sense) or incubation in the presence of additives which apply limited osmotic stress.

With very few exceptions, cryoprotection is essential. As a general guide, cryoprotectant mixtures including dimethyl sulphoxide (DMSO), glycerol or a sugar, and sometimes a third component are most appropriate for unorganized systems, whereas DMSO alone is usually adequate for organized systems. Cryoprotectants are usually prepared in culture medium which itself contains 2 per cent or 3 per cent sucrose and minerals (preparation in water can greatly reduce the cryoprotective effect; Withers, 1980b), and applied at a temperature close to 0°C. Recent work has shown, however, that application at such a low temperature may not be necessary (e.g. Grimsley and Withers, 1983).

Slow or step-wise freezing appears to be suitable for most material but there are important exceptions wherein injury is avoided by the use of very rapid freezing and thawing. Washing to remove cryoprotectants does not seem to be necessary, although it is commonly used. Experiences with cell suspension cultures and some embryos would suggest that washing may even be harmful (Withers, 1979; Withers and King, 1979). Damaging effects of washing may be alleviated by the use of relatively warm medium (Finkle *et al.*, 1982).

Recovery is normally carried out under standard conditions, except that semi-solid rather than liquid culture medium is often used during the initial recovery period. To elicit the desired pattern of growth in some organized cultures, it may be necessary to modify the composition of the recovery medium.

Recommended procedures

In this section, a model cryopreservation procedure will be given for each type of specimen, citing the most relevant reference but not necessarily the exact source as the method may be a composite. Suggested variations and a note of particular difficulties or points of interest follow. Fig. 11.1 shows apparatus which is suitable for many of the procedures.

Protoplasts

Model procedure: *Pregrowth* – harvest cells in exponential growth; isolate protoplasts. *Cryoprotection* – 5% v/v DMSO + 10% w/v glucose; dispense into ampoules. *Freezing* – 1 or 2°C min^{-1} to -35°C; transfer to liquid nitrogen. *Thawing* – rapidly in warm water at +40°C. *Post-thaw treatments* – wash in liquid medium. *Recovery* – on standard semi-solid medium. (Takeuchi *et al.*, 1982.)

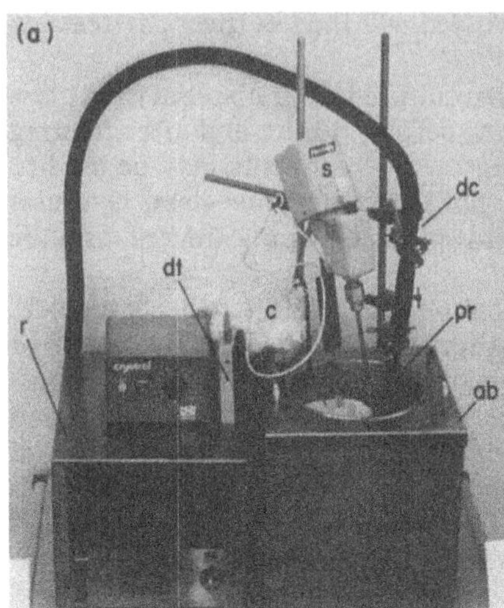

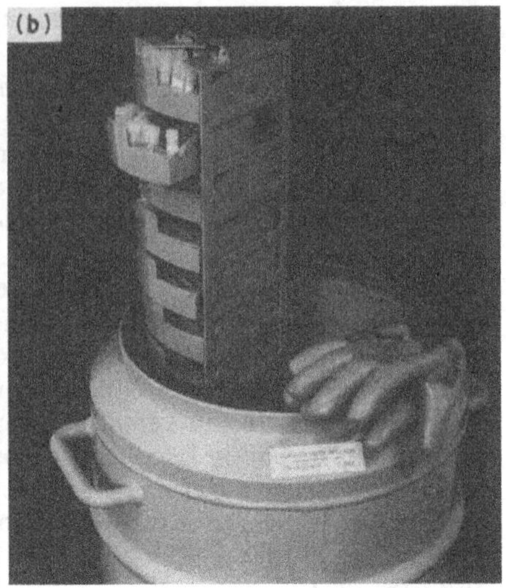

Fig. 11.1 Cryopreservation apparatus suitable for carrying out many of the storage procedures detailed in the text. (a) Improvised freezing unit, based on Withers and King (1980). Key: r = thermostated refrigeration unit; dc = dip cooler; ab = insulated alcohol bath; s = stirrer (spark-free); pr = polystyrene raft carrying specimen ampoules; dt = digital thermometer recording specimen temperature; c = cover for bath consisting of a bag of polystyrene beads. (b) Liquid nitrogen cooled storage refrigerator. Specimen ampoules are held in stacks of drawers. Note insulated safety gloves for use when handling instruments and specimens cooled by liquid nitrogen.

To date, seven higher plant species have been cryopreserved as protoplasts. In some cases recovery has not extended beyond the recording of a positive viability test. The most widely investigated, *Daucus carota*, has responded best in terms of recovery growth and plant regeneration.

Preliminary investigations suggest that prospects are good for the wide use of cryopreservation for protoplasts. The technique should find applications in genetic manipulation, and although not immediately relevant to genetic conservation in the more conventional sense, it may provide a means of storing genotypes which fail to respond to techniques involving larger, more organized tissues.

Pollen

Model procedure I, Cryopreservation: *Pretreatment* – harvest pollen from dehisced anthers; dry to *c*. 30 per cent moisture content; enclose in a vial. *Cryoprotection* – none. *Freezing* – plunge directly into liquid nitrogen. *Thawing* – rapidly in warm water at +40°C. *Recovery* – by germination under standard test conditions or fertilization. (Barnabas and Rajki, 1976.)

Model procedure II, Freeze-drying: *Pretreatment* - harvest pollen from dehisced anthers and enclose in a vial. *Cryoprotection* - none. *Freezing* - freeze-dry for *c.* 30 minutes. *Store* - at −20°C or in liquid nitrogen. *Thawing and rehydration* - in a hygrostat at *c.* 90 per cent relative humidity. *Recovery* - by germination under standard test conditions or fertilization. (Akihama *et al.*, 1978.)

Conservation of pollen has a dual role to play in plant breeding and associated disciplines. Firstly, pollen can embody a significant portion of the crop genotype in a compact, readily stored form. The genes may be stored for long periods of time, spanning seasons or even just days or weeks to overcome the problems caused by asynchrony in flowering. Secondly, isolated pollen can be a useful source of haploid plants to be used in genetic manipulation studies or for the rapid development of homozygous diploid plants to introduce into breeding programmes.

Either approach to storage would appear to be appropriate for fertilization requirements. Comparative data are scant but it would appear that a higher percentage survival might be expected from material which has been cryopreserved. Storage periods of up to 4 years have been recorded for cryopreserved pollen stored at −76°C (Frank *et al.*, 1982; no doubt storage in liquid nitrogen would be at least as safe), and 9 years for freeze dried pollen (Akihama *et al.*, 1978).

One possible caution in relation to genetic stability has been raised by Grout and Crisp (personal communication) who found differences in the storage quality of seed produced from fertilization with cryopreserved pollen of *Brassica* sp. Gamete selection may be occurring.

As yet, no-one has attempted to cryopreserve or freeze-dry isolated pollen and then induce androgenesis; it is anticipated that cryopreservation would be more appropriate.

Anthers and pollen embryos

Model procedure: *Pregrowth* - 3 to 4 weeks under standard conditions to induce androgenesis; express embryos where appropriate. *Cryoprotection* - 7% v/v DMSO; dispense into ampoules. *Freezing* - 2°C min^{-1} to *c.* −100°C; transfer to liquid nitrogen. *Thawing* - rapidly in warm water at +40°C. *Post-thaw treatment* - wash in liquid medium. *Recovery* - on standard semi-solid medium. (Bajaj, 1977.)

Five species have been cryopreserved as anther cultures or pollen embryos. However, survival levels are very low (e.g. 1–5 per cent) and it is clear that much further work will be necessary before a satisfactory, routine procedure emerges. Possible modifications of the above procedure include the use of DMSO plus sucrose (both at 7 per cent) instead of DMSO alone and rapid freezing. In view of the size of anthers and possible difficulties in achieving cryoprotectant penetration, pollen-derived embryos may be the more suitable subject for cryopreservation.

Cell suspension cultures

> **Model procedure:** *Pregrowth* – culture for 4–7 days in medium containing 6 per cent mannitol; harvest during rapid growth. *Cryoprotection* – 0.5M DMSO + 0.5M glycerol + 1M sucrose; dispense into ampoules. *Freezing* – 1°C min^{-1} to −35°C; hold for 40 minutes; transfer to liquid nitrogen. *Thawing* – rapidly in warm water at +40°C; *Post-thaw treatments* – none. *Recovery* – the cells suspended in cryoprotectant solution are layered over standard semi-solid medium; any liquid not taken up after *c.* 7 days is drained away. (Withers and King, 1980.)

Cell suspension cultures have received more attention than any other system. To date, over 30 species have been cryopreserved successfully as cells, the majority achieving good recovery growth from a high proportion (up to 80 per cent) of the cell population.

A cryopreservation procedure can be recommended here without difficulty, but some additional points may aid the widest possible application (Kartha *et al.*, 1982a; King, personal communication; Maddox *et al.*, 1983; Withers and King, 1980). Sorbitol, DMSO or proline may be substituted for mannitol at the pregrowth stage. Sucrose may be omitted from the cryoprotectant procedure or replaced by proline. Gradual dilution of the cryoprotectant-containing medium during the post-thaw period and use of a liquid medium appropriate for growth from low cell densities may aid recovery.

Although detail cannot be given here, it is appropriate to mention that cell suspension cultures have been the subjects of electron microscopical and physiological studies which aid our understanding of the nature of freezing injury in this and other biological systems (e.g. Cella *et al.*, 1982; Pritchard *et al.*, 1982; Withers and Davey, 1978).

The role of cell suspension cultures other than in genetic manipulation is very limited. However, they do provide valuable model systems in which to study genetic stability in storage by cryopreservation. Studies have been carried out in at least two laboratories to monitor stability after freezing and thawing in defined biochemical mutant lines (Hauptmann and Widholm, 1982; King, personal communication). In all cases examined, frozen and thawed cells perform in the same way as control cultures under test conditions. These findings are corroborated by observations of retention of morphogenic potential in stored material (e.g. Nag and Street, 1973). Stability in relation to time in extended storage has yet to be examined.

Moss protonemal cultures

> **Model procedure:** *Pregrowth* – culture for *c.* 12 days on semi-solid medium supplemented with 0.5M mannitol. *Cryoprotection* – 10% w/v glucose + 5% v/v DMSO dispense into ampoules. *Freezing* – 1°C min^{-1} to −35°C; transfer to liquid nitrogen. *Thawing* – rapidly in warm water at +40°C. *Post-thaw treatments* – tissue is drained free of cryoprotectant solution but not washed. *Recovery* – culture on a standard semi-solid medium. (Grimsley and Withers, 1983.)

The moss *Physcomitrella patens* provides a valuable model system for the study of developmental processes in plants. The many auxotrophic and developmentally abnormal strains which are available could be used to assess intergenotypic responses to cryopreservation and stability of genotypes over long periods of storage in liquid nitrogen. The system is as suitable, if not superior to, cell suspensions for these purposes and should provide useful information relevant to the conservation of all biological material under conditions of cryopreservation.

Callus cultures

Model procedure: *Pregrowth* - harvest actively growing callus. *Cryoprotection* - 10% v/v DMSO + 10% w/v polyethylene glycol (Carbowax 6000) + 8% w/v glucose; dispense into ampoules. *Freezing* - 1°C min^{-1} to −30°C; transfer to liquid nitrogen. *Thawing* - rapidly in warm water at +40°C. *Post-thaw treatments* - wash in warm (+22°C) liquid medium. *Recovery* - on standard semi-solid medium. (Finkle *et al.*, 1982.)

Callus cultures have been relatively neglected and must be recognized as among the most difficult subjects for cryopreservation. This reflects their relative bulk, slow rate of growth, cellular heterogeneity and often large mean cell size. Whilst fragmented callus may be handled as a cell suspension, the procedure recommended here is based upon the most extensive study of callus cultures conducted to date and involving eight species (Finkle *et al.*, 1982). These species include *Phoenix dactylifera* (datepalm) the callus of which, in fact, consists of a mass of pro-embryo like structures. Some intergenotypic differences were found in the survival of callus of this species and of *Oryza sativa* (rice).

Callus cultures are considered to be particularly prone to genetic instability. There are, however, examples of satisfactory cloning via callus which do not incur excessive variation. In view of this and their relative ease of handling, callus cultures may be favoured in some conservation work. Further development of cryopreservation methodology is therefore essential.

Shoot-tips

Model procedure: *Pregrowth* - dissect shoot-tips from parent plant/culture and pregrow for 2 days on medium supplemented with 5% v/v DMSO. *Cryoprotection* - 5% v/v DMSO dispense into ampoules. *Freezing* - 0.5°C min^{-1} to −40°C; transfer to liquid nitrogen. *Thawing* - rapidly in warm water at +40°C. *Post-thaw treatments* - wash in liquid medium. *Recovery* - on standard semi-solid medium. (Kartha *et al.*, 1979).

Although the above procedure or one very similar has been demonstrated to have a fairly wide applicability, a considerable degree of qualification is required. This reflects several factors including heterogeneity of the experimental material in terms of its origin and mode of culture, acute unresponsiveness of certain species and different experimental strategies adopted by research workers. However, in view of the central importance of shoot-cultures to conservation work and as means of clonally propagating elite varieties of root and tuber crops and many fruits, it is essential that good methods be developed.

The magnitude of the conservation problem is illustrated by the fact that for potato, over 6000 clones are currently maintained in field plantations at the International Potato Centre (CIP) in Peru, and for cassava, over 2000 at the International Centre for Tropical Agriculture (CIAT) in Colombia. These could all usefully be transferred to *in vitro* conservation as shoot cultures. A further application of *in vitro* culture of shoot-tips is in the elimination of pathogens and provision of disease-free stocks for germplasm exchange. By such means, transportation and quarantine procedures can greatly be simplified.

Pregrowth is usually necessary but the inclusion of DMSO would not appear to be essential. The author (unpublished observations) has found that the critical factor may be promotion of healing of the excision wound whilst inhibiting excessive growth, rather than a specific DMSO effect.

The use of a higher level of DMSO (e.g. 10 or 15% v/v) and freezing step-wise through, for example, −5°C, −10°C, −15°C, −20°C (Grout and Henshaw, 1978; Kartha *et al.*, 1982b; Sakai and Uemura, 1982) have also been successful, bringing the total number of species succumbing to a cryo-preservation method based on extracellular freezing to more than ten. Several of these species, including *Solanum* spp. (potato) and *Fragaria* sp. (strawberry) respond well to both this approach and ultra-rapid freezing but one example, *Brassica napus*, appears to be far more amenable to ultra-rapid freezing (Withers, 1982c; see Fig. 11.2). Ultra-rapid freezing was first used successfully for *in vitro* material by Grout and Henshaw (1978) studying the cryopreservation of potato genotypes. The cryoprotected shoot-tip was placed on a hypodermic needle and plunged into liquid nitrogen. Thawing was carried out by plunging the needle into warm liquid medium. The practical disadvantages are obvious but in view of its evident efficacy with certain material, efforts are underway to try to combine ultra-rapid freezing with some means of protecting and enclosing the specimen (Withers, unpublished observations).

Another approach to freezing (Kartha *et al.*, 1982b) has led to some success with the particularly difficult species *Manihot esculenta* (cassava). Shoot-tips in droplets of cryoprotectant solution are placed on a metal foil sheet in a Petri dish for cooling. The supporting foil is then dropped into liquid nitrogen. This method also has practical disadvantages but it should be open to improvement.

In a number of cases, cryopreserved shoot-tips have failed to undergo organized recovery growth. Electron microscopical studies indicate that damage to the shoot apex may be massive (Grout and Henshaw, 1980; Haskins and Kartha, 1980). Strong intergenotypic variations in total recovery and extent of callusing versus shoot development have been found in potato (Towill, 1984). To some extent this defeats the object of using

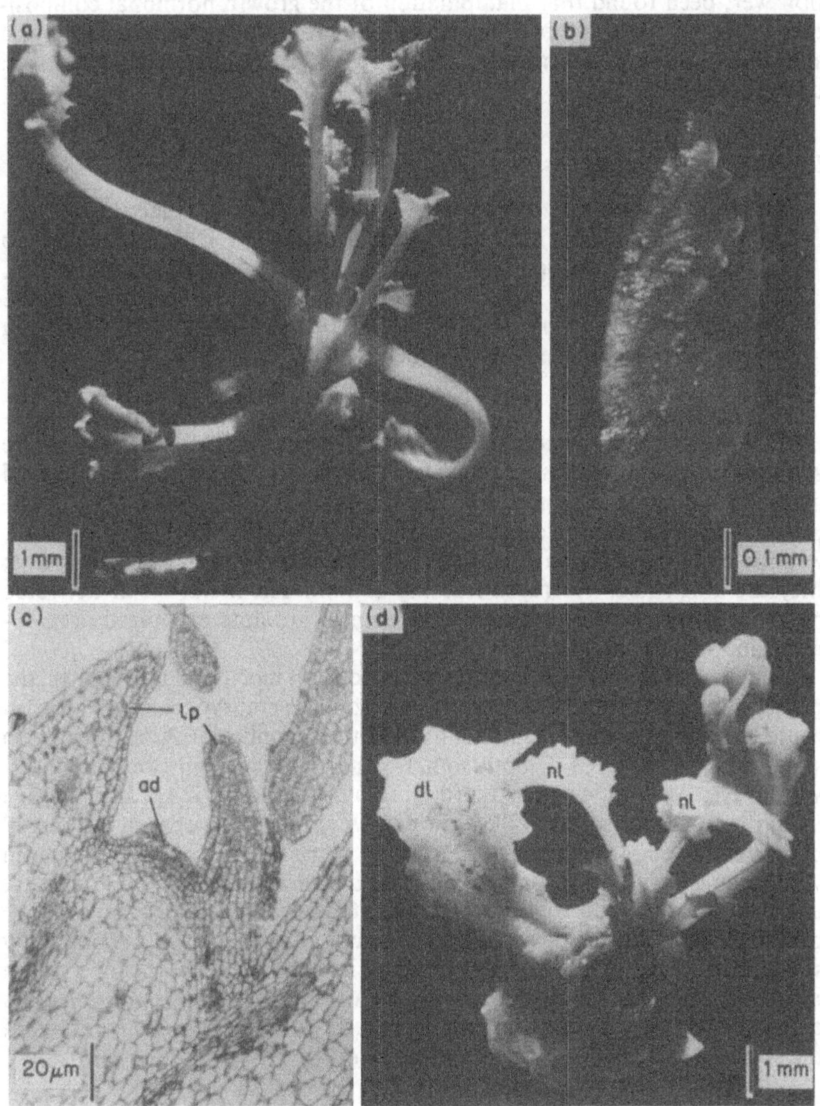

Fig. 11.2 The cryopreservation of shoot-tips of *Brassica napus*. (a) Shoot culture. (b) Dissected shoot-tip consisting of apical dome and leaf primordia/leaflets. (c) Vertical section through shoot-tip showing apical dome (ad) and leaf primordia (lp). (d) Cryopreserved shoot-tip (ultra-rapidly frozen), photographed 3 weeks after thawing. Leaf primordia which were present at the time of freezing have developed into deformed leaves (dl); leaves formed after thawing (nl) are of normal morphology. Pregrowth in the presence of DMSO promotes development of the latter. (Withers and Marshall, unpublished observations.)

shoot-tips for genetic conservation by cryopreservation as the desired genetic stability may be lost by the failurė of non-adventitious development. It has, however, been found that manipulation of the growth hormonal composition of the post-thaw recovery medium may promote organized growth and suppress callusing (Henshaw *et al.*, 1985a; Withers and Benson; unpublished).

In vitro *cultured embyros*

> **Model procedure:** *Pregrowth* - select early stage embryos globular/heart-shaped). *Cryoprotection* - 10% v/v DMSO; transfer to aluminium foil envelope (without suspending cryoprotectant solution). *Freezing* - 1°C min^{-1} to −40°C; transfer to liquid nitrogen. *Thawing* - slowly in air. *Post-thaw treatments* - none. *Recovery* - on standard semi-solid medium. (Withers, 1979.)

The method above has been used successfully for somatic embryos regenerated from a cell suspension of *Daucus carota* (carrot) and immature zygotic embryos of a number of cereals (Withers, 1979; 1982c; unpublished observations). In the former but not the latter case, slow freezing in the presence of a suspending cryoprotectant solution was entirely unsuccessful. Zygotic embryos may, in fact, have very imprecise freezing requirements; they are able to survive rapid freezing as well. The inclusion of activated charcoal in the recovery medium may help to promote organized development and reduce callusing.

Immature embryos are important sources of inoculum material in the genetic manipulation of some crops, and somatic embryogenesis is a useful route to mass propagation. In addition to these applications, the facility to store cultured embryos of both somatic and zygotic origin can aid conservation. Firstly, zygotic embryos produced by wide crosses may fail to survive *in vivo* due to endosperm failure. Rescue *in vitro* and cryopreservation will permit their long-term storage. Secondly, some recalcitrant seeds contain very large embryos (e.g. palms and *Theobroma cacao* - cocoa) which are likely to be difficult to cryopreserve. If secondary embryos could be produced from these and then selected for cryopreservation at a small, early stage, both cloning to replicate the accession and its genetic conservation might be achieved. The true recalcitrance of oil palm (*Eldeis guineensis*) is questioned but it is interesting that cryopreservation of somatri embryos of the species has been achieved (Engelmann *et al.*, 1985).

Techniques for the ultra low temperature storage of seeds

Areas requiring attention

In considering the limitations of conventional seed storage, four problems can be identified:

(i) quantitative loss of orthodox seeds as a result of distribution, viability testing and deterioration;

(ii) genetic instability in orthodox seeds;
(iii) genetic variation resulting from periodic regeneration and replenishment of seed stocks;
(iv) rapid loss of viability in recalcitrant seeds.

Clearly, efforts could be directed towards the conversion of all materials to an *in vitro* state and subsequent cryopreservation. However, this would not be an intelligent approach. Wide genetic diversity could not be conserved in any practical way and in any case, unnecessary technical complexity would be introduced. *In vitro* storage has a part to play in the resolution of problems associated with seed recalcitrance but should not be considered as a first option for orthodox seeds where cryopreservation of the seed itself is recommended. (At this time, it is not possible to make specific recommendations for the cryopreservation of recalcitrant seeds.)

Cryopreservation of dry, orthodox seeds

Model procedure: *Preparation* – dehydrate seeds to below *c.* 10 per cent water content; enclose in an aluminium foil or paper envelope. *Freezing* – immerse in liquid nitrogen. *Thawing* – slowly in air at room temperature. *Post-thaw treatments* – none. *Recovery* – standard germination procedure. (Stanwood and Roos, 1979.)

Over 40 species have now been cryopreserved as seeds (Sakai and Noshiro, 1975; Stanwood and Bass, 1978; Stanwood and Roos, 1979; Styles *et al.*, 1982). They range from very small to relatively large (e.g. several legumes), indicating that unit size is more important than in the case of *in vitro* cultures. However, water content is critical, and the concensus finding appears to be that at moisture levels in excess of 15 per cent, viability will be seriously affected by freezing damage. (There are a very few specific exceptions such as *Papaver somniferum* [poppy] [Stanwood and Roos, 1979] which will tolerate 17.5 per cent water content.)

In surveying the literature on the ultra-low temperature storage of orthodox seeds, there appear to be few problems and attention is increasingly given to the logistics and costs involved in this approach to conservation. Possible difficulties should, however, be mentioned. Mechanical injury may occur during the freezing of some seeds. Styles *et al.*, (1982) recommend that packets, as described above, should be used rather than ampoules for any seeds which are brittle (e.g. as a result of excessive dryness), fragile or elongate, in order to avoid mechanical damage.

Cracking of the seed coat can, however, be beneficial (Jordan *et al.*, 1982). An example is found in seeds of *Setaria lutescens* in which one cycle of freezing and thawing improved germination rates by facilitating water uptake. Repeated freezing and thawing led to a decline in viability, attributed to cytological damage, but this is unlikely to be a problem in routine storage.

In cases where the level of physical damage is intolerable and jeopardizes germination, slow freezing may provide a solution. Mumford and Grout (1978) found that seeds of *Manihot esculenta* (cassava) with a very low water content of 2–6 per cent were seriously damaged by direct immersion in liquid nitrogen. In contrast, slow freezing at *c.* 100°C min^{-1} (this is in fact rapid in comparison with methods for cultures) maintained viability levels at 100 per cent of controls. Some manipulation of thawing rates may also be beneficial, although detailed comparative data are lacking.

Storage periods of up to 600 days in liquid nitrogen are quoted (Styles *et al.*, 1982) and much longer terms could be anticipated, offering a considerable improvement in genetic stability and viability compared with conventional cold storage. Cryopreservation thus offers the potential of resolving the three problem areas defined in the storage of orthodox seeds.

Cryopreservation of imbibed orthodox seeds and recalcitrant seeds

It is known that orthodox seeds suffer injury when frozen with a relatively high water content, hence the need to desiccate prior to storage. However, a study of the behaviour in freezing of partially imbibed material may provide insights into the nature of avoidance of freezing injury in recalcitrant seeds which cannot be desiccated.

Sakai and Noshiro (1975) have frozen seeds of several species at a range of water contents and shown that at a moisture content in the range of 16–20 per cent, care is needed to avoid exposure to temperatures in the region of − 30°C during thawing (following rapid freezing). In the work of Grout (1979) using imbibed seeds of *Lycopersicon esculentum* (tomato), it was demonstrated that material with a water content typical of recalcitrant seeds (25–40 per cent) could survive slow cooling to − 20°C (*c.* 3°C min^{-1}) or rapid freezing by direct immersion in liquid nitrogen, provided that the cryoprotectant DMSO was applied at 10–15% v/v. An upper limit to hydration, compatible with survival was found between 40 per cent and 70 per cent moisture content. However, shoot meristems dissected from seeds which had been imbibed to *c.* 73 per cent moisture content, cryoprotected with 15% v/v DMSO and frozen rapidly in liquid nitrogen were capable of recovery in culture (*c.* 35 per cent survival).

Clearly, the tomato seed system is far removed from true recalcitrant seeds, but an important lesson to be learned from it is that if recalcitrant seeds are treated as tissue cultures which themselves are usually 'recalcitrant' in the same sense, there is scope for translation of all of the expertise from *in vitro* work. Furthermore, once the transition to an *in vitro* system is taken, there is no need to preserve the whole seed; the embryo or just a meristem will suffice. The embryo is unlikely to be as strict in its environmental requirements as the whole seed and being much smaller will provide a more suitable specimen for cryopreservation work. Investigations (Grout, 1986) into the storage of embryos of *Elaeis guineensis* (oilpalm) support these remarks.

A final comment to be made in respect of this type of material is that true recalcitrance should be established. There are examples of seeds, for example those of *Citrus limon* (lemon), which can be physically manipulated (in this case by removing the testa; Mumford and Grout, 1979), thereby permitting desiccation to low levels and storage by either conventional methods or cryopreservation.

Plant genetic conservation in practice – matching problems and solutions

At the outset, the seriousness of the problem in plant genetic conservation was emphasized. Some possible solutions to persistent difficulties in germplasm storage and means of improving existing technical approaches were then offered. However, as in any discipline, efforts in developing techniques will only be justified if the techniques are eventually applied to real problems. Is this the case for plant genetic conservation?

Plant germplasm is widely undervalued but as a result of the foresight and initiative of a few dedicated scientists an organizational framework has been set up to oversee the development of an international network of genetic resource centres (Plucknett *et al.*, 1983; Wilkes, 1983). Responsibility for promotion of the network lies with the International Board for Plant Genetic Resources (IBPGR) which is an autonomous, international, scientific organization under the aegis of the Consultative Group on International Agricultural Research (CGIAR). The IBPGR was created in 1974 and is based at the Food and Agriculture Organization of the United Nations (FAO) in Rome who provides its Secretariat.

The IBPGR operates by identifying and putting in order of priority crops requiring conservation, and co-ordinating a rolling programme of appropriate work at institutes including International Agricultural Research Centres. For all seed crops, it was anticipated that the global network of centres would be operational by 1986 (Wilkes, 1983). This, of course, relates to those seed crops which produce orthodox seeds amenable to conventional storage. New technology in the form of cryopreservation is available to extend the facilities of orthodox seed-banks, although implementation will probably depend upon economic factors and an assessment of the genuine need for cryopreservation.

For recalcitrant seed-producing and vegetatively propagated species, it is recognized that the development of storage technology must continue so that material already in sub-optimal storage can safely be conserved and so that current efforts to collect and evaluate valuable germplasm from diminishing natural environments are not wasted.

During 1981 and 1982 the IBPGR agreed to the convening of expert advisory committees with responsibility for seed storage and *in vitro* storage. The Advisory Committee on *In Vitro* Storage in particular faces a daunting task in that all vegetatively propagated crops and recalcitrant seed-producing crops are potential subjects for consideration. In its first report (IBPGR, 1983), such crops which carry a high IBPGR priority rating were addressed in terms of the status of conservation techniques applicable to them. Accordingly, it was recommended that *in vitro* methods suitable for application in gene-banks should be developed for potato, cassava, sweet potato, yam, aroids, banana and plantains, coconut, sugarcane, cacao, citrus fruits, temperate fruits and grape-vine. Several other crops awaited expert advice before inclusion in the list.

Among the recommendations of the committee there are clear acknowledgements of cryopreservation as the only satisfactory method for *in vitro* base collections, and the crucial importance of monitoring genetic stability in culture, genetic characterization, including the use of biochemical techniques, disease indexing and the collection and dissemination of relevant information. A programme of research is now being promoted to develop the necessary crop-specific technology. (Much previous work has, regrettably, concentrated upon model systems.)

It is hoped that in the near future cryopreservation will be implemented as a routine gene-banking procedure alongside the already widespread use of *in vitro* techniques for pathogen eradication, clonal propagation and short-term storage.

Recent progress

Progress has been recorded recently in a number of areas of plant genetic conservation. The IBPGR and its *In Vitro* Advisory Committee (Holden and Williams, 1984) have continued to promote information exchange (Wheelans and Withers, 1984) and have considered the use of *in vitro* techniques for germplasm collection (IBPGR, 1984; Yidana *et al.*, 1986) and the design of *in vitro* genebanks (IBPGR, 1986).

Slow growth storage of shoot cultures is being developed for more crops with genuine conservation problems (e.g. Staritsky *et al.*, 1986; Wanas *et al.*, 1986). However, instability in slow growth of unorganized cultures is apparent in the work of Hiraoka and Kodama (1984). Conversely, retention of secondary product synthetic capacity in cryopreserved cultures has been reported (Chen *et al.*, 1984; Seitz *et al.*, 1983; Watanabe *et al.*, 1983; Ziebolz and Forche, 1985). The latter two works involve treatment of small pieces of callus as cell suspensions. Further details of stability in cryopreserved cultures may be found in Withers (1985a, 1986).

The development and wider application of cryopreservation procedures for shoot-tips are documented in reports by Henshaw *et al.*, (1985b), Katano *et al.*, (1984) and Moriguchi *et al.*, (1985) (see also Kartha, 1985; Withers 1986). Attention has continued to be given to furthering our understanding of the process of cryopreservation of cell suspensions (Pritchard *et al.*, 1986) and to the improvement of conservation procedures for orthodox and recalcitrant seeds (Grout, 1986; Stanwood, 1985).

Acknowledgement

The author gratefully acknowledges receipt of a Science and Engineering Research Council Advanced Fellowship.

References

Akihama, T., Omura, M. and Kozaki, I. (1978). Further investigation of freeze-drying for deciduous tree pollen. In *Long Term Preservation of Favourable Germplasm*, pp. 1–7. Edited by Akihama, T. and

Nakajima, K. Fruit Tree Research Station, MAF, Japan.

Bajaj, Y.P.S. (1977). Survival of *Nicotiana* and *Atropa* pollen embryos frozen at − 196°C. *Current Science (India)* **46**, 305–7.

Barnabas, B. and Rajki, E. (1976). Storage of maize (*Zea mays* L.) pollen at − 196°C in liquid nitrogen. *Euphytica* **25**, 747–52.

Bridgen, M.P. and Staby, G.L. (1981). Low pressure storage and low oxygen storage of plant tissue cultures. *Plant Science Letters* **22**, 177–86.

Cella, R., Colombo, R., Galli, M.G., Nielsen, E., Rollo, F. and Sala, F. (1982). Freeze-preservation of *Oryza sativa* L. cells: a physiological study of freeze-thawed cells. *Physiologia Plantarum*, **55**, 279–84.

Chen, T.H.H., Kartha, K.K., Leung, N.L., Kurz, W.G.W., Chatson, K.B. and Constabel, F. (1984). Cryopreservation of alkaloid producing cell cultures in periwinkle (*Catharanthus roseus*). *Plant Physiology* **75**, 726–31.

Cocking, E.C. (1986). The tissue culture revolution. In *Plant Tissue Culture and its Agricultural Applications*, pp. 3–20. Edited by Withers, L.A. and Alderson, P.G. Butterworth, London.

Cocking, E.C., Davey, M.R., Pental, D. and Power, J.B. (1981). Aspects of plant genetic manipulation. *Nature* **293**, 265–70.

CIAT (1978). Annual Report, Genetic Resources Unit, International Centre for Tropical Agriculture, Cali, Colombia.

Dale, P.J., Cheyne, V.A. and Dalton, S.J. (1980). Pathogen elimination and *in vitro* plant storage in forage grasses and legumes. In *Tissue Culture Methods for Plant Pathologists*, pp. 119–24. Edited by Ingram, D.S. and Helgeson, J.P. Blackwell, Oxford.

Engelmann, F. Duval, Y. and Derenddre, J. (1985). Survival and proliferation of oilpalm (Eldeis guineensis) somatic embryos after freezing in liquid nitrogen, C.R. *Acad. Sci. Ser III Sci. de la vie* **301**, 111–16.

Finkle, B.J,, Ulrich, J.M. and Tisserat, B. (1982). Responses of several lines of rice and datepalm callus to freezing at − 196°C. In *Plant Cold Hardiness and Freezing Stress: Volume 2: Mechanisms and Crop Implications*, pp. 643–60. Edited by Li, P.H. and Sakai, A. Academic Press, New York.

Frank, J., Barnabas, B., Gal, E. and Farkas, J. (1982). Storage of sunflower pollen. *Zeitschrift für Pflanzenzüchtung* **89**, 341–3.

Grimsley, N.H. and Withers, L.A. (1983). Cryopreservation of cultures of the moss *Physcomitrella patens*. *Cryoletters* **4**, 251–8.

Grout, B.W.W. (1979). Low temperature storage of imbibed tomato seeds: a model for recalcitrant seed storage. *Cryoletters* **1**, 71–6.

Grout, B.W.W. (1986). Embryo culture and cryopreservation for the conservation of genetic resources of species with recalcitrant seed. In *Plant Tissue Culture and its Agricultural Applications*, pp. 303–309. Edited by Withers, L.A. and Alderson, P.G. Butterworth, London.

Grout, B.W.W. and Henshaw, G.G. (1978). Freeze-preservation of potato shoot-tip cultures. *Annals of Botany* **42**, 1227–9.

Grout, B.W.W. and Henshaw, G.G. (1980). Structural observations on the growth of potato shoot-tip cultures after thawing from liquid nitrogen. *Annals of Botany* **46**, 243–8.

Haskins, R.H. and Kartha, K.K. (1980). Freeze-preservation of pea meristems: cell survival. *Canadian Journal of Botany* **58**, 833–40.

Hauptmann, R.M. and Widholm, J.M. (1982). Cryostorage of cloned amino acid analog-resistant carrot and tobacco suspension cultures. *Plant Physiology* **70**, 30–34.

Henshaw, G.G., O'Hara, J.F. and Stamp., J.A. (1985a). Cryopreservation of potato meristem. In *Cryopreservation of Plant Cells and Organs*, pp. 159–70. Edited by Kartha, K.K. CRC Press, Boca, Raton.

Henshaw, G.G., Keefe, P.D. and O'Hara, J.F. (1985b). Cryopreservation of potato meristems. In In Vitro *Techniques – Propagation and Long-term Storage*, pp. 155–60. Edited by Schäfer-Menukr, A. Dordrecht, Nijhoff/Junk.

Henshaw, G.G., O'Hara, J.F. and Westcott, R.J. (1980a). Tissue culture methods for the storage and utilization of potato germplasm. In *Tissue Culture Methods for Plant Pathologists*, pp. 71–6. Edited by Ingram, D.S. and Helgeson, J.P. Blackwell, Oxford.

Henshaw, G.G., Stamp, J.A. and Westcott, R.J. (1980b). Tissue culture and germplasm storage. In *Developments in Plant Biology, Volume 5, Plant Cell Cultures: Results and Perspectives*, pp. 277–82. Edited by Sala, F., Parisi, B., Cella, R. and Cifferi, O. Elsevier – North Holland, Amsterdam.

Hiraoka, N. and Kodama, T. (1984). Effects of non-frozen cold storage on the growth, organogenesis and secondary metabolism of callus cultures. *Plant Cell, Tissue and Organ Culture* **3**, 349–57.

Holden, J.H.W. and Williams, J.T. (1984). *Crop Genetic Resources – Conservation and Evaluation*. George Allen and Unwin, London.

IBPGR, (1983). *IBPGR Advisory Committee on* In Vitro *Storage – Report of the first meeting*. International Board for Plant Genetic Resources Publication AGP: IBPGR/82/84. IBPGR, Rome.

IBPGR. (1984). The potential for using *in vitro* techniques for germplasm collection. International Board for Plant Genetic Resources Publication AGP:IBPGR/83/108. IBPGR, Rome.

IBPGR. (1986). Design, planning and operation of *in vitro* genebanks. International Board for Plant Genetic Resources Publication AGP: IBPGR/85/154. IBPGR, Rome.

Ingle, J. (1982). Genetic manipulation and plant improvement. *Span* **25**, 50–3.

Jones, L.H. (1974). Long term survival of embryoids of carrot (*Daucus carota* L.). *Plant Science Letters* **2**, 221–4.

Jordan, J.L., Jordan, L.S. and Jordan, C.M. (1982). Effects of freezing to −196°C and thawing on *Setaria lutescens* seeds. *Cryobiology* **19**, 435–42.

Kartha, K.K (1985). Meristem culture and germplasm preservation. In *Cryopreservation of Plant Cells and Organs*, pp. 115–34. Edited by Kartha, K.K. CRC Press, Boca Raton.

Kartha, K.K., Leung, N.L. and Gamborg, O.L.(1979). Freeze-preservation of pea meristems in liquid nitrogen and subsequent plant regeneration. *Plant Science Letters* **15**, 7–15.

Kartha, K.K., Leung, N.L., Gaudet-LaPrairie, P. and Constabel, F. (1982a). Cryopreservation of periwinkle, *Catharanthus roseus* cells cultured *in vitro*. *Plant Cell Reports* **1**, 135–8.

Kartha, K.K., Leung, N.L. and Mroginski, L.A. (1982b). *In vitro* growth responses and plant regeneration from cryopreserved meristems of cassava (*Manihot esculenta* Crantz). *Zeitschrift für Pflanzenphysiologie* **107**, 133–40.

Katano, M., Ishihara, A. and Sakai, A. (1984). Survival of apple shoot-tips cultured *in vitro* after immersion in liquid nitrogen. *Japanese Journal of Breeding* **34**, 212–13.

King, M.W. and Roberts, E.H. (1979). *The storage of recalcitrant seeds - achievements and possible approaches*. International Board for Plant Genetic Resources Publication AGP: IBPGR/79/44. IBPGR, Rome.

Maddox, A., Gonsalves, F. and Shields, R. (1983). Successful preservation of plant cell cultures at liquid nitrogen temperatures. *Plant Science Letters* **28**, 157–62.

Mix, G. (1984). *In vitro* preservation of potato germplasm. In *Efficiency in Plant Breeding*, pp. 194–5. Edited by Lange, W., Zeven, A.C. and Hogenboom, N.G. Pudoc Wageningen.

Moriguchi, T., Akihama, T. and Kozaki, I. (1985). Freeze preservation of dormant pear shoot apices. *Japanese Journal of Breeding* **35**, 196–9.

Mumford, P.M. and Grout, B.W.W. (1978). Germination and liquid nitrogen storage of cassava seed. *Annals of Botany* **42**, 255–7.

Mumford, P.M. and Grout, B.W.W. (1979). Desiccation and low temperature (−196°C) tolerance of *Citris limon* seed. *Seed Science and Technology* **7**, 407–10.

Nag, K.K. and Street, H.E. (1973). Carrot embryogenesis from frozen cultured cells. *Nature* **245**, 270–2.

Nitzsche, W. (1978). Erhaltung der Lebensfühigkeit in Getrocknetern Kallus. *Zeitschrift für Pflanzenphysiologie* **87**, 469–72.

Plucknett, D.L., Smith, N.J.H., Williams, J.T. and Anishetty, N.M. (1983). Crop germplasm conservation and developing countries. *Science* **220**, 163–9.

Pritchard, H.W., Grout, B.W.W., Reid, D.S. and Short, K.C. (1982). The effects of growth under water stress on the structure, metabolism and cryopreservation of cultured sycamore cells. In *The Biophysics of Water*, pp. 315–18. Edited by Franks, F. and Mathias, S. Cambridge, John Wiley.

Pritchard, H.W., Grout, B.W.W. and Short, K.C. (1986). Osmotic stress as a pregrowth procedure for cryopreservation: 3. Cryobiology of sycamore and soybean cell suspensions. *Annals of Botany* **57**, 379–87.

Roberts, E.H. (1975). Problems of long-term storage of seed and pollen for genetic resources conservation. In *Crop Genetic Resources for Today and Tomorrow*, pp. 269–96. Edited by Frankel, O.H. and Hawkes, J.G. Cambridge University Press, Cambridge.

Roberts, E.H. and King, M.W. (1982). Storage of recalcitrant seeds. In *Crop Genetic Resources - the Conservation of Difficult Material*, pp. 39–48.

Edited by Withers, L.A. and Williams, J.T. IUBS/IBPGR, Paris, (IUBS Series B-42).

Sakai, A. and Noshiro, M. (1975). Some factors contributing to the survival of crop seeds cooled to the temperature of liquid nitrogen. In *Crop Genetic Resources for Today and Tomorrow*, pp. 317–26. Edited by Frankel, O.H. and Hawkes, J.G. Cambridge University Press, Cambridge.

Sakai, A. and Uemura, M. (1982). Recent advance of cryopreservation of apical meristems. In *Plant Cold Hardiness and Freezing Stress. Volume 2: Mechanisms and Crop Implications*, pp. 635–41. Edited by Li, P.H. and Sakai, A. Academic Press, New York.

Seitz, U., Alferman, A.W. and Reinhard, E. (1983). Stability of biotransformation capacity in *Digitalis lanata* cell cultures after cryogenic storage. *Plant Cell Reports* **2**, 273–6.

Stanwood, P.C. (1985). Cryopreservation of seed germplasm for genetic conservation. In *Cryopreservation of Plant Cells and Organs*, pp. 199–226. Edited by Kartha, K.K. CRC Press, Boca Raton.

Stanwood, P.C. and Bass, L.N. (1978). Ultracold preservation of seed germplasm. In *Plant Cold Hardiness and Freezing Stress*, pp. 361–71. Edited by Li, P.H. and Sakai, A. Academic Press, New York.

Stanwood, P.C. and Roos, E.E. (1979). Seed storage of several horticultural species in liquid nitrogen (−196°C). *HortScience* **14**, 628–30.

Staritsky, G., Dekkers, A.J., Louwaars, N.P. and Zandvoort, E.A. (1986). *In vitro* conservation of aroid germplasm at reduced temperatures. In *Plant Tissue Culture and its Agricultural Applications*, pp. 277–83. Edited by Withers, L.A. and Alderson, P.G. Butterworth, London.

Styles, E.D., Burgess, J.M., Mason, C. and Huber, B.M. (1982). Storage of seed in liquid nitrogen. *Cryobiology* **19**, 195–99.

Takeuchi, M., Matsushima, H. and Sugawara, Y. (1982). Totipotency and viability of protoplasts after long-term freeze-preservation. In *Plant Tissue Culture, 1982*, pp. 797–8. Edited by Fujiwara, A. Japanese Association for Plant Tissue Culture, Tokyo.

Towill, L.E. (1984). Survival at ultralow temperatures of shoot-tips from *Solanum tuberosum* groups andigena, phureja, stenotomum, tuberosum and other tuber-bearing *Solanum* species. *Cryoletters* **5**, 319–26.

Wanas, W.H., Callow, J.A. and Withers, L.A. (1986). Growth limitation for the conservation of pear genotypes. In *Plant Tissue Culture and its Agricultural Applications*, pp. 285–90. Edited by Withers, L.A. and Alderson, P.G. Butterworth, London.

Watanabe, K., Mitsuda, H. and Yamada, Y. (1983). Retention of metabolic and differentiation potentials of green *Lavandula vera* callus after freeze preservation. *Plant and Cell Physiology* **24**, 119–22.

Westcott, R.J. (1981). Tissue culture storage of potato germplasm. 2. Use of growth retardants. *Potato Research* **24**, 343–52.

Wheelans, S.K. and Withers, L.A. (1984). The IBPGR International Database on *In Vitro* Conservation. *Plant Genetic Resources Newsletter* **60**, 33–8.

Wilkes, G. (1983). Current status of crop plant germplasm. *CRC Critical Reviews* **1 (2)**, 131–81.

Withers, L.A. (1979). Freeze-preservation of somatic embryos and clonal plantlets of carrot (*Daucus carota* L.). *Plant Physiology* **63**, 460–7.

Withers, L.A. (1980a). *Tissue Culture Storage for Genetic Conservation*. International Board for Plant Genetic Resources Publication AGP: IBPGR/82/30. IBPGR, Rome.

Withers, L.A. (1980b). The cryopreservation of higher plant tissue and cell cultures - an overview with some current observations and future thoughts. *Cryoletters* **1**, 239–50.

Withers, L.A. (1982a). Storage of plant tissue cultures. In *Crop Genetic Resources - the Conservation of Difficult Material*, pp. 49–82. Edited by Withers, L.A. and Williams, J.T. IUBS/IBPGR, Paris, (IUBS Series B-42).

Withers, L.A. (1982b). *Institutes Working on Tissue Culture for Genetic Conservation*. International Board for Plant Gentic Resources, Rome. (Revision of 1981 Edition; 1983 and 1985 Editions take the form of a computer data-base).

Withers, L.A. (1982c). The development of cryopreservation techniques for plant cell, tissue and organ cultures. In *Plant Tissue Culture, 1982*, pp. 793–4. Edited by Fujiwara, A. Japanese Association for Plant Tissue Culture, Tokyo.

Withers, L.A. (1985a). Cryopreservation of cultured cells and meristems. In *Cell Culture and Somatic Cell Genetics of Plants, Volume 2*, pp. 254–316. Edited by Vasil, I.K. Academic Press, New York.

Withers, L.A. (1985b). Cryopreservation of cultured plant cells and protoplasts. In *Cryopreservation of Plant Cells and Organs*, pp. 243–67. Edited by Kartha, K.K. CRC Press, Boca Raton.

Withers, L.A. (1986). Cryopreservation and Genebanks. In *Plant Cell Culture Technology*, pp. 96–140. Edited by Yeoman, M.M. Blackwell, Edinburgh.

Withers, L.A. and Davey, M.R. (1978). A fine-structural study of the freeze-preservation of plant tissue cultures. I The frozen state. *Protoplasma* **94**, 207–19.

Withers, L.A. and King, P.J. (1979). Proline - a novel cryoprotectant for the freeze-preservation of cultured cells of *Zea mays* L. *Plant Physiology* **64**, 675–8.

Withers, L.A. and King, P.J. (1980). A simple freezing unit and cryopreservation method for plant cell suspensions. *Cryoletters* **1**, 213–20.

Yidana, J.A., Withers, L.A. and Ivins, J.D. (1986). Development of a simple method for the collection and propagation of cocoa *in vitro*. *Acta Hort.* (In press).

Ziebolz, B. and Forche, E. (1985). Cryopreservation of plant cells with special attributes. In In Vivo *Techniques - Propagation and Long-term Storage*, pp. 181–3. Edited by Schäfer-Menukr, A. Dordrecht, Nijhoff/Junk.

12

The preservation of organisms responsible for parasitic diseases

E. James
Department of Ophthalmology
Medical University of South Carolina, SC, USA

Introduction

The problem

The number of people in the world affected with diseases caused by parasites is considerable, and is increasing. Current estimates are that 24 million individuals are infected with South American and 45 million with African trypanosomiasis, 200 million with schistosomiasis, 215 million with chronic malaria and several hundred million with filarial worm infections (WHO, 1982). The intestinal nematode parasites are even more prevalent; *Enterobius* is in the Guinness Book of World Records as the most common human infection after the common cold and periodontal disease (McWhirter, 1983). Many hundreds of millions more people are also exposed to the risk of contracting these infections.

Parasitic diseases maim, disfigure, debilitate and kill many millions each year. They are also a major contributory factor in many deaths ascribed to other causes. In particular they affect or interact with the nutrition of individuals and can directly cause malnutrition, undernutrition and malabsorption and are thus an important factor in the 41 000 deaths each day from hunger (Hunger Project, 1982). They affect the ability of the infected individual to contribute to the local economy and can even cause towns and villages to cease to function as economic units, as is the case with onchocerciasis where 30 per cent of the male workforce in an infected area may be incapacitated through blindness. This may result in mass migration away from the region and the disuse of valuable agricultural land.

Many of the diseases, such as malaria and hookworm, were endemic in what have now become the developed nations, for example malaria occurred as far north as Finland until the late 1940s and was only eradicated from continental Europe in the years between 1945 and 1964 (Bruce-Chwatt and De Zulueta, 1980). Schistosomiasis occurred in Portugal until only a few years ago. The frequency with which cases of malaria and other parasitic diseases are appearing among the populations of developed nations, however, is increasing again as intercontinental travel becomes commonplace and, all too often, infected individuals are being misdiagnosed or diagnosed too late by doctors who are not acquainted with parasitic infections – for example there were 1909 cases of malaria in the U.K. in 1978 and 10 deaths (Bruce-Chwatt and De Zulueta, 1980).

It is, however, the peoples of the Third World, those below the Brandt Commission's (1980) north–south divide, who suffer most from parasitic diseases, and whose economic and social progress is thwarted by their continued existence. An increase in the national economies of those affected regions is the key to eventual sustained control of these diseases. In the developed nations standards of sanitation, hygiene and overall health care have improved along with economic advancement and, as a result, diseases such as plague, cholera and typhoid fever have ceased to exist together with many of the parasitic diseases.

The importance of parasitic diseases in the cycle of deprivation, poverty and hunger has long been recognized, but in 1977 the World Health Organization, in collaboration with the World Bank and the United Nations Development Programme, instigated a Special Programme of Research and Training in six of the major diseases (five parasitic diseases and leprosy). Many governmental research establishments, among both the developed and underdeveloped nations, and private Foundations and Trusts have also contributed much towards the understanding and control of these diseases. The developed nations have always been strong on committing resources to basic research, but the emphasis of their direct contribution has been shifting more in this direction and away from field-orientated programmes of epidemiology, treatment and control as personnel in affected countries become better trained in these latter skills.

The parasites

Although viruses, bacteria, rickettsias, etc. are parasitic, parasitology does not include these groups of organisms and is concerned chiefly with the protozoa, helminths (worms), the blood-sucking arthropods and those invertebrates which act as disease vectors.

Four groups of protozoa are parasitic to man: the amoebae (e.g. *Entamoeba*, *Iodamoeba*, *Naegleria*); the ciliata (e.g. *Balantidium*); the sporozoa (e.g. *Plasmodium*, *Babesia*, *Coccidia*, *Eimeria*, *Toxoplasma*); and the flagellata (e.g. *Trypanosoma*, *Leishmania*, *Giardia*).

The major groups of parasitic helminths are the nematoda or roundworms (e.g. *Ascaris*, *Enterobius*, *Trichuris*, the hookworms, the filariae, *Trichinella*, *Capillaria*) and the platyhelminths which are divided into the cestoda (tapeworms e.g. *Taenia*, *Echinococcus*) and the trematoda (flukes or flatworms, e.g. *Schistosoma*, *Fasciola*, *Opisthorchis*).

The blood-sucking or parasitic arthropods include mites and ticks (arachnids) and many groups of insects (the flies - mosquitoes, blackflies, midges, tsetse flies, gad flies, sandflies, myiasis flies - and fleas, chiggers and lice). It is perhaps not suprising that almost all these blood-sucking insects act as vectors for parasitic diseases, and also for many viral, rickettsial and bacterial infections. The other main group of intermediate hosts is the molluscs, which transmit the trematode (and some nematode) parasites. Animals such as freshwater fish and crabs, which are used by man as food, also act as intermediate hosts for some trematodes, cestodes and nematodes (Muller, 1975).

Other groups of parasites which are of importance to humans are those affecting livestock and, as a result, human livelihood, nutrition and health. These include the protozoa *Babesia*, *Anaplasma*, *Theileria*, *Trypanosoma* and *Trichomonas*, and of the helminths, the trematodes (Dunn, 1978) *Fasciola*, *Fasciolopsis* and *Schistosoma*, cestodes such as *Taenia*, *Echinococcus*, and a large number of nematode parasites which include *Trichostrongylus*, *Ostertagia*, *Haemonchus*, *Cooperia*, *Nematodirus* and *Dictyocaulus*.

Parasites and low temperatures

Groups preserved

Low temperature preservation of parasites now plays a crucial role in both field-orientated and basic studies. Cryopreservation techniques have been successfully applied to all the groups of parasitic protozoa except the ciliata. Perhaps the relatively small number of people affected with balantidiasis, the general low level of pathogenicity of *Balantidium* and the paucity of research being carried out with this parasite has not yet provided the impetus for its cryopreservation to be investigated.

Helminthologists have generally been slow to perceive the benefits of cryopreservation, but most of the important intestinal nematode parasites of domestic animals, some of those of humans and many of the filarial species have been successfully cryopreserved. Among the many other species of helminths and invertebrate vectors, however, none has been cryopreserved successfully except the schistosomula stage of *Schistosoma* spp. (see below). There is one report of a viable cestode (*Taenia crassiceps*) being recovered after cooling to −35°C (Ham, 1982) and of viable eggs of the blackfly (*Simulium*) being recovered after cooling to −196°C (Ham, 1983). Studies on the genetics of vector species such as mosquitoes and *Simulium* would also benefit if they could be cryopreserved.

Later in this chapter the protozoa and helminths will be discussed separately. Both protozoa and helminths respond to the stresses imposed by low temperatures in a manner which is similar to most other cell types, although they differ from each other in two important respects. Most protozoa reproduce asexually within their vertebrate host; the helminths, however, reproduce sexually, and if an asexual phase occurs, this is often in the intermediate host. For many applications of cryopreservation, low levels of survival are therefore adequate for the protozoa but for the helminths the aim is to achieve the highest possible level of survival. This does not mean that low survival of protozoa should be tolerated as there is likely to be a degree of selection which could alter the genetic characteristics of the cryopreserved isolate.

As different cell types have frequently been shown to require different cryopreservation schedules (Leibo and Mazur, 1971), high levels of survival of helminths, which contain cells differentiated for many different functions, are theoretically more difficult to achieve; the best protocol is likely to represent a compromise between the different requirements of the different cell types within the individual helminth.

Preservation for routine maintenance and field collection

The conventional methods for the routine maintenance of parasites involve sequential passage through laboratory animals or *in vitro* culture. These methods are frequently very expensive, tedious, impractical and time consuming. There is considerable risk of genetic drift or selection or loss of the organisms through contamination of cultures or human error. Cryopreservation overcomes these problems and, although many of the techniques used are suboptimal, for the protozoa at least, cryopreservation is now an essential research tool.

Cryopreservation ought to have become an essential tool also for the helminths and the vectors but has only become such a tool for the intestinal nematodes of domestic animals (where perhaps the costs of maintenance in large animals are higher) and for some of the filarial nematodes where laboratory life cycle maintenance is difficult or impossible. Almost every laboratory working with parasitic protozoa uses a liquid nitrogen storage system; many of the pharmaceutical companies developing potential anthelmintics similarly store the larvae of the domestic animal nematodes.

In the development of cryopreservation methodology the amount of material to be cryopreserved at any one time is usually small and the emphasis has generally been on simple techniques. These two factors have allowed most of the techniques so far devised to be suitable for field use as well as laboratory use; many species of protozoa and helminth can now be collected in the field and transported in liquid nitrogen back to the laboratory for further study.

Preservation for genetic conservation

The trypanosomes were the first parasites to be investigated extensively with regard to cryopreservation, largely because during an infection a succession of antigenic variants of the parasites occurs and a stable method was required for preserving these variants for study. The pioneering work with trypanosomes was carried out by Lumsden, Cunningham and others at Tororo, Uganda during the 1960s. They coined the term 'stabilate' to describe the discrete population or clone of organisms with specific biological characteristics preserved on a unique occasion (Lumsden and Hardy, 1965). Stabilation has provided for co-ordination of international research through the availability of defined strains and species to workers throughout the world.

Cryopreservation techniques are also used to store variants of the different species of malaria and are particularly useful for the collection, and assessment of susceptibility to new chemotherapeutic agents, of drug resistant strains. Biological and biochemical comparisons of the large number of *Leishmania* and *Entamoeba* zymodemes has been possible with the assistance of cryopreservation; this has resulted in revised species classifications for these species and better understanding of the chemotherapeutic regimens required (Sargeaunt and Williams, 1979; Le Blanq, 1983).

Helminths do not vary antigenically during the course of an infection in the same way as protozoa, although different stages in the life cycle may express different surface (Philip *et al.*, cited by Wakelin and Denhem, 1983) or somatic antigens, and different geographical isolates of particular species often have quite different biological characteristics (e.g. *Onchocerca volvulus*, Anderson *et al.*, 1974; *Trichinella* spp., Dick, 1983). Drug resistance also exists amongst some species of nematodes. Gastro-intestinal nematodes of domestic animals are being cryopreserved to conserve drug resistant strains (Coles *et al.*, 1980), and work is in progress to conserve different species and strains of *Onchocerca* and *Schistosoma* and several species of filarial nematode larvae. Also at the National Institutes of Health, Bethesda, USA a large number of strains of *S. mansoni* with varying virulence have been preserved in liquid nitrogen (Stirewalt and Lewis, personal communication).

Cryopreservation of parasites for genetic conservation is in its infancy and there are, as yet, few co-ordinated programmes for preserving type strains or for setting up reference centres as has been done for the free-living nematode *Coenorhabditis elegans* (Anon, 1980).

Destruction

Low temperatures are occasionally employed to kill parasites. The use of a specific cooling schedule to favour the survival of bull sperm and kill the venereally transmitted protozoan *Trichomonas foetus* (Joyner, 1954) has been discontinued following conflicting results (Smith, 1961). However, it has been suggested that selective destruction of early or late intra-erythrocytic stages of malaria might be possible by using a specific cryo-preservation protocol (Wilson *et al.*, 1977). Recently (Mutetwa and James, unpublished data), it has been shown that there are quantitative and qualitative differences in the survival of the different intra-erythrocyte stages of *Plasmodium chabaudi*.

Mandatory exposure of meat or fish to low temperatures is required in those countries where there is a significant risk of transmission of certain helminth parasites. The Netherlands Institute of Fishery Products requires that lightly salted or green herring be frozen at −20°C for 24 hours (Gustafson, 1953) to kill the larvae of *Anisakis marina* (an aberrant parasite of humans normally found in seals and dolphins; Ruitenberg, 1970). Freezing is also recommended for the control of beef tapeworm and *Trichinella* (Smith, 1975), although recent reports suggest that there is considerable variability in the tolerance to freezing of *Trichinella* isolates obtained from the arctic, temperate regions and the tropics (Dick, 1983).

Preservation for vaccination

Currently, the only anti-parasite vaccines in commercial production are against the cattle and sheep lung-worms, *Dictyocaulus viviparus* and *D. filaria* respectively. Live third stage larvae obtained from faecal culture are X-irradiated and distributed through the normal first class mail service for oral administration by a local veterinary surgeon. The larvae have an approved shelf life of up to 45 days so there has been little incentive for cryopreserving these vaccines.

Other radiation-attenuated vaccines produced along similar lines have been developed against dog hookworm (*Ancylostoma caninum*), malaria and schistosomiasis. The dog hookworm vaccine was produced in the USA (Miller, 1978) and then withdrawn on grounds of cost. Experimental development of the malaria sporozoite and schistosome schistosomula vaccines reached an advanced stage, although recently their further development appears to have been shelved pending the results of studies aimed at producing defined antigenic 'dead' vaccines through gene cloning technology. Cryopreservation techniques have been developed for both the malaria sporozoite (Leef *et al.*, 1979) and the schistosome schistosomula (James, 1981) vaccines which would provide them with a shelf-life and a convenient method of delivery.

It is likely that the putative live vaccine against schistosomiasis may be rescucitated in the future but in the meantime parellel studies on a cryopreserved radiation-attenuated vaccine against the cattle schistosome *S. bovis* in the Sudan are continuing. Cattle schistosomiasis is endemic in large parts of Africa and causes considerable morbidity, mortality and loss of potential beef and milk yields. Recent cost-benefit analyses (McCaulay *et al.*, 1983) have indicated that the economic advantages of such a vaccine would be considerable. The prototype vaccine delivered in one shot of 10 000 organisms irradiated at 3 krad induced approximately 60–70 per cent protection as assessed by reductions in worm burden, faecal egg output and tissue egg burden following field-challenge (Majid *et al.*, 1980). Subsequent studies (Taylor and Bickle, personal communication) indicate that an irradiation dose of 10–20 krad may give greater protection.

As schistosome cercariae or schistosomula have a normal shelf-life of only a few hours or days, it is envisaged that the attenuated schistosomula to be used for vaccination would be cryopreserved (Fig. 12.1). This would also facilitate batch quality control prior to distribution and administration in the field. Cryopreserved radiation-attenuated *S. mansoni* schistosomula have been shown to be highly protective against challenge in mice (Bickle and James, 1978; Murrell *et al.*, 1979), although cryopreservation interacts with the irradiation step (James and Dobinson, 1984) and alters the dose required for induction of optimum protection. The logistics of producing, storing, transporting and delivering an anti-schistosome vaccine have been demonstrated (James *et al.*, in preparation) using *S. mansoni* schistosomula harvested, irradiated and cryopreserved in England and administered to baboons in Kenya. A similar approach is being taken in vaccinating cattle and water buffalo against *S. japonicum* in China (Hsu *et al.*, 1983) where this parasite exists as a zoonosis. Vaccines against both *S. bovis* and *S. japonicum* would indirectly affect man by improving the local economies and the level of

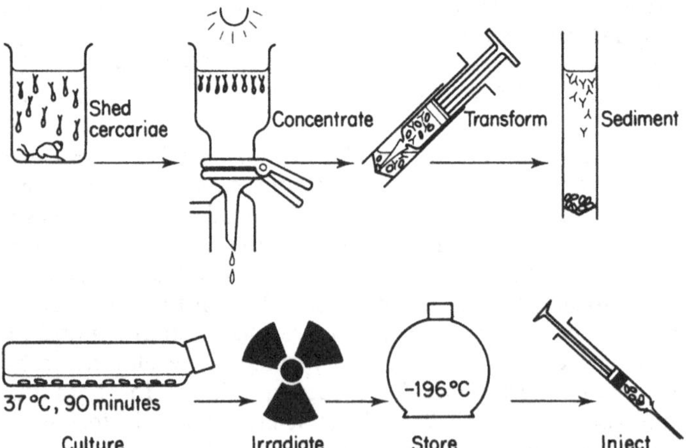

Fig. 12.1 Protocol for producing live cryopreserved radiation-attenuated vaccine. Cercariae are harvested from snails subjected to an increase in water temperature and illumination; they are concentrated on a millipore filtration apparatus, transformed by shear-force separation of the heads from the tails by 10 passages through a 21G needle and allowed to sediment to obtain a relatively tail-free pellet of schistosomula. The schistosomula are then incubated in culture medium (lactalbumin hydrolysate with Earles salts) at 37 °C for 90 minutes, exposed to ^{60}Co-irradiation and cryopreserved (the cryopreservation technique is outlined in Fig. 12.4). Following transportation to the site of administration the schistosomula are thawed, counted and their viability assessed, and they are given by intramuscular injection.

human nutrition, and a vaccine against *S. japonicum* would reduce the level of contamination of the environment with parasite eggs and hence reduce the level of transmission of oriental schistosomiasis to humans.

Cryopreservation also plays an important role in the research and development of vaccines against two cattle protozoal diseases, East Coast Fever and babesiosis. The most effective vaccination strategy developed against *Theileria* spp., the agent of East Coast Fever, appears to be the 'infection-treatment' technique (Radley *et al.*, 1975; Robson *et al.*, 1977; Uilenberg *et al.*, 1977; Paling and Geysen, 1981) which involves the injection of infective particles obtained from the tick vector followed by treatment with oxytetracycline. There is also another approach using cultured attenuated schizonts (Pipano, 1981) which has actually been used in limited field trials against *T. annulata* (Stepanova *et al.*, 1977). Low temperature preservation of stabilates and of material to be used for vaccination appears most often to be based on the cryopreservation technique described by Cunningham *et al.*, (1974).

For babesiosis, a live 'avirulent' vaccine produced by passage of *Babesia bovis* infected blood through splenectomized calves has been in use in Australia since 1964 (Callow, 1975). The low temperature preservation of this parasite has been well studied (Dalgliesh and Mellors, 1974; Dalgliesh *et al.*, 1976). More recently, inactivated merozoite and merozoite surface coat vaccines derived from tissue culture have been produced which can be stored lyophilized. The results obtained with these inactivated vaccines appear to be superior in some circumstances to those obtained with the live vaccine (Ristic and Levy, 1981), although it is probable that there will be a place for both types of vaccine in future control programmes.

Cutaneous leishmaniasis is a disease producing an ulcerative lesion in infected individuals, often on exposed portions of the lower limbs, forearms and face. After the lesion has healed the patient is left immune to further infection but with a disfiguring scar. A 'vaccine' comprising normal unattenuated parasites can be administered at a less visible body location which will induce temporary ulceration and good immunity. A frozen stored 'vaccine' of this type has been field tested (Green *et al.*, 1983) and excellent success has been claimed, although the ability of the parasites to produce lesions with high ulceration rates appeared to decline with prolonged (18 months) storage.

Preservation for biological control

There are a number of parasitic organisms which might be used to control disease intermediate hosts such as insects and molluscs, although none of these appears to be close to development as control agents yet. Studies on the cryopreservation of these organisms have not been reported, although data concerning the cryopreservation of *Neoaplectana* spp. larvae, which are being considered as possible biological control agents for Triatomine bugs, the vectors of South American trypanosomiasis, are imminent.

Cryopreservation of protozoa

Techniques

The cryopreservation of protozoa has been reviewed by Weinman and McAllister (1947). Diamond (1964), Dalgliesh (1972), Miyata (1975), James (1980a), Leef *et al.*, (1981), and James (1984). There are many techniques in use and these reflect both the requirements of the parasites and the facilities and conditions available to the workers.

Two techniques in particular have been adapted to a number of different species. The first technique involves slow cooling at approximately 1 °C min^{-1} to −70 °C followed by placing the sample in liquid nitrogen for storage. This technique was derived initially for *Trypanosoma* spp. by careful experimentation by workers at the East African Trypanosomiasis Research Organization, Tororo, Uganda between 1960 and 1969 (EATRO Annual Reports). The second technique, which may well have been first used by Laveran and Mesnil (1904) also for *Trypanosoma*, was outlined by Coggeshall (1939) for malaria, and simply involves plunging the organisms in a suspension, contained in a large glass tube, to produce a rapid cooling rate. Variations on this technique using capillary tubes or depositing a droplet of the suspension on a mica sliver (Luyet and Gehenio, 1954) and plunging into liquid nitrogen to increase the cooling velocity, have been reported.

Most parasitic protozoa have been, and often still are, cryopreserved by either or both of these techniques or by further variations on them. Different workers have also varied the type and concentration of cryoprotectant (although almost invariably glycerol or DMSO has been used), and the incubation time and temperature. Thawing has been virtually universally carried out by agitating the sample container in a water bath at 37 °C or 40 °C (James, 1984).

There are, of course, exceptions; for example it was found that 82 °C min^{-1} and 265 °C min^{-1} were the optimum cooling rates respectively for *Babesia bigemina* and *B. rodhaini* (Dalgliesh and Mellors, 1974; Dalgliesh *et al.*, 1976), and for *Plasmodium falciparum* trophozoites and *P. berghei* sporozoites 300 °C min^{-1} and 50 °C min^{-1} respectively were optimal (Mutetwa, 1983; Leef *et al.*, 1979). A two-step cooling procedure (rapid cooling to −31 °C, hold for 30 minutes, plunge into liquid nitrogen), developed initially for lymphocytes (Walter *et al.*, 1975) has been successfully adapted for the erythrocytic stages of *P. knowlesi* (Wilson *et al.*, 1977).

It cannot, however, be emphasized strongly enough that a cryopreservation protocol which is optimal for one species of organism may be inappropriate for another species. For example, the cooling rate requirements for the intra-erythrocytic stages of three species of malaria, *P. chabaudi*, *P. falciparum*, and *P. galinaceum* differ markedly, being 3600 °C min^{-1}, 300 °C min^{-1} and 1 °C min^{-1} respectively (Mutetwa, 1983; Mutetwa and James, in preparation).

Intracellular protozoan parasites

Differences in the intrinsic tolerance of the species of host red cell (rodent, human and avian respectively) may account for the different cooling rate requirements of the three species of malaria mentioned above. The optimum cryopreservation schedule for an intracellular organism is likely to represent a compromise between the requirements of the parasite and those of its host cell. Leef *et al.*, (1981) made a valid distinction in reviewing the intra- and extracellular protozoa separately, although in practice the techniques used often do not differ much.

The integrity of the host red cell is crucial to the survival of the intra-erythrocytic malaria trophozoite, in as far as parasite survival was generally inversely related to red cell haemolysis (Mutetwa and James, 1984a) (Fig. 12.2). This association may seem obvious but it has been questioned in the past (Wilson *et al.*, 1977) and many workers are still in the habit of using slow cooling protocols for intra-erythrocytic human and rodent malarias which undoubtedly are very damaging to the host red cells.

What is not known yet is how much the requirements of parasitized red cells differ from non-parasitized red cells. It is conceivable that, with the *P. chabaudi*/mouse red cell model at least, parasitized red cells survive cooling better at slower rates than do unparasitized cells and that survival levels of approximately 20 per cent, the highest achieved so far (Mutetwa and James, 1984a, b), may be the maximum possible. The use of techniques to enrich the populations of parasitized cells, such as on discontinuous density gradients, should resolve this.

With intracellular parasites, and particularly those of erythrocytes, special care must be taken during the post-thawing and the dilution procedures otherwise the cells are lysed. This is most important for protocols using glycerol as the cryoprotectant. Several procedures have been devised for deglycerolysing erythrocytes (Rowe *et al.*, 1980) but for malaria a single step dilution into 15% w/v glucose (0.9 M) for the removal of 10% glycerol gives

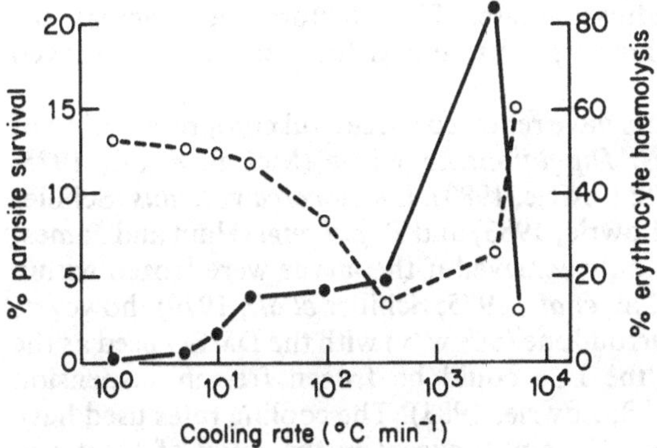

Fig. 12.2 Relationship between survival of *Plasmodium chabaudi* intra-erythrocytic stages and red cell haemolysis following cryopreservation.

Percentage survival (●) of *P. chabaudi* trophozoites and percentage haemolysis (○) cryopreserved using different cooling rates and in the presence of 10% v/v glycerol. Samples were thawed and deglycerolysed using medium containing 15% w/v glucose. (From Mutetwa and James, 1984b.)

good results (Mutetwa and James, 1984a) – this is basically 'osmotically squeezing' the cells (Lovelock, 1952), and the glycerol:glucose ratio is almost the same as that given by Smith (1961).

Cryopreservation of helminths

Tissue nematodes

There have been relatively few reports concerning the cryopreservation of helminths, and almost all of these reports have been devoted to the nematodes. Helminth preservation has been reviewed recently by James (1980a). The first successes in any controlled studies appear to have been with the microfilariae of *Dirofilaria immitis* and *Litomosoides carinii* which successfully survived (motility 38–83 per cent) fast cooling to −70°C in citrated blood without any cryoprotectant (Weinman and McAllister, 1947).

Subsequent studies have generally reported the use of slow cooling rates around 1°C min^{-1} for microfilariae, although these larvae appear to be especially suited to the double-incubation ethanediol technique (see below). Cryoprotectant has been omitted (Weinman and McAllister, 1947) or used at concentrations up to 40% ethanediol (Ham *et al.*, 1981). Other authors have variously reported the use of DMSO (Ogunba, 1969; Obiamiwe and MacDonald, 1971), glycerol (Chinery and Atiemo, 1973), methanol (Ham *et al.*, 1979) and the extracellular protectant hydroxyethyl starch (Minjas and Townson, 1980).

The microfilariae, perhaps because of their small size (*c.* 280 μm × 10 μm) appear to tolerate the stresses of cryopreservation relatively well. The infective third-stage larvae L_3 which are somewhat larger (0.5–2.0 mm × 25 μm) are less tolerant of these stresses. This may be because the surface cuticle is thicker or because L_3's contain more cells or simply because the organism's cross-sectional radius is greater. These factors would increase the time taken for cryoprotectants to penetrate and for water to be removed during freezing.

Nevertheless, several authors have reported successful cryopreservation of L_3's of a number of species: *Dipetalonema viteae* (McCall *et al.*, 1975; McCall and Weathersby, 1978; Lowrie, 1983), *Onchocerca volvulus* (Schiller *et al.*, 1979), *Brugia malayi* (Lowrie, 1983) and *B. pahangi* (Ham and James, 1983). Initially, survival was only achieved if the larvae were frozen within the tissues of the vector (McCall *et al.*, 1975; Schiller *et al.*, 1979), however, by incorporating polyvinylpyrrolidone (5% w/v) with the DMSO used as the intracellular cryoprotectant the L_3's could be frozen free in suspension (McCall and Weathersby, 1978; Lowrie, 1983). The cooling rates used have all been slow (approximately 1°C min^{-1}) except in the study of Ham and James (1983) who used 20% v/v methanol as the cryoprotectant and a cooling rate of 5°C min^{-1} to an intermediate temperature of −21°C followed by plunging into liquid nitrogen.

Gastro-intestinal nematodes

Slow cooling has also been used successfully for *Nematodirus* spp. *Trichostrongylus* spp. and *Trichonema* spp. (Parfitt, 1971; Isenstein and Herlich, 1972; Bemrick, 1978), while other species, *Ancylostoma* spp., *Haemonchus contortus*, *Ostertagia* spp., *Cooperia* spp. and *Oesophagostomum* spp. survived better at faster cooling rates (Weinman and McAllister, 1947; Campbell *et al.*, 1972, 1973).

In a comprehensive study 15 separate species of cattle and sheep nematodes were cryopreserved successfully (Van Wyk *et al.*, 1977) but little detail was given of the volumes of the samples or of the cryopreservation technique and it is therefore not possible to estimate the cooling rate used, although this was probably relatively slow. In a further study, using eight species, a consistently high survival (75–95 per cent) following slow cooling (1 °C min^{-1}), but variable levels of survival (0–80 per cent) following rapid cooling (rate unspecified) have been obtained (Coles *et al.*, 1980).

The suspending medium used for all these gastro-intestinal nematode larvae has been distilled water or tap water, or occasionally, 0.9% saline. The addition of cryoprotectants does not appear to confer any protection to these parasites (Coles *et al.*, 1980). This suggests that the larvae may already contain cryoprotective compounds. These infective larvae do spend a considerable time on pasture which is sometimes frozen to as low as −26°C during the winter months (Kates, 1950), and other animals such as terrestrial arthropods, intertidal molluscs and polar fishes similarly exposed to low temperatures do synthesize their own cryoprotectants. There also appears to be an increase, at least for *D. viviparus*, in tolerance to cryopreservation of larvae harvested during the winter (James, unpublished observation).

The crucial step in successful cryopreservation is the prior exsheathment of the larvae. Ensheathed larvae survive poorly, if at all (Campbell and Thompson, 1973; Van Wyk *et al.*, 1977). This is also the case for infective larvae of the cattle lungworm *D. viviparus* (Fig. 12.3). As *D. viviparus* larvae do not survive slow cooling below −40°C without being exsheathed, the sheath must be preventing the dehydration which occurs during freezing and hence the organisms are likely killed by intracellular ice formation. The sheath is a barrier to water movement and its normal function is to prevent desiccation during the time the larvae are lying exposed on pasture. The circumstantial evidence thus points to these larvae probably producing their own cryoprotective substances.

The infective larvae of hookworms, which are found in warm and damp climatic regions of the world, still need to be exsheathed in order to survive cryopreservation (Miller and Cunningham, 1965; Kelly *et al.*, 1976), but the survival of these species is improved (from a mean of 47 per cent to 89 per cent) with the addition of 10% DMSO (Kelly *et al.*, 1976).

Trematodes

Only one species of trematode, *Schistosoma*, has been cryopreserved. After numerous unsuccessful experiments, the breakthrough came with a two-step

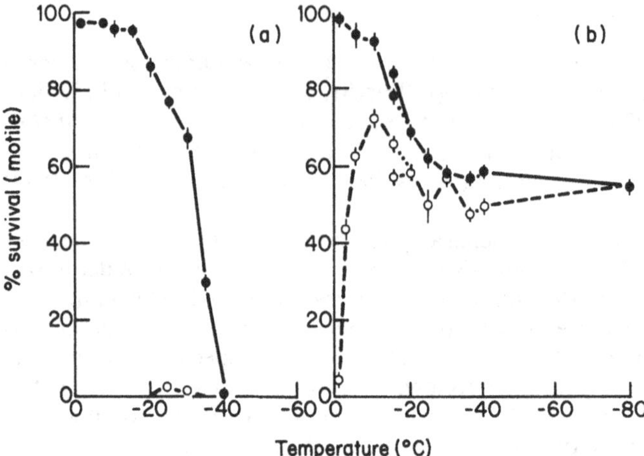

Fig. 12.3 Effect of exsheathment on tolerance to freezing of *Dictyocaulus viviparus* (nematoda).

Third-stage larvae of *D. viviparus* were cooled at 1 °C min^{-1} to various intermediate temperatures (horizontal axis) before plunging (cooling rate approximately 10 000 °C min^{-1}) into liquid nitrogen (○). No-plunge controls (●). (a) The survival of ensheathed larvae. (b) The survival of larvae exsheathed by exposure to 2% sodium hypochlorite for 10 minutes at 20 °C. (From James, in preparation.)

cooling schedule using 17.5% v/v methanol as the cryoprotectant. The suspension of schistosomula, distributed in 20 μl droplets onto the surface of an aluminium block pre-cooled to 0 °C, was cooled at 0.7 °C min^{-1} to −28 °C and then plunged into liquid nitrogen (cooling rate approximately 10 000 °C min^{-1}) (James and Farrant, 1977). Survival was 0.4 per cent as assessed by the ability of the organisms to develop into adult worms relative to unfrozen controls.

From calculations of the proportions of the solutes and of the frozen and unfrozen solvent fractions present in the sample at −28 °C, it was deduced that the residual unfrozen water surrounding the schistosomula contained approximately 40% v/v methanol. A new technique was then devised in which the methanol was added at 0 °C to the schistosomula at a final concentration of 40% and the sample was plunged directly into liquid nitrogen. Methanol is toxic to many cell types and also to schistosomula, but the toxicity is time dependent. Schistosomula were able to tolerate the methanol and the rapid cooling if the pre-freeze exposure time was approximately 10 seconds (James, 1980b). This technique improved the level of survival to 6.1 per cent.

A further significant improvement in survival was achieved by substituting ethanediol for the methanol. Ethanediol is less toxic to schistosomula and is more viscous, it is therefore easier to achieve a glass transition during rapid cooling. However, the increased viscosity also means that permeation of the ethanediol into the schistosomula takes longer. A two-step cooling technique with ethanediol at a starting concentration of 12.5% v/v gave optimum survival using an intermediate temperature of −22.5°C. At this temperature the residual unfrozen water was calculated to contain approximately 35% v/v ethanediol. To enhance permeation of the ethanediol, the incubation temperature needed to be increased to 37°C, but at this temperature the additive was toxic. At 0°C it took too long to permeate and the schistosomula were being damaged by dehydration. The cryoprotectant was therefore added in two separate steps (James, 1981). The first addition of 10% was made at 37°C and the ethanediol was allowed to penetrate over 10 minutes. Following cooling to 0°C and addition of more ethanediol to bring the concentration to 35% v/v the schistosomula effectively became partially dehydrated by the equivalent of 25% ethanediol which was tolerable at 0°C (Fig. 12.4). Survival using this technique was boosted to 47 per cent (James, 1981).

This basic double-incubation ethanediol technique has subsequently been adapted for several other helminths. Good survival (71–79 per cent) of *Onchocerca* spp. microfilariae was obtained using ethanediol concentrations of 10% and 40% at 37°C for 15 minutes and at 0°C for 10–30 seconds respectively (Ham *et al.*, 1981). The same basic protocol can be used for *Brugia* sp. microfilariae by extending the second incubation time to 40–45 seconds

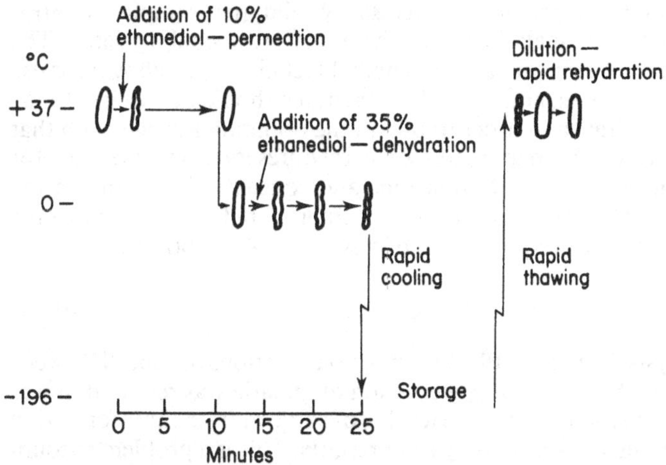

Fig. 12.4 Stresses imposed on schistosome schistosomula by cryopreservation using the ethanediol double-incubation technique.

Addition of 10% v/v ethanediol to schistosomula suspended in culture medium (lactalbumin hydrolysate with Earles salts) at 37°C initially causes dehydration and shrinkage. Schistosomula regain their normal shape over 10 minutes as the ethanediol permeates. The sample is then cooled to 0°C and after 5 minutes an equal volume of 60% v/v ethanediol is added to bring the final concentration to 35% v/v. Over the next 10 minutes the schistosomula dehydrate and shrink and are then distributed in 20 μl droplets onto glass slivers and plunged (cooling rate 5100°Cmin^{-1}) into liquid nitrogen. Samples are thawed and simultaneously diluted to remove the ethanediol by dropping into 2 ml of medium pre-warmed to 42°C. Warming rate is approximately 12 000°Cmin^{-1}.

(Ham, personal communication) and for *Wuchereria bancrofti* microfilariae by further extension to 60 seconds (Owen and Anantaraman, 1982). Small numbers of the infective larvae of *Neoaplectana* spp. and the newborn larvae of *Trichinella* spp. also can be cryopreserved by variations of this technique (James and Minter; Dobinson and James, unpublished observations).

This technique requires the minimum of apparatus and is eminently suitable for field collection of parasites. Large numbers of microfilariae of *O. volvulus* (El Sheikh and Ham, 1982) and of *Wuchereria bancrofti* (Owen and Anantaraman, 1982) have been collected in the Sudan and rural India this way. The technique appears to be readily adaptable but for application to other species consideration will have to be given to possible variations in tolerance to cryoprotectants and to dehydration and to the speeds with which cryoprotectants and water permeate the organisms.

Conclusions

Cryopreservation techniques

The techniques of parasite cryopreservation, once established, have tended to become taken for granted as another tool of research. This is largely as it should be, except that rather too often the assumption is made that a technique which gives good survival with one species of parasite will be ideal for other species. This assumption is not necessarily valid; for example, the intra-erythrocytic species of malaria require very different cooling rates. The evolution of the double-incubation ethanediol technique (see above) for use with different parasites demonstrates how each step should be checked for its effect on a new organism in order to obtain the optimum survival with that species. As many of the parameters of a cryopreservation schedule, for example cooling rate, cryoprotectant type and concentration, temperature and time of addition, warming rate, etc, are interrelated, altering one of the parameters will alter the requirements of any or all of the others.

Viability assays

In a recent analysis (Mutetwa, 1983) of the 41 publications produced between 1939 and 1980 in which the cryopreservation of malaria was reported, it was possible to estimate the approximate level(s) of survival from the information given by the various workers in only eight reports. This is a problem encountered throughout the literature on cryopreservation of parasites. All too often survival is only reported in the form of subjective statements or is impossible to quantify as unfrozen controls were not included. Anyone considering conducting a study on parasite cryopreservation should pick quantitative viability assay and include the relevant control(s), otherwise the significance of their findings will be lost.

Future

There are still many organisms for which a cryopreservation technique has yet to be devised and whose cryopreservation would produce considerable benefit to researchers. In particular, the cryobiology of trematodes and cestodes and the arthropod vectors (mosquitoes, blackflies, etc.) is virtually unknown. Low temperature preservation has become fairly organized for the protozoa and as more strains are preserved it will become necessary to establish libraries of type strains and strains with special characteristics to be made available to interested workers. This could be of service to those involved in disease diagnosis particularly where uniformity of the antigens used is important. Recently, one small step in this direction has been made by liquid nitrogen storage of *S. mansoni* eggs for use in the circumoval precipitin test (Ismail *et al.*, 1983). Preservation of helminth eggs may be another way that life cycle maintenance programmes also can be made easier (Uga *et al.*, 1983).

Certain helminths could be useful for investigating cellular interactions in cryobiology as at least some of the cells of these organisms are likely to have been damaged during cryopreservation and it would be useful to determine how much damage the organism can tolerate, how damage is repaired and which types of cells are most crucial to survival of the whole organism. Information gained here may be applicable to other metazoans and to tissue and organ freezing.

Acknowledgements

I would gratefully like to acknowledge financial support from the International Atomic Energy Agency and the Wellcome Trust.

References

Anderson, J., Fuglsang, H., Hamilton, P.J.S., Marshall, T.F. de C. (1974). Studies on onchocerciasis in the United Cameroon Republic II. Comparison of onchocerciasis in rain-forest and Sudan-savanna. *Transactions of the Royal Society of Tropical Medicine and Hygiene* **68**, 209–22.

Anonymous. (1980). Coenorhabditis genetics center. *Nature* **284**, 513.

Bemrick, W.J. (1978). Tolerance of equine stronglylid larvae to desiccation and freezing. *Cryobiology* **15**, 214–18.

Bickle, Q.D. and James, E.R. (1978). Resistance against *Schistosoma mansoni* induced by immunization of mice with cryopreserved schistosomula. *Transactions of the Royal Society of Tropical Medicine and Hygiene* **72**, 677–8.

Brandt, W. (1980). North-South: A program for survival. *International Commission on International Development Issues*, Chairman Willy Brandt, MIT Press, Cambridge, USA.

Bruce-Chwatt, L.J. and De Zulueta, J. (1980). *The Rise and Fall of Malaria in Europe, a Historico-epidemiological Study*. Oxford University Press, Oxford.

Callow, L.L. (1975). Vaccination against bovine babesiosis. In *Immunity to*

Parasitic Animals, pp. 121–49. Edited by Miller, L.H., Pino, J.A. and McKlevey, J.J. Jr. Plenum Press, New York and London.

Campbell, W.C. and Thompson, B. (1973). Survival of nematode larvae after freezing over liquid nitrogen. *Australian Veterinary Journal* **49**, 110–11.

Campbell, W.C., Blair, L.S. and Egerton, J.R. (1972). Motility and infectivity of *Haemonchus contortus* larvae after freezing. *Veterinary Record* **91**, 13.

Campbell, W.C., Blair, L.S. and Egerton, J.R. (1973). Unimpaired infectivity of the nematode *Haemonchus contortus* after freezing for 44 weeks in the presence of liquid nitrogen. *Journal of Parasitology* **59**, 425–7.

Chinery, W.A. and Atiemo, N.A. (1973). A preliminary observation on the preservation of microfilariae of *Acanthochilonema perstans* (Manson 1891) at sub-zero temperature. *Ghana Medical Journal* **12**, 203–204.

Coggeshall, L.T. (1939). Preservation of viable malaria parasites in the frozen state. *Proceedings of the Society of Experimental Biology and Medicine* **42**, 499–501.

Coles, G.C., Simpkin, K.G. and Briscoe, M.G. (1980). Routine cryopreservation of ruminant nematode larvae. *Research in Veterinary Science* **28**, 391–2.

Cunningham, M.P., Brown, C.G.D., Burridge, M.J. and Purnell, R.E. (1974). Cryopreservation of infective particles of *Theileria parva*. *International Journal of Parasitology* **3**, 583–7.

Dalgliesh, R.J. (1972). Theoretical and practical aspects of freezing parasitic protozoa. *Australian Veterinary Journal* **48**, 233–9.

Dalgliesh, R.J. and Mellors, L.T. (1974). Survival of the parasitic protozoan, *Babesia bigemina*, in blood cooled at widely different rates to – 196°C. *International Journal for Parasitology* **4**, 169–72.

Dalgliesh, R.J., Swain, A.J. and Mellors, L.T. (1976). Bioassay to measure effects of cooling and warming rates and protection by dimethylsulphoxide on survival of frozen *Babesia rodhaini*. *Cryobiology* **13**, 631–7.

Diamond, L.S. (1964). Freeze-preservation of protozoa. *Cryobiology* **1**, 95–102.

Dick, T.A. (1983). Species and intraspecific variation. In *Trichinella and trichinosis*, pp. 31–73. Edited by Campbell, W.C. Plenum Press, New York and London.

Dunn, A.M. (1978). *Veterinary Helminthology*, second edition. William Heinemann Medical Books, London.

East African Trypanosomiasis Research Organization, *Annual Reports (1960–69)*, Tororo, Uganda.

El Sheikh, H. and Ham, P.J. (1982). Human onchocerciasis: Cryopreservation of isolated microfilariae. *Lancet* **Feb. 20th**, 450.

Green, M.S., Kark, J.D., Witztum, E., Greenblatt, C.L. and Spira, D.T. (1983). Frozen stored *Leishmania tropica* vaccine: The effects of dose, route of administration and storage on the evolution of the clinical lesion. Two field trials in the Israel Defense Forces. *Transactions of the*

Royal Society of Tropical Medicine and Hygiene **77**, 152–9.

Gustafson, P.V. (1953). The effect of freezing on encysted *Anisakis* larvae. *Journal of Parasitology* **39**, 585–8.

Ham, P.J. (1982). Recovery of *Taenia crassiceps* cysticerci (ERS ToI derived strain) from sub-zero temperatures. *Journal of Helminthology* **56**, 131–3.

Ham, P.J. (1983). Cryopreservation of blackfly eggs (Diptera-Simulidae). *Cryobiology* **20**, 729.

Ham, P.J. and James, E.R. (1983). Successful cryopreservation of *Brugia pahangi* third-stage larvae in liquid nitrogen. *Transactions of the Royal Society of Tropical Medicine and Hygiene* **77**, 815–19.

Ham, P.J., James, E.R. and Bianco, A.E. (1979). *Onchocerca* spp. cryopreservation of microfilariae and subsequent development in the insect host. *Experimental Parasitology* **47**, 384–91.

Ham, P.J., Townson, S., James, E.R. and Bianco, A.E. (1981). An improved technique for the cryopreservation of *Onchocerca* microfilariae. *Parasitology* **83**, 139–46.

Hunger Project. (1982). *The End Hunger Briefing Workbook*. The Hunger Project, 77 Cromwell Road, London.

Hsu, S.Y.Li, Hsu, H.F., Xu, S.T., Shi, F.H., He, Y.X., Clarke, W.R. and Johnson, S.C. (1983). Vaccination against bovine schistosomiasis japonica with highly X-irradiated schistosomula. *American Journal of Tropical Medicine and Hygiene* **32**, 367–70.

Isenstein, R.S. and Herlich, H. (1972). Cryopreservation of infective third-stage larvae of *Trichostrongylus axei* and *T. colubriformis*. *Proceedings of the Helminthological Society of Washington* **39**, 140–42.

Ismail, S.A., Stek, M.Jr., and Leef, J.L. (1983). Circumoval precipitin (COP) test in schistosomiasis with frozen *Schistosoma mansoni* eggs. *Transactions of the Royal Society of Tropical Medicine and Hygiene* **77**, 809–11.

James, E.R. (1980a). Protozoa and helminth parasites of man and animals. In *Low Temperature Preservation in Medicine and Biology*. Edited by Ashwood-Smith, M.J. and Farrant, J. Pitman Medical, Tunbridge Wells.

James, E.R. (1980b). Cryopreservation of *Schistosoma mansoni* schistosomula using 40% (10 M) methanol and rapid cooling. *Cryoletters* **1**, 535–44.

James, E.R. (1981). *Schistosoma mansoni*: Cryopreservation of schistosomula by two-step addition of ethanediol and rapid cooling. *Experimental Parasitology* **52**, 105–16.

James, E.R. (1984). Cryopreservation of parasitic protozoa. In *Maintenance of Microorganisms*. Edited by Kirsop, B.E. Academic Press, London and New York.

James, E.R. and Farrant, J. (1977). Recovery of infective *Schistosoma mansoni* schistosomula from liquid nitrogen: A step towards storage of a live schistosomiasis vaccine. *Transactions of the Royal Society of Tropical Medicine* **71**, 498–500.

James, E.R. and Dobinson, A.R. (1984). *Schistosoma mansoni*: The interactive effects of irradiation and cryopreservation on parasite maturation and immunization of mice. *Experimental Parasitology* **57**, 279–86.

Joyner, L.P. (1954). The elimination of *Trichomonas foetus* from infected semen by storage in the presence of glycerol. *Veterinary Record* **66**, 727–31.

Kates, K.C. (1950). Survival on pasture of free-living stages of some common gastrointestinal nematodes of sheep. *Journal of the Helminthological Society of Washington* **17**, 39–58.

Kelly, J.D., Campbell, W.C. and Whitlock, H.V. (1976). Infectivity of *Ancylostoma caninum* larvae after freezing over liquid nitrogen. *Australian Veterinary Journal* **52**, 141–3.

Laveran, A. and Mesnil, F. (1904). *Trypanosomes and Trypanosomiasis*. Baillière, London.

Leef, J.L., Strome, C.P.A. and Beaudoin, R.L. (1979). Low-temperature preservation of sporozoites of *Plasmodium berghei*. *Bulletin of the World Health Organization* **57**, 87–91.

Leef, J.L., Hollingdale, M.R. and Beaudoin, R.L. (1981). Principles of cryopreservation of protozoan parasites and erythrocytes. *World Health Organization report*, WHO/MAL/81.940.

Le Blanq, S.M. (1983). *An epidemiological and taxonomic study of old world* Leishmania *using isoenzymes*. PhD Thesis, London University.

Leibo, S.P. and Mazur, P. (1971). The role of cooling rates in low-temperature preservation. *Cryobiology* **8**, 447–52.

Lovelock, J.E. (1952). Resuspension in plasma of human red blood cells frozen in glycerol, *Lancet*, **i**, 1238.

Lowrie, R.C. Jr. (1983). Cryopreservation of third-stage larvae of *Brugia malayi* and *Dipetalonema viteae*. *American Journal of Tropical Medicine and Hygiene* **32**, 767–71.

Lumsden, W.H.R. and Hardy, G.J.C. (1965). Nomenclature of living parasite material. *Nature* **205**, 1032.

Luyet, B.J. and Gehenio, P.M. (1954). Effect of ethylene glycol in protecting various amoeboid organisms against freezing injury. *Journal of Protozoology* **1**, (suppl), 7.

McCall, J.W., Jun, J. and Thompson, P.E. (1975). Cryopreservation of infective larvae of *Dipetalonema viteae*. *Journal of Parasitology* **61**, 340–42.

McCall, J.W. and Weathersby, A.B. (1978). In *4th International Congress of Parasitology, 19–16 August, 1978, Warsaw*, Short Communications, Section C II 9, 155, Warsaw, Poland.

McCaulay, E.H., Majid, A.A., Tayeb, A. and Bushara, H.O. (1983). Clinical diagnosis of schistosomiasis in sudanese cattle. *Tropical Animal Health and Production* **15**, 129–36.

McWhirter, N. (1983). *The Guinness Book of Records 1984*. Guinness Superlatives Ltd., 2 Cecil Court, London Road, Enfield.

Majid, A.A., Bushara, H.O., Saad, A.M., Hussein, M.F., Taylor, M.G., Dargie, J.D., Marshall, T.F. deC. and Nelson, G.S. (1980). Observations on cattle schistosomiasis in the Sudan, a field study in comparative medicine III. Field testing of an irradiated *Schistosoma bovis* vaccine.

American Journal of Tropical Medicine and Hygiene **29**, 452–5.

Miller, T.A. (1978). Industrial development and field use of the canine hookworm vaccine. *Advances in Parasitology* **16**, 333–44.

Miller, T.A. and Cunningham, M.P. (1965). Freezing of infective larvae of the dog hookworm *Ancylostoma caninum*. *East African Trypanosomiasis Research Organization Annual Report, 1965*, 19–22.

Minjas, J.N. and Townson, H. (1980). The successful cryopreservation of microfilariae with hydroxyethyl starch as cryoprotectant. *Annals of Tropical Medicine and Parasitology* **74**, 571–3.

Miyata, A. (1975). Cryo-preservation of the parasitic protozoa. *Japanese Journal of Tropical Medicine and Hygiene* **3**, 161–200.

Muller, R.A. (1975). *Worms and Disease*. William Heinemann Medical Books, London.

Murrell, K.D., Stirewalt, M.A. and Lewis, F.A. (1979). *Schistosoma mansoni*: Vaccination of mice with cryopreserved irradiated schistosomules. *Experimental Parasitology* **50**, 265–71.

Mutetwa, S.M. (1983). *Studies on the cryopreservation of erythrocytic stages of malaria parasites*. PhD Thesis, London University.

Mutetwa, S.M. and James, E.R. (1984a). Cryopreservation of *Plasmodium chabaudi* I: Protection by glycerol and dimethylsulphoxide during cooling and by glucose following thawing. *Cryobiology* **21**, 329–39.

Mutetwa, S.M. and James, E.R. (1984b). Cryopreservation of *Plasmodium chabaudi* II: Cooling and warming rates. *Cryobiology* **21**, 552–8.

Obiamiwe, B.A. and MacDonald, W.W. (1971). The preservation of *Brugia pahangi* microfilariae at sub-zero temperatures and their subsequent development to the adult stage. *Annals of Tropical Medicine and Parasitology* **65**, 547–54.

Ogunba, E.O. (1969). Preservation of frozen *Brugia pahangi* using dimethylsulphoxide. *Journal of Parasitology* **55**, 1101–102.

Owen, D.G. and Anantaraman, M. (1982). Successful cryopreservation of *Wuchereria bancrofti* microfilariae. *Transactions of the Royal Society of Tropical Medicine and Hygiene* **76**, 232–3.

Parfitt, J.W. (1971). Deep freeze preservation of nematode larvae. *Research in Veterinary Science* **12**, 488–9.

Paling, R.W. and Geysen, D. (1981). Observations on Rwandan strains of *Theileria parva* and the value of *T. parva* nyakizu as a possible vaccine strain. In *Advances in the Control of Theileriosis*, pp. 238–41. Edited by Irvin, A., Cunningham, M.P. and Young, A.S. Martinus Nijhoff, The Hague.

Pipano, E. (1981). Schizonts and tick stages in immunization against *Theileria annulata*. In *Advances in the Control of Theileriosis*, pp. 242–52. Edited by Irvin, A., Cunningham, M.P. and Young, A.S. Martinus Nijhoff, The Hague.

Radley, D.E., Brown, C.G.D., Cunningham, M.P., Kimber, C.D., Musisi, F.L., Payne, R.C., Purnell, R.E., Stagg, S.M. and Young, A.S. (1975). East Coast fever: 3. Chemoprophylactic immunization of cattle using oxytetracycline and a combination of theilerial strains. *Veterinary Parasitology* **1**, 51–60.

Ristic, M. and Levy, M.G. (1981). A new era of research toward solution of bovine babesiosis. In *Babesiosis*, pp. 509–44. Edited by Ristic, M. and Kreier, J.P.

Robson, J., Pedersen, V., Odeke, G.M., Kamya, E.P. and Brown, C.G.D. (1977). East coast fever immunization trials in Uganda: Field exposure of zebu cattle immunized with isolates of *Theileria parva*. *Tropical Animal Health and Production* **9**, 219–31.

Rowe, A.W., Lenny, L.L. and Mannoni, P. (1980). Cryopreservation of red cells and platelets. In *Low Temperature Preservation in Medicine and Biology*. Edited by Ashwood-Smith, M.J. and Farrant, J. Pitman Medical, Tunbridge Wells.

Ruitenberg, J. (1970). Anisakis: *Pathogenesis, serodiagnosis and prevention*. PhD Thesis, University of Utrecht, Holland.

Sargeaunt, P.G. and Williams, J.E. (1979). Electrophoretic isoenzyme patterns of the pathogenic and non-pathogenic intestinal amoebae of man. *Transactions of the Royal Society of Tropical Medicine and Hygiene* **73**, 225–32.

Schiller, E.L., Turner, V.M., Marroquin, H.F. and D'Antonio, R. (1979). The cryopreservation and *in vitro* cultivation of larval *Onchocerca volvulus*. *American Journal of Tropical Medicine and Hygiene* **28**, 997–1009.

Stepanova, N., Zablotsky, V., Mutuzkina, Z., Rasulov, J., Umarov, J. and Tuhtaev, B. (1977). Vaccination against theileriasis using live vaccines. *Veterinarya Moscow* **3**, 69–70.

Smith, A.U. (1961). *Biological Effects of Freezing and Supercooling*. Edward Arnold, London.

Smith,H.J. (1975). An evaluation of low temperature sterilization of trichinae infected pork. *Canadian Journal of Comparative Medicine* **39**, 316–20.

Uga, S., Araki, K., Matsumura, T. and Iwamura, N.I. (1983). Studies on the cryopreservation of eggs of *Angiostrongylus cantonensis*. *Journal of Helminthology* **52**, 297–303.

Uilenberg, G., Silayo, R.S., Moangala, C., Tondeur, W., Tatchell, R.J. and Sanga, H.J.N. (1977). Studies on Theileriidae (Sporozoa) in Tanzania. X. A large-scale field trial on immunization against cattle theileriosis. *Zeitschrift fur Tropenmedizin und Parasitologie* **28**, 499–506.

Van Wyk, J.A., Gerber, H.M. and Van Aardt, W.P. (1977). Cryopreservation of the infective larvae of the common nematodes of ruminants. *Onderstepoort Journal of Veterinary Research* **44**, 173–94.

Wakelin, D. and Denham, D.A. (1983). The immune response. In *Trichinella and trichinosis*. Edited by Campbell, W.C. Plenum Press, London and New York.

Walter, C.A., Knight, S.C. and Farrant, J. (1975). Ultrastructural appearance of freeze-substituted lymphocytes frozen by interrupting rapid cooling with a period at −26°C. *Cryobiology* **12**, 103–109.

Weinman, D. and McAllister, J. (1947). Prolonged storage of human pathogenic protozoa with conservation of virulence: Observations on the storage of helminths and leptospiras. *American Journal of Hygiene* **45**, 102–121.

World Health Organization. (1982). *Special Programme for Research and Training in Tropical Diseases Newsletter*, number 18, May 1982, UNDP/World Bank/WHO.

Wilson, R.J.M., Farrant, J. and Walter, C.A. (1977). Preservation of intra-erythrocytic forms of malarial parasites by one-step and two-step cooling procedures. *Bulletin of the World Health Organization* **55**, 309–315.

13

Low temperature preservation in medicine and veterinary science

B.J. Fuller
Academic Department of Surgery
Royal Free Hospital School of Medicine

Introduction

Low temperatures may have profound effects on mammalian systems, experiments on storage of meat at low temperatures and on the numbing effect of cold being dsecribed as early as the seventeenth century (Boyle, 1683). During the succeeding period many studies have been reported by scientists stimulated by the concept that it might be possible to preserve cells, tissues or even whole organisms by the application of low temperatures. This early work is reviewed by Keilin (1959). A reduction in temperature leads to a lowering of metabolic activity and this has been used in medicine to combat the harmful cellular effects which accompany ischaemia (cessation of blood flow), for example during surgical procedures on heart (Kirklin *et al.*, 1961), and liver (Bernhard *et al.*, 1957), and in protecting organs, tissues and cells which have been removed from the body for transplantation. The following discussion will consider:

(*a*) the application of lowered temperatures in an attempt to overcome the harmful effects of ischaemia;
(*b*) the rationale for techniques used in tissue preservation, which fall into two broad categories – (*i*) storage in the liquid state at temperatures above freezing point (hypothermic storage) and (*ii*) preservation in the frozen state at deep subzero temperatures (cryopreservation);
(*c*) examples of current methods of preservation employed in clinical or animal science.

Detailed technical information concerning specific storage protocols are available elsewhere (Karow and Pegg, 1981; Ashwood-Smith and Farrant, 1980).

The harmful effects of ischaemia

The circulatory system bathes the tissues and organs of the body with a nutrient-rich plasma in which salts, dissolved gases and energy substrates such as glucose are maintained within closely defined limits. The hydrogen ion concentration is regulated such that pH remains within the range 7.31–7.43, and the osmolality of the plasma remains close to 300 mosmol kg^{-1}. The circulatory system also acts as a heat exchanger, maintaining a constant body temperature close to 37°C. Within this controlled environment cells of tissues and organs are able to function and replicate (where necessary) with a high degree of efficiency. To carry out these vital processes, a continuous supply of energy is required, which is derived from metabolic breakdown of substrates (glucose, fatty acids, ketone bodies, amino acids) supplied from the plasma. Dissolved oxygen is also supplied in the bathing plasma. The substrates are oxidized via a series of enzymic transformations to carbon dioxide, with a concomitant reduction of the cellular pool of nicotinamide adenine dinucleotides (NAD, NADP).

Energy is liberated from the substrate by the enzymic transformations, with the greatest energy yield being produced when the reduced nucleotides are reconverted to the oxidized form via the electron transfer mechanism of the mitochondria. This process requires molecular oxygen to produce the final end product, water. The cell is able to harness and store the energy in the phosphate bonds of adenosine triphosphate (ATP) and this energy can be used as required, for example for synthetic processes.

To perform enzymatic catabolic and synthetic processes with high efficiency, the internal environment of the cell must be maintained within closely defined limits of ionic composition, pH, osmolality, etc. Most cells are in osmotic equilibrium with the plasma, but the ionic content of the intracellular milieu may be very different from that of the plasma (Table 13.1). Plasma contains relatively high concentrations of sodium and chloride and low concentrations of potassium and magnesium. Conversely the

Table 13.1 Composition of typical extracellular and intracellular fluids (millimoles/litre)

	Extracellular fluid	Intracellular fluid
Na^+	140	15
K^+	4	150
Mg^{++}	1	5
Cl^-	110	10
HCO_3^-	25	10

intracellular medium is rich in potassium and magnesium and has lower sodium and chloride concentrations. Calcium distribution also varies between plasma and intracellular solutions, the relationship being complicated as calcium can be 'bound' intracellularly to proteins. These concentration differences result in a tendency for sodium to diffuse into the cell across the membrane, whilst potassium and magnesium diffuse out. The cell will resist these ionic redistributions, which would eventually destroy the essential characteristics of the intracellular milieu, by actively transporting ions against concentration gradients, using the sodium–potassium ATPase or 'sodium pump' located in the plasmalemma. Sodium is translocated out of the cell, potassium into the cell (on a stoichiometric basis of three sodium ions per two potassium ions), and energy is expended in the form of hydrolysis of ATP.

The cell contents normally have a negative charge with respect to extracellular fluid because as the sodium and potassium ions tend to diffuse back down their concentration gradients, potassium diffuses out at a faster rate than that achieved by sodium ions diffusing in. Chloride ions distribute freely across the cell membrane, and the negative charge dictates that less chloride will partition into the intracellular fluid, given that some intracellular anion results from charges on protein. Under conditions of normal cellular metabolism these diffusional movements are self limiting in that if the negative charge increases (too great a cation loss), the further diffusion of positively-charged potassium ion out of the cell is opposed. The net result of (Na^+-K^+) ATPase activity is the movement of sodium and chloride out of the cell. Water distributes freely across the cell membrane and its movement is dictated by osmotic forces across the membrane. Thus the (Na^+-K^+) ATPase, by controlling sodium chloride expulsion from the cell, also controls internal osmolality and consequently cell volume.

When the blood supply to a tissue is arrested, the mechanisms for maintaining intracellular homeostasis, and consequently vital cell processes, are immediately jeopardized because the continuing supply of energy substrates and dissolved oxygen is halted. Many cells have metabolite reserves which can be mobilized to provide energy substrates, for example glucose from glycogen, but the absence of supplies of molecular oxygen is most immediately noticeable. At normal body temperatures the available oxygen, present as dissolved oxygen or as oxygen bound to haemoglobin in the red corpuscles in the vascular bed, is used up very quickly once blood flow is interrupted. At a gross level this is manifest as a transition in colour of an ischaemic organ, for example kidney from 'pink' (normal content of oxyhaemoglobin) to 'purple' (deoxyhaemoglobin predominant), once ischaemia exceeds 2–3 minutes. The consumption of the available oxygen means that the electron transport process can no longer function. Reduced nucleotides (NADH, NADPH) accumulate and these inhibit many of the enzymatic processes of both catabolism and anabolism. Without participation of the electron transport system, energy transduction and ATP formation are blocked. Many cells have the capability to perform some anaerobic metabolism, for example, glycolysis, which can liberate energy from substrates. This is an inefficient process, liberating only 3 per cent of the chemical energy available in glucose compared to about 50 per cent released by normal aerobic glucose catabolism. Anaerobic metabolism also results in the accumulation of hydrogen ions, and the resultant pH shift itself eventually inhibits the glycolytic process.

The abolition of all available methods for cell energy trapping results in rapid depletion of ATP concentrations, such that, using ischaemic rat liver as an illustration, ATP concentrations have reduced by > 80 per cent in 15 minutes. As discussed above, ATP is essential for operation of the sodium pump in the cell membrane, and ischaemia thus results in a loss of normal intracellular homeostasis because the tendency for ions to diffuse across the membrane can no longer be balanced by active transport processes. Sodium ions diffuse into, and potassium ions diffuse out of, the cell. Furthermore, the presence of intracellular impermeant anions results in more sodium entering than potassium leaving the cell. This leads to a loss of membrane negative charge allowing even more chloride to enter. Net movement of sodium and chloride into the cell will result in water uptake (and cell swelling) by osmosis. In the early stages of ischaemia these changes are readily reversed when normal blood supply to the tissue is re-established. However, as the ischaemia period increases the imbalances that result become lethal to the tissue, through a series of events which at present are only partially understood. It has been postulated that a reduction in intracellular pH and alteration in ionic composition results in release of autolytic enzymes from lysozomes in the cell (Rangel *et al.*, 1969). Biochemical and spatial disorganizations caused by ischaemia may result in abnormal enzymic transformations which can release free radicals (Thaw *et al.*, 1982). The free radicals react autocatalytically with membrane lipids causing degradation. There may also be a redistribution of Ca^{2+} within the cell causing activation of membrane-bound phospholipases, leading to membrane degradation (Chien *et al.*, 1977). The end result is a tissue which cannot regain normal internal homeostasis when blood supply is restored. For example at 37°C mammalian kidney may tolerate approximately 60 minutes of ischaemia (Rikukawa and Lindsey, 1968) and liver approximately 30 minutes (Abouna, 1974). The vascular bed of tissue and organs may also be a site of ischaemic damage. If the vasculature has been affected such that reperfusion of the tissue is compromised when the blood supply is restored, further metabolic deprivation will occur. Red blood cells, themselves rendered ischaemic when trapped within a tissue deprived of normal circulation, become less deformable (Weed *et al.*, 1969) and may physically block tissue blood supply during reperfusion. Swelling of the endothelial cells lining small vessels may also contribute to a reduced blood supply. Endothelial cells participate in vascular homeostasis by synthesizing and releasing a range of pro- and anti-coagulants (Thorgeirsson *et al.*, 1978) and derangement of this control may result in vascular stasis. Cumulative toxicity resulting from ischaemia must be halted or at least slowed in an acceptable low temperature preservation regimen.

To prevent lethal ischaemic damage in mammalian tissues, the tissue or organ may be supplied with substrates, vitamins, cofactors, etc. essential for maintenance of viability at physiological temperature. This is tissue or organ culture in which lowered temperatures are rarely used (Jakoby, 1979).

This approach is not routinely used in clinical transplantation. However, normothermic culture of islets of Langerhans, prior to transplantation, has suggested that tissue culture may be beneficial in reducing the immunogenicity of the tissue (Bowen *et al.*, 1980), making host rejection of the graft less likely, and it may therefore play a greater role in the future.

Alternatively, ischaemic damage can be reduced if the chemical and physical processes involved are slowed by reduction in temperature. Ideally, this should result in a reduction of metabolic activity, slowed depletion of important metabolites, cofactors and high-energy compounds and minimized accumulation of harmful products. The problem of differential disorganization of metabolic pathways under such conditions may limit the duration of hypothermic storage.

Hypothermic techniques for short-term storage of tissues and organs, or prevention of ischaemic damage during surgery may be successful; for example kidneys removed and stored in ice for up to 12 hours functioned successfully after transplantation (Calne *et al.*, 1963). However, reduced metabolism may itself produce deleterious effects which accumulate with time. For example, the reduction in the activity of the (Na^+-K^+) ATPase in many tissues is such that by 5 °C its activity is *c.* 1 per cent of that measured at 37 °C. The results of the reduced activity of the (Na^+-K^+) ATPase is similar to that seen in warm ischaemia. This has been demonstrated using isolated rat hepatocytes (Berthon *et al.*, 1980) which were incubated in tissue culture medium for 60 minutes at 38 °C, then for 90 minutes at 1 °C and finally for 60 minutes at 38°C (Fig. 13.1). During the hypothermic period there was a passive redistribution of sodium, potassium, chloride and water across the cell membrane, resulting in a change in membrane potential. Rewarming of the cells to 38°C resulted in almost complete restoration of normal ion balance. The rapidity of this reversal was due to cellular concentrations of necessary substrates and high energy compounds being maintained by hypothermic conditions such that normal (Na^+-K^+) ATPase function was quickly restored on rewarming.

After 90 minutes at 1 °C there was only a small reduction in membrane potential of isolated rat hepatocytes (Fig. 13.1), and this negative charge still provided a significant driving force for potassium ions to move out of the cells. There was also a gradual loss of cellular high-energy adenine nucleotides, possibly via enzymatic breakdown (Buhl and Jensen, 1979).

Two strategies have been developed to improve hypothermic storage of tissues and organs.

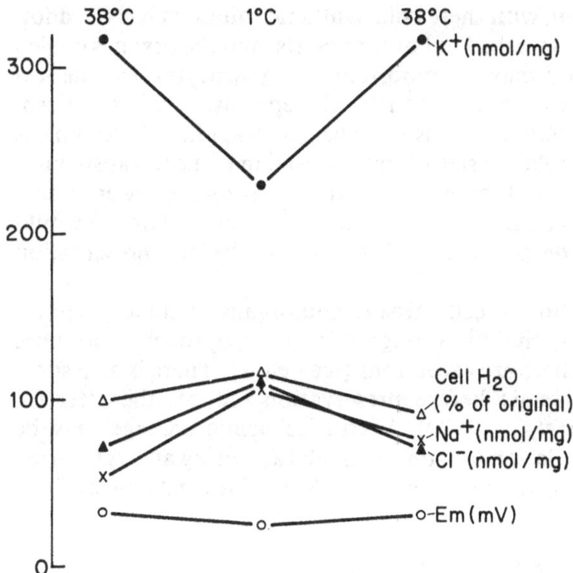

Fig. 13.1 The content of Na^+, K^+, Cl^-, water and the membrane charge of isolated rat hepatocytes measured at 38°C, after cooling to 1°C for 90 minutes and subsequent return for 60 minutes (redrawn from the data of Berthon *et al.*, 1980).

(1) *Preservation in solutions of 'intracellular' composition*

In this method the tissue is suspended in, or an organ is perfused with, a solution of an ionic composition adjusted to reduce the passive diffusions induced by hypothermia. Successful preservation of rat kidneys was enhanced when they were perfused with a solution of elevated potassium and magnesium content (Keeler *et al.*, 1966). A further development was the use of a high magnesium, high potassium solution made hyperosmolar with respect to intracellular fluid using glucose which has low tissue permeability at reduced temperatures (Collins *et al.*, 1969). This extended kidney storage in experimental systems to about 30 hours at 0°C.

Other solutions based on the same premise but with differences in relative concentrations of ions or species of impermeant molecules have been found to be similarly successful (Sacks *et al.*, 1973; Ross *et al.*, 1976). Interpretation of much of the early work on the importance of particular constituents is difficult because of variations in technique between the studies. However, when different constituents of storage solutions were examined in a standardized model of kidney preservation (Green and Pegg, 1979) certain conclusions could be drawn. The most important factor appears to be the prevention of cellular oedema by inclusion of impermeant solutes (sucrose, glucose, sulphate). In addition, maintenance of pH by inclusion of phosphate (providing a buffer system) or organic buffers (e.g. HEPES) also seems to be important. The efficacy of other additives and drugs remains to be proved.

For solid organs flushing with these cold synthetic solutions has the additional benefits that blood is washed out of the vessels, and the organ is cooled rapidly. Removal of blood may be important as erythrocytes become less deformable during cold entrapment and block reperfusion of the microcirculation once normal blood flow is established. Rapidity of cooling is important because of the relative size of organs used in clinical transplantation. The core temperatures of large organs such as livers are lowered only slowly by simple surface cooling, and thus warm ischaemic damage would be experienced by the deeper parts of the organ before preservation commenced.

Typically, for preservation of cells, tissues and organs of clinical interest the limit of successful hypothermic storage using this approach is no more than 1–2 days, although there are exceptions (see below). There is also some evidence that warm ischaemia before preservation reduces the effective storage period (Johnson *et al.*, 1972). Warm ischaemic damage may be reduced by gaseous persufflation of kidney vasculature, using an oxygen-rich gas, during hypothermic storage (Fischer *et al.*, 1978). The mechanism of this protection is unknown.

(2) *Continuous hypothermic perfusion*

This method is applicable only to large tissues or organs, and involves continuous perfusion of the vascular bed, typically with a solution similar to plasma in ionic composition. The storage temperature, 6–10°C, is higher than that used for simple solution storage (0°C). At these temperatures metabolism is assumed to proceed sufficiently to maintain cellular integrity and homeostasis. Substrates are added and the solution is oxygenated prior to perfusion. Compounds such as short-chain fatty acids can be utilized by kidneys during hypothermic perfusion (Huang *et al.*, 1971), and so provide energy for vital cell processes. Under such conditions kidneys are also able to resynthesize tissue reserves of high-energy adenine nucleotides depleted by prior warm ischaemia (Pegg *et al.*, 1981). Continuous perfusion of rat liver also similarly provides partial maintenance of ATP and cellular ion content (Attenburrow *et al.*, 1981). The perfusion pressures employed (usually 40–60 mmHg) are much lower than normal blood pressure and are designed to be the minimum for provision of good overall tissue perfusion. To balance this hydrostatic pressure and prevent interstitial oedema, oncotic agents such as albumin are usually added (Belzer *et al.*, 1967). Perfusion is advantageous in that accumulation of noxious agents (e.g. hydrogen ions) within the tissue may be prevented, and residual blood can be removed, so clearing the microcirculation. By contrast, prolonged flushing may not always clear the vascular bed of the organ (Foreman *et al.*, 1982), resulting in problems of re-establishing adequate blood supply after storage.

Continuous hypothermic perfusion of kidneys is preferred in the clinical situation if the organs have been damaged by warm ischaemia. Experimentally, the most consistent technique for prolonged storage of kidneys has been continuous perfusion (3 days: Belzer *et al.*, 1967). However, the method is practically more complex and expensive than simple storage after flushing

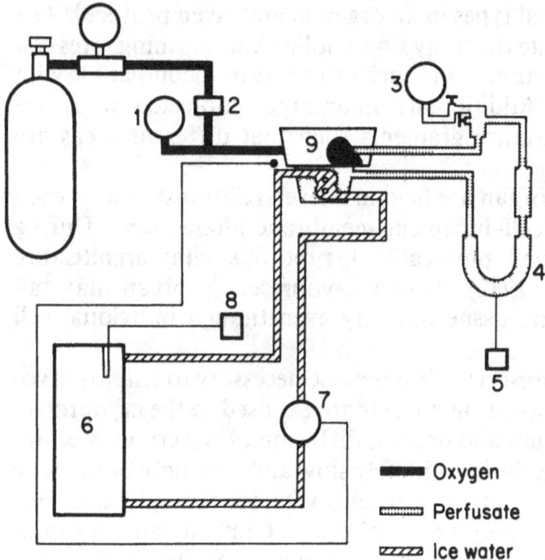

Fig. 13.2 A diagram of a perfusion machine for hypothermic kidney preservation. The apparatus is cooled by circulating chilled water (6). Perfusate is supplied to the kidney (9) via the renal artery. The perfusate pump is adjusted to provide the required inflow pressure which is monitored by a pressure gauge (3). Oxygenation may be provided by blowing sterile-filtered gas mixtures across the surface of the perfusate reservoir.

(1 = gasflow; 2 = gasfilter; 3 = arterial pressure gauge; 4 = roller pump; 5 = flow meter; 6 = ice and water; 7 = pump; 8 = temperature monitor; 9 = kidney in organ chamber.)

because of the use of large volumes of perfusate and perfusion machinery. A typical kidney perfusion circuit is depicted in Fig. 13.2. The use of sterile, non-toxic and largely disposable tubing, with oxygenators and membrane filters, in addition to a reliable pump, contribute to the cost and complexity of the circuitry. When perfusion is attempted by inexperienced personnel, organ damage can be caused by bad technique (Cerra *et al.*, 1977). An interesting experimental approach to prolong successful perfusion has been to reconnect the kidney to a donor animal for a short period of normal blood circulation in the middle of a period of cold perfusion (Wijk *et al.*, 1980). This has been found to allow consistent 6-day survival of canine kidneys. At present, however, such a method is not clinically applicable, and the mechanisms of this 'resuscitation' remain unclear.

Cryopreservation

Organs of clinical interest, for example kidney, consist of a variety of cell types (e.g. endothelial cells of the vasculature, parenchymal cells, and ductular cells). Different cell types require different regimens of cooling and warming for successful cryopreservation, and it may be difficult to match the optima for all the different cell types in an organ in any given protocol. The size and shape of organs dictate that only slow cooling and warming rates can be achieved by surface conductance, and such rates may be incompatible with survival of a given cell type. Additionally, in an organ subjected to surface cooling there will be temperature gradients such that different areas are cooled at different rates.

Individual cells within an organ are held in a fixed relationship with each other through cell–cell and cell–basement membrane interactions. During freezing extracellular ice may physically disrupt this vital architecture (Fig. 13.3) (Jacobsen *et al.*, 1982). Thus a cryopreserved organ may fail because of breakdown of the tissue integrity even though individual cell viability has been retained.

In attempts at successful preservation it may be necessary to employ cryoprotectant additives. In a suspension all cells are exposed to the cryoprotectant simultaneously. For tissues and organs, diffusion of the cryoprotectant after simple immersion would be inordinately slow and may only be of value where tissues comprise a limited number of cell layers, for example, skin. For larger tissues or organs, the only practical way of introducing cryoprotectants is by perfusion of the vascular bed. Even with perfusion, addition and removal must be carried out in a slow, controlled fashion as the additives, in the required concentrations, often exert osmotic forces which prove

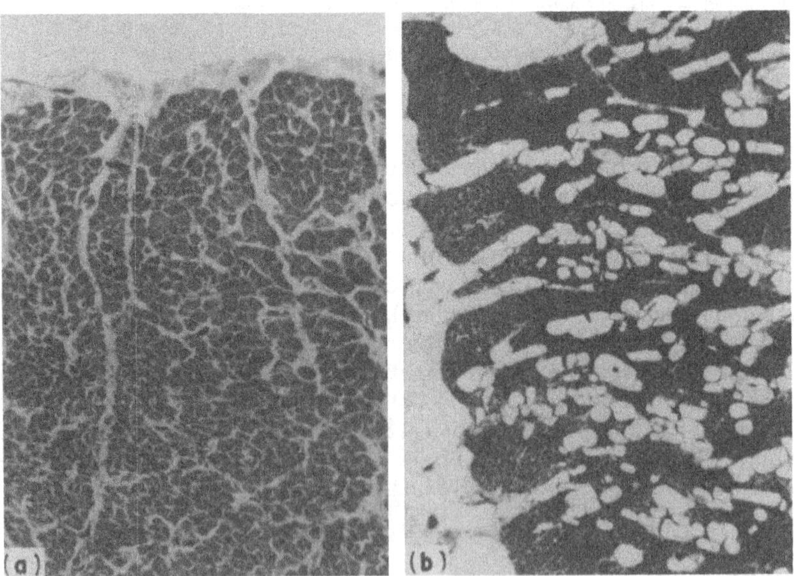

Fig. 13.3 Photomicrographs of smooth muscle samples cooled to $-21\,°C$. (a) Sample cooled without freezing at $0.3\,°C\ min^{-1}$ and taken through freeze substitution (Mag. × 1110). (Hunt *et al.*, 1982). (b) Sample cooled with freezing at the same rate to $-21\,°C$ and freeze substituted. The large cavities correspond to sites of ice crystal formation which have disrupted the tissue. (Mag. × 725). (Provided by Dr C. Hunt).

damaging *per se*. Increasing the concentration of cryoprotectants will also increase the problems of achieving equilibrium within the cell mass. In cryopreservation of cell suspensions the cells are usually present at a low density (< 10 per cent packed cell volumes). However, at higher cell densities (50–80 per cent) survival is reduced (Nei, 1968), by mechanisms which are not fully understood. Cell densities in organs are obviously high and this type of damage may be important in organ cryopreservation.

It will be apparent from the above descriptions that only by careful experimentation will the problems of organ cryopreservation be overcome; empirical methods have little chance of success, and considerable scientific effort is being applied to these problems (Jacobsen 1979; Karow, 1981). Techniques such as persufflation of the vascular bed of an organ with a cooled gas, for example pre-cooled helium, have been developed to allow greater flexibility and control of cooling rates. The use of microwave heating to allow rapid warming of large organs has also been advocated, and there is interest in the use of high concentrations of cryoprotectants to prevent ice formation. However, such approaches have yet to be fully developed. To date there have been few reports of successful organ cryopreservation (Dietzman *et al.*, 1974; Guttman *et al.*, 1977), and these successes have been difficult to reproduce.

As a result of the complex techniques required for whole organ transplantation, there is some interest in isolating the important cell types from the donor organ and implanting them as a suspension. For example, experimentally-induced diabetes in animals has been reversed by implanting islets of Langerhans isolated by enzyme digestion of the donor pancreas (Ballinger and Lacy, 1972), and certain types of liver diseases have been controlled by implanting suspensions of hepatocytes, again under experimental conditions (Matas *et al.*, 1976). Studies have been undertaken on cryopreservation of these organ-derived cells to produce a bank of material for transplantation. As might be expected from the preceding discussion, the results are much more encouraging than for attempted cryopreservation of the complete organ, although many problems remain to be overcome. The clinical manifestations of diabetes mellitus have been controlled to some degree in the rat (Taylor *et al.*, 1983), dog (Rajotte *et al.*, 1983) and pig (Wise *et al.*, 1983) using cryopreserved islets of Langerhans. Prolonged survival of cryopreserved hepatocytes after transplantation in rat has been demonstrated, with cells retaining at least some metabolic and ultrastructural characteristics of normal hepatocytes (Fuller *et al.*, 1982; Fuller *et al.*, 1983). It remains to be demonstrated whether this approach to tissue transplantation can be successfully applied to clinical problems in the future.

Involvement of oxygen free-radicals in hypothermic organ storage

Recently, there has been increasing evidence of the involvement of oxygen-derived free radical damage in stored and transplanted tissues. It has not yet been possible to measure the radicals in complex biological systems *in vivo* because they are so highly reactive, but evidence of their involvement has been found by looking at the products of radical interaction with biological substrates (polyunsaturated fatty acids, proteins, etc; Pryor, 1976). In hypothermic kidney preservation it has been shown that poor preservation protocols allowed development of this type of damage (Green *et al.*, 1986), even before blood supply was restored to the kidney after transplantation. Other studies have shown that addition of particular scavenging agents to renal preservation solutions improved subsequent organ function (Halasz *et al.*, 1985). Much remains to be learnt about this type of aberrant metabolism.

Some examples of current uses of low temperatures in clinical or animal science

Storage of blood and blood components

In the UK whole blood is routinely stored at 4°C after dilution with acid-citrate-dextrose solution (Beutler, 1972). The maximum permitted storage for subsequent transfusion is 21 days. Platelets, as a platelet-rich fraction prepared from anticoagulated blood, are often stored at 22°C. Many laboratories resuspend the platelets in autologous plasma and maintain a gentle agitation in the samples. Seventy-two hours is thought to be the maximum useful storage period.

Erythrocytes can be successfully cryopreserved, and frozen storage has been used for many years, particularly in the USA. Several cryopreservation methods have been employed using the cryoprotective additive glycerol (Rowe and Lenny, 1981). The protocols used for the successful recovery of frozen/thawed erythrocytes also, fortuitously, remove leucocytes and plasma proteins which are responsible for haemolytic transfusion reactions. These reactions can be a major problem in patients requiring multiple transfusions over a prolonged period, for example those suffering severe anaemia in thalassaemia major. Cryopreservation also allows banking of erythrocytes of rare blood groups for emergency use.

Gamete preservation

Spermatoza from a large number of mammalian and avian species have been successfully cryopreserved. This has revolutionized artificial insemination for agricultural and clinical purposes. Most often the cryoprotective additive employed is glycerol, whilst other additives such as egg yolk have been found to be important. The sperm samples are often packaged in small volumes (0.2 ml) in plastic straws, which are cooled by direct exposure to liquid nitrogen vapour. Samples are rewarmed by immersing the straws in warm water. The detailed theory and application of present methods are covered in recent reviews (Watson, 1979; Polge, 1980).

Cryopreservation of unfertilized ova or early stage embryos (1 cell to blastocyst) has been attempted in a variety of species. The technique is being used increasingly in agricultural science for meatstock production, and it may have a role to play in species conservation (Leibo, 1979). Clinically, there is interest in applying the method as part of the programme for extra-corporeal fertilization in overcoming female infertility. Success rates achieved with unfertilized oocytes have generally been lower than those obtained with embryos. As early as 1952 it was demonstrated that a small percentage of rabbit embryos could continue to divide in culture after cooling to −79°C with glycerol as cryoprotectant (Smith, 1952). The embryos of a variety of mammalian species including cattle, sheep, goats, mice and rabbits have since been successfully cryopreserved. Most often the penetrating cryoprotectants glycerol or dimethyl sulphoxide (DMSO) are used. A variety of regimens have been developed, characteristically, using slow cooling rates ($< 1°C\,min^{-1}$), to low subzero temperatures (e.g. −80°C) before immersion in liquid nitrogen and slow rates of warming (Leibo, 1979). There have been only a few reports of clinical embryo cryopreservation (Trounson and Conti, 1982), the ethical and legal implications of which are to be examined.

Improvements in the understanding of cryobiological damage, and modern techniques for evaluating this, have led to advances on several fronts of gamete preservation. In earlier indications that unfertilized ova were difficult to cryopreserve have to be re-evaluated in the light of more recent studies showing that in the mouse it has been possible to improve recovery of unfertilized ova by attention to detail of the preservation protocol (Fuller and Bernard, 1984). Methods allowing the storage of cells at low temperatures in a solid, vitrified form, have been sought over several years because it was recognized that the harmful effects of ice crystallization could be avoided. The work of Rall and Fahy (1985) has shown that this is possible for mouse embryos using a special combination of cryoprotectants. The successful use of cryopreservation to store embryos in human *in vitro* fertilization programmes has become more frequent (Cohen *et al.*, 1985).

Bone marrow storage

Long-term storage of autologous bone marrow in patients undergoing high-dosage radiation therapy or chemotherapy for malignant disease is attractive because the patient's bone marrow can be reconstituted by infusing thawed material after the treatment. For some leukaemias marrow cells taken in remission periods can be introduced when the disease reappears. With new technology, for example use of monoclonal antibodies, it may be possible to selectively remove leukaemic cells from a marrow sample and again store the marrow for the patient's future need. Bone marrow banks for allogeneic transplantation could be set up, and recent advances in immunosuppressive therapy have increased optimism concerning allogeneic grafting. The technique commonly employed is to use DMSO as cryoprotectant, and to cool the marrow in a thin layer in special plastic bags, between metal plates, at a slow rate. Rewarming is rapid, in a water bath at 37°C (Weiner *et al.*, 1976).

Skin preservation

Skin can be stored in cold medium in the liquid state close to 0°C for several days, but there is a progressive loss of viability. For prolonged storage cryopreservation is employed. The penetrating cryoprotectants glycerol or DMSO have been commonly used.

Equilibration of even thin tissues with cryoprotectant is slow, and in fact exposure of skin rolls to cryoprotectant for 2 hours at 4°C is common. The rolls of skin are cooled slowly (1–5°C min^{-1}), and warming, by placing in sterile saline at 37°C, is rapid (Bondoc and Burke, 1971).

Corneal storage

Corneas are transplanted clinically to reverse blindness caused by loss of transparency of the cornea. Storage in a specialized medium at 4°C for up to approximately 5 days is compatible with successful grafting. Several regimens for corneal cryopreservation have been advocated, involving the use of DMSO, cooling at a slow controlled rate (1–5°C min^{-1}) and rapid rewarming (Mueller *et al.*, 1964; Capella, 1978). These methods have yet to be widely accepted, although in some centres results for successful graft using cryopreserved cornea have been reported to be equal to those for liquid-stored tissue (Van Horn and Schultz, 1977).

Bone and cartilage

Bone and cartilage grafts are required for many orthopaedic procedures. Non-living bone grafts can be stored either deep-frozen or freeze-dried. These may provide a lattice for new bone deposit and eventually be replaced by host lamellar bone. Low-temperature storage of viable articular cartilage has been attempted, but present methods result in considerable loss of chondrocyte viability, and further research is required (Langer, 1981).

Kidney, liver, pancreas and heart

The most common method for storing such organs for clinical transplantation is an initial perfusion of the vascular bed using a chilled synthetic solution of elevated potassium content and increased osmolality, followed by ice storage as described above. For kidneys, Collins' solution or Marshall's Citrate solution are often used, and ice storage for between 24 and 48 hours is clinically acceptable provided the organ has not experienced warm ischaemia prior to preservation. Some centres which have suitable expertise perform continuous hypothermic perfusion with an albumin-based solution, and this is the method of choice if warm ischaemic damage has been incurred. Organs are viable for up to 3 days.

In clinical liver grafting, only ice storage after flushing has been used extensively. Great care is taken to avoid warm ischaemic damage, and the acceptable length of preservation is at present not much more than 12 hours. Another point of note is that the potassium-rich perfusate must be voided from the liver before re-establishment of normal circulation, for example by flushing the liver with plasma protein fraction just before the anastomoses are complete. If this is not done the potassium-rich solution passes directly from the liver via a short venous drainage back to the heart, and may cause sudden cardiac arrest.

Conversely, in the case of the heart, the cardioplegic actions of high potassium-content perfusates are considered beneficial because they inhibit cardiac muscle activity immediately the perfusion is commenced and thus prevent ischaemic damage during the early stages of flushing. Clinical heart preservation using initial perfusion and low temperature storage has not been used for periods of greater than approximately 6 hours.

For clinical preservation of the pancreas great care is taken to avoid warm ischaemia during removal of the organ, and initial perfusion with a potassium-rich hyperosmolar solution is employed, followed by ice storage. Pancreas transplantation is a relatively new field and clinical preservation has not exceeded 24 hours.

External application of low temperatures

Reduced temperatures cause arteriolar vasoconstriction and thus reduce blood supply (Smith, 1970). This is one beneficial action of cold packs applied to reduce oedema or 'swelling' in traumatized joints. More recently, this knowledge has been used to alleviate alopecia, a side-effect of some chemotherapeutic regimens used for cancer patients. During the period of drug treatment, the scalp temperature of the patient is reduced to about 20°C by application of pre-cooled gel packs or by use of a specialized 'cap' through which a coolant is circulated. This results in good conservation of the patient's hair (Dean *et al.*, 1979) as the vasoconstriction and the reduced temperature ensure that little of the circulatory drug reaches the hair follicles.

References

Abouna, G.J. (1974). Liver. In *Organ Preservation for Transplantation*, 349–71. Edited by Karow, A.M., Abouna, G. and Humphries, A. Little Brown and Co., Boston.

Ashwood-Smith, M.J. and Farrant, J. (1980) *Low Temperature Preservation in Medicine and Biology*. (Eds). pp. 323. Pitman Medical, Tunbridge Wells.

Attenburrow, V.P., Fuller, B.J. and Hobbs, K.E.F. (1981). Effects of temperature and method of hypothermic preservation on hepatic energy metabolism. *Cryoletters* **2**, 15–20.

Ballinger, W.F. and Lacy, P.E. (1972). Transplantation of intact pancreatic islets in rats. *Surgery* **72**, 175–83.

Belzer, F.O., Ashby, B.S. and Dunphy, J.E. (1967). 24-hour and 72-hour preservation of canine kidneys. *Lancet* **2**, 536–9.

Bernhard, W., Cahill, G. and Curtis, G. (1957). The rationale of surgery under hypothermia in certain patients with severe hepatocellular disease. *Arch. Surg.* **145**, 289–96.

Berthon, B., Claret, M., Mazet, J. and Poggioli, J. (1980). Volume and temperature dependent permiabilities in isolated rat liver cells. *Journal of Physiology* **305**, 267–77.

Beutler, E. (1972). Preservation of erythrocytes – liquid storage. In

Haematology, pp. 1299–1300. Edited by Williams, W.J., Beutler, E., Erslev, A. and Rundles, R. McGraw-Hill, New York.

Bondoc, C.C. and Burke, J.F. (1971). Clinical experience with viable frozen human skin and a frozen skin bank. *Annals of Surgery* **174**, 371–82.

Bowen, K. Andruc, L. and Lafferty, K. (1980). Successful allotransplantation of mouse pancreatic islets to non-immunosuppressed recipients. *Diabetes* **29**, 98–104.

Boyle, R. (1683). *New experiments and observations touching cold.* R. Davis, London.

Buhl, M. and Jensen, M. (1979). The role of 5′ nucleotidase in purine depletion of ischaemic renal tissue. In *Organ Preservation 11*, pp. 239–58. Edited by Pegg, D.E., Jacobsen, I.A. Churchill Livingston, Edinburgh.

Calne, R.Y. Pegg, D.E., Pryse-Davis, J. and Leigh-Brown, F. (1963). Renal preservation by ice cooling. An experimental study relating to kidney transplantation from cadavers. *British Medical Journal* **2**, 651–5.

Capella, J.A. (1978). Techniques for corneal preservation. In *Corneal Preservation*, pp. 308–17. Edited by Capella, J.A., Edelhauser, H.F. and Van Horm, D. Springfield, Charles C. Thomas.

Cerra, F.B., Raza, S., Andres, G.A. and Siegal, J. (1977). The endothelial damage of pulsatile renal perfusion and its relationship to perfusion pressure and colloid osmotic pressure. *Surgery* **81**, 534–41.

Chien, K., Abrams, J., Pfau, R. and Farber, J. (1977). Prevention by chlorpromazine of ischaemic liver cell death *American Journal of Pathology* **88**, 539–53.

Cohen, J., Simons, R., Fehilly, C., Fischel, S., Edwards, R., Hewitt, J., Rowland, G., Steptoe, P. and Webster, M. (1985). Birth after replacement of hatching blastocyst cryopreserved at expanded blastocyst stage. *Lancet* **I**, 647.

Collins, G.M., Bravo-shugarman, M. and Terasaki, P. (1969). Kidney preservation for transportation. Initial perfusion and 30 hr. ice storage. *Lancet* **2**, 1219–22.

Dean, J., Salmon, S. and Griffith, K. (1979). Prevention of doxorubian - induced hair loss with scalp hypothermia. *New England Journal of Medicine* **301**, 1427–9.

Dietzman, R.H., Robelo, A.E., Graham, E.F., Crabo, B.G. and Lillehei, R.C. (1974). Long term functional success following freezing of canine kidneys. *Surgery* **74**, 181–9.

Elford, B.C. and Solomon, A.K. (1974). Temperature dependence of cation permeability of dog red cells. *Nature* **248**, 522–4.

Fischer, J.H., Czerniak, A., Hauer, V. and Isselhard, W. (1978). A new simple method for optimal storage of ischaemically damaged kidneys. *Transplantation* **25**, 43–8.

Foreman, J., Wusteman, M.C. and Pegg, D.E. (1982). Washout of red blood cells from kidneys damaged by warm ischaemia. In *Organ Preservation; Basic and Applied Aspects*, pp. 183–6. MIT Press, Lancaster.

Fuhrman, F.A. (1956). Oxygen consumption of mammalian tissues at reduced temperature. In *The Physiology of Induced Hypothermia*, pp. 50–51. Pub. 451. Nat. Acad. Sci. Nat. Res. Council: Washington.

Fuller. B.J. and Attenburrow, V.D. (1979). The effects of hypothermic storage of liver by continuous perfusion and simple portal flushing on hepatic protein synthesis and urea production. In *Organ Preservation 11.*, pp. 278-92. Edited by Pegg, D.E. and Jacobsen, I.A. Churchill Livingstone, Edinburgh.

Fuller, B.J. and Bernard, A. (1984). Successful *in vitro* fertilization of mouse oocytes after cryopreservation using glycerol. *Cryoletters* **5**, 307–12.

Fuller, B.J., Lewin, J. and Sage, L. (1983). Ultrastructural assessment of cryopreserved hepatocytes after prolonged ectopic transplantation. *Transplantation* **35**, 15–18.

Fuller, B.J., Woods, R.J., Nutt, L.H. and Attenburrow, V.D. (1982). Survival of hepatocytes upon thawing from − 196°C: functional assessment after transplantation. In *Organ Preservation: Basic and Applied Aspects*, pp. 381–83. Edited by Pegg, D.E., Jacobsen, I.A. and Halasz, N. MIP Press, Lancaster.

Green, C.J., Healing, G., Lunec, J., Fuller, B.J. and Simpkin,S. (1986). Evidence of free-radical induced damage in rabbit kidneys after simple hypothermic preservation and auto-transplantation. *Transplantation* **41**, 161–5.

Green, C.J. and Pegg, D.E. (1979). The effect of variation in electrolyte composition and osmolality of solutions for infusion and hypothermic storage of kidneys. In *Organ Preservation 11*, pp. 86–101. Edited by Pegg, D.E. and Jacobsen, I.A. Churchill Livingstone, Edinburgh.

Guttman, E.M., Lizin, J., Robitaille, P., Blanchard, H. and Turgeon-Knaack, C. (1977). Survival of canine kidneys after treatment with dimethyl sulphoxide, freezing at − 80°C, and thawing by microwave illumination. *Cryobiology* **14**, 559-67.

Halasz, N., Bennett, J., Bry, W. and Collins, G.M. (1985). Protection of preserved and re-perfused kidneys against free-radical injury. *Cryobiology* **22**, 614–20.

Huang, J.S., Downes, G.L. and Belzer, F.O. (1971). Utilization of fatty acids in perfused hypothermic dog kidneys. *Journal of Lipid Research* **12**, 622–8.

Hunt, C.J., Taylor, M.J. and Pegg, D.E. (1982). Freeze-substitution and isothermal freeze fixation studies to elucidate the pattern of ice formation in smooth muscle at 252K. *Journal of Microscopy* **124**, 177–86.

Jacobsen, I.A. (1979). Steps towards long-term preservation of kidneys at sub-zero temperatures. *Scandinavian Journal of Urology and Nephrology*, Supplementum **52**.

Jacobsen, I.A., Pegg, D.E., Starklut, H., Chemnitz, C., Hunt, C., Barfort, P. and Diaper, M. (1982). Effect of cooling and warming rate on glycerolized rabbit kidneys. *Cryobiolgy* **19**, 668.

Jakoby, W. and Pastan, H. (1979). *Cell Culture. Methods in Enzymology Vol LVIII*, Academic Press, London.

Johnson, R.W., Anderson, M., Morley, A.R., Taylor, R. and Swinney, J. (1972). Twenty-four hour preservation of kidneys injured by prolonged warm ischaemia. *Transplantation* **13**, 174–9.

Karow, A.M. and Pegg, D.E. (1981). *Organ Preservation for Transplantation, 2nd. ed.* Marcel Dekker, New York.

Karow, A.M. Jr. (1981). Problems of Organ Cryopreservation. In *Organ Preservation for Transplantation*, pp. 517–52. Edited by Karow, A.M. and Pegg, D.E. Marcel Dekker, New York.

Keeler, R., Swinney, J., Taylor, R. and Uldall, M. (1966). The problem of renal preservation. *British Journal of Urology* **38**, 653–6.

Keilin, D. (1959). The problems of anabiosis or latent life: history and current concept. *Proceedings of the Royal Society* B. **150**, 159–191.

Kirklin, J., Dawson, B., Davloo, R.A. and Theye, R.A. (1961). Open intracardiac operations; use of circulatory arrest during hypothermia induced by blood cooling. *Annals of Surgery* **154**, 796–76.

Langer, F. (1981). Bone and cartilage. In *Organ Preservation for Transplantation. 2nd. Ed.*, pp. 443–50. Edited by Karow, A.N. and Pegg, D.E. Marcel Dekker, New York.

Leibo, S.P. (1979). Fundamental cryobiology of mouse ova and embryos. In *The Freezing of Mammalian Embryos, CIBA Foundation Symposium 52*, pp. 69–91. Excerpta Medical, Amsterdam.

Matas, A.J., Sutherland, D.E., Steffes, M.W., Mauer, S.M., Lowe, A., Simmons, R.L. and Najarian, J.S. (1976). Hepatocellular transplantation for metabolic deficiencies: decrease of plasma bilirubin in Gunn rats. *Science* **192**, 892–94.

Mueller, F.O., Casey, T.A. and Trevor-Roper, P.D. (1964). Use of deep-frozen human cornea in full thickness grafts. *British Medical Journal* **2**, 473–5.

Morris, J.G. (1968). The kinetics of enzyme-catalyzed reactions. In *A Biologists' Physical Chemistry*, pp. 261–300, Edward Arnold, London.

Nei, T. (1968). Mechanisms of haemolysis of erythrocytes by freezing at near subzero temperatures. 11 Investigations of factors affecting haemolysis by freezing. *Cryobiology* **4**, 303–11.

Pegg, D.E., Jacobsen, I.A. and Halasz, N. (1982). *Organ Preservation: Basic and Applied Aspects*. MTP Press, Lancaster.

Pegg, D.E., Wusteman, M.C. and Foreman, J. (1981). Metabolism of normal and ischaemically injured kidneys during perfusion for 48 hours at 10°C. *Transplantation* **32**, 437–43.

Polge, C., Smith, A.U. and Parkes, A.S. (1949). Revival of spermatozoa after vitrification and dehydration at low temperatures. *Nature* **164**, 666.

Polge, C. (1980). Freezing of spermatozoa. In *Low Temperature Preservation in Medicine and Biology*, pp. 45–64. Edited by Ashwood-Smith, M.J. and Farrant, J. Pitman Medical, London.

Popovic, V., Popovic, P. and Karow, A.M. (1974). Hypothermia and hibernation. In *Organ Preservation for Transplantation*, pp. 11–22. Edited by Karow, A.M., Abouna, G.J. and Humphries, A.L. Little Brown and Co., Boston.

Pryor, W.A. (1976). The role of free-radical reactions in biological systems. In *Free Radicals in Biology Vol. 1*, pp. 1–49. Edited by Pryor, W.A. Academic Press, New York.

Rajotte, R.U., Warnock, G.L., Bruch, L.C. and Procyshyn, A.W. (1983) Transplantation of cryopreserved and fresh rat islets and canine pancreatic fragments: comparison of cryopreservation protocols. *Cryobiology* **20**, 169–84.

Rall, W.F. and Fahy, G. (1985). Ice free cryopreservation of mouse embryos at −196°C. *Nature* **313**, 573–5.

Rangel, E.M., Bruchner, W.L., Byfield, J., Dinber, A., Yakeishi, Y., Stevens, G.H. and Fonkalsrud, E.W. (1969). Enzymatic evaluation of hepatic preservation using cell stabilizing drugs. *Surgery, Gynaecology and Obstetrics* **129**, 963–72.

Rikukawa, Y. and Lindsey, E. (1968). Furosemide as a protective agent in renal ischaemia. I. Preliminary Investigations in Rats. In *Organ Perfusion and Preservation*, pp. 131–5. Edited by Normal, J., Appleton Century Crofts, New York.

Ross, H., Marshall, V.C. and Escott, M. (1976). 72-hour canine kidney preservation without continuous perfusion. *Transplantation* **21**, 498–501.

Rowe, A. and Lenny L. (1981). Red blood cells. In *Organ Preservation for Transplantation*, pp. 285–322. Edited by Karow, A.M., and Pegg, D.E. Marcel Dekker, New York.

Sacks, S.A., Petritsch, P.H. and Kaufman, J.J. (1973). Canine kidney preservation using a new perfusate. *Lancet*, **1**, 1024–8.

Sen, A.K. and Woddas, W.F. (1962). Determination of the temperature and pH dependence of glucose transfer across the human erythrocyte membrane measured by glucose exit. *Journal of Physiology* **160**, 392–403.

Smith, A.U. (1952). Behaviour of fertilized rabbit eggs exposed to glycerol at low temperatures. *Nature* **170**, 374–5.

Smith, A.U. (1970). Frostbite, hypothermia and resuscitation after freezing. In *Current Trends in Cryobiology*, pp. 181–208. Plenum Press, New York.

Taylor, M.J., Duffy, T.J., Hunt, C.J., Morgan, S.R. and Davisson, P.J. (1983) Transplantation and *in vitro* perifusion of rat islets of Langerhans after slow cooling and warming in the presence of either glycerol or dimethyl sulfoxide. *Cryobiology* **20**, 185–204.

Thaw, H.H., Forsberg, J.O., Del Maestro, R.F., Gerdin, B., McKenzie, F.N. and Afors, K.E. (1982). A free-radical approach to tissue injury. In *Organ Preservation: Basic and Applied Aspects*, edited by Pegg, D.E., Jacobsen, I.A., Halasz, N. MTP Press, Lancaster.

Thorgeirsson, G. and Robertson, A.L. (1978). The vascular endothelium – pathobiologic significance. *American Journal of Pathology* **93**, 803–48.

Trounson, A.O. and Conti, A. (1982). Research in human *in vitro* fertilization and embryo transfer. *British Medical Journal* 244–8.

Van Horn, D.L. and Schultz, R.D. (1977). Corneal Preservation: Recent advances. *Survey of Ophthalmology* **21**, (4), 301–12.

Watson, P.F. (1979). In *Oxford Reviews of Reproductive Biology*, Vol. 1, pp. 283–350. Edited by Finn, C. Oxford University Press, Oxford.

Weed, R.I., La Celle, P. and Merrill, E.W. (1969). Metabolic dependence of red cell deformity. *Journal of Clinical Investigation* **48**, 795–809.

Weiner, R.S., Tobias, J.S. and Yankee, R.A. (1976). The processing of human bone marrow for cryopreservation and reinfusion. *Biomedicine* **24 (4)**, 226–31.,

Wijk, J. Van Der, Sloof, M. Rijkmans, B. and Koostra, G. (1980). Successful 96- and 144-hour experimental kidney preservation: a combination of standard machine preservation and newly-developed normothermic *ex-vivo* perfusion. *Cryobiology* **17**, 473–77.

Wise, M.H., Yates, A., Gordon, C. and Johnson, R.W. (1983). Subzero preservation of mechanically-prepared procine islets of Langerhans: response to a glucose challenge *in vitro*. *Cryobiology* **20**, 211–18.

Zimmerman, F.A., Dietz, H.G., Kohler, Ch., Kilian, N., Kosterhom, J. and Scholz, R. (1982). Effects of hypothermia on anabolic and catabolic processes and on oxygen consumption in perfused rat livers. In *Organ Preservation: Basic and Applied Aspects*, pp. 121–6. Edited by Pegg, D.E., Jacobsen, I.A., Halasz, N. MTP Press, Lancaster.

14

Cryotherapy

C.J. Green
Surgical Research Group
MRC Clinical Research Centre

Fundamental considerations and general principles

Historical development

That the injurious effects of subzero temperatures could be usefully harnessed for the deliberate destruction of unwanted tissues or to render localized areas insensible has been known for many years. Indeed, the use of cold in medicine is probably as ancient as its use for storage. The simple observation that exposure to extreme cold led to frost-bite of extremities accompanied by loss of sensation must have been made by primitive hominids. There are many reports of its analgesic properties – for example, more than 2500 years ago the Ancient Egyptians recorded the value of cold for the treatment of trauma and inflammation, whilst in about 400 BC Hippocrates recommended ice and snow packs for the local relief of pain prior to surgery and in certain diseases of the bones and joints (Jones, 1931). The stories of such early physicians as Avicenna of Persia (980–1070), related by Gruner (1930), and Severino of Naples (1580–1656) as reported by Bartholin of Copenhagen (1661) reveal that cold proved useful for pre-operative analgesia. Later, in describing his experience of operating on soldiers during Napoleon's disastrous retreat from Moscow in 1812, the military surgeon Baron Larré made the important observation that amputations could be done painlessly and without haemorrhage if the limbs were covered with ice or snow during the operation (Larré, 1832). Indeed, such simple cooling with local icepacks was still in use for limb surgery as recently as 1941 (Allen, 1943).

The destructive effects of subzero temperatures have also long been recognized (Boyle, 1683). The earliest known experimental work was carried out in 1777 by John Hunter, who noted that local cooling of cocks' combs resulted in vascular stasis and tissue necrosis followed by excellent healing and recovery (Palmer, 1835). At about the same time, Spallanzini (1784) observed that some cells were able to survive extremely low temperatures. Cryotherapy then went through one of its fashionable periods during the nineteenth century and was promoted for both its analgesic and destructive properties by James Arnott (1797–1883). This unusual man devoted the last 50 years of his life to the advancement of cryotherapy and to a campaign against the dangers of inhalational anaesthesia and narcotics. Arnott used cold for such varied conditions as headache, erysipelas, neuralgia and cancer (Bird, 1949) and, in 1851 he described the treatment of ulcerating breast cancers by application of freezing mxture packs and claimed 'some improvement' in the patients' general condition (Arnott, 1848).

In 1890, a vacuum-insulated container was designed by Dewar for the storage of liquid air and the way was opened for further exploitation of freezing techniques. In 1899, Campbell White reported treating small cutaneous lesions with swabs dipped in liquid air and this simple technique is still used by some dermatologists. Solid CO_2 was used later for a similar purpose (Juliusberg, 1903), and this remained the most commonly employed technique until Rowbotham *et al.*, (1959) designed a cannula filled with CO_2 in alcohol to freeze gliomas.

The first sophisticated cryoprobe was developed by Cooper and Lee (1961) and consisted of a vacuum insulated tube through which liquid nitrogen was pumped until it underwent a phase change as it emerged to cool a probe to $-196°C$. In fact, this is the basic design of most modern probes and is correctly regarded as the springboard for modern cryosurgery. Subsequently, Amoils, who wanted a small easily controlled instrument for ophthalmic surgery, introduced an enclosed CO_2-expansion cryoprobe (Amoils, 1967) which reduced the temperature to $-50°C$. This model has since been converted to use nitrous oxide to achieve temperatures of $-89°C$ at its orifice.

Mechanisms of damage in cryotherapy

Cellular events during freezing and thawing

The physical and biochemical damage caused by freezing and thawing has been discussed in detail elsewhere in this volume. Most of the manoeuvres described to prevent this damage so that tissues can be stored are reversed in cryosurgery where the aims are to kill as efficiently as possible. It must, however, be emphasized that the mechanisms of damage have usually been studied in single cell systems, often in low *in vitro* concentrations, and rarely in tissues of any great bulk or mixed cell species. In cryosurgery, on the other hand, a crude assault is made on relatively large masses of tissue consisting of mixed populations of cells packed closely together, and to which blood returns after thawing. It is therefore likely that damage results from many factors (Grout and Morris, this volume).

The formation of intracellular ice has usually been considered the most damaging stress during freezing but in a solid tumour it is likely that extracellular ice will be equally disruptive to cells packed closely together.

Damage after thawing

Perhaps in the cryosurgical situation, the damage which occurs *after* thawing is as important as the injuries resulting from freezing. It is clear that freezing and thawing *per se* cannot be held entirely responsible for the extent of damage which develops after thawing (Fraser and Gill, 1967; Smith and Fraser, 1974). It is most likely that the microcirculation is compromised by freezing and that cells surviving non-lethal temperatures are later killed by post-thaw ischaemia (restriction of blood supply) and infarction.

What vascular changes are likely during cryosurgery? During the initial cooling, arterioles and venules constrict particularly at temperatures between +11°C and +3°C (Kreyberg, 1957). Although this has not been demonstrated, it is possible that the degree of vasoconstriction is dependent on cooling rate and, as vasoconstriction may be a physiological cryoprotective mechanism (Rothenborg, 1977), it could shield some tumour cells from death. What is more certain is that the circulation in a cryolesion shuts down very rapidly after thawing (Bellman and Adams-Ray 1956). The endothelium is particularly sensitive to freeze-thaw damage and a cascade of thrombogenic events is triggered with initial reflow of blood. This involves increased permeability of vascular walls, increased blood viscosity, loss of erythrocyte deformibility, extravascular oedema associated with lowered intracapillary hydrostatic pressure, sludging and a decrease in blood flow. Release of serotonin from platelets damaged by freezing, and activation of contact factors results in platelet aggregation and the formation of microthrombi.

Other mechanisms of damage may come into play immediately after thawing. Reports that partial oxygenation may be more damaging than complete ischaemia (Hossman and Kleihues, 1973) and that most ischaemic damage occurs during the early phases of recirculation (Hearse *et al.*, 1975) suggest that oxidative processes and toxic free radicals may also play a significant part in ischaemic damage. Normally, this potentially runaway process is controlled and scavenged by efficient enzyme systems such as superoxide dismutase which converts O_2 to H_2O_2, and catalase and glutathione peroxidase which reduces H_2O_2 to H_2O. In addition, a range of endogenous cellular anti-oxidants including vitamin E, glutathione, ascorbic acid, and selenium salts are important. It is possible that, after freezing the cellular activity of these scavengers will have been damaged thus allowing free radicals such as singlet oxygen, hydroxides and peroxides to accumulate to toxic levels during the immediate recirculatory phase.

The importance of post-thaw ischaemia has been demonstrated in a series of experiments (Le Pivert, 1981). In one study, using a transplantable carcinoma in rats, it was found that if frozen tumour cells were retransplanted immediately after thawing, they grew in another host but if transfer was delayed by 48 hours they did not grow.

It is also possible that auto-antibodies are generated in response to altered 'self' antigens and these may assist in killing not only the primary tumour but also metastatic cells.

Cell population effects

The ability of different cell populations to tolerate cold should be emphasized. When preserving tissues, solid organs or tumours containing mixed cell populations, it is difficult to achieve optimum cooling rates for survival of all species of cell; similarly, it is difficult to achieve cooling rates capable of killing all cells during cryotherapy.

Sensitivity to cold can also vary in cells of the same species, for example in rapidly dividing neoplatic cells water content is directly proportional to mitotic index which decreases toward the centre of a tumour. Hence, cells on the periphery which are young and dividing rapidly may be more susceptible to freezing injury.

It is therefore clear that cells should be exposed to as many different freezing and thawing hazards as possible to ensure their death. As shown later, this can only be achieved realistically by subjecting the whole tumour to several freeze-thaw cycles or by using a roving spray technique which will provide rapid fluctuations in tissue temperature. The importance of more than one freeze-thaw regimen has been demonstrated in several animal studies (Myers *et al.*, 1969). When the cure rates of virally-induced tumours in mice were compared after excision, or one, two or three freeze-thaw treatments, it was found that excision resulted in 100 per cent cure, but one freeze-thaw treatment cured only 30 per cent and even three treatments allowed 10 per cent regrowth. These workers also showed that cell suspensions from mice treated only once were capable of inducing tumours in 34 per cent of virgin mice whilst no tumours developed if the original tumour had been frozen three times. Furthermore, to illustrate just how difficult it is to kill all cells with existing cryotherapeutic instruments, they showed that probe tip temperatures of $-60°C$ were inadequate and 100 per cent cure rates were only achieved once the primary tumours had been subjected to three freeze-thaw cycles at a probe temperature of −180°C. It has also been found that a fast cooling rate of −260°C min^{-1} was more effective than −100°C min^{-1} (Neel *et al.*, 1971).

As cryotherapy is often used to kill unwanted tissues which may have been virally induced or are chronically ulcerating, it is worth emphasizing at this stage that viruses and bacteria are relatively resistant to low temperature. The possibility that viruses released from frozen tumours will survive and could then cause further malignant change cannot be discounted. Surgeons should not assume that lowering the temperature to −180°C sterilizes either the probe or the treated tissues; it certainly does neither. In a series of experiments (Green, unpublished data) in which chronically infected ulcerations were swabbed immediately after spraying with liquid nitrogen spray it was possible to grow gram-positive and gram-negative bacteria by normal plating techniques. Furthermore, when those colonized plates were also subjected to three 3-minute freeze-thaw cycles by liquid nitrogen spray, it proved possible to transfer thawed colonies to second plates and obtain subsequent regrowth.

Conclusions

To maximize kill and avoid 'escaped' malignant cells, liquid nitrogen should be used to achieve measured temperatures of at least −60°C beyond the limits of the tumour. Rapid freezing at cooling velocities more than −100°C min^{-1} followed by slow rewarming is likely to achieve the best results. The time of application, once the requisite ice-ball (see below) is formed, should be at least 3 minutes, and the freeze-thaw cycle shoud be repeated at least twice. In practice, these conditions may be difficult to achieve, and it is fortunate that post-thaw ischaemia probably completes destruction beyond what could theoretically be expected of freezing alone.

Physical production of the cryolesion

There are three methods which can be used to reduce the temperature of tissues to lethal levels during cryotherapy. The first techniques utilizes the cooling effect resulting from vaporization of a liquid, either within a metal probe tip or by spraying a liquid and vapour mixture directly onto a tissue. The second technique lowers temperature by the adiabatic expansion of a gas or liquid as it passes through a narrow orifice or valve under pressure – the Joule–Thomson effect. The third method makes use of the thermoelectric Peltier effect in which a direct current passing through a junction of dissimilar conductors leads to a fall in temperature. In practice, most modern cryoprobes are based on the first two phenomena. Nitrous oxide is the gas most commonly used in equipment based on the Joule–Thomson effect and this achieves a temperature of approximately −89°C at the orifice, whilst liquid nitrogen is at −196°C. However, the lowest temperatures achieved at the probe to tissue junction will be approximately −70°C and −185°C respectively.

When a cryosurgical probe or a spray has been applied to vascularized tissue for some time, an ice-ball forms. Cells close to the cold source are cooled to the lowest temperature *and* at the greatest velocity whereas those cells at the periphery of the ice-ball are necessarily at the freezing point of the tissue fluid, at approximately −2°C. Eventually, equilibrium will be reached between heat energy substracted and heat brought to the area by conduction and vascular convection, and a thermal profile which is approximating that illustrated in Fig. 14.1 will be produced (Zacharian, 1971). The sharp temperature gradient developed between the probe tip and the ice interface will be approximately 10°C mm^{-1}.

The *final* temperature of each cell within this ice-ball will therefore depend on its distance from the probe, the temperature of the refrigerant (hence its capacity for heat extraction), on the vascularity of the tissue, and on the geometry of the probe itself. However, before this occurs, the temperature at any point within the lesion will be a function of duration of application as well as of the other four variables. Generally, it appears that most, if not all, cells require cooling to below −20°C to ensure death and some tumour cells may escape even if cooled below −60°C. Hence, many cells in the outer isothermic shells of the lesion are likely to survive the immediate and direct effects of freezing.

As suggested earlier, the rate of cooling is important. If the probe tip is brought down to its final temperature rapidly, the rate of cooling of adjacent cells will be much faster than the rate of cooling of cells on the periphery of the cryolesion. Freeze substitution studies have confirmed these differential rates of cooling and demonstrated the presence of intracellular ice at the centre of the lesion and of extracellular crystals at the periphery (Whittaker, 1974). Cells close to the probe are therefore subjected to the most lethal conditions, whilst those cells further away may survive. It is thought that the velocity of cooling should be at least 100°C min^{-1} for maximum destruction.

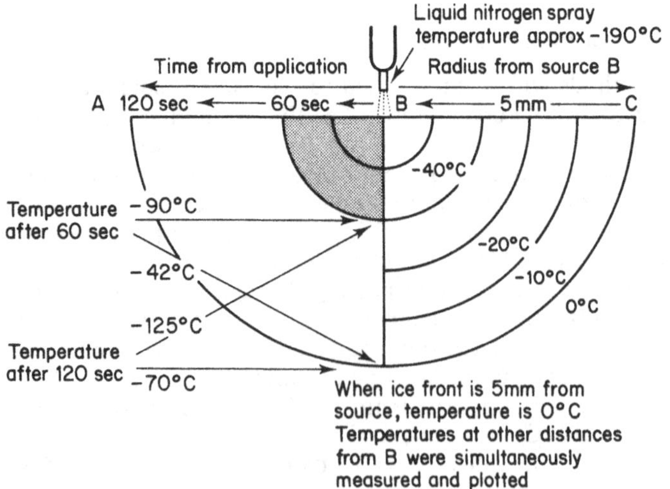

Fig. 14.1 Temperature profile in tissue subjected to a source of refrigerant at −190°C (After Zacharian, 1971).

In practice, this is only likely to occur to cells within 1mm of a probe at −190°C and cells within the remainder of the ice-ball will be cooled much more slowly.

What other factors should be considered? The *final* radius of the ice-ball will be limited by the probe temperature and by the vascularity of the tissue under attack. It will also depend on the diameter and efficiency of probe to tissue contact, or alternatively on the area over which refrigerant is sprayed. Clearly, the larger the mass, the slower the innermost cells cool. Although these innermost cells may eventually be cooled to low temperatures, the rate may be slow enough for them to remain supercooled and survive. This is not just a theoretical possibility as it has been demonstrated that normal and neoplastic cells can survive at the centre of a mass cooled to −253°C (Klinke, 1939).

The ice-ball characteristics vary still further during a second or subsequent freeze. Not only does the specific heat of the tissue alter but, more importantly, impaired microcirculation after the first freeze-thaw cycle allows a marked increase in cooling velocity at any given point in the ice-ball. Hence, if the freeze-thaw cycle is repeated 2–3 times, a cell at that point is likely to have been subjectd to several different cooling and rewarming velocities.

The shape of the cryolesion will depend on the shape of the probe or on the spray pattern adopted, and on the proximity of blood vessels traversing the tumour mass. As shown in studies using thermography to map different isotherms within a tissue (Bradley, 1977), the ice-ball is rarely hemispherical. Hence, the radius of the visible surface will not necessarily provide an accurate guide to the depth of the lesion. The depth is more likely to be about two-thirds of the surface radius *if* the probe or spray is applied at one point only. Multiple probes or a roving spray will produce very different lesions: with a spray, for example, it is possible to create a plaque of frozen tissue say 1mm thick but 5cm in diameter.

To ensure killing all malignant cells, the −60°C isotherm should theoretically be extended beyond the estimated periphery of the tumour for, although −20°C and −40°C have been considered lethal temperatures, work by Gage, (1979) suggests that this is not the case. It follows that the visible ice-ball must extend well beyond the tumour edge.

In general, cryosurgical units based on Freon, CO_2 or N_2O are only adequate for benign conditions and for blocking peripheral nerves in the treatment of intractable pain. At best, the refrigerant temperature of −89°C is only capable of reducing cells closest to the probe to about −65°C and this will cool only a very limited mass of tissue below the −20°C necessary to kill benign tumours. The extreme cold attained by liquid nitrogen is far more destructive and, of course, potentially more dangerous if used carelessly. The simplest approach to cryotherapy is to cool brass rods in liquid nitrogen in an insulated flask and then apply these directly to the tissue. Otherwise, this refrigerant can be used to cool specifically manufactured probes or it can be sprayed directly onto the tissue. A spray is the only way of effectively freezing large tumour masses and is preferred to multi-headed probes. It has the major advantage that the base of the tumour can be frozen first by spraying around a carefully delineated area thus rapidly immobilizing potentially malignant cells and shutting down the vascular supply early in the freezing sequence. This perhaps reduces the risk of cells being pushed along in advance of a cold-front and released into the systemic circulation. The remainder of the tumour within this frozen stockade can then be treated by spraying in ever-decreasing circles. This spray technique reduces the temperature of a cell mass far more quickly than is possible with a probe, yet by delicate application the surgeon is able to form solid frozen plaques on the surface without damaging deeper structures. However, great care must be exercised, and wherever large tumours are sited over other soft tissues, particularly glandular structures, deep temperatures should be monitored by several needle thermocouples placed at strategic intervals. If the mass to be treated overlies cartilage or bone, it can simply be frozen down to periosteal level without monitoring. Finally, to ensure that no cells escape destruction, the lesion should be subjected to at least two freeze-thaw cycles.

Local tissue reactions: the cryolesion

Burning pain is the immediate reaction to application of a cryoprobe, particularly to fingers, toes or soles of the feet. The pain becomes more intense the deeper the ice-ball develops below the surface, and becomes even more pronounced during thawing, lasting for several minutes afterwards.

The tissue may swell and become oedematous immediately after thawing, especially in very young or elderly patients. Occasionally, an urticarial response also develops immediately afterwards. Within 2–4 hours, bullae containing serous or serosanguinous fluid or even blood are formed. These may be large and alarming for the patient. As the blisters resolve an eschar forms and this may take anything from 7–21 days to separate away and reveal healthy granulation tissue below. Scarring is minimal and the cosmetic result is usually excellent. Hypertrophic scars develop only very occasionally but hypopigmentation lasting several months is commonly encountered in coloured people.

This general healing pattern is exaggerated by the more intense and prolonged freezing employed to cure malignant lesions. For example, blistering is more likely to be haemorrhagic, and cryolesions around the eye or over the forehead may swell enormously and become dependent especially in the elderly (Elton, 1977).

Prolonged freezing produces lesions which may take many weeks to heal and, in poorly vascularized areas such as the extremities of old people, delayed healing may occur. Typically, a serosanguinous exudate occurs in 2–3 days and oozing may continue for 2–3 weeks to be gradually replaced by a yellowish necrotic membrane. This can easily be mistaken for an infected wound but, in fact, secondary infection is rare. After resolution of the crust, milia are often seen in the wound and even pseudo-carcinomatous hyperplasia develops on rare occasions. This often looks like a recurrence of the original tumour but should not be confused with it. Permanent alopecia is to be expected after severe freezing. Contracted and hypertrophic scarring may occur but true keloid formation has never been reported.

Other complications may arise. Cryolesions on the nose and ears cause deficits or even perforations in the cartilage and tenderness may persist for months. Particular care should be taken with eyelid lesions as freezing may produce conjunctivitis and blockage of the naso-lacrimal duct. Headaches are commonly encountered after freezing lesions on the temple and forehead. Because large blood vessels are relatively resistant to freezing damage, post-operative haemorrhage is rare but it can occur. Similarly, if nerves such as the facial nerve are incorporated in a deep ice-ball, paralysis may result but this is nearly always temporary because nerve tissue regenerates well after freezing.

The variable susceptibility of different tissues is of practical importance. Bone, fascia, tendon sheath, perineurium and the walls of large blood vessels are relatively resistant. Hence, tumour masses may be killed leaving behind those structures either intact or with recoverable function. Similarly, although the axons of peripheral nerves will degenerate distal to a site at which an ice-ball has formed, regeneration occurs without neuroma because the collagenous matrix and perineurium remain intact.

The histological appearance of the cryolesion and its immediate tissue environment varies, of course, with the anatomical sites (Walder, 1971). In skin, the epidermis regenerates very rapidly, and the lesion is surrounded by a cellular infiltrate of macrophages, lympocytes and eosinophilic granulocytes. In the connective tissue only limited fibrinogenesis takes place and in a random manner so that parallel bundles are not laid down to cause contracture. There is minimal scarring in the oral cavity. In bone, cellular elements are killed but skeletal support is maintained. Large blood vessels seem particularly resistant to cold: it has been reported that even after freezing for 5–10 minutes at $-150°C$, no macroscopic changes were observed and normal blood flow was resumed after thawing. Nevertheless, at a microscopic level, intima did detach from the internal layer and elastic fibres became straightened so, clearly, there must be some danger of stenosis through intimal proliferation and thrombosis, and this is more likely to occur in veins. The effect of freezing nervous tissue, whether in the brain or in peripheral nerves, is also interesting. In the brain, a pale peripheral zone develops around an infarct. Myelin sheaths are swollen and ruptured, and vacuolation occurs in glial cytoplasm. Oedema invariably develops. Little damage is seen in the vessel walls and, although endothelial cells are damaged, there is minimal disturbance to the basement membrane, reticulin, collagen or elastin. Hence, haemorrhage by rhexis is rarely encountered. The damage within the lesion becomes more noticeable until about the twelfth day and it then heals with minimal shrinkage by fibrosis. In peripheral nerves, the axon degenerates away from the myelin sheath along with distortion of the myelin leaflets and damage to Schwann cells. The absence of scarring and neuroma formation allows rapid axonal regeneration at a rate of approximately 1.2 mm day^{-1}.

Systemic reactions: immunological factors

The body would be expected to respond to such an insult with a non-specific cellular infiltrate around the cryolesion just as it would to necrosis resulting from any other causative agency and, indeed, this is the case. However, this particular form of destruction raises some interesting theoretical and practical possibilities. First, consider the likely events after other forms of treatment. Whenever a primary tumour is excised or totally destroyed by cautery such that antigenicity cannot be expressed on the surface of cells, there is always the likelihood that, in removing what may be a slowly developing growth in near-equilibrium with the host and continuously emitting weak antigenic 'signals' latent cancer cells may be released from a last constraint. Proliferation may then be accelerated both locally and at metastatic sites. Another alternative would be radiotherapy. In this situation, the primary tumour is left to die *in situ* but the cell nucleus is the first to suffer, hence coding control over surface antigens will be lost.

On the other hand, when a tumour is subjected to extreme cold and left to die *in situ*, membrane lipoprotein complexes and hence antigen–antibody complexes and receptor sites are inevitably disrupted or altered, but are unlikely to be totally destroyed. Furthermore, the nucleus often appears relatively intact. For a limited time after treatment it is not unreasonable to expect antigenicity to be enhanced and enough antigen released systemically to produce a strong specific humoral response which hopefully will kill escaped cells of the same tumour species. The problem is finding a way of sustaining that response.

How does this speculation accord with clinical and experimental evidence? Many surgeons have reported regression of malignant tumours beyond their expectations – either disappearance of a tumour mass beyond the limits of the observed ice-ball, or regression of secondary nodules which were separate from the main primary and which had not been frozen. It is easy to dimiss these observations as anecdote but their potential importance merits serious investigation in experimental models.

A review of the literature suggests that humoral responses do occur in response to the freezing of *some* tumour species, but that response is relatively short lived (Faraci *et al.*, 1975). For example, in experimental mice with induced sarcomas, those treated by cryotherapy withstood secondary challenge significantly better than those in which the tumours were ligated or excised (Neel *et al.*, 1973). The cytotoxic activity of lymphocytes and sera taken from mice with cryosurgically treated tumours was significantly greater than that from tumour-bearing or tumour-amputated mice *but* this activity was elevated for only 12–19 days after treatment (Faraci *et al.*, 1975). On the other hand, in experiments using a different tumour system, circulating antibodies could not be detected after cryosurgery (Grace, 1977). The amount of tissue frozen may also be important. In rats with two transplanted tumours in which one was treated by cryosurgery, it was found that if a large mass of frozen tissue was left *in situ*, then the other tumour either regressed very slowly or not at all, whereas when only a small mass was left behind, regression of the other tumour was very rapid indeed (Blackwood and Cooper, 1972). The specificity of the response is a moot point. Antibody production of a non-specific nature has been demonstrated by fluorescent staining following freezing damage inflicted on rat livers (Li *et al.*, 1977). Similarly, in human patients, a comparison of five serum proteins before and after traditional prostate surgery and cryosurgical treatment indicated a wide variation in immunoglobulin response, particularly in IgG, and suggested a marked immunogenic response to cryosurgery (Drylie *et al.*, 1968). More recently, it has been claimed that in a rabbit gastric tumour model freezing induced a specific immunity which was transferable to other animals using lymphocytes to establish passive immunity, and that enhancement was transferrable to another body by injection of serum (Yamasaki *et al.*, 1978). In the author's laboratory, pilot experiments have been carried out to try to demonstrate specificity, using either transplanted tumours or weakly antigenic skin allografts differing only at a single histocompatibility locus: in each case there have been significant immunological responses in cryotreated groups expressed either by lower rates of tumour metastasis, or by enhanced rejection of secondary allografts and raised cytotoxic activity but these were inconclusive for specificity (Green, unpublished data).

It is too early to draw conclusions as in so much experimental tumour immunology the results are often contradictory and it is difficult to extrapolate from mouse to man. However, it is felt that cryotherapy does elicit an immune response in the host and that this could be exploited with adjuvant manoeuvres and by using frozen primary tumour homogenates as sequential vaccine injections, to attempt to boost host immunity over a long period.

Surgical applications

General considerations

Cryotherapy has now been used extensively in humans and other animals both as an agent of destruction and for discrete blockade of peripheral nerves to prevent post-operative pain and relieve chronic intractable pain. Any assessment of its value must be made against conventional excision surgery, electrocautery, radiofrequency lesion making, laser treatments, radiotherapy and, of course, conservative treatment. As an analgesic, cryotherapy has to be compared with blockade using injectable local analgesics and with other means of deliberately interrupting axons by surgical division, crushing, or lytic agents such as alcohol or phenol. Comparisons must be made of the improvement in early return of patient mobility, subjective decrease in pain perception and decreased requirement for systemic analgesics such as morphine.

Before considering the specific areas of human and veterinary surgery in which cryotherapy has proved useful it is worth examining the general advantages commonly ascribed to it. First, it is often claimed that general anaesthesia can be avoided because intense cold acts as a local analgesic. This statement may need modifying for many human patients do indeed report only a tingling sensation and slight pain if they have been given weak analgesics by mouth some 30 minutes before application but other patients require high dosages of pethidine and inhalation of nitrous oxide–oxygen gas mixtures, and a few need a general anaesthetic, particularly when extensive freezing of tumours involving the external genitalia are to be carried out. As other animals are unlikely to tolerate any degree of pain with stoic indifference, they should at least be given neuroleptanalgesics, and for some cases light surgical anaesthesia is required. For example, fentanyl–fluanisone narcosis may be used in canines and most other mammals, xylazine in equines and bovines, and ketamine catalepsis in cats, birds, and reptiles (Green, personal data). It is true that once the pain (often intense) associated with thawing has subsided then post-operative pain is rarely a problem, even though the cryolesion is superficially similar to a burn and results in blistering. Indeed, because of destruction of local nerve receptors, palliation of chronic pain, particularly in cases of extensive and incurable cancer is often the strongest indication for cryotherapy. This is particularly valuable in elderly patients who are poor anaesthetic risks and for whom hospital admission can be avoided. Secondly, statements that post-surgical bleeding is rarely encountered because small blood vessels to the lesion contract and thrombose are only partially correct – it is possible for large vessels within a tumour to start bleeding 30–60 minutes after freezing at a time when post-operative attention may have been relaxed. Thirdly, healing is usually excellent by second-intention granulation with a minimum of cosmetic distortion through scarring, but occasionally the lesion becomes infected and healing is then delayed. However, colour changes in skin and hair will occur and these are most noticeable in pigmented people. Finally, biopsy of the frozen lesion is possible without anaesthesia or suturing, and histological interpretation is satisfactory after a piece of solid frozen tumour has been simply removed by scalpel or bone forceps.

Surgical specialties

Dermatology and plastic surgery

Some workers consider cryotherapy the method of choice for adenoma sebaceum, basal cell epitheliomas, dermatofibromas, warts, keratoses (actinic, arsenical, radiation and seborrheic), verrucas (plantar and vulgaris), and valuable for angiomas, carbuncles, epitheliomas, fibromas, granuloma annulare, herpes simplex and herpes zoster, keloids, kerato acanthomas, leishmaniasis, lupus vulgaris, lupus erythematosus, nevus, pustules, verrucae, acne vulgaris, myxoid cysts, haemangiomas and epitheliomas of the eyelids (Zacharian, 1977). Pre-malignant and malignant tumours, including melanotic freckle (Dawber and Wilkinson, 1979) and even metastatic melanomas (Brietbart, 1981) have also been treated, and excellent cure or palliation of the condition claimed. Clearly, these cases must be carefully selected.

Perhaps the safest conclusion to draw from the many reports in the literature is that cryotherapy is a useful tool in the management of benign and malignant conditions of the integument to be used instead of, together with, or after, excision surgery, radiotherapy and chemotherapy. It is particularly valuable where a good cosmetic outcome is required, and for sites such as the forehead, nose, ears, eyelids and over bone and cartilage which are otherwise difficult to manage with irradiation.

Gynaecology

Cryotherapy has been used for some years for the treatment of benign cervical lesions (Lortat-Jacobs and Solente, 1930). Carbon dioxide snow has been used to treat a series of 325 patients with chronic cervicitis and the technique reported to be superior to the other methods then available: follow-up at 2 years revealed a 70 per cent cure rate after a single treatment (Weitzner, 1940). Even better results were obtained with a specially designed probe utilizing Freon; 58 to 60 patients responded satisfactorily (Hall, 1942). However, cryotherapy did not become really popular until a liquid nitrogen probe had been designed and proved suitable for neoplastic conditions (Cahan, 1967; Townsend *et al.*, 1967). Since that time, cryotherapy has been used extensively for precursor and invasive neoplastic conditions of the vulva, vagina, cervix and endometrium (Chamberlain, 1975) but is probably best restricted to benign conditions such as endocervical polyps, condyloma acuminata and benign leukoplakia. When compared with electrocautery (Ostergard and Townsend, 1969), it proved quicker and more acceptable to the patients and there is little evidence that it causes stenosis or interferes with subsequent pregnancy (Chamberlain, 1975). In another series the immediate cure rates were very much better with cryotherapy than with other modalities (Ostergard *et al.*, 1969). However, for cases of cervical malignancy it should only be used as palliation in otherwise inoperable cases since residual growth in the lymphatics or cervix have been reported following cryotherapy (Crisp, 1972).

A cryoprobe passed through the cervix has also been used to destroy endometrium to cure dysfunctional uterine haemorrhage (Schenker and Polishuk, 1972) and has since been adopted as a reasonably successful procedure. Attempts as passing cryoprobes down a laparoscope have been less successful however in closing fallopian tubes to sterilize women or in treating endometriosis.

It can be concluded that cryotherapy is useful for treating benign lesions and dysfunctional uterine haemorrhage but can only be a palliative in otherwise inoperable malignancy.

Neurosurgery

In the pioneering work of Fay (1940), a primitive cooling cannula was implanted into the substance of certain malignant brain tumours. Various freezing mixtures were passed to lower the temperature of the area to modest subzero levels and histological examination revealed destruction of tumours adjacent to the probe but tumours on the periphery were unaffected. Later, an improved probe consisting of a double-lumen cannula, through which concentrated saline solutions circulated, was designed and used in a series of malignant brain tumours (Rowbotham *et al.*, 1959). The limitations of this treatment were again illustrated by histological evidence of necrosis where the temperature had fallen to $-20°C$ but neoplastic tissue surviving peripherally.

It was not until cryoprobes were insulated along their length by silk, Teflon or by vacuum insulation (Tytus and Ries, 1961; Cooper and Lee, 1961) that more effective coolants such as Freon 12, carbon dioxide or liquid nitrogen could be used. Indeed, development of these probes and the ability to produce a measured temperature of −196°C with liquid nitrogen opened the way for cryotherapy to spread into many other disciplines.

Cryosurgery has since been used for treatment or palliation in several conditions of the central nervous system (CNS). First, it is valuable in making deliberate brain lesions. Soon after Tytus and Ries (1961) had demonstrated in laboratory animals that discrete lesions could be created in the brain, Cooper and Lee (1961) reported that a similar probe could be used for stereotactic thalamectomy in treating Parkinson's disease. The technique has since been used successfully in a large series of patients at the same treatment centre; relief of tremor and rigidity has been claimed in 90 per cent of cases, with only a 1 per cent mortality and 0.25 per cent permanent hemiplegia risk after freezing the ventrolateral nucleus (Cooper, 1969). It has also been reported that the improvement in tremor is stable in 40 per cent of patients (Ojemann and Ward, 1971). Although patients with primarily unilateral involvement are most likely to benefit from ventrolateral thalamotomy, more than half those with severe bilateral symptoms can expect improvement in contralateral tremor and rigidity (Markham and Rand, 1963). Thalamotomy has little effect on bradykinesia and none on the bulbar and autonomic symptoms (Rand, 1981).

In another series of 57 patients where previous treatment of the ventrolateral nucleus of the thalamus had failed, unilateral and bilateral lesions have been created in the pulvinar area of the thalamus to treat such clinical syndromes as dystonia musculorum deformans, cerebral palsy, spastic hemiplegia following cerebrovascular accidents or trauma, and intractable pain (Cooper *et al.*, 1973). In some patients, the treatment resulted in dramatic relief of spasticity along with a return of motor functions, and the treatment was attended by a minimum of complications.

Perhaps the greatest advantage of cold lesion making in the CNS over chemical, mechanical and thermal methods is that a preliminary assessment of the projected lesion sites can be made. Temporary modest temperature reduction abolishes symptoms if the site is ideal. Conversely, it produces unwanted side-effects which get worse as the freeze progresses but are reversed when the probe is rewarmed. It has been suggested that the cryolesion is superior to other techniques because of reversibility, reproducibility, sharp definition, avascularity, controlled variability in size, safety, simplicity and speed of application (Cooper, 1962a, b).

Treating patients with Parkinsonian symptoms with L-dopa for 12–18 months initially and then using cryothalamotomy in those whose tremor is not improved and then continuing L-dopa medication as appropriate is now recommended (Rand, 1981).

Discrete cryolesions have also been made in the brain for the treatment of psychiatric disorders (Kelly *et al.*, 1973; Richardson, 1973) as alternatives to heat or radioactive yttrium. Richardson (1973) using a N_2O cooled probe achieving temperatures of −70°C reported an improvement in 67 per cent of 40 patients without complications arising.

It has also been used with varying success for destroying brain tumours. Two methods for killing large neoplasms were originally employed. Where the tumours were relatively inaccessible, they were destroyed by a semi-freehand stereotactic method but where they were on or near the brain surface, a large cryoprobe was used to freeze the tumour (Cooper and Stellar, 1963). It was claimed that attachment to the probe was a useful aid to traction allowing a clearly defined plane of cleavage between tumour and healthy tissue, with minimal bleeding from the tumour itself. Others agree that this combined freezing and excision technique has a place when removing small vascularized surface tumours (Richardson, 1975). However, a controlled jet of liquid nitrogen directed at the tumour via a fine nozzle is claimed to provide a quicker and safer congelation of an en plaque menangioma accompanied by rapid coagulation of surface haemorrhage and minimal damage to the surrounding brain (Hamlin, 1971).

Deeply-placed brain tumours in the thalamus, third ventricle, pituitary and pineal regions have also been destroyed by combining stereotactic biopsy and cryoprobe technique in which three freeze-thaw cycles were employed at temperatures down to −80°C (Conway, 1973).

How does this compare with radiofrequency lesion making or laser? Rand (1981) feels that cryotherapy is of no advantage over other procedures in the primary management of benign or malignant brain tumours with the exception of pituitary tumours but concedes that it is a useful adjunct, either by taking advantage of the adhesive property of the probe for retraction or for destroying residual tumour after excision.

Finally, cryotherapy has been used for treating migraine, to produce vascular lesions, and for destruction of the pituitary.

The observation that injury to the sphenopalatine area during a nasal operation produced prolonged relief in a case of hemicranical migraine, (Cook, 1973) led to freezing of the region of the occipital artery, the superficial temporal artery and sphenopalatine arteries in a series of 322 migraine cases. The probe had a tip temperature varying from −120°C to −160°C and it was not placed in contact with the artery but simply positioned near it. It was noted that pain or headache returned during the initial freeze but decreased until almost abolished after repeated freeze-thaw cycles. The author reported that approximately 45 per cent of the patients had considerable relief from their migraines 6 months later, whilst 13 per cent enjoyed complete relief. Only speculation can be made concerning the reason for this as freezing was unlikely to cause permanent changes in the vessel walls themselves. It is more likely that intra- and extra-arterial sympathetic supply to the vessels degenerates after freezing and this would tally with observations that central effects of pain could be produced during the freeze and that stellate block relieved symptoms in cold-induced paresthesia of the upper limb.

When blood vessels are frozen they are not usually occluded on rewarming but enthothelial detachment and thickening of the walls leads to stenosis and even complete obliteration of the lumen after several weeks. This property was utilized in treating a series of 20 arteriovenous anomalies or angiomas (Walder *et al.*, 1970). After multiple freeze-thaw applications with an open cryogenic technique, vessels appeared completely blocked at the lesion site in 13 cases examined by serial follow-up angiography, and the remainder were reduced in size. In subsequent reports (Walder, 1972; Peeters and Walder, 1973) more details of the technique were given and it was concluded that the best results were likely if cryotherapy was combined with orthodox procedures.

Adenohypophyseal hyperfunction and pituitary tumours have been treated over many years by total or partial excision, by external irradiation and by transphenoidal implantation of radioactive yttrium seeds. However, as each of these techniques has some disadvantages, Cooper (1963) attempted cryogenic destruction of the gland by passing a cryoprobe through a transfrontal approach. This also had limitations in that the pituitary could only be partially destroyed and there was a distinct hazard of damage to the frontal lobe, as well as to the optic nerves and chiasma during insertion (Cooper and Hirose, 1966).

For this reason, other workers have used a transphenoidal approach, placing the cryoprobe either in the midline of the sella turcica or to the right or left at accurately measured distances by stereotactic instrumentation (Rand, 1964). The large number of cases dealt with provides adequate material for review and evaluation (Rand, 1981).

The signs and symptoms of pituitary hypersecretion including acromegaly and Cushing's syndrome have been controlled by cryotherapy (Maddy *et al.*, 1969; Levin *et al.*, 1974). Initially, temperatures of −50°C to −70°C were used but later these were reduced to −180°C for 8–10 minutes. It has also been demonstrated that cryolesions made 3–4mm from the midline were more effective than those made in the midline (Conway and Garcia, 1970).

Pituitary gland cryoablation has also been employed successfully for the palliation of endocrine-dependent lesions such as metastatic and prostatic cancer and progressive proliferative diabetic retinopathy (Dashe *et al.*, 1966; Bleasel and Lazarus, 1965). Early failures with freezing were attributed to insufficient reduction in temperature (Rand, 1968) and therefore temperatures down to −180°C are employed to obtain maximum kill.

It may be that significant long-term improvement will be seen in approximately only 50 per cent of cases (Franco *et al.*, 1973), although others feel that the lower complication rate with cryotherapy renders it the method of choice in selected cases (Richardson, 1975).

Oral surgery

Generally, cryolesions in the oral cavity heal very well and cryotherapy is useful for several conditions. It affords better results in treating white patches and pre-malignant leukoplakia (Poswillo, 1975), and is especially valuable for diffuse lesions involving the palate, cheek, tongue, buccal floor and vermillion border of the lip. As in other specialties, however, cryotherapy should only be used as a palliative for carcinoma and then only when other factors contra-indicate orthodox treatment. Nevertheless, where all else fails, it may provide tumour control, debulking, good relief of intractable pain and control of chronic haemorrhage (Gage, 1969; Leopard and Poswillo, 1974).

Benign lesions such as papillomas, fibroepithelial polyps, giant cell and pregnancy epulids, mucus-retention cysts and pseudopyogenic granulomata have all been effectively treated with cryoprobes. Cryotherapy has also proved valuable for vascular naevi, haemangiomas, fibroangiomas and lymphangiomas, as well as haemostasis in bleeding diatheses. Patients with painful ulcers such as those which occur in aphthous ulceration, herpes simplex, lichen planus and Behcet's disease can at least be relieved of their pain after short periods of freezing.

Otorhinolaryngology

Perhaps the most profitable use of cryotherapy in this speciality stems from its haemostatic property. For example, vascular lesions of the upper air passages may be removed, particularly capillary haemangiomata of the nose or supraglottis. It has also been used successfully for the control of severe epistaxis (Bluestone, 1965) and for removal of nasal polyps (von Leden and Rand, 1967). Cryonecrosis is also an acceptable method for destroying diseased tonsillar tissue (Hill, 1965). For malignant tumours of the head and neck, cryotherapy is probably not suited to primary treatment but it is certainly valuable for palliation (Hill, 1965; Holden and McKelvie, 1972).

Cryotherapy has proved useful for the selective destruction of the vestibular neuroepithelium in treating Menière's disease. In one series (Wolfson and Cutt, 1971), long-term follow-up for 1–4 years revealed that vertigo was either completely eliminated or substantially reduced in 76 per cent of the cases, and hearing was not harmed. Facial nerve complications occurred in a few patients but all recovered within 6 weeks.

Orthopaedic surgery

In a series of experimental studies (Kuylenstierna *et al.*, 1980) it has been shown that viable osteocytes disappear from cortical bone 2–7 days after freezing the whole intact bone but marrow may survive. Nevertheless, the earlier work using dogs' femora and mandibles (Gage *et al.*, 1966) suggested that neoplastic and other pathological tissue could be eradicated from bone by freezing without radical excision. Cryotherapy has since been used for treating giant cell tumours (Marcove *et al.*, 1973) and chondrosarcoma (Marcove *et al.*, 1977), and may find a place in treating bony lesions in the maxillo-facial region (Bradley, 1978).

General surgery

Cryotherapy is commonly used for benign and locally invasive tumours such as leukoplakia, basal cell carcinoma and Bowen's tumour. The cure rates are comparable to those of other modalities and cosmetic results are usually superior. In Gorlin's syndrome, it is probably the treatment of choice (Lloyd-Williams and Holden, 1975). In malignant conditions, it is usually a last resort in all cases except squamous cell carcinoma of the skin where it may be the primary treatment. As a palliative, it provides pain relief, marked reduction in the bulk of tumour and secondary infection, and a decrease in serosanguinous discharge.

It is particularly valuable fρr anorectal conditions such as second and third degree haemorrhoids, anal warts and rectal polyps, and has more recently found favour for the radical treatment of rectal carcinoma. Complications such as anal sepsis, fistulea, bleeding, incontinence, stenosis or perforations are only occasionally encountered afterwards (Osborne *et al.*, 1978). It is also useful for treating pilonidal disease (O'Connor, 1979).

Although not widely accepted for thoracic surgery, cryotherapy has proved useful for primary bronchopulmonary lesions (Carpenter *et al.*, 1977) and pulmonary metastases (Uhlschmid *et al.*, 1979).

Urology

Although there are scattered reports on the use of cryotherapy for genital warts in men (Green *et al.*, 1978), bladder neoplasms (Reuter, 1972) and verrucous carcinoma of the penis (Hughes, 1979), it is only in the management of prostatic disease that it has been extensively employed (Soanes *et al.*, 1966; Peterson *et al.*, 1978). The transurethral approach has been recommended for benign hyperplasia (Hansen and Wanstrup, 1973) but an open perineal incision is preferred for prostatic cancer (Peterson *et al.*, 1978). Whether or not an immunological response is elicited after cryotherapy of the prostate is still debatable (Jones, 1974) in spite of the many publications favouring this hypothesis (Ablin *et al.*, 1973).

Ophthalmic surgery

Cryotherapy is used extensively in ophthalmology for three main reasons. Firstly, the fact that a cold probe adheres strongly to moist tissue enables it to be used for holding the lens during cataract operations, for holding the globe during enucleation for choroid melanoma, and for holding and haemostatis during surgery for orbital haemangiomata. Secondly, the cryoprobe provides a way of locally and selectively damaging intra-ocular tissues without damaging the scleral wall of the globe. It is in routine use for retinal detachment surgery and is used prophylactically when retinal holes have appeared prior to detachment. Cryosurgery is also very effective for destroying the small peripheral haemangiomata occurring in the Von Hippel Lindan syndrome, and for destroying part of the ciliary body to treat thrombotic glaucoma (Fison, personal communication). Thirdly, cryotherapy is valuable for periocular conditions such as trichiasis and eyelid tumours. Eyelash follicles are destroyed at temperatures between −15°C and −20°C with little permanent damage to other structures except melanocytes. Cryotherapy cannot cure entropion but can be used in conjunction with surgery to treat distichiasis. As an alternative to excision surgery or radiotherapy for eyelid basal or squamous cell carcinomas, cryotherapy is valuable for small lesions near the lacrimal puncture since this structure and its canaliculus are remarkably resistant to freezing yet easily occluded by other therapy. It is also very effective for treating squamous papillomas and seborrhoeic keratoses in this region (Collins, personal communication).

Cryoanalgesia

General considerations

When peripheral nerve fibres are progressively cooled, a conduction block similar to that produced by local analgesics develops. This blockade is complete at temperatures between 0°C and 10°C but prolonged interruption of conduction after thawing does not occur until the nerve has been frozen at temperatures between − 5°C and − 20°C (Evans *et al.*, 1980). The axons are not killed predictably until cooled to − 20°C or below and recovery then becomes a function of axonal regeneration. The axons and myelin sheaths disintegrate but, although the connective tissue is devitalized, the perineurium and epineurium remain intact to provide guides for subsequent axonal regeneration. So long as axonal growth is constant, then the reappearance of sensory and motor activity becomes a function of the distance between cryolesion and end-organ (Evans *et al.*, 1980). As this has been applied clinically to the relief of both acute and chronic pain, it is now generally termed cryoanalgesia.

Clinical applications

The fact that such low temperature blockade is both prolonged and reversible offers great advantages over other methods of nerve destruction such as section or crushing. The technique has proved popular in several clinical situations.

Early reports of the use of cryoanalgesia for the management of chronic intractable pain claimed that 81 per cent of patients studied gained some benefit from the procedure (Lloyd *et al.*, 1976). The median duration of pain relief was 11 days but some patients gained prolonged relief for up to 224 days. The cryoprobe was applied via an open or by a percutaneous approach, and it is probably significant that the percutaneous was less successful - in fact as these patients only had temporary relief, it is likely that the nerve was incompletely damaged through failure to reduce the temperature below − 20°C. This group has used the technique for patients with low back pain, facial nerve pain, tic doloreux, atypical facial neuralgia and malignant neuralgia. In cases where the nerve was first exposed, symptomatic relief was achieved in more than 90 per cent of patients (Barnard *et al.*, 1978). Interestingly sensory function often returned long before the reappearance of intractable pain.

The post-operative value of cryoanalgesia has now been demonstrated (Katz *et al.*, 1980; Glynn *et al.*, 1980). Patients undergoing thoracotomy who received intercostal nerve freezes by direct application at the termination of surgery had less post-operative pain than control groups and their requirement for analgesics was markedly reduced. In another study (Wood *et al.*,

1979), ilio-inguinal nerve blocks were performed on patients admitted for herniorraphy, and muscle pain was relieved.

It can be concluded that freezing is effective if applied to the exposed nerve. It produces a reversible blockade, does not lead to neuroma formation, does not cause neuritis and can be safely repeated without ill effect once sensation has returned (Evans, 1981).

Veterinary applications

Cryotherapy has become widely accepted in veterinary practice since 1970 (Borthwick, 1970) and is used in all classes of animal for similar conditions to those outlined above (Rickards, 1981). In our laboratory cryotherapy has been of value in treating small exotic birds such as parakeets, reptiles (including tortoises), lizards and snakes, and in small mammals such as hamsters and gerbils where application of a probe to kill neoplastic tumour is often more convenient and carries less risk of bleeding than a scalpel.

The list of skin conditions which have been treated by cryotherapy includes both benign and malignant neoplasms. Rodent ulcers, lick granulomas, interdigital cysts, chronic dermatitis accompanied by dermal hyperplasia, chronic ulcers, feline eosinophilic lip granulomas, papillomas, equine sarcoid (Lane, 1977), tumours of the eyelid, equine cutaneous habronemiasis (Migiola *et al.*, 1978) and anal furunculosis in dogs (Borthwick, 1971; Lane and Burch, 1975) have each yielded to destruction by low temperature. More recently, malignant tumours such as periocular squamous cell carcinomas of cattle (Farris and Fraunfelder, 1976) and horse (Hilbert *et al.*, 1977), melanocytic tumours, and squamous cell carninomas of the feline nose (Rickards, 1981) have been treated and good results claimed.

In the buccal cavity, cryotherapy has been found to be effective in the treatment of the chronic ulcerative stomatitis commonly encountered in cats and snakes, and also in the treatment of benign tumours such as polyps or epulids in dogs and cats (Green *et al.*, 1977; Green, 1978). Oropharyngeal malignancies are particularly difficult to treat in dogs and rapid effective treatment of the primary tumour is essential to avoid extensive local invasion of bone. In a recent series, (White, 1981 personal communication) cryotherapy was preferred to fractionated radiotherapy, and adamantinomas, squamous cell carcinomas and fibrosarcomas carried the best prognosis.

In horses, exuberant granulation tissue is a common sequel to wounds of the lower limb, and cryotherapy is the method of choice for these cases (Farris, 1981). Squamous cell carcinoma of the penis is also quite common in this species and cryotherapy can be recommended for this also. Cryoanalgesia has been employed in horses (Farris, 1981) and a limited number of neurectomies have been carried out successfully by freezing the exposed palmar digital nerves for relief of chronic pain. As this has been recommended elsewhere (Tate, 1977), it is clearly a valuable technique which should avoid the problems of neuroma formation associated with conventional surgical neurectomy and probably deserves more widespread use.

Conclusions

If used properly and only in carefully selected cases, cryotherapy is clearly a valuable tool for human and veterinary surgery. Unfortunately, it is often maligned as inadequate when in fact the operator has not taken the trouble to learn the techniques properly. All too often it is apparent that the probe has been unable to reduce the temperature sufficiently at a high enough velocity and then the operator has failed to apply it for long enough or in repeat cycles. It must be stated firmly that equipment cooled by nitrous oxide will produce a superficial lesion which cannot possibly cope with malignant lesions. Liquid nitrogen cools a probe far more efficiently. Liquid nitrogen used as a spray through fine needles is even more flexible and will produce very accurately defined lesions controlled both for area and depth. In this laboratory sprays are used virtually to the exclusion of other techniques for all surface work.

References

Ablin, R.J., Soanes, W.A. and Gonder, M.J. (1973). Elution of *in vivo* bound antiprostatic epithelial antibodies following multiple cryotherapy of carcinoma of prostate. *Urology* **2**, 276–9.

Allen, F. (1943). Refrigeration anaesthesia and treatments. *Anaesthesia and Analgesia* **22**, 264–73.

Amoils, S.P. (1967). The Joule-Thompson cryoprobe. *Archives of Ophthalmology* **78**, 201–207.

Arnott, J. (1851). *The treatment of cancer by the regulated application of an anaesthetic temperature*, pp. 1–32. Churchill, London.

Arnott, J. (1848). On severe cold or congelation as a remedy of disease. *London Medical Gazette* Dec 1. 1848.

Barnard, J.D.W., Lloyd, J.W. and Glynn, C.J. (1978). Cryosurgery in the management of intractable facial pain. *British Journal of Oral Surgery* **16**, 135–42.

Bartholin, T. (1661). "*De Nivis Uso Medico*", pp. 132–3. Hanbold, Copenhagen.

Bellman, S. and Adams-Ray, J. (1956). Vascular reactions after experimental cold injury. *Angiology* **7**, 339–43.

Bird, H.M. (1949). James Arnott M.D. 1797–1883. *Anaesthesia* **4**, 10–17.

Blackwood, C.E. and Cooper, I.S. (1972). Response of experimental tumor systems to cryosurgery. *Cryobiology* **9**, 508–15.

Bleasel, K. and Lazarus, L. (1965). Cryogenic hypophysectomy. *Medical Journal of Australia* **2**, 148–50.

Bluestone, C.D. (1965). Intranasal freezing for severe epistaxis: preliminary report of an experimental and clinical study. *Transactions of the American Academy of Ophthalmology and Otolaryngology* **69**, 310–17.

Borthwick, R. (1970). Cryosurgery in veterinary practice: a preliminary report. *Veterinary Record* **86**, 683–6.

Borthwick, R. (1971). The treatment of perianal sinuses in the dog by cryosurgery. *Journal of the American Animal Hospital Association* **7**, 45–71.

Boyle, R. (1683). "*New Experiments and Observations Touching Cold*". Robert Davis, London.

Bradley, P.F. (1977). Thermography as an aid to cryosurgery. *Acta Thermographic* **2**, 83–90.

Bradley, P.F. (1978). Modern trends in cryosurgery of bone in the maxillofacial region. *International Journal of Oral Surgery* **7**, 405–15.

Brietbart, J. (1981). Personal communication.

Cahan, W.G. (1967). Cryosurgery of the uterus: description of technique and potential application. *American Journal of Obstetrics and Gynecology* **99**, 138–55.

Carpenter, R.J., Neel, H.B. and Sanderson, D.R. (1977). Cryosurgery of bronchopulmonary structures. An approach to lesions inaccessible to the rigid bronchoscope. *Chest* **72**, 279–84.

Chamberlain, G. (1975). Cryosurgery in gynaecology. *British Journal of Hospital Medicine* **14**, 26–37.

Conway, L.W. (1973). Stereotaxic diagnosis and treatment of intracranial tumours including an initial experience with cryosurgery for pinealomas. *Journal of Neurosurgery* **38**, 453–60.

Conway, L.W. and Garcia, J.H. (1970). Cryohypophysectomy. Postmortem findings in 16 cases. *Journal of Neurosurgery* **32**, 435–42.

Cook, N. (1973). Cryosurgery of migraine. *Headache* **12**, 143–50.

Cooper, I.S. (1962a). A cryogenic method for physiologic inhibition and production of lesions in the brain. *Journal of Neurosurgery* **19**, 853–838.

Cooper, I.S. (1962b). Cryogenic surgery of the basal ganglia. *Journal of the American Medical Association* **181**, 600–604.

Cooper, I.S. (1963). Cryogenic surgery: a new method of destruction or extirpation of benign or malignant tissue. *New England Journal of Medicine* **268**, 743–9.

Cooper, I.S. (1969). *Involuntary Movement Disorders*, pp. 54–77, 317–400. Hoeber, New York.

Cooper, I.S., Amin, I., Chandra, R. and Waltz, J.M. (1973). A surgical investigation of the clinical physiology of the LP-pulvinar complex in man. *Journal of the Neurological Sciences* **18**, 89–110.

Cooper, I.S. and Hirose, T. (1966). Application of cryogenic surgery to resection of parenchymal organs. *New England Journal of Medicine* **274**, 15–18.

Cooper, I.S. and Lee, A.S. (1961a). Cryostatic congelation: a system for producing a limited controlled region of cooling or freezing of biological tissues. *Journal of Nervous and Mental Disorders* **133**, 259–63.

Cooper, I.S. and Lee, A.J. (1961b). Cryothalamectomy-hypothermic congelation: a technical advance in basal ganglia surgery. *Journal of the American Geriatric Society* **9**, 714–18.

Cooper, I.S. and Stellar, S. (1963). Cryogenic freezing of brain tumours for excision or destruction *in situ*. *Journal of Neurosurgery* **20**, 921–30.

Crisp, W.E. (1972). Cryosurgical treatment of neoplasia of the uterine cervix. *Obstetrics and Gynecology* **39**, 495–9.

Dashe, A.M., Solomon, D.H., Rand, R.W., Frasier, S.D., Brown, J. and Spears, I. (1966). Stereotaxic hypophyseal cryosurgery in acromegaly and other disorders. *Journal of the American Medical Association* **198**, 591–6.

Dawber, R.P.R and Wilkinson, J.D. (1979). Melanotic freckle of Hutchinson: treatment of mascular and nodular phases with cryotherapy. *British Journal of Dermatology* **101**, 47–9.

Drylie, D.M., Jordan, W.P. and Robbins, J.B. (1968). Immunologic consequences of cryosurgery. I. serum proteins. *Investigative Urology* **5**, 619–26.

Elton, R.F. (1977). The course of events following cryosurgery. *Journal of Dermatologic Surgery and Oncology* **3**, 448–51.

Evans, P.J.D. (1981). Cryoanalgesia. The application of low temperatures to nerves to produce anaesthesia or analgesia. *Anaesthesia* **36**, 1003.

Evans, P.J.D., Lloyd, J.W. and Green, C.J. (1980). Cryoanalgesia technique. *Lancet* **i**, 1188–9.

Faraci, R.P., Bagley, D.H., Marrone, J.A.C. and Breazley, R.M. (1975). *In vitro* demonstration of cryosurgical augmentation of tumor immunity. *Surgery* **77**, 433–8.

Farris, H.E. (1981). Veterinary medicine. In *Handbook of Cryosurgery*, pp. 315–40. Edited by Ablin, R.J. Marcel Dekker, New York and Basel.

Farris, H.E. and Fraunfelder, F.T. (1976). Cryosurgical treatment of ocular squamous cell carcinoma of cattle. *Journal of the American Veterinary Medical Association* **168**, 213–16.

Fay, T. (1940). Observations on prolonged human refrigeration. *New York Journal of Medicine* **40**, 1351–4.

Franco, P.S., Hershman, J.M. and Galbraith, J.G. (1973). Treatment of acromegaly by stereotaxic cryohypophysectomy. *Southern Medical Journal* **66**, 747–53.

Fraser, J. and Gill, W. (1967). Observations on ultra-frozen tissue. *British Journal of Surgery* **54**, 770–76.

Gage, A.A. (1969). Cryosurgery for oral and pharyngeal carcinoma. *American Journal of Surgery* **118**, 669–72.

Gage, A.A. (1979). What temperature is lethal for cells? *Journal of Dermatologic Surgery and Oncology* **5**, 459–60.

Gage, A.A., Greene, G.W., Neiders, M.E. and Emmings, F.G. (1966). Freezing bone without excision. *Journal of the American Medical Association* **196**, 770–74.

Glynn, C.J., Lloyd, J.W. and Barnard, J.D.W. (1980). Cryoanalgesia in the management of pain after thoracotomy. *Thorax* **35**, 325–7.

Grace, D.M. (1977). Comparison of cryotherapy and surgery for treatment of spontaneous mouse tumour. *Canadian Journal of Surgery* **20**, 249–54.

Green, A.N., Smith, N.H. and Balsdon, M.J. (1978). Testing the effectiveness of cryosurgery for genital warts in men. *Nursing Mirror* **147**, 26–7.

Green, C.J. (1978). The scientific basis of cryotherapy. *Veterinary Dermatology Newsletter* **3**, 34–45.

Green, C.J., Cooper, J.E. and Jones, D.M. (1977). Cryotherapy in the reptile. *Veterinary Record* **101**, 529.

Gruner, O.C. (1930). A treatise on the canon of medicine of Avicenna. Luzac, London.

Hall, F.E. (1942). The use of quick freezing methods in gynecologic practices. *American Journal of Obstetrics and Gynecology* **43**, 105–111.

Hamlin, H. (1971). Abeyance of haemorrhage by 'cryospray' during meningioma removal. *Neurochirurgia* **14**, 115–17.

Hansen, R.I. and Wanstrup, J. (1973). Cryoprostatectomy. Histological changes elucidated by serial biopsies. *Scandinavian Journal of Urology and Nephrology* **7**, 100–104.

Hearse, D.J., Humphrey, S.M., Nayler, W.G., Slade, A. and Border, D. (1975). Ultrastructural damage associated with re-oxygenation of the anoxic myocardium. *Journal of Molecular Cardiology* **7**, 315–24.

Hilbert, B.J., Farrell, R.K. and Grant, B.D. (1977). Cryotherapy of periocular squamous cell carcinoma in the horse. *Journal of the American Veterinary Medical Association* **170**, 1305–1308.

Hill, C.L. (1965). Cryosurgery in otolaryngology. *Rhode Island Medical Journal* **48**, 544–5.

Hippocrates. "*Heracleitus on the Universe*" Aphorisms Vol. 4. Translated by W.H.S. Jones. (1931). 5:165, 7:201. Heinemann, London.

Holden, B.H. and McKelvie, P. (1972). Cryosurgery in the treatment of head and neck neoplasia. *British Journal of Surgery* **59**, 709–12.

Hossman, K.A. and Kleihues, P. (1973). Reversibility of ischaemic brain damage. *Archives of Neurology* **29**, 375–84.

Hughes, P.S.H. (1979). Cryosurgery of verrucous carcinoma of the penis (Buschke-Lowenstein Tumor). *Cutus* **24**, 43–5.

Jones, W.H.S. (1931). *Hippocrates, Heracleitus on the Universe, Aphorisms*, Volume 4: Ch 5, p. 165; Ch 7, p. 201. (Translation). Heinemann, London.

Jones, L.W. (1974). Cryosurgery for prostatic carcinoma. *Urology* **4**, 499–503.

Juliusberg, M. (1903). Gefrierbehandlung bei Hautkrankheiten. *Berliner Klinische Wochenschrift* **42**, 260–63.

Katz, J., Nelson, W., Forest, R. and Bruce, D. (1980). Cryoanalgesia for post thoracotomy pain. *Lancet* **i**, 512–13.

Kelly, D., Richardson, A. and Mitchell-Heggs, N. (1973). Stereotactic limbic leucotomy: a preliminary report on forty patients. *British Journal of Psychiatry* **123**, 133–40.

Klinke, J. (1939). Direct proof that cancer and normal cells live after freezing at temperatures down to minus 253°C. *Growth* **3**, 169–73.

Kreyberg, L. (1957). Local freezing. *Proceedings of the Royal Society of Medicine* **147**, 546–7.

Kuylenstierna, R., Anniko, M., Lundquist, P.G. and Nathanson, A. (1980). Experimental cryobiology of bone: a light and electron microscopical investigation. *Cryobiology* **17**, 563–70.

Lane, J.G. (1977). The treatment of equine sarcoids by cryosurgery. *Equine Veterinary Journal* **9**, 127–33.

Lane, J.G. and Burch, D.G.S. (1975). The cryosurgical treatment of canine anal furunculosis. *Journal of Small Animal Practice* **16**, 387–92.

Larré, D.J. (1832). "*Memoirs de Chirurgie Militaire et Campagnes .... Paris 1812–1817*". Translated by J.C, Mercer (1832). Carey and Lea, Philadelphia.

Leopard, P.J. and Poswillo, D.E. (1974). Practical cryosurgery for oral lesions. *British Dental Journal* **136**, 185–96.

Le Pivert, P.J. (1981). Basic considerations of the cryolesion. In *Handbook of Cryosurgery*, pp. 15–68. Edited by Ablin, R.J. Marcel Dekker, New York, Basel.

Levin, S.R., Hofeldt, F.D., Schneider, V., Becker, N., Karam, J.H., Seymour, R.J., Adams, J.E. and Forsham, P.H. (1974). Cryohypophysectomy for acromegaly: Factors associated with altered endocrine function and carbohydrate metabolism. *American Journal of Medicine* **57**, 526–35.

Li, A.K.C., Trencher, P.S., Holborow, E.J., Newsome, C. and Wynne, A.T. (1977). Experimental smooth muscle antibodies. *Clinical and Experimental Immunology* **27**, 273–7.

Lloyd, J.W., Barnard, J.D.W. and Glynn, C.J. (1976). Cryoanalgesia: a new approach to pain relief. *Lancet* **ii**, 932–4.

Lloyd-Williams, K. and Holden, H.B. (1975). Cryosurgery in general and ENT surgery. *British Journal of Hospital Medicine* **14**, 14–25.

Lortat-Jacobs, G. and Solente, C. (1930). *La Cryotherapie*. Maisson et Cie, Paris.

Maddy, J.A., Winternitz, W.W., Norrel, H., Quillen, D. and Wilson, C.B. (1969). Acromegaly: treatment by cryoablation. *Annals of Internal Medicine* **71**, 497–505.

Marcove, R.C., Lyden, J.R., Huvos, A.G. and Bullough, P.B. (1973). Giant-cell tumors treated by cryosurgery. A report of twenty-five cases. *Journal of Bone and Joint Surgery* **55**, 1633–44.

Marcove, R.C., Stovell, P.B., Huvos, A.G. and Bullough, P.B. (1977). The use of cryosurgery in the treatment of low and medium grade chrondrosarcoma. A preliminary report. *Clinical Orthopaedics* **122**, 147–56.

Markham, C.H. and Rand, R.W. (1963). Stereotaxic surgery in Parkinson's disease. *Archives of Neurology* **8**, 621–31.

Migiola, S., Blanton, A.B. and Davenport, J.W. (1978). Cryosurgical treatment of equine cutaneous habronemiasis. *Veterinary Medicine and Small Animal Clinics* **73**, 1073–6.

Myers, R.S., Hammond, W.G. and Ketcham, A.S. (1969). Tumor-specific transplantation immunity after cryosurgery. *Journal of Surgery and Oncology* **1**, 241.

Neel, H.B., Ketcham, A.S. and Hammond, W.G. (1971). Requisites for successful cryogenic surgery of cancer. *Archives of Surgery* **102**, 45–8.

Neel, H.B., Ketcham, A.S. and Hammond, W.G. (1973). Experimental evaluation of *in situ* oncocide for primary tumour therapy: comparison of tumor-specific immunity after complete excision, cryonecrosis and ligation. *Laryngoscope* **23**, 376–87.

O'Connor, J.J. (1979). Surgery plus freezing as a technique for treating pilonidal disease. *Diseases of the Colon and Rectum* **22**, 306–307.

Ojemann, G.A. and Ward, A.A. (1971). Present indications for L-dopa and thalamectomy in the treatment of Parkinson's disease. *Northwest Medicine* **70**, 101–104.

Osborne, D.R., Higgins, A.F. and Hobbs, K.E. (1978). Cryosurgery in the management of rectal tumours. *British Journal of Surgery* **65**, 859–61.

Ostergard, D.R. and Townsend, D.E. (1969). Comparison of electrocauterization and cryosurgery for the treatment of benign disease of the uterine cervix. *Obstetrics and Gynecology* **33**, 58–63.

Ostergard, D.R., Townsend, D.E. and Hirose, F.M. (1969). The long-term effects of cryosurgery of the uterine cervix. *Journal of Cryosurgery* **2**, 17–22.

Palmer, J.F. (1835). *The Works of John Hunter.*

Peeters, F.L.M. and Walder, H.A.D. (1973). Intraoperative vertebral angiography in arteriovenous malformations. *Neuroradiology* **6**, 169–73.

Petersen, D.S., Millerman, L.A., Rose, E.F., Bonney, W.W., Schmidt, J.D., Hawtrey, C.E. and Culp, D.A. (1978). Biopsy and clinical course after cryosurgery for prostatic cancer. *Journal of Urology* **120**, 308–311.

Poswillo, D.E. (1975). Cryosurgery in oral surgery. *British Journal of Hospital Medicine* **14**, 47–54.

Rand, R.W. (1968). Cryogenic techniques in stereotactic neurosurgery. Cryohypophysectomy and cryothalmectomy. *International Surgery* **49**, 212.

Rand, R.w. (1981). Neurosurgery. In *Handbook of Cryosurgery*, pp. 153–93. Edited by Ablin, R.J. Marcel Dekker, New York, Basel.

Rand. R.W., Dashe, A.M., Paglia, D.E., Conway, L.W. and Solomon, D.H. (1964). Stereotactic cryohypophysectomy. *Journal of the American Medical Association* **189**, 255–9.

Reuter, H.J. (1972). Endoscopic cryosurgery of prostate and bladder tumours. *Journal of Urology* **107**, 389–93.

Richardson, A. (1973). Stereotactic limbic leucotomy: surgical technique. *Postgraduate Medical Journal* **49**, 860–64.

Richardson, A. (1975). Cryosurgery in neurosurgery. *British Journal of Hospital Medicine* **14**, 39–46.

Rickards, D.A. (1981). Companion animal practice. In: *Handbook of Cryosurgery*, pp. 341–74. Edited by Ablin, R.J. Marcel Dekker, New York and Basel.

Rothenberg, H.W. (1977). Cryoprotective properties of vasoconstriction. *Cryobiology* **14**, 349–61.

Rowbotham, G.F., Haigh, A.L. and Leslie, W.G. (1959). Cooling cannula for use in the treatment of cerebral neoplasms. *Lancet* **1**, 12–15.

Schenker, J.G. and Polishuk, W.Z. (1972). Regeneration of rabbit endometrium after cryosurgery. *Obstetrics and Gynecology* **40**, 638–45.

Smith, J.J. and Fraser, J. (1974). An estimation of tissue damage and thermal history in the cryolesion. *Cryobiology* **11**, 139–47.

Soanes, W.A., Gonder, M.J. and Shulman, S. (1966). Apparatus and technique for cryosurgery of the prostate. *Journal of Urology* **96**, 508–11.

Spallanzani, A. (1784). '*Translation of Dissertation Relative to the Natural History of Animals and Vegetables*'. London.

Tate, L.P. (1977). Cryosurgery in partial and complete neurectomy in the horse. *Proceedings of the 23rd Annual Convention of the American Association of Equine Practitioners, Vancouver, 1977.*

Townsend, D.E., Ostergard, D.R. and Hirose, F.M. (1967). The effect of cryosurgery on the cervix uteri. In *Cryosurgery*. Edited by Rand, R.W. Charles C. Thomas, Springfield, Illinois.

Tytus, J.S. and Ries, L. (1961). Further observations on rapid freezing and its possible application to neurosurgical techniques. *Bulletin of the Mason Clinics* **15**, 51–61.

Uhlschmid, G., Kolb, E. and Largiader, F. (1979). Cryosurgery of pulmonary metastases. *Cryobiology* **16**, 171–8.

Von Leden, H. and Rand, R.W. (1967). Cryosurgery of the head and neck. *Archives of Otolaryngology (Chicago)* **85**, 93–8.

Walder, H.A.D. (1971). Experimental cryosurgery. In *Cryogenics in Surgery*, pp. 150–81. Edited by Leden, H. von and Cahan, W.G. Medical Examination Publishing Co., New York.

Walder, H.A.D. (1972). Cryosurgery as a technical method, its indication and results. *Acta Chirurgica Belgica* **71**, 203–19.

Walder, H.A.D., Jaspar, H.H. and Meijer, E. (1970). Application of cryotherapy in cerebrovascular anomalies. An experimental and clinical study. *Psychiatra, neurologia, neurochirurgia* **73**, 471–86.

Weitzner, K. (1940). The treatment of endocervicitis with carbon dioxide snow (dry ice). *American Journal of Surgery* **48**, 620–24.

White, A.C. (1899). Liquid air in medicine and surgery. *Medical Records* **56**, 109–12.

Whittaker, D.K. (1974). Ice crystals formed in tissues during cryosurgery. *Cryobiology* **11**, 192–201.

Wolfson, R.J. and Cutt, R.A. (1971). Long-term results with cryosurgery for Menière's disease. *Archives of Otolaryngology* **93**, 483–6.

Wood, G.J., Lloyd, J.W., Evans, P.J.D. and Bullingham, R.E.S. (1979). Cryoanalgesia and day care herniorraphy. *Lancet* **ii**, 479.

Yamasaki, T. (1978). Immunological survey for endoscopic cryosurgery, and its basic study. *Cryobiology* **15**, 702.

Zacharian, S. (1971). Cryosurgery for cancer of the skin. *Cancro* **24**, 349–55.

Zacharian, S.A. (1977). *Cryosurgical Advances in Dermatology and Tumours of the Head and Neck*. Charles C. Thomas, Springfield, Illinois.

15

The freezing of food tissues

D.S. Reid
Department of Food Science and Technology
University of California

Introduction

In a comparison of freezing of biological tissue as a food resource to freezing in attempts to preserve viability (cryopreservation) the most obvious difference is that of scale. The vast tonnages required to satisfy the needs of the frozen food market are orders of magnitude greater than the quantities typical of cryopreservation and storage in, for example, a clinical situation. For this reason, if no other, it is clearly inappropriate to employ storage at the temperature of liquid nitrogen for frozen foods, as the large quantities of cryogen required would be excessively costly. Also, the provision of frozen food to the public at such low temperatures would constitute a major hazard, as handling these materials would require extreme safety precautions inappropriate to both commercial food storage and the home environment. For reasons of both safety and economy the temperatures utilized for frozen food storage are therefore those of large scale readily available mechanical refrigeration, typically ranging between −20°C and −40°C. In contrast the storage of cryopreserved material is usually accomplished at temperatures between −80°C and −196°C. Any changes resulting from chemical activity that might take place in frozen foods during storage at higher subzero temperatures are likely to occur at rates significantly faster than those encountered during storage for cryopreservation. Another, related consequence of difference in storage temperature is the amount of unfrozen water in the system. Whilst at the lower temperatures as little as 12 per cent of the water in a tissue system may not be formed into ice, at the higher temperatures of frozen food storage significantly more unfrozen water will be present (Riedel, 1961; Duckworth, 1981). It should be noted that cryoprotectant may change the amount of ice formed at very low temperatures (See Taylor, this volume).

A further important difference between frozen foods and cryopreserved materials is a consequence of sample size. In general, the size of the food unit being frozen is more nearly akin to the size of entire organs that might be the subject of cryopreservation, rather than the size of single cells or small cell clusters. In cryopreservation, as the size of the individual unit to be preserved increases, serious practical problems arise. Particularly, accurately controlled cooling throughout the tissues becomes less feasible due to restrictions of internal heat transfer, and it is partly as a result of this problem that organ preservation is difficult and often unsuccessful. Elaborate perfusion techniques for protective solutions have been developed in attempts to minimize tissue damage resulting from this essentially non-controlled cooling and to achieve prolonged viable organ storage (Fuller, this volume). Most food commodities which are subjected to freezing for storage also have unit sizes which do not permit controlled rapid rate freezing, and it is therefore not surprising to find freezing of food tissues is also accompanied by significant tissue damage (Reeve and Brown, 1966; Partmann, 1975).

In the case of food tissues with effective cryoprotectant additives perfusion is generally inappropriate and is often either not possible or inpracticable. Freezing damage, and associated cell death may therefore be inevitable for food tissues and may have to be accepted. It has to be borne in mind that the objective of the food freezer is not the same as that of the cryobiologist. The cryobiologist aims to prepare and freeze living cells and tissues in such a way that, on thawing, an acceptable proportion of viable cells are recovered and in the case of organs and whole organisms that integrated function is restored. The food freezer, in contrast, aims to freeze material in such a way that, after thawing, a material of the best possible nutritional and eating quality is obtained. The concept of eating quality is somewhat ill-defined, yet everyone can tell when frozen food is of unacceptable quality. Prefreezing treatments employed to achieve these aims may actually result in cell and tissue damage.

These different objectives of the cryobiologist and the food freezer may not always be incompatible, for successful cryopreservation might produce a material of good eating quality. The preservation of viability by a technique designed specifically for food processing is, however, less likely. Furthermore, cryopreservation can make use of a wide range of additives to help minimize freezing damage but many of these will be inappropriate for use when freezing food. In addition, it should be remembered that not all food tissues are 'alive' even when 'fresh'. For example, muscle does not become meat until after a complex series of biochemical changes have taken place following the slaughter of the animal, and it is not considered 'alive'. Many vegetables, however, when freshly harvested must be considered 'alive' (e.g. peas seeds, potato tubers).

The characteristic behaviour of different biological tissues subjected to frozen preservation as food commodities is extremely interesting, with animal-derived and plant-derived tissues differing in many respects. Each will be considered separately below. There are observations made during investigations of food preservation by freezing that may be of value to the cryobiologist who wishes to have a better understanding of the variety of routes by which freezing damage might occur and the consequences of such damage.

Freezing of plant tissues

Food tissues derived from plants comprise both fruits and vegetables which, for purposes of this discussion, will be distinguished from a food science, rather than a botanical viewpoint.

Fruits

Botanically, fruits consist of the seed-bearing organs of plants and form an important food commodity. Fruits are usually consumed after they have reached physiological maturity and they may even have entered senescence. Fruits in the commercial food sense may exclude botanical fruits such as those of pea, bean. and cucumber which are considered to be vegetables, and may be harvested and eaten whilst still at an immature stage. Those foods which are classed as fruits are frequently consumed raw (i.e. as fresh tissues). As a consequence, a major contribution to their accepted eating quality is their texture. This arises to a considerable extent from their cellular turgor pressure, which provides a resistance to compression, followed by a sudden yielding as the cell wall complexes burst. The contribution of the cell walls themselves to texture in fruits is usually not of great importance. The important component of the complex is the plasma membrane, which regulates water permeability. Anything which alters the cell permeability of the plasma membrane will affect the turgor pressure of the cell. Freezing, unfortunately, often results in a large increase in the permeability of this membrane and on thawing, tissues become soft and lacking in turgor (Brown, 1979). A frozen, thawed strawberry is a most striking illustration of the effect of freezing and consequential cell membrane damage on the texture of fruits. This is demonstrated most clearly, when the quality of strawberries is evaluated using a combination of mechanical testing, sensory analysis and histological examination (Szezesniak and Smith, 1969). Frozen, thawed strawberries were much less firm than fresh strawberries and showed extensive cell plasmolysis, and even cell rupture. Much more juice was released, presumably as a consequence of membrane damage. Another similar illustration of the effect of freezing on the permeability of cell wall membranes, and its consequences in terms of texture, is provided by the application of microscopy to tomato tissues (Mohr and Stein, 1969; Mohr, 1971).

Methods exist which may help to reduce the impact of this freezing damage, for example the addition of sugars. Sugars act by partially dehydrating the cells of the fruit, thus making it more difficult for these cells to freeze. As in the addition of cryoprotective additives during cryopreservation, this may help protect cells. Unfortunately, as already indicated, it is difficult to obtain deep tissue penetration, and hence high effectiveness of such additives, because the necessary vascular systems may not exist. Furthermore, the addition of sugars may produce unacceptable textural changes within the tissues. In general, when cryoprotective compounds are used, careful control of freezing conditions is also employed to maximize their effectiveness. Fruits are typically of such a size that the required degree of control is often not possible. Addition of sugars, therefore may do little in the way of preserving the texture of the fruit.

Frozen fruits are often eaten whilst only partially thawed, as in this way the ice present helps provide some extra degree of firmness. Utilizing fruits whilst only partially thawed can have other advantages, for in addition to textural defects, freezing damage often produces a dislocation of the normal enzymic processes of the cell (this is discussed below when considering vegetable freezing). As a consequence, further changes may occur that lower fruit quality. In strawberries, the mushy texture of the fully thawed material is, at least in part, a consequence of enzymic degradation processes, as well as membrane damage. The use of partially thawed fruit may help to make such undesirable alterations less noticeable to the consumer.

Vegetables

Vegetables, do not lend themselves to as simple a classification as fruits. In a generalized plant, all parts are, in some particular instance consumable as a vegetable. Examples of the botanical variety of vegetables are listed in Table 15.1. As previously noted, even botanical fruits may be considered as vege-

Table 15.1 Some examples of the botanical variety of edible plant tissues used as vegetables

Part of plant eaten	Example
Root	carrot, parsnip
Tuber	potato, yam
Young shoot	asparagus
Stem	celery
Leaf	lettuce, spinach
Flower bud	broccoli, cauliflower
Immature fruit	green bean, cucumber
Mature fruit	tomato
Seed	pea

tables. As a result of this diversity of tissue type, generalizations about preservative freezing for vegetables are even more difficult to make than those concerning fruits. Many vegetables are consumed after cooking which will soften cell walls and degrade cell membranes, rendering them fully permeable. Cooking, therefore, causes turgor loss and therefore other losses of turgor pressure that are a direct consequence of freezing will be of lesser significance. Vegetables which are consumed raw, such as lettuce, cannot always be satisfactorily frozen, for turgor pressure is commonly an important component of their eating quality.

Although loss of turgor may be of less importance in frozen vegetables that are to be cooked than in frozen fruits, other freezing related problems exist. Vegetables tissues consist of cells with fairly rigid cell walls, and a complex, compartmentalized interior. Membranes within the cells maintain the separate identities of these compartments. During freezing both membranes and cells walls may be damaged such that, in addition to the loss of turgor already mentioned, the inner compartmentalization of the cells may be disrupted. Substances which are normally kept apart may come together and the consequential reactions lead to undesirable changes in flavour, colour and odour. Texture also may be further affected by degradation of structural polymers within the cell walls. Such degradation processes are partially responsible for texture loss in frozen, thawed strawberries also.

Enzymically catalyzed reactions are of particular significance to the frozen food industry, as they can lead to a variety of undesirable changes.

Studies of the microstructure of frozen plant products indicate clearly why such reactions become important (Mohr and Cocking, 1968.; Reeve, 1970.; Mohr, 1974.; Brown, 1977). Freezing causes disruption of the internal cell structures, bringing together enzymes and substrates which would normally be well separated. These react to produce degradation products, often with easily detectable, unnatural flavours. Pectolytic enzymes may degrade cell walls, and undesirable colour changes may occur.

A brief heat treatment may inactivate such enzymes, and render the tissue more suitable for long-term frozen storage (Joslyn and Cruess, 1929). The thermal inactivation of enzymes eliminates many of the undesirable reactions, at the cost of destroying cell membrane integrity, together with a partial solubilization of cell wall material. A reaction of particular importance to texture is the gelatinization of the starches within the cell. This is considered at length by Reeve (1970). Blanching, as this treatment is called, is akin to a partial cooking process and is, therefore, appropriate for vegetables which are normally consumed cooked. It may not, however, be appropriate for tissues normally consumed raw, particularly in those cases where a noticeable flavour change occurs during heating.

In order to determine whether an adequate blanch has been performed, it is necessary to identify either the enzymes which cause the detrimental changes or suitable marker enzymes whose inactivation can readily be monitored (Finkle, 1971.; Richardson, 1976). The most frequently monitored enzymes are catalase and peroxidase, which lend themselves readily to simple assay procedures. It has been shown that vegetable tissues blanched to inactivity of these enzymes are suitable for freezing without further enzymically produced quality deterioration. The enzymes lipoxygenase, pectin esterase and peridoxase have all been implicated in the loss of quality of certain unblanched plant materials during freezing and frozen storage (Joslyn, 1966). The use of blanching to improve the quality characteristics of some frozen vegetables illustrates a major difference between food freezing and cryobiological freezing. Blanching kills the tissue so that the food freezer is, in fact, freezing a dead tissue. The end result, however, will be a more acceptable food product than would frozen, unblanched material. The food processor, where possible, will choose blanching and freezing conditions which optimize the eating quality of the final product.

A variety of enzymically catalyzed reactions are possible in frozen, unblanched plant tissue, and in frozen plant tissues which have been inadequately blanched (Powrie, 1973,; Burnette, 1977 and Table 15.2.). These reactions may also be responsible for some of the damage that occurs in frozen materials as a result of suboptimal cryopreservative procedures. Frozen storage does not suspend these reactions, it merely slows them down. Indeed, at higher sub-freezing temperatures the kinetics of some enzymically catalyzed processes in frozen systems may be accelerated (Fennema *et al.*, 1973.; Singh and Wang, 1977). This may be a consequence of the increased concentration of essential solutes in the liquid phase as a result of ice seperation outweighing the reduction in reaction constant as temperature is lowered.

In tissues which cannot be blanched prior to freezing other means must be sought to minimize the enzymically induced changes during frozen storage. In some cases chemical inhibition of the process may be appropriate, and in this area in particular important cross-fertilization between cryopreservation and food freezing preservation might be expected. However, the toxicity of many effective plant cryoprotectants prevents their use in foods. Similarly, the methods popular in food preservation such as the addition of sulphur dioxide to prevent some enzyme reaction (van Arsdel *et al.*, 1969) would be detrimental to live tissue. Nevertheless, this could be an important area of cooperation.

Table 15.2 Enzymic reactions important in damaged plant tissues

Enzyme	Substrate	Product
Lipoxygenase	Unsaturated fatty acid	Fatty acid hydroperoxide Carbonyl compounds
Catalase	Hydroperoxide	Oxygen
Pectic enzymes	Pectins	Pectic acid, uronides, galacturonic acid
Lipase	Triglycerides	Fatty acids, monoglycerides, glycerol
Polyphenol oxidase	Phenols	Melanins (brown pigments)
Peroxidase	Peroxides + electron donors	Oxidized donors

Freezing of muscle tissues

The freezing of muscle-derived food tissues involves a different set of problems to the freezing of plant-derived food tissues (Powrie, 1973b). Muscle tissue is quite uniform in composition, and has few major components. In contrast, plant tissues are quite heterogeneous, containing a wide variety of cells, enzyme systems, storage organs, etc; consequently, a wide variety of reaction sequences and damage events are possible. In muscle tissues, the proteins are predominantly of two or three main types, with the actin/myosin complex being at high concentration. Also the enzyme systems are specialized rather than diverse. In addition, muscles lack the rigid cell walls of the plants; their mode of freezing is therefore somewhat different. Two main categories of muscle foods should be considered, meat and fish. The problems of meat freezing and fish freezing also vary in many respects, as a result of differences in the stabilities of many of their components.

Meats

Meat is usually post-rigor animal muscle tissue. The processes of conversion of living muscle to meat are complex, and involve a large number of biochemical changes. After slaughter, the muscle undergoes rigor. The adenosine triphosphate (ATP) in the muscle is gradually used up by continuing metabolic processes, but while it is still present in significant quantity, the muscle is capable of contraction. Once the ATP is exhausted, the muscles become rigid, until subsequent autolysis causes them to soften again (Hultin, 1976). The freezing of fully aged meat presents few problems. If, however, the tissue is frozen before all the ATP has exhausted the regulatory mechanisms of the muscle are disrupted. On thawing, the ATP still present causes an immediate, uncontrolled contraction of the muscle, producing a tough, inedible tissue mass. This problem can be controlled in at least two ways, other than simply waiting until the ATP is exhausted. The muscle can be handled, prior to freezing, in a way which will produce more rapid exhaustion of the ATP store. A technique for electrical stimulation can achieve this. Alternatively, the frozen material may be held at $-3°C$ to $-6°C$ for several hours prior to complete thawing. At these temperatures, the ATP becomes exhausted, whilst the ice framework within the tissues prevents contraction of the fibres. Furthermore, workers concerned with meat freezing must also be aware of the phenomenon of cold shortening. Again, this is a problem which is manifested only in the presence of ATP when, during cooling and prior to freezing, an uncontrolled shortening of the muscle fibres may occur, producing a tough product. This contraction is thought to result from a change in membrane permeabilities (see Morris, this volume).

The most noticeable effect of freezing on meat is a loss of fluid (drip loss) on thawing. Studies have been carried out in order to relate this drip loss to the freezing conditions. By controlling freezing rates, and measuring drip loss in conjunction with histological examination it has been shown that maximum drip loss (and hence presumably the maximum cell membrane damage) occurred at freezing rates which correspond to the changeover in freezing pattern from intracellular freezing to extracellular freezing (Bevilacqua *et al.*, 1977; Anon and Calvelo, 1980). This result is consistent with findings on cod freezing which will be discussed below (Love, 1966).

During frozen storage of meat, the main changes which occur involve oxidation of lipids, together with some loss in protein solubility. There are not, however, any major enzymic problems, apart from those already mentioned as being important if ATP has not been depleted. Consequently, blanching, or heat inactivation of the muscle enzymes, is not necessary. Meat is therefore frozen raw (unlike most vegetables).

Fish

Fish tissues are much more susceptible to change during freezing and frozen storage than meat tissues, though once again the major component being frozen is muscle. The proteins of the fish muscle complex appear to be less stable to freezing than those of meat; during freezing and frozen storage of fish muscle significant amounts of denaturation can occur (Shenonda, 1980) and the proteins typically become much less soluble. Myofibrillar proteins can be extracted from fish muscle using a range of salt solutions, but, as storage time increases, the amount of extractable protein decreases. Much of the solubility loss is a result of crosslinking of the proteins. One undesirable aspect associated with crosslinking is the production of formaldehyde in certain species. Lipids, moisture content, and many other factors are also involved. The use of additives to help maintain the myofibrillar proteins in soluble, functional form has been studied; this approach has close parallels with the use of protectant substances in cryopreservation. Some of the additives which have a useful effect are monosodium glutamate, certain polyalcohols and carboxylic acids (Matsumoto, 1979). Although useful as a probe to investigate the mechanism of freezing denaturation, it is unlikely that many of the effective additives would have a practical use in frozen food systems.

The chemical composition of many fish is such that compounds of disagreeable taste or colour may be formed during freezing or frozen storage, formaldehyde is only one example. Many of these compounds are produced as a result of enzymic activity (Martin *et al.*, 1982), but thermal inactivation of the enzymes is not appropriate because of the extensive, deleterious effects of thermal processing on fish muscle structure. A lower temperature for storage is more appropriate, as the detrimental changes take place more slowly at −30°C than at −20°C, or −10°C. Some of the spoilage products typically found in frozen fish are listed in Table 15.3

Table 15.3 Some spoilage products found in fish

Trimethylamine	Hypoxanthine
Dimethylamine	Formaldehyde
Cadaverine	Tryosine
Putrescine	Ammonia
Histamine	Pyridoxal-5′-phosphate

Table 15.4 Relationship between freezing rate and cell damage in fish muscle, as indicated by nuclear material in drip fluid (from Love, 1966)

Freezing time 0°C to −5°C	Nuclear material in drip fluid	Ice characteristics
5	low	Many small crystals in cell
25	high	Several large crystals in cell
50	low	One crystal filling in cell
100	high	Changeover from intracellular to extracellular
200	low	Extracellular

An insight into circumstances of freezing damage in muscle tissues of cod has been demonstrated (Love, 1966). Slow freezing has been shown to result in extracellular ice, causing dehydration, and rapid freezing to produce a large number of intracellular ice crystals. Intermediate freezing rates, interestingly, at the point where the pattern of ice formation was intermediate between extra- and intracellular (Table 15.4) produced maximum damage. Research on beef has produced similar findings (Bevilacqua *et al.*, 1979; Anon and Calvelo, 1980).

It should be pointed out that there is a significant difference between the method of harvesting fish and meat. Fish are hunted, and their glycogen stores may often be depleted and severely stressed at their time of death. This can affect muscle pH, which affects the final quality of the product. Also, the nutritional status of the fish is uncertain and their biochemical status unknown and variable. This will affect the freezing behaviour of individual fish. Meat animals are usually farmed, and at the time of slaughter they might be less severely stressed than fish. Their nutritional status is known and their biochemical status is well defined. The freezing behaviour of individuals is, therefore, more predictable.

Conclusions

There are many lessons to be learned by studying the freezing of tissues for food use. As much higher storage temperatures have to be used in food freezing than in cryopreservation, the potential for change and damage within the tissues is greater.

The characteristics of plant tissues are different from those of animal tissues, the cellulosic cell walls playing an important role in freezing characteristics. Plant cells also contain many enzymes which, if delocalized during the freezing process, cause problems associated with undesirable enzyme reactions. Plants are typically complex, and a wide range of cell types have to be frozen simultaneously and successfully.

Animal tissues that are frozen as food are, in general, muscle tissues. These tissues are more uniform, have a high concentration of one major component, and pose fewer enzymatic degradation problems (except perhaps in the case of fish) than plant tissues.

The role of cryoprotectants in food freezing is limited, as a result of problems of toxicity, flavour and food legislation.

A general conclusion might be that the preservation of edibility by freezing encounters the same phenomena as seen when freezing to preserve viability but that the responses of the investigator or processor are, of necessity, very different.

References

Anon, M.C. and Calvelo, A. (1980). Freezing rate effects on the drip loss of frozen beef. *Meat Science* **4**, 1–14.

Ashwood-Smith, M.J. (1980). Low temperature preservation of cells, tissues and organs. In *Low Temperature Preservation in Medicine and Biology*, pp. 19–44. Edited by Ashwood-Smith, M.J. and Farrant, J. Pitman, London.

Bevilacqua, A., Zaritzky, N.E. and Calvelo, A. (1979). Histological measurements of ice in frozen beef. *Journal of Food Technology* **14**, 237–51.

Brown, M.S. (1977). Texture of frozen fruits and vegetables. *Journal of Texture Studies* **7**, 391–404.

Brown, M.S. (1979). Frozen fruits and vegetables: their chemistry, physics and cryobiology. *Advances in Food Research* **25**, 181–235.

Burnette, F.S. (1977). Peroxidase and its relationship to food flavour and quality. *Journal of Food Science* **42**, 1–6.

Duckworth, R.B. (1981). Differential thermal analysis of frozen food systems. I Determination of unfreezable water. *Journal of Food Technology* **6**, 317–27.

Fennema, O., Powrie, W.D. and Marth, E. (1973). *The Low Temperature Preservation of Food and Living Matter*. Dekker, New York.

Finkle, B.J. (1971). Freezing preservation. In *Biochemistry of Fruits and their Products*, Vol. 2, pp. 635–85. Edited by Hulme, A.C. Academic Press, New York.

Hultin, H.O. (1976). Characteristics of muscle tissue. In *Principles of Food Science. I Food Chemistry*, pp. 577–617. Edited by Fennema, O.R. Dekker, New York.

Joslyn, M.A. (1966). The freezing of fruits and vegetables. In *Cryobiology*, pp. 565–607. Edited by Meryman, H.T. Academic Press, London.

Joslyn, M.A. and Creuss, W.V. (1929). Freezing of fruits and vegetables for retail distribution in paraffined paper containers. *Fruit Products Journal* **8**, 9–12.

Love, R.M. (1966). Freezing of animal tissue. In *Cryobiology*, pp. 317–405. Edited by Meryman, H.T. Academic Press, London.

Martin, R.E., Flick, G.J., Hebard, C.E. and Ward, D.R. (1982). *Chemistry and Biochemistry of Marine Food Products*. Avi Publishing, Westport, CT.

Matsumoto, J.J. (1979). Denaturation of fish muscle proteins during frozen storage. In *Proteins at Low Temperatures*, pp. 205–224. Edited by

Fennema, O. American Chemistry Society, Washington.
Mohr, W.P. (1971). Freeze thaw damage to protoplasmic structure in high moisture edible plant tissue. *Journal of Texture Studies* **2**, 316–27.
Mohr, W.P. (1974). Freeze-thaw (and blanch) damage to vegetable ultrastructure. *Journal of Texture Studies* **5**, 13–27.
Mohr, W.P. and Cocking, E.C. (1968). A method of preparing highly vacuolated, sensescent and damaged plant tissue for ultrastructural study. *Journal of Ultrastructure Research* **21**, 171–181.
Mohr, W.P. and Stein, M. (1969). Effect of different freeze-thaw regimes on ice formation in tomato fruit parenchyma tissue. *Cryobiology* **6**, 15–21.
Partmann, W. (1975). The effects of freezing and thawing on food quality. In *Water Relations of Foods*, pp. 505–37. Edited by Duckworth, R.B. Academic Press, London.
Pegg, D.E. (1973). *Organ Preservation*. Churchill Livingstone, London.
Powrie, W.D. (1973b). Characteristics of food phytosystems and their behaviour durng freeze preservation. In *Low Temperature Preservation of Foods and Living Matter*, pp. 354–380. Edited by Fennema, O., Powrie, W.D., Marth, E. Dekker, New York.
Powrie, W.D. (1973a). Characteristics of food myosystems and their behaviour during freeze preservation. In *Low Temperature Preservation of Foods and Living Matter*, pp. 282–353. Edited by Fennema, O., Powrie, W.D. and Marth, E. Dekker, New York.
Reeve, R.M. (1970). Relationships of histological structure to texture of fresh and processed fruits and vegetables. *Journal of Texture Studies* **1**, 247–84.
Reeve, R.M. and Brown, M.S. (1966). Some structural and biochemical changes related to frozen fruits and vegetables. *Cryobiology* **3**, 214–23.
Richardson, T. (1976). Enzymes. In *Principles of Food Science. I Food Chemistry*, pp. 285–345. Edited by Fennema, O.R. Dekker, New York.
Riedel, L. (1961). In the problem of bound water in meats. *Kaltetechnik* **13**, 122–8.
Shenouda, S.Y.K. (1980). Theories of protein denaturation during frozen storage of fish flesh. *Advances in Food Research* **26**, 275–311.
Singh, R.P. and Wang, C.Y. (1977). Quality of frozen foods - a review. *Journal of Food Process Engineering* **1**, 97–127.
Szczesniak, A.S. and Smith, B.J. (1969). Observations on strawberry texture, a three pronged approach. *Journal of Texture Studies* **1**, 65–89.
van Arsdel, W.D., Copley, M.J. and Olson, R.L. (1969). *Quality and Stability of Frozen Foods*. Wiley-Interscience, New York.

Indices

Species index

Subject index